Inderbir Singh's

Textbook of
HUMAN HISTOLOGY

Late Professor Inderbir Singh
(1930–2014)

Tribute to a Legend

Professor Inderbir Singh, a legendary anatomist, is renowned for being a pillar in the education of generations of medical graduates across the globe. He was one of the greatest teachers of his times. He was a passionate writer who poured his soul into his work. His eagle's eye for details and meticulous way of writing made his books immensely popular amongst students. He managed to become enmeshed in millions of hearts in his lifetime. He was conferred the title of Professor Emeritus by Maharishi Dayanand University, Rohtak.

On 12th May 2014, he has been awarded posthomously with Emeritus Teacher Award by National Board of Examination for making invaluable contribution in teaching of Anatomy. This award is given to honour legends who have made tremendous contribution in the field of medical education and their work had vast impact on the education of medical graduates. He was a visionary for his times and the legacies he left behind are his various textbooks on gross anatomy, histology, neuroanatomy, and embryology. Although his mortal frame is not present amongst us, his genius will live on forever.

Inderbir Singh's

Textbook of HUMAN HISTOLOGY

with Colour Atlas and Practical Guide

Seventh Edition

Revised and Edited by

NEELAM VASUDEVA MBBS MD
Director Professor and Head, Department of Anatomy
Maulana Azad Medical College, New Delhi

SABITA MISHRA MBBS DNB PhD (AIIMS)
Professor, Department of Anatomy
Maulana Azad Medical College, New Delhi

The Health Sciences Publishers

New Delhi | London | Philadelphia | Panama

Jaypee Brothers Medical Publishers (P) Ltd

Headquarters

Jaypee Brothers Medical Publishers (P) Ltd
4838/24, Ansari Road, Daryaganj
New Delhi 110 002, India
Phone: +91-11-43574357
Fax: +91-11-43574314
Email: jaypee@jaypeebrothers.com

Overseas Offices

J.P. Medical Ltd
83 Victoria Street, London
SW1H 0HW (UK)
Phone: +44 20 3170 8910
Fax: +44 (0)20 3008 6180
Email: info@jpmedpub.com

Jaypee Medical Inc
The Bourse
111 South Independence Mall East
Suite 835, Philadelphia, PA 19106, USA
Phone: +1 267-519-9789
Email: jpmed.us@gmail.com

Jaypee Brothers Medical Publishers (P) Ltd
Bhotahity, Kathmandu, Nepal
Phone: +977-9741283608
Email: kathmandu@jaypeebrothers.com

Jaypee-Highlights Medical Publishers Inc
City of Knowledge, Bld. 237, Clayton
Panama City, Panama
Phone: +1 507-301-0496
Fax: +1 507-301-0499
Email: cservice@jphmedical.com

Jaypee Brothers Medical Publishers (P) Ltd
17/1-B Babar Road, Block-B, Shaymali
Mohammadpur, Dhaka-1207
Bangladesh
Mobile: +08801912003485
Email: jaypeedhaka@gmail.com

Website: www.jaypeebrothers.com
Website: www.jaypeedigital.com

Inderbir Singh's Textbook of Human Histology

First Edition	*: 1987*	*Fifth Edition*	*: 2006*
Second Edition	*: 1992*	*Reprint*	*: 2008*
Third Edition	*: 1992*	*Reprint*	*: 2009*
Fourth Edition	*: 2002*	*Sixth Edition*	*: 2011*
Reprint	*: 2005*	*Seventh Edition*	*: **2014***

ISBN 978-93-5152-322-2

Printed at Sanat Printers, Kundli

Preface

Textbook of Human Histology by Professor Inderbir Singh has remained an authoritative and standard textbook for the past many decades and it is our proud privilege to revise this book and bring out the 7th edition. The strength and popularity of this textbook has been its simple language and comprehensiveness that has essentially remained unchanged since its inception. Professor Singh's eye for details and his meticulous writing style has always been popular amongst the generations of medical students. Although all the chapters have been revisited and thoroughly revised, we have taken special care to retain the basic essence of the book.

To make this standard textbook fulfill the needs of today's generation of students, some new features have been introduced in this edition. A new chapter on Light Microscopy and Tissue Preparation has been added to acquaint the students with the basics of histology. Every student of histology is expected to identify the slides and differentiate amongst them in a perfect manner. To make the students familiar with the various slides, Histological Plates have been added in each chapter that include a photomicrograph, line drawing, and salient features that are visible while examining under the microscope.

Each chapter has been rearranged to provide sequential learning to the students. All the diagrams have been redrawn and many new illustrations have been added for easy comprehension of the basic concepts. Clinical and Pathological Correlations have been added at relevant places for creating an interest of the students in the understanding of pathologies associated with various tissues.

For providing an overview of histology to the student and for quick recall, an atlas has been provided at the beginning of the book. The atlas includes more than 80 slides of histological importance along with their important features.

As envisioned by Professor Inderbir Singh, this textbook is of utmost utility not only for the undergraduate students but also for the students pursuing postgraduation in Anatomy. Keeping this in mind, advanced information on various topics has been included as Added Information to cater to the needs of postgraduate students.

The revision of this book was a team effort. We are thankful to our colleagues for their constant encouragement throughout our venture. We extend our heartfelt thanks to our staff in the Histology laboratory for preparing the slides for photography. We are thankful to Dr Sawti Tiwari for her important contribution in drawing some of the figures.

We are grateful to Professor Ivan Damjanov, an esteemed teacher and expert in the field of pathology well known across the globe, for allowing us to use some of the slides from his collection. We gratefully acknowledge Professor Harsh Mohan, a well known surgical pathologist of India, for providing pathological correlations in the book. We are thankful to Dr Sunayna Misra [M.D (Path.), PGI Chandigarh] for her valuable suggestions and inputs especially in the pathological correlations.

We extend our heartfelt thanks to Shri Jitendar P Vij (Group Chairman) and Mr Ankit Vij (Group President) for providing us the opportunity to revise Text of Human Histology and for their persistent support in publication of this book.

Dr Sakshi Arora (Chief Development Editor), the driving force of this endeavour, deserves a special thanks for her tireless efforts. She has perservered throughout this venture with a smile on her face. We are thankful to her entire development team comprising Dr Mrinalini Bakshi, Dr Swati Sinha, and Ms Nitasha Arora (Editors), and Mr Prabhat Ranjan, Mr Neeraj Choudhary, Mr Ankush Sharma, Mr Phool Kumar, Mr Deep Dogra and Mr Sachin Dhawan (Designers and Operators) for providing insights and creative ideas that helped in polishing this book to best meet the needs of students and faculty alike.

We present the 7th edition of this most popular textbook to the medical fraternity as our tribute to a legendary anatomist, Professor Inderbir Singh for being a pillar in the education of generations of doctors throughout the world.

NEELAM VASUDEVA
SABITA MISHRA

Contents

Chapter 5: General Connective Tissue

Chapter 6: Cartilage

Chapter 7: Bone

Chapter 8: Muscular Tissue

Chapter 9: Lymphatics and Lymphoid Tissue

Chapter 10: The Blood and the Mononuclear Phagocyte System

Chapter 11: Nervous System

Chapter 12: Skin and its Appendages

Chapter 13: The Cardiovascular System

Chapter 14: The Respiratory System

Chapter 15: Digestive System: Oral Cavity and Related Structures

Chapter 16: Digestive System: Oesophagus, Stomach and Intestines

Chapter 17: Hepatobiliary System and Pancreas

Chapter 18: The Urinary System

Chapter 19: Male Reproductive System

Chapter 20: Female Reproductive System

Chapter 21: Endocrine System

Chapter 22: Special Senses: Eye

Chapter 23: Special Senses: Ear

Colour Atlas

HISTOLOGY & ITS STUDY

The study of histology is very important for the understanding of the normal functioning of the human body. It also forms the essential basis for the study of the changes in various tissues and organs in disease. (This is the science of pathology). From these points of view the study of histology is best done taking one organ system at a time. That is the approach most teachers prefer to take in practical classes of histology. It is also the basis on which the chapters of this book have been organised.

However, in practical examinations, the emphasis is on the ability of the student to recognise a tissue or organ that is being viewed through a microscope. Here it becomes necessary to know how to distinguish between similar looking tissues or organs belonging to different systems. This atlas has been organised to serve this objective. Tissues and organs that have a similar appearance are considered in one lot. For example, if a slide presents something that looks like a tube, whether it be an artery or the ureter or the ductus deferens, these are considered together. This makes the grouping unusual, but this is exactly what the student needs at the time of an examination.

At the same time it is true that an organ can be composed of several tissues, (or layers), and the ability to recognise them can go a long way is arriving at a correct diagnosis of the organ being seen. We will, therefore, first try to study and identify the various tissues that make up different organs. We will then have a good basis for identifying any organ that we are required to recognise.

BASIC TISSUES THAT CAN BE RECOGNISED IN HISTOLOGICAL SECTIONS

EPITHELIA

The outer surface of the body, and the luminal surfaces of cavities (big or small) lying within the body are lined by one or more layers of cells that completely cover them. Such layers of cells are called epithelia. Epithelial tissue forms the lining of the general body surfaces, passages and cavities within the body. Basement membrane connects the epithelium to the underline subepithelial tissues.

Classification of epithelial tissue is based on shape of the cells, number of cell layers and special modifications seen on the cells. Epithelia may be ***simple***, when they consist of only one layer of cells, or ***stratified*** when there are several layers of cells. Epithelial cells may be flat (or squamous), cuboidal, columnar etc.

Several types of epithelia can be recognised. Learning to identify an epithelium can be of considerable help in finding out what organ you are seeing.

Simple Squamous Epithelium

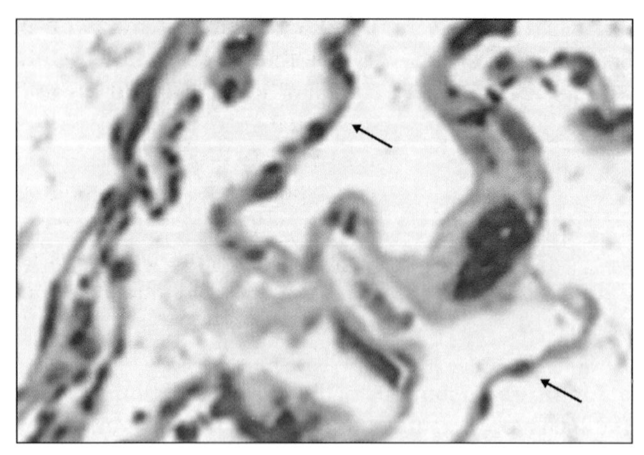

Fig. A1.1: An alveolus of the lung showing a lining of simple squamous epithelium (arrows)

- The cells of this epithelium are flattened.
- In sections they appear so thin that bulgings are produced on the surface by nuclei.
- In surface view (Fig. A1.3) the cells have polygonal outlines that interlock with those of adjoining cells.
- A simple squamous epithelium lines the alveoli of the lungs, the free surfaces of peritoneum, pleura and pericardium. Here it is given the name **mesothelium**. It also lines the inside of blood vessels,where it is called **endothelium**, and of the heart where it is called **endocardium**.

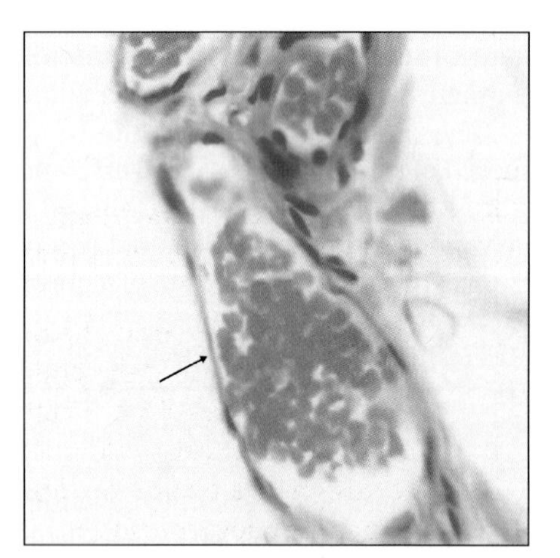

Fig. A1.2: A capillary lined by endothelium (arrow)

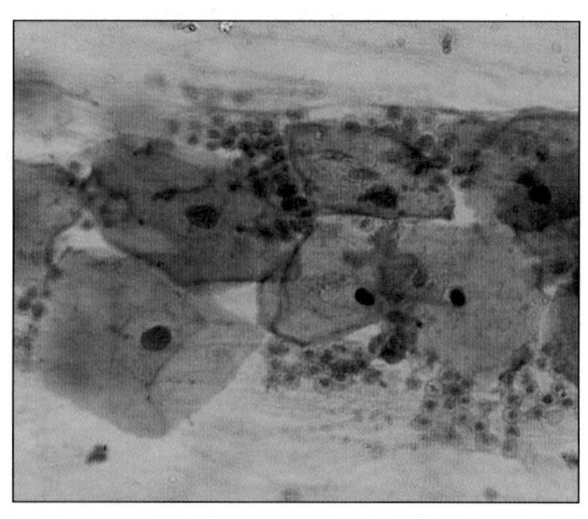

Fig. A1.3: Simple squamous epithelium (surface view)

Simple Cuboidal Epithelium

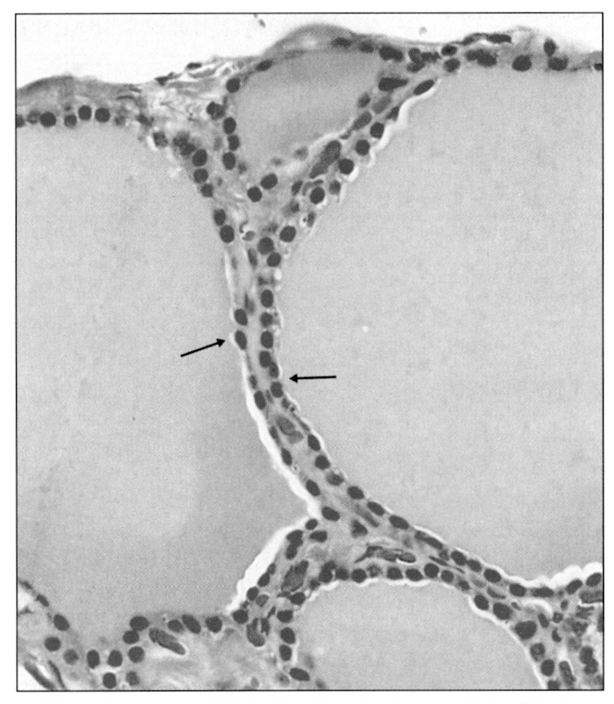

Fig. A1.4: A thyroid follicle lined by simple cuboidal epithelium (arrow)

- The epithelium is made up of cells that look like squares (in which the length and breadth is equal).
- Nuclei are rounded.
- A typical cuboidal epithelium lines follicles of the thyroid gland, kidney tubules, germinal layer of ovary and ducts of various glands.

Simple Columnar Epithelium

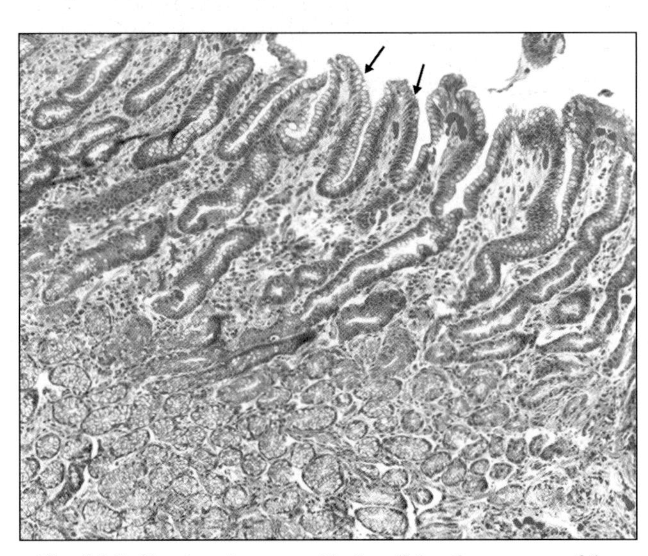

Fig. A1.5: Simple columnar epithelium lining the mucosa of the stomach (arrow)

- In this epithelium the height of the cells is much greater than their width.
- The nuclei are oval being elongated in the same direction as the cells. They lie near the bases of the cells. Because of this we see a zone of clear cytoplasm above the nuclei.
- A simple columnar epithelium lines the mucous membrane of the stomach and of the large intestine.

Columnar Epithelium showing Striated Border

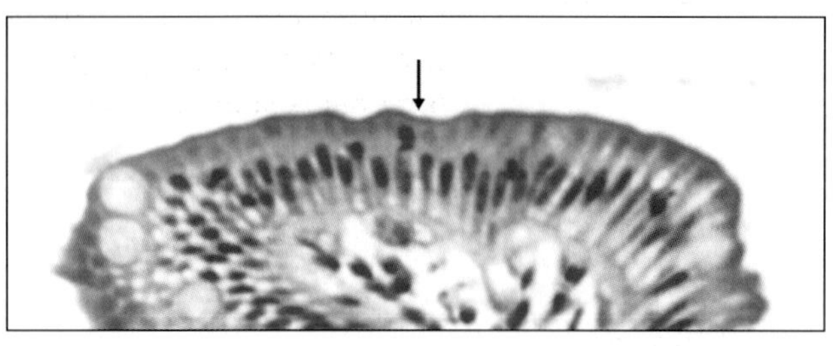

Fig. A1.6: Columnar epithelium with a striated border in the small intestine (arrow)

- In some regions the free surfaces of the cells of columnar epithelium show a thickening with vertical striations in it: this is called a striated border.
- This is seen typically in the small intestine

Pseudostratified Ciliated Columnar Epithelium

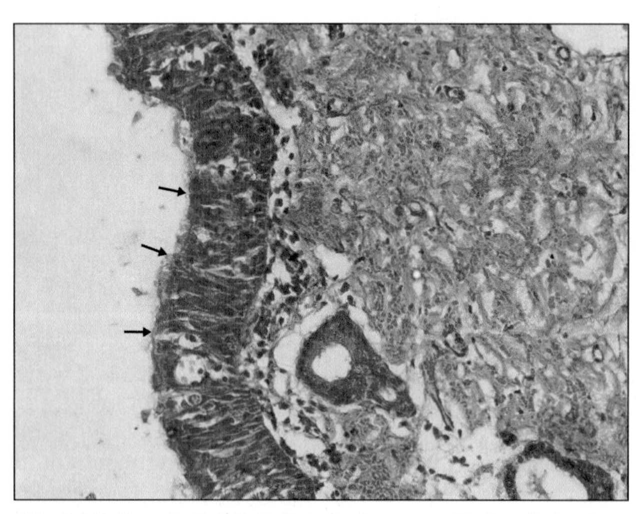

Fig. A1.7: Pseudostratified ciliated columnar epithelium in trachea (arrow)

- Pseudostratified epithelium differs from simple columnar epithelium in that it appears to be multi-layered. However, there is actually only one layer of cells. The multi-layered appearance is due to the fact that the nuclei lie at different levels in different cells. Such an epithelium is seen in the ductus deferens.
- In some situations, pseudostratified columnar epithelium bears hair-like projections called cilia.
- Pseudostratified ciliated columnar epithelium is seen in trachea and in large bronchi.

Pseudostratified Columnar Epithelium with Stereocilia

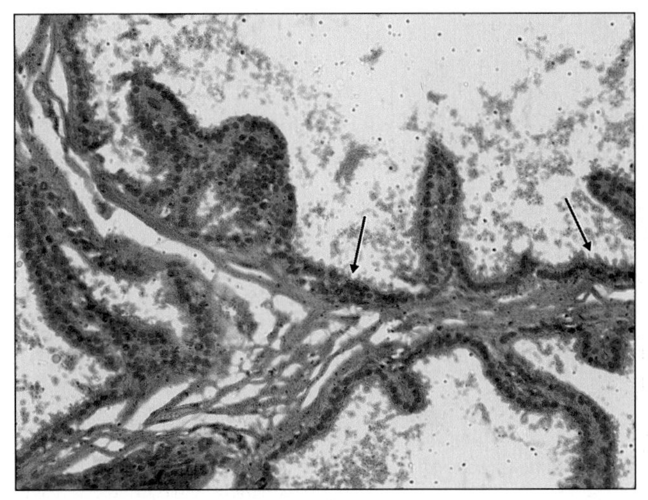

Fig. A1.8: Pseudostratified columnar epithelium with stereocilia in epididymis (arrow)

- In some situations, the pseudostratified columnar epithelium bears stereocilia as seen in epididymis.
- Stereocilia are actually long microvilli and not cilia.

Transitional Epithelium

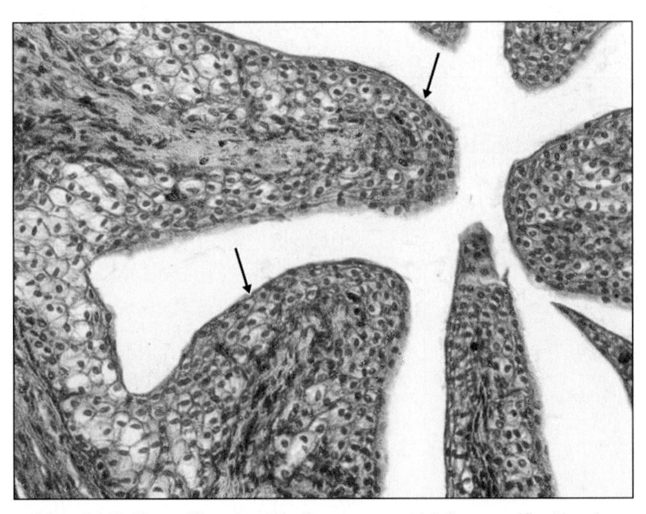

Fig. A1.9: Transitional epithelium seen at high magnification in ureter (arrow)

- In this type of epithelium we see several layers of cells with round nuclei.
- The deepest cells are columnar or cuboidal. The middle layers are made up of polyhedral or pear-shaped cells.
- The cells of the surface layer are large and often shaped like an umbrella
- This epithelium lines many parts of the urinary tract.

Stratified Squamous Epithelium (Non-keratinised)

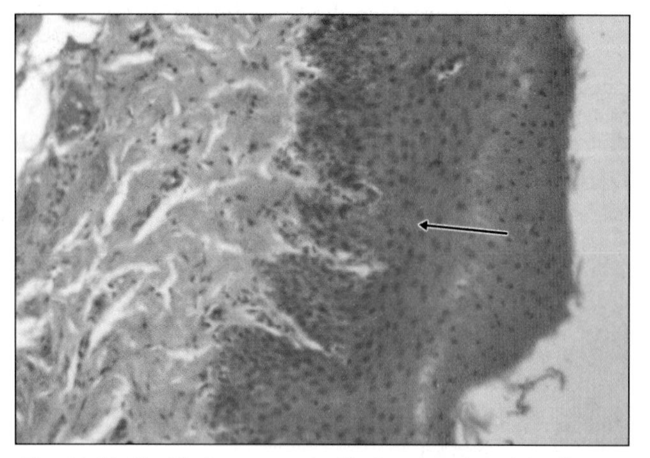

Fig. A1.10: Stratified squamous epithelium (non-keratinised) seen in oesophagus (arrow)

- The stratified epithelium is made up of several layers of cells.
- Although this is called stratified squamous epithelium, only the most superficial cells are squamous (flattened).
- The cells in the deepest (or basal) layer are columnar. In the middle layers they are polyhedral, while the more superficial layers show increasing degrees of flattening.
- The nuclei are oval in the basal layer, rounded in the middle layer, and transversely elongated in the superficial layers.
- There is no superficial keratinized zone; flattened nuclei are seen in the topmost layer.
- This kind of epithelium is seen lining some internal organs like the oesophagus and the vagina.

Stratified Squamous Epithelium (Keratinised)

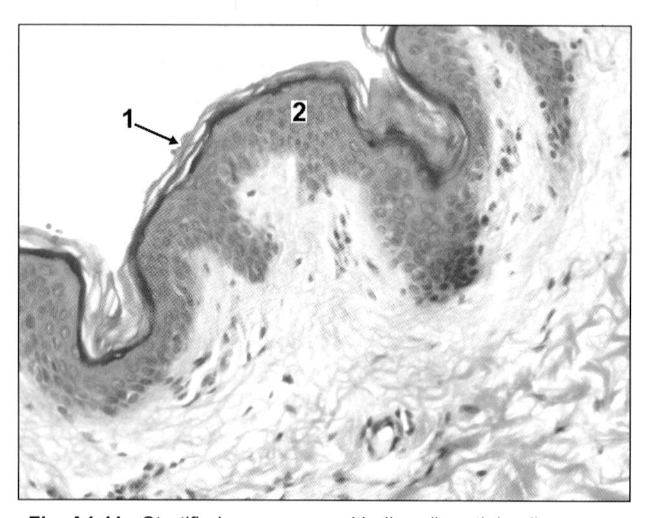

Fig. A1.11: Stratified squamous epithelium (keratinised) as seen in skin

- Here the deeper layer are covered by additional layers that represent stages in the conversion of cells into non-living fibres. This process is called keratinisation (or cornification).
- The surface layer is made up of keratin which appears as fibres. No cellular outline or nuclei can be seen.
- It is seen typically in epidermis of the skin.

Key

1. Keratin
2. Stratified squamous epithelium

CONNECTIVE TISSUE

In most organs there are areas filled in by fibres that are described as connective tissue. The main constituent of connective tissue is collagen fibres that stain pink. In stretch preparations they are seen as wavy bundles. Other fibres present (elastic, reticular) can be seen with special stains. Connective tissue also contains many cells but only their nuclei can be made out.

Irregular Connective Tissue

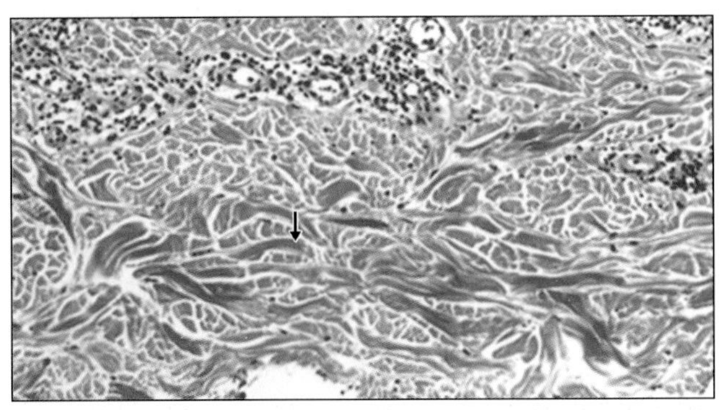

Fig. A2.1: Irregular connective tissue as seen in dermis of skin (arrow)

- Irregular connective tissue is typically seen in dermis of skin.
- It consists of compactly packed bundles of collagen fibres that are not arranged in orderly fashion.
- Thin elastic fibres are present, but are not seen with H & E stain.

Regular Connective Tissue

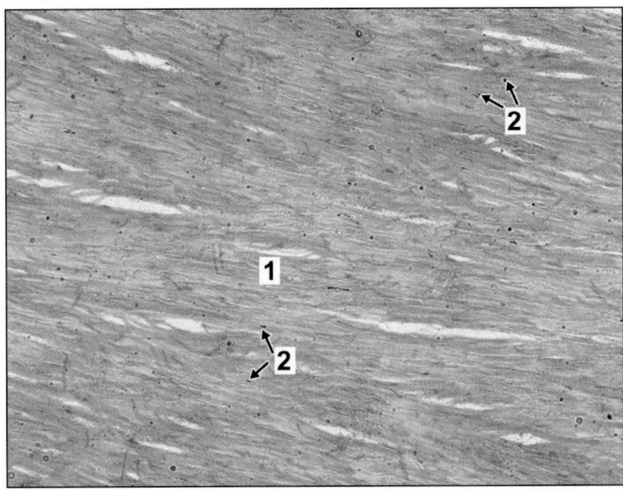

Fig. A2.2: Regular connective tissue as seen in tendon

- Tendons are also made up of collagen fibres, but here the fibres (or fibre bundles) are arranged in orderly fashion parallel to each other.
- Nuclei of some cells (mainly fibroblasts) are seen between the bundles of collagen. They are elongated (elliptical).

Key

1. Collagen fibres
2. Nuclei of fibroblasts

Adipose Tissue

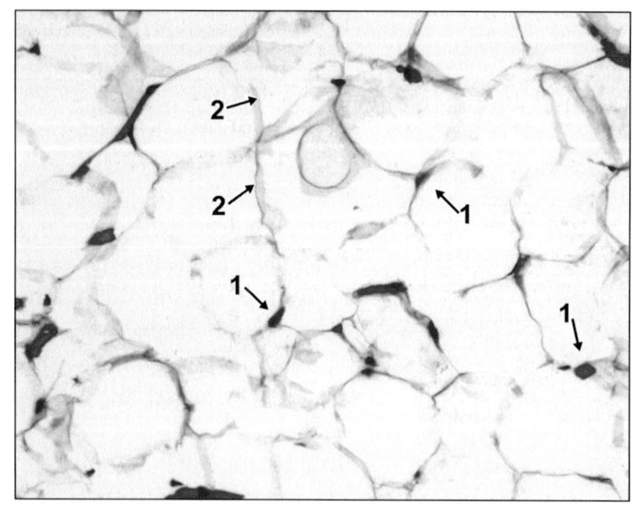

Fig. A2.3: Adipose tissue

- Adipose tissue is made up mainly of compactly arranged fat cells.
- In routine sections the cells appear empty as the fat gets dissolved during preparation of the section.
- The cytoplasm of each cell is seen as a pink 'rim'.
- The nucleus is flat and lies to one side.

Key

1. Nuclei of adipocyte
2. Thin rim of cytoplasm

SPECIALISED CONNECTIVE TISSUE

CARTILAGE

Unlike connective tissue, that can be deformed easily, cartilage is a special form of connective tissue that is firm, and retains its shape.

Hyaline Cartilage

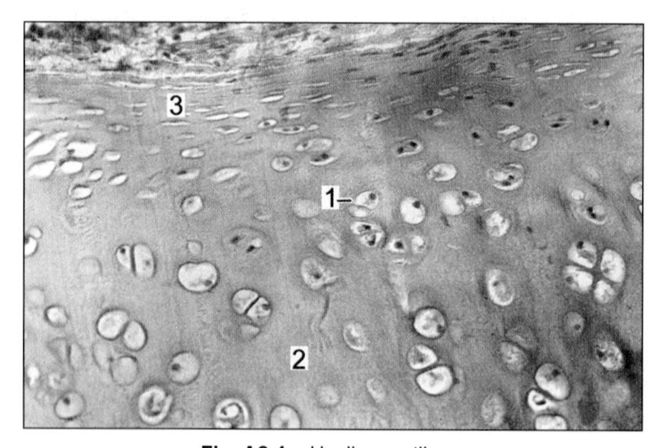

Fig. A2.4: Hyaline cartilage

- It is characterised by groups of cartilage cells (chondrocytes) surrounded by a homo-geneous matrix which separates the cells widely.
- Near the surface of the cartilage the cells are flattened and merge with the cells of the overlying connective tissue. This connective tissue forms the perichondrium.
- Costal cartilage and articular cartilage of synovial joint are example of hyaline cartilage.

Key

1. Chondrocytes
2. Homogenous matrix
3. Perichondrium

Elastic Cartilage

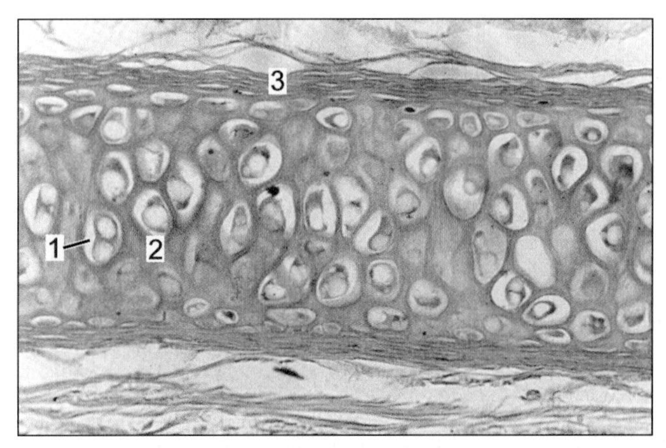

Fig. A2.5: Elastic cartilage

- In elastic cartilage chondrocytes are surrounded by matrix containing many elastic fibres
- Perichondrium covering is present over the cartilage
- It is seen typically in auricle and epiglottis

Key

1. Chondrocytes
2. Elastic fibres
3. Perichondrium

Fibrocartilage

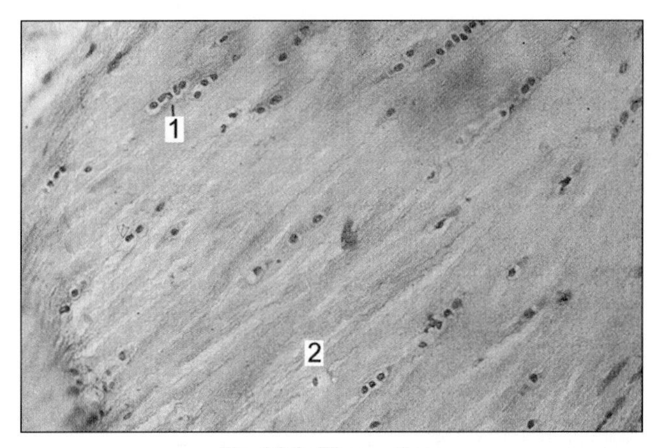

Fig. A2.6: Fibrocartilage

- Fibrocartilage is characterized by presence of collagen fibres arranged in bundles with rows of chondrocytes intervening between the bundles
- Perichondrium is absent
- Fibrocartilage is seen typically in pubic symphysis and manubrio sternal joint

Key

1. Chondrocytes
2. Bundles of collagen fibres

Fibrocartilage can be confused with the appearance of a tendon. Note that chondrocytes in fibrocartilage are rounded, but the cells in a tendon (fibrocytes) are flattened and elongated.

BONE

Compact Bone

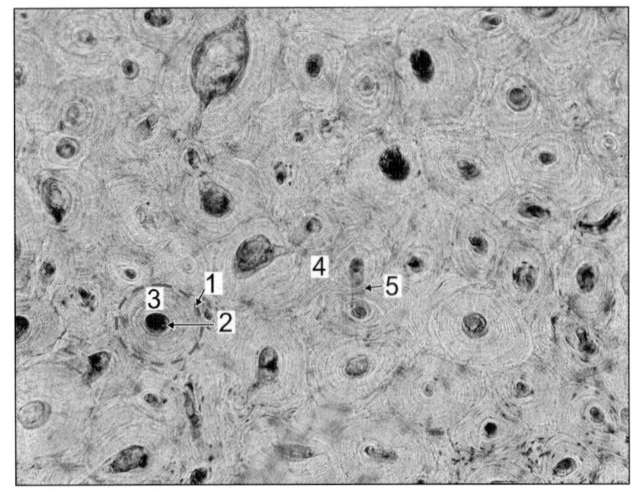

Fig. A2.7: Compact bone; transverse section

- A transverse section through the compact bone shows ring-like osteons (or Haversian systems).
- Haversian canal is seen at the centre of each osteon.
- Around the canal there are concentric lamellae of bone amongst which there are small spaces called lacunae in which osteocytes are present.
- Delicate canaliculi radiate from the lacunae containing processes of osteocytes.
- Interstitial lamellae fill intervals between Haversian systems.
- Volkmanns canal interconnecting the adjacent haversian canal may be seen

Note: The appearance of compact bone is so characteristic that you are not likely to confuse it with any other tissue.

Key

1. Haversian system (osteon)
2. Haversian canal
3. Concentric lamellae
4. Interstitial lamellae
5. Volkmann's canal

Spongy (Cancellous) Bone

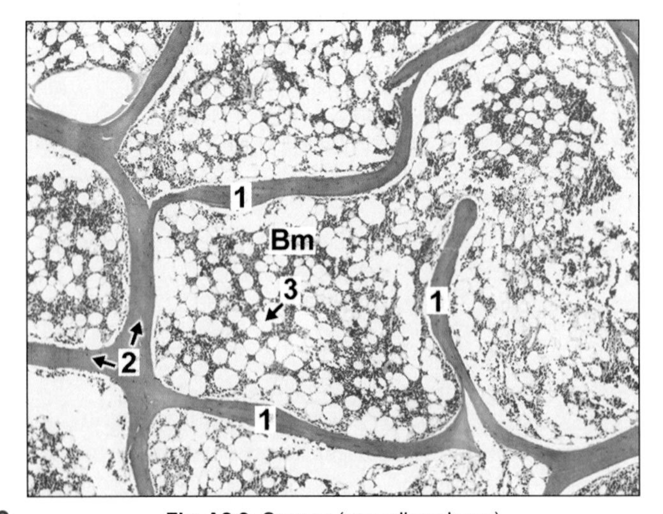

Fig. A2.8: Spongy (cancellous bone)

- It is made up of a network of bony trabeculae (pink) in which the nuclei of some osteocytes can be seen.
- The spaces of the network are filled in by bone marrow in which numerous fat cells are present.
- The spaces between the fat cells are occupied by numerous blood forming cells (only nuclei of which are seen).

Key

1. Trabeculae
2. Osteocytes
3. Fat cells

Bm = Bone marrow

MUSCLE

Skeletal Muscle

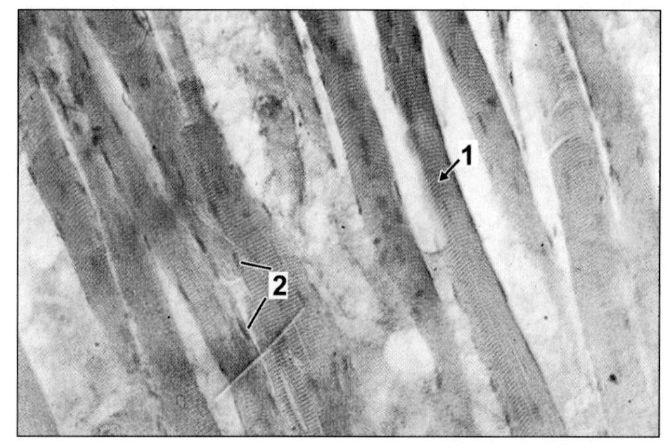

Fig. A3.1: Longitudinal section through skeletal muscle

- In a longitudinal section through skeletal muscle the fibres are easily distinguished as they show characteristic transverse striations.
- The fibres are long and parallel without branching.
- Many flat nuclei are placed at the periphery.
- The muscle fibres are separated by some connective tissue.

Key

1. Muscle fibres with transverse striations
2. Peripherally placed nuclei

Smooth Muscle

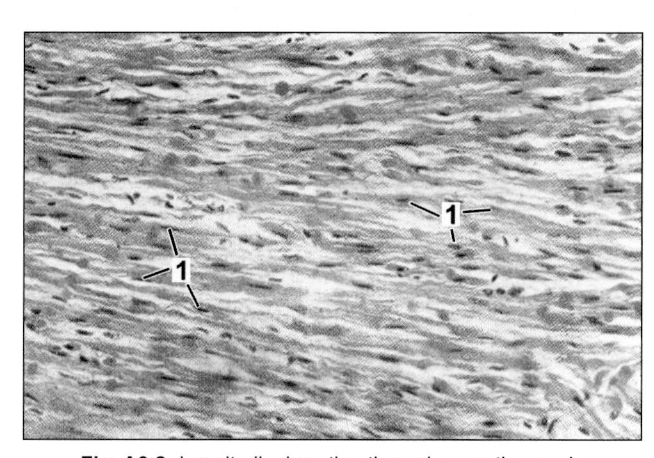

Fig. A3.2: Longitudinal section through smooth muscle

- In a longitudinal section through smooth muscle elongated spindle shaped cells without straitions are seen.
- A single elongated (oval) centrally placed nucleus can be identified.
- Smooth muscle is present in the walls of parts of the alimentary canal, in the urogenital tract etc.

Key

1. Oval centrally placed nuclei

Cardiac Muscle

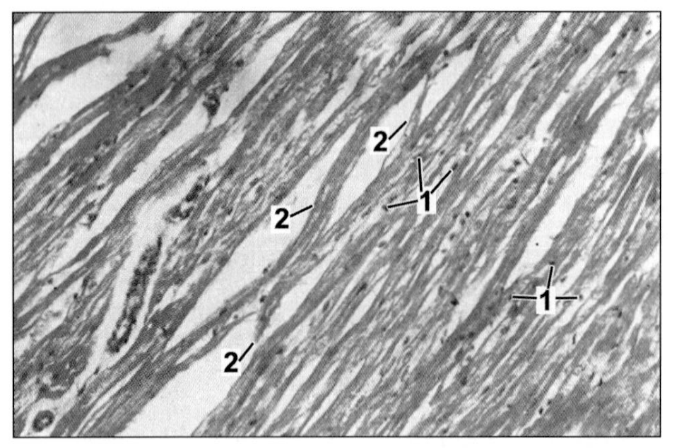

Fig. A3.3: Longitudinal section through cardiac muscle

- The fibres are made up of 'cells' each of which has a centrally placed nucleus and transverse striations.
- Adjacent 'cells' are separated from one another by transverse lines called intercalated discs.
- Fibres show branching.

Key

1. Centrally placed nuclei
2. Branching and anastomosing of fibres

Nervous Tissue

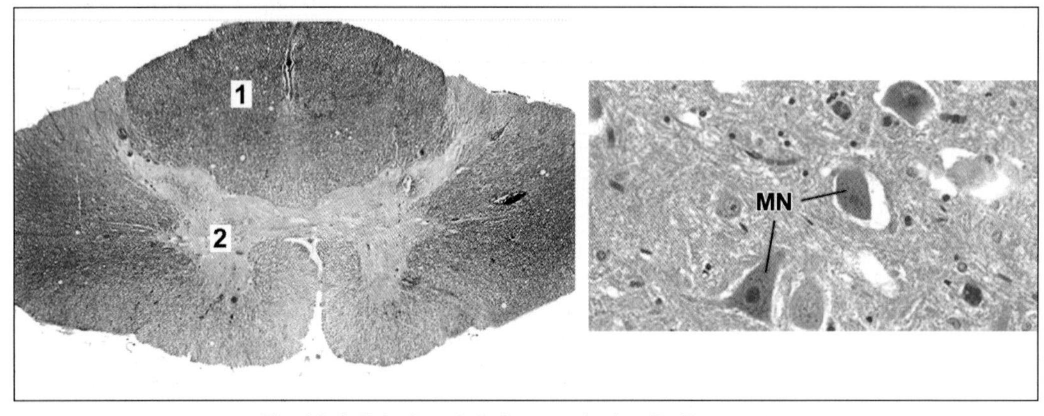

Fig. A3.4: Spinal cord. **A.** Panoramic view **B.** Grey matter

- Grey matter contains cell bodies of neurons and neuroglia
- White matter contains axons of neurons and neuroglia

Key

1. White matter
2. Grey matter

MN. Multipolar neurons

STRUCTURES THAT ARE USUALLY SEEN AS SINGLE TUBES

ARTERIES

The structure of an artery varies greatly with its size. Each artery shows three layers, the tunica intima, tunica media and the tunica adventitia (in internal to external order). The lumen is lined by endothelium (flattened cells).

Elastic Artery

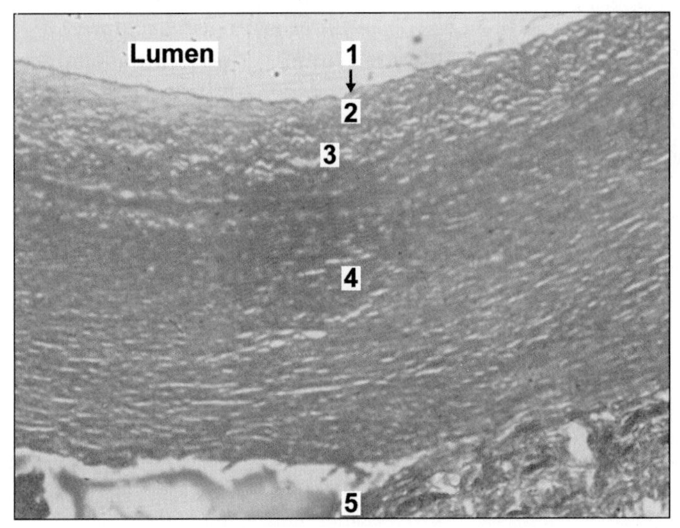

Fig. A4.1: Elastic artery

Elastic artery is characterised by presence of

- Tunica intima consisting of endothelium, subendothelial connective tissue and internal elastic lamina.
- The first layer of elastic fibres of tunica media is considered the internal elastic lamina.
- Thick tunica media with many elastic fibres and some smooth muscle fibres.
- Tunica adventitia containing collagen fibres and vasa vasorum.

Key

1. Endothelium
2. Subendothelial connective tissue
3. Internal elastic lamina

(1–3: Tunica intima)

4. Tunica media
5. Tunica adventitia

Muscular Artery

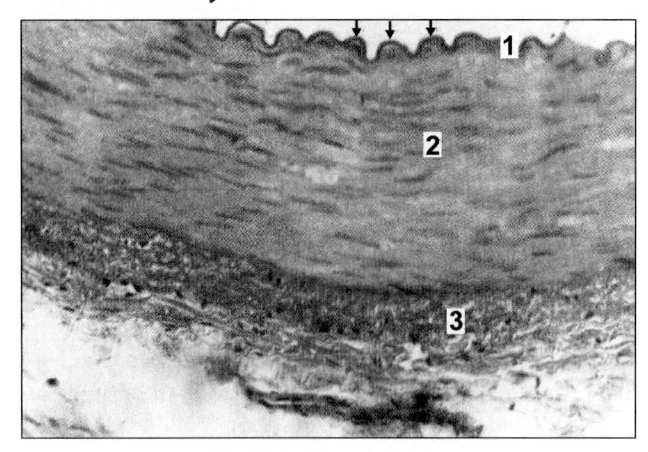

Fig. A4.2: Muscular artery

- In muscular arteries, the tunica intima is made of endothelium and internal elastic lamina (arrow) which is thrown into wavy folds due to contraction of smooth muscle in the media.
- Tunica media-is composed mainly of smooth muscle fibres arranged circularly.
- Tunica adventitia-contains collagen fibres.

Key

1. Tunica intima
2. Tunica media
3. Tunica adventitia

Vein

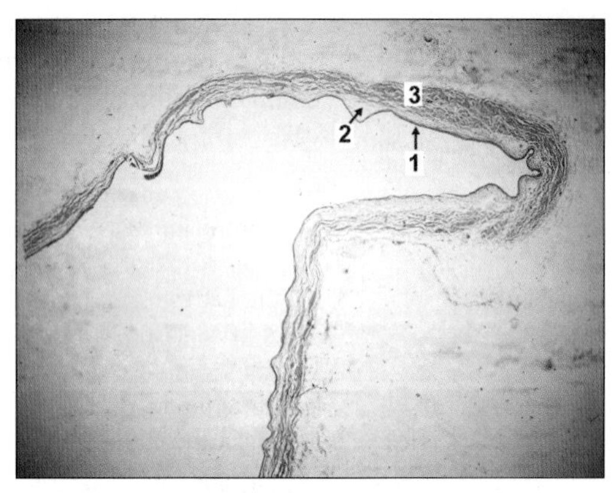

Fig. A4.3: Large vein

The vein has a thinner wall and a larger lumen than the artery. The tunica intima, media and adventitia can be made out, but they are not sharply demarcated. The media is thin and contains only a small quantity of muscle. The adventitia is relatively thick. Note again that the luminal surface appears as a dark line, with an occasional nucleus along it.

Key

1. Tunica intima
2. Tunica media
3. Tunica adventitia

VERMIFORM APPENDIX

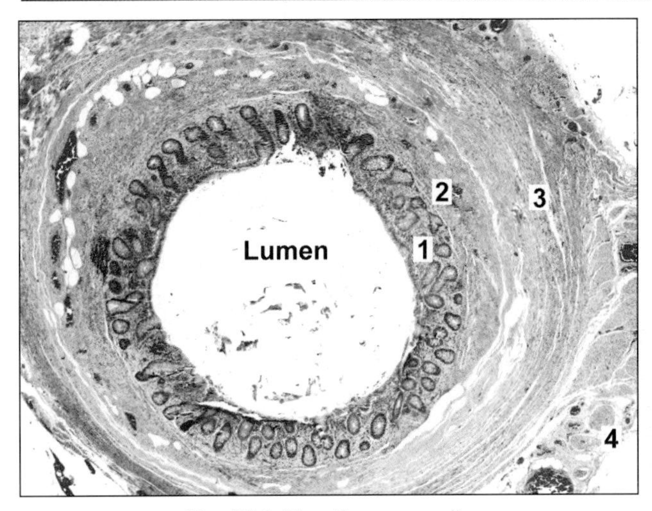

Fig. A5.1: Vermiform appendix

- The vermiform appendix is the narrowest part of the gastrointestinal canal and is seen as a tubular structure.
- The innermost layer of the mucosa, is lined by simple columnar epithelium with goblet cells and lymphocytes.
- Crypts are seen in lamina propria.
- The next layer, the submucosa may show a variable number of lymphatic nodules.
- The submucosa is surrounded by smooth muscle layer (muscularis externa) followed by serosa.
- The longitudinal muscle coat is complete and equally thick all round. Taenia coli as seen in colon are not present.

Key

1. Mucosa
2. Submucosa
3. Muscularis externa
4. Serosa

Vermiform appendix can be easily recognised due to its tubular form bearing resemblance to the colon with presence of lymphoid tissue.

URETER

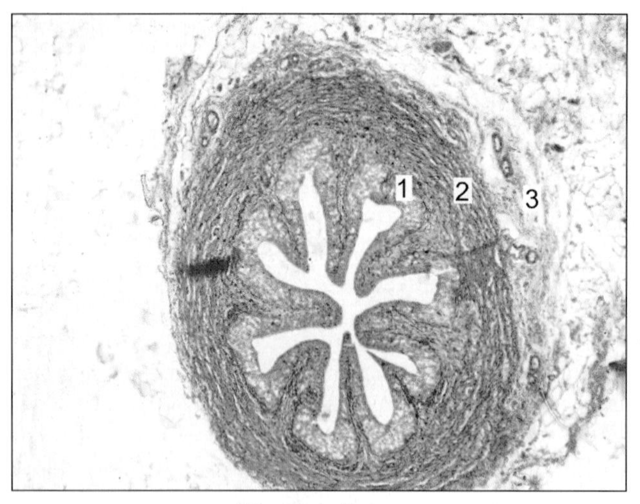

Fig. A6.1: Ureter

- The ureter can be recognised because it is tubular and its mucous membrane is lined by transitional epithelium.
- The epithelium rests on a layer of connective tissue (lamina propria).
- The mucosa shows folds that give the lumen a star shaped appearance.
- The muscle coat has an inner layer of longitudinal fibres and an outer layer of circular fibres. This arrangement is the reverse of that in the gut.
- The muscle coat is surrounded by adventitia made of fibroelastic connective tissue in which blood vessels and fat cells are present.

Key

1. Mucosa comprising of transitional epithelium and lamina propria
2. Muscle coat
3. Adventitia

DUCTUS DEFERENS

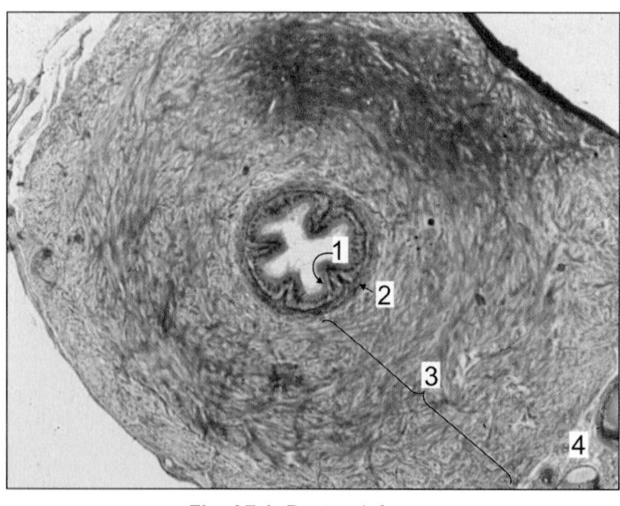

Fig. A7.1: Ductus deferens

- This is a tube that is distinguished from the ureter as its mucous membrane is lined by pseudostratified columnar epithelium.
- The muscle coat is very thick. Three layers, inner longitudinal, middle circular and outer longitudinal are seen.
- The muscle coat is surrounded by adventitia containing blood vessels and nerves.

Key

1. Small lumen lined by pseudostratified columnar epithelium
2. Submucosa
3. Muscle layer
4. Adventitia

UTERINE TUBE

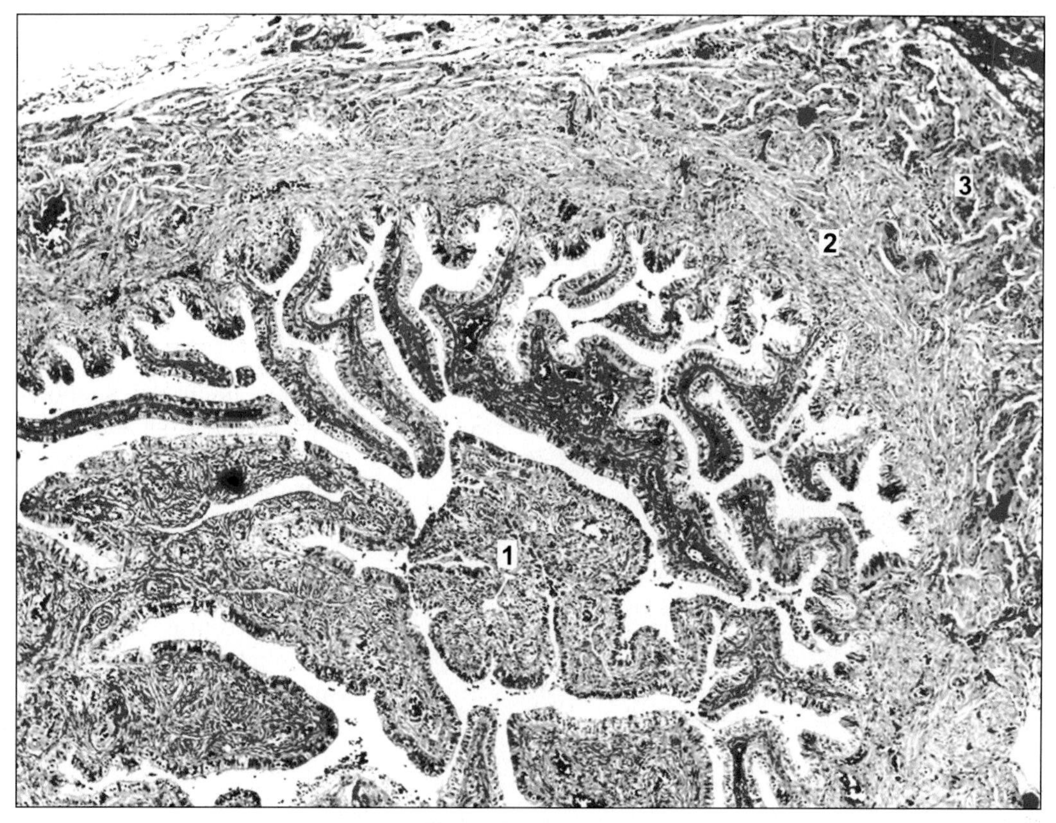

Fig. A8.1: Uterine tube

- The uterine tube is characterized by presence of numerous branching mucosal folds that almost fill the lumen of the tube.
- The mucosa is lined by ciliated columnar epithelium.
- The uterine tube has a muscular wall with an inner circular and outer longitudinal muscle layer.

Key

1. Mucous membrane with numerous branching folds
2. Inner circular muscle layer
3. Outer longitudinal muscle layer

STRUCTURES MADE UP MAINLY OF LYMPHOID TISSUE

LYMPHOID TISSUE

Lymphocytes are one variety of cells of blood. Collections of them are frequently seen in many tissues. Such aggregations constitute lymphoid tissue. Such tissue is seen in the form of aggregations of dark staining nuclei. Some organs (lymph nodes, spleen) are made up almost entirely of such tissue. At some sites lymphoid tissue shows nodules where the lymphocytes are more densely packed than elsewhere. The nodule may show a central area that is lighter staining (because the cells are less densely packed)

LYMPH NODE

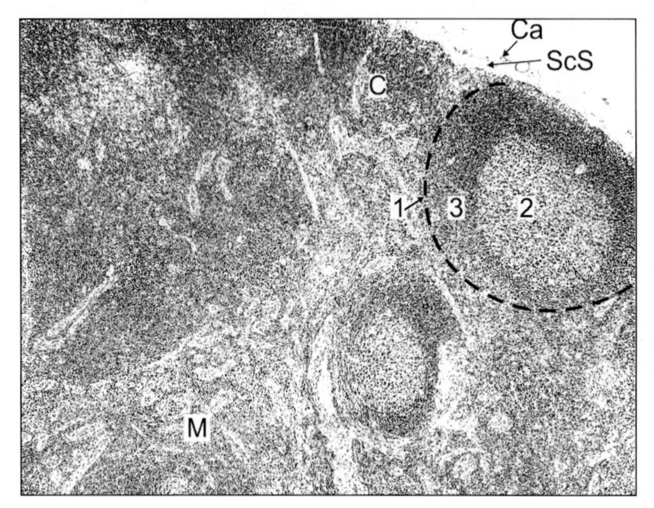

Fig. A9.1: Lymph node

- A lymph node has an outer cortex and an inner medulla.
- The cortex is packed with lymphocytes. A number of rounded lymphatic follicles (or nodules) are present. Each nodule has a pale staining germinal centre surrounded by a zone of densely packed lymphocytes.
- Within the medulla the lymphocytes are arranged in the form of anastomosing cords. Several blood vessels can be seen in the medulla.
- A thin capsule surrounds the lymph node, sending trabeculae into the cortex.
- Beneath the capsule is a clear space called subcapsular sinus.

Note: All lymphoid tissue are easily recognised due to presence of aggregation of dark staining nuclei. The nuclei belong to lymphocytes.

Key

1. Lymphatic nodule
2. Germinal centre
3. Zone of dense lymphocytes

C. Cortex
M. Medulla
Ca. Capsule
ScS. Subcapsular sinus

SPLEEN

It is characterized by–

- A thick capsule with trabecula extending from it into the organ (not shown in photomicrograph)
- The substance of the organ is divisible into the red pulp in which there are diffusely distributed lymphocytes and numerous sinusoids; and the white pulp in which dense aggregations of lymphocytes are present. The latter are in the form of nodules surrounding arterioles.

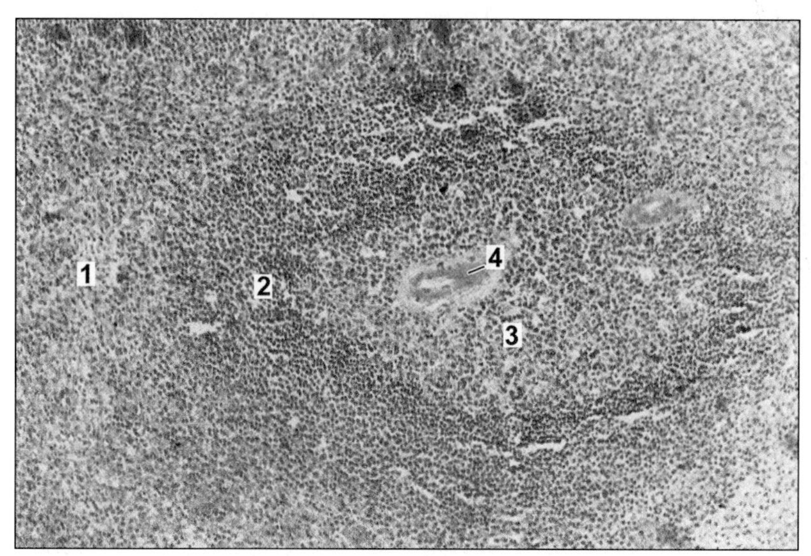

Key

1. Red pulp
2. White pulp
3. Germinal centre
4. Arteriole present in an eccentric position

Fig. A10.1: Spleen

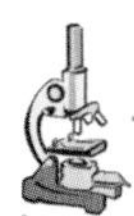

When cut transversely the cords of spleen resemble the lymphatic nodules of lymph nodes, and like them they have germinal centres surrounded by rings of densely packed lymphocytes. However, the nodules of the spleen are easily distinguished from those of lymph nodes because of the presence of an arteriole in each nodule. The arteriole is placed eccentrically at the margin of the germinal centre.

THYMUS

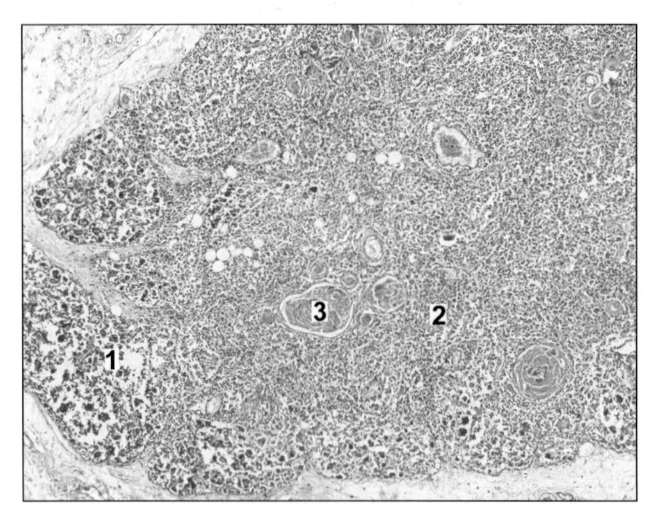

Fig. A11.1: Thymus

Key

1. Cortex
2. Medulla
3. Corpuscle of Hassall

In the slide it can be seen that

- The thymus is made up of lymphoid tissue arranged in the form of distinct lobules. The presence of this lobulation enables easy distinction of the thymus from all other lymphoid organs.
- The lobules are partially separated from each other by connective tissue.
- In each lobule an outer darkly stained cortex (in which lymphocytes are densely packed); and an inner lightly stained medulla (in which the cells are diffuse) are present.
- The medulla contains pink staining rounded masses called the corpuscles of Hassall.

PALATINE TONSIL

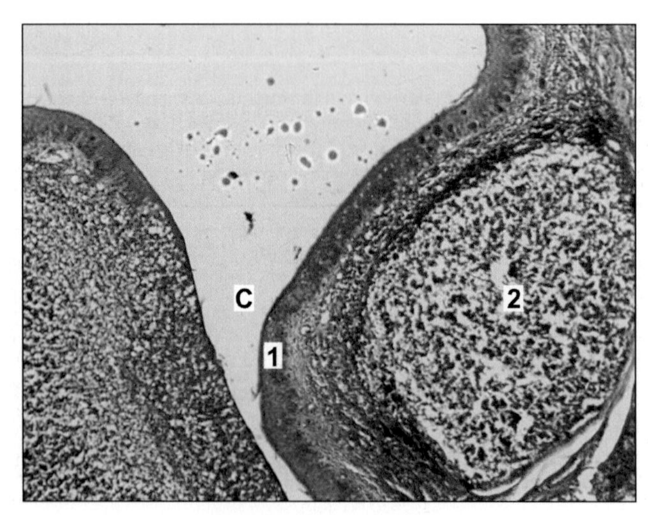

Fig. A12.1: Palatine tonsil

- Palatine tonsil is an aggregation of lymphoid tissue that is readily recognized by the fact that it is covered by a stratified squamous epithelium on its oral surface.
- At places the epithelium dips into the tonsil in the form of deep crypts.
- Deep to the epithelium there is diffuse lymphoid tissue in which typical lymphatic nodules can be seen.

Key

C. Crypt
1. Stratified squamous epithelium
2. Lymphatic nodule

SOME STRUCTURES COVERED BY STRATIFIED SQUAMOUS EPITHELIUM

SKIN

The skin consists of two layers. The most superficial layer is the epidermis which consists of stratified squamous epithelium (keratinised). The epidermis rests on a thick layer of connective tissue which is called the dermis.

Thick Skin

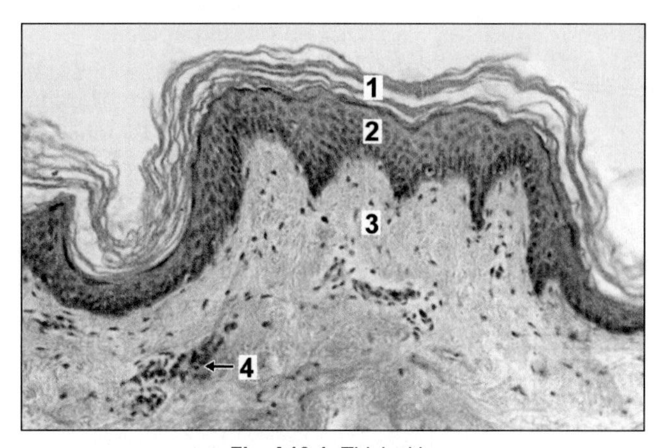

Fig. A13.1: Thick skin

Thick or glabrous skin is characterized by

- Presence of thick epidermis made up of keratinized stratified squamous epithelium (stratum corneum is very thick)
- Absence of hair follicles and sebaceous glands
- Presence of sweat glands in the dermis.
- Thick skin is found in palm of hand and sole of foot.

Key

1. Keratin
2. Epidermis (stratified squamous epithelium)
3. Dermis
4. Sweat glands

Thin Skin

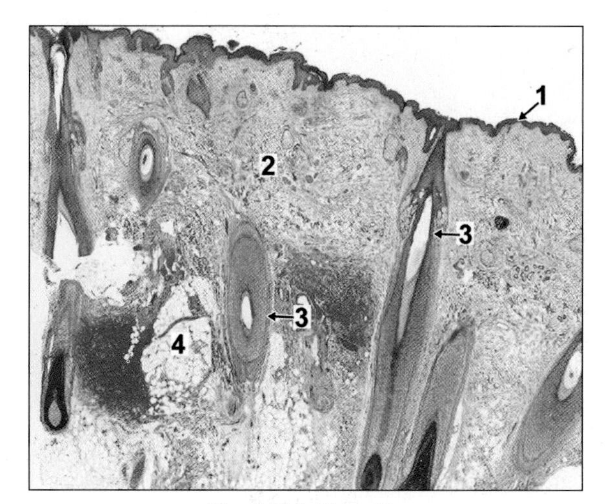

Fig. A13.2: Thin skin

Thin skin or hairy skin is characterized by

- Presence of thin epidermis made up of keratinized stratified squamous epithelium (stratum corneum is thin)
- Presence of hair follicles, sebaceous glands and sweat glands in the dermis.
- Thin skin is found in all parts of body except palm of hand and sole of foot.

Key

1. Epidermis
2. Dermis
3. Hair follicle
4. Sebaceous gland

TONGUE

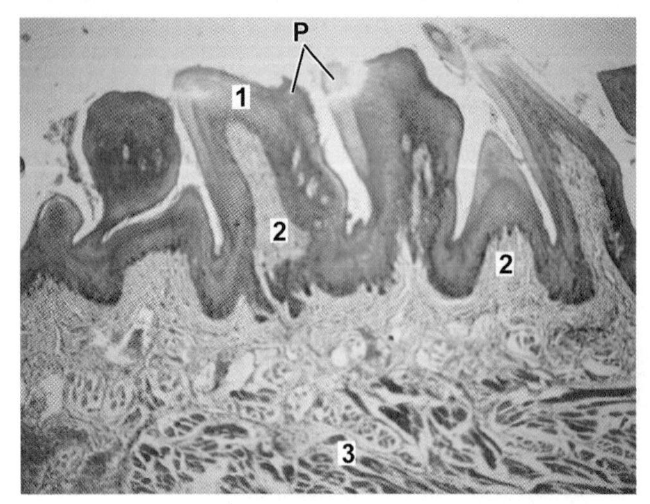

Fig. A14.1: Tongue

- The tongue is another structure covered on both surfaces by stratified squamous epithelium (non-keratinised).
- The undersurface of the tongue is smooth, but on the dorsum the surface shows numerous projections or papillae.
- Each papilla has a core of connective tissue (lamina propria) covered by epithelium. Some papillae are pointed (filiform), while others are broad at the top (fungiform). A third type of papilla is circumvallate, the top of this papilla is broad and lies at the same level as the surrounding mucosa.
- The main mass of the tongue is formed by skeletal muscle seen below the lamina propria. Muscle fibres run in various directions so that some are cut longitudinally and some transversely. Numerous serous glands and mucous glands are present amongst the muscle fibres.

Key

P. Papillae
1. Stratified squamous epithelium
2. Lamina propria
3. Skeletal muscle

OESOPHAGUS

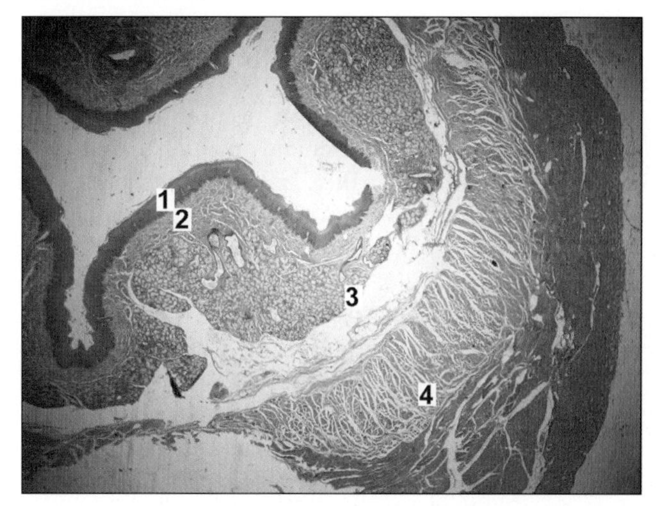

Fig. A15.1: Oesophagus (Low power)

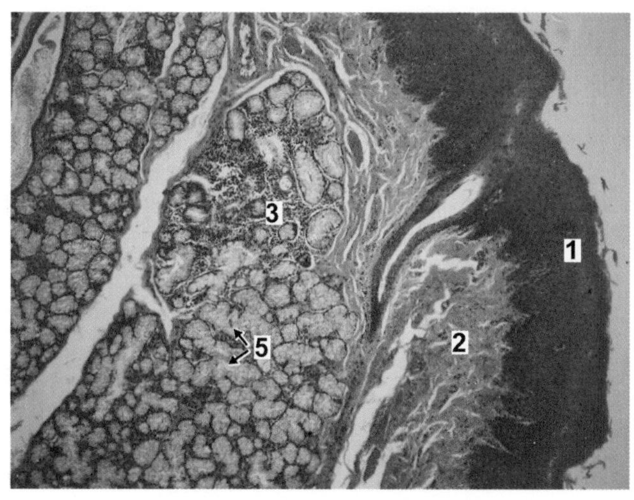

Fig. A15.2: Oesophagus (High power)

In transverse section the oesophagus shows the following layers (from within outwards):

- Lining of non-keratinised stratified squamous epithelium.
- The underlying connective tissue of the lamina propria.
- The muscularis mucosae in which the muscle fibres are cut transversely.
- The lining epithelium, lamina propria and muscularis mucosa collectively constitute the mucosa.
- The submucosa having oesophageal glands (mucous secreting).
- The layer of circular muscle, and the layer of longitudinal muscle constituting the muscularis externa.

Note: In muscularis externa the muscle is of the striated variety in the upper one third of the oesophagus, mixed in the middle one third, and smooth in the lower one third.

Key

1. Muscosa lined by stratified squamous epithelium
2. Lamina propria
3. Submucosa
4. Muscularis externa
5. Submucous glands

VAGINA

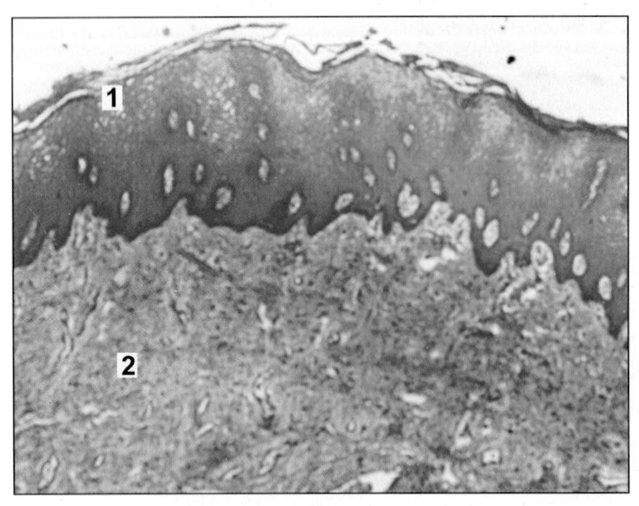

Fig. A22.1: Vagina

Key

1. Stratified squamous non-keratinised epithelium
2. Lamina propria

- The vagina is a fibromuscular structure consisting of an inner mucosa, a middle muscular layer and an outer adventitia.
- The mucosa consists of Stratified squamous non-keratinised epithelium.
- Loose fibroelastic connective tissue (lamina propria) with many blood vessels and no glands.
- The mucosa of vagina is rich in glycogen and hence stains palely which distinguishes it from oesophagus.
- Muscular layer consists of smooth muscle fibres.

The structure of the vagina has a superficial resemblance to that of the urinary bladder. However, the two can be distinguished by the fact that the mucosa of the vagina is lined by stratified squamous epithelium (non-keratinised).

CORNEA

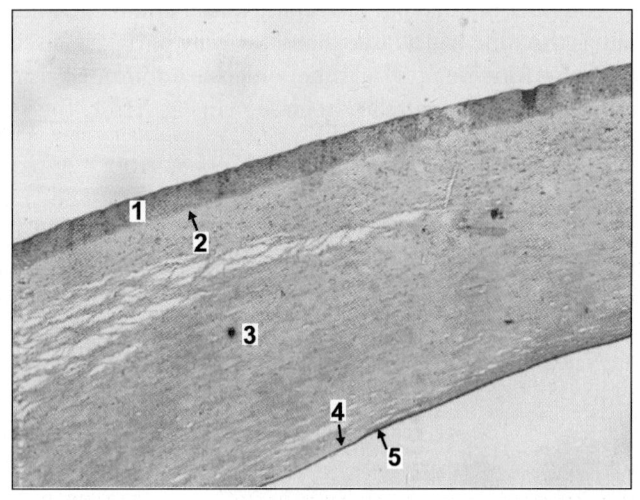

Fig. A16.1: Cornea

The cornea is made up of five layers

- The outermost layer is of non-keratinised stratified squamous epithelium (corneal epithelium).
- The corneal epithelium rests on the structureless anterior limiting lamina (also called Bowman's membrane)
- Most of the thickness of the cornea is formed by the substantia propria (or corneal stroma) made up of collagen fibres embedded in a ground substance.
- Deep to the substantia propria there is a thin homogeneous layer called the posterior limiting lamina.
- The posterior surface of the cornea is lined by a single layer of flattened or cuboidal cells.

Note: The structure of the cornea is fairly distinctive and its recognition should not be a problem.

Key

1. Stratified sqaumous corneal epithelium
2, Anterior limiting membrane (Bowman's)
3. Substantia propria
4. Posterior limiting membrane (Descemet's)
5. Posterior cuboidal epithelium

SOME ORGANS IN WHICH TISSUES ARE ARRANGED IN PROMINENT LAYERS

In this group we will consider organs that have a thick wall and a fairly large lumen. Some of these are tubular, but as the tube has a large diameter, only part of it is seen in a section. The wall in most of these organs is made up of an inner mucosa, a submucosa and layers of muscle. One such organ, the oesophagus, has already been seen in Fig. A15.1. The vermiform appendix (Fig. A5.1) also has a similar structure.

STOMACH

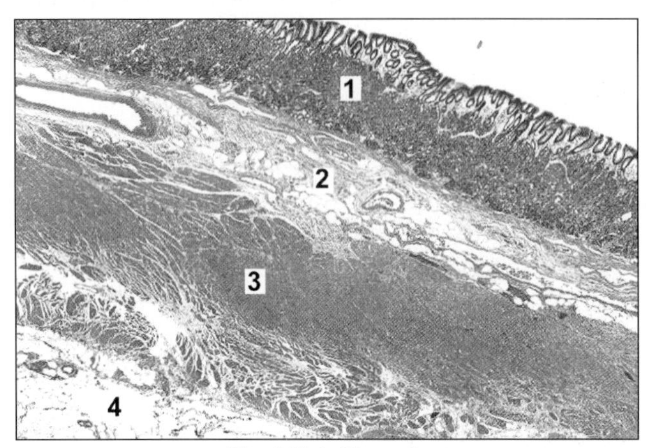

Fig. A17.1: Stomach

The basic structure of stomach is similar to oesophagus i.e. it is composed of (from within outwards):

- Mucosa
- Submucosa
- Muscularis externa
- Serosa

- Mucosa is lined by simple tall columnar epithelium. It shows invaginations called gastric pits.
  - *Lamina propria* contains gastric glands.
  - *Muscularis mucosa* is made of layers of smooth muscle.
- Submucosa consists of fibroelastic connective tissue, blood vessels and meissner's nerve plexus.
- Muscularis externa is composed of three layers of smooth muscle–inner oblique, middle circular and outer longitudinal.
- Serosa is visceral peritoneum (Simple squamous epithelium) over a layer of loose connective tissue.

Key

1. Mucosa
2. Submucosa
3. Muscularis externa
4. Serosa

SMALL INTESTINE

The basic structure of small intestine is similar to oesophagus and stomach i.e. it is composed of:

- Mucosa
- Submucosa
- Muscularis externa
- Serosa

- Mucosa is made of simple columnar absorptive epithelium with goblet cells. The epithelium and the underlying lamina propria shows finger-like evaginations called *intestinal villi.* Epithelium also shows tubular invagination from the base of the villi into the lamina propria known as *crypts of Lieberkuhn* (intestinal glands). These crypts are lined by columnar and goblet cells.
  - Lamina propria consisting of connective tissue
  - Muscularis mucosa is made of smooth muscle fibres. This layer is responsible for movement and folding of mucosa
- Submucosa shows presence of *Brunner's gland in duodenum* and *Peyer's patches in ileum*
- Muscularis externa and serosa corresponds exactly with stomach.

JEJUNUM

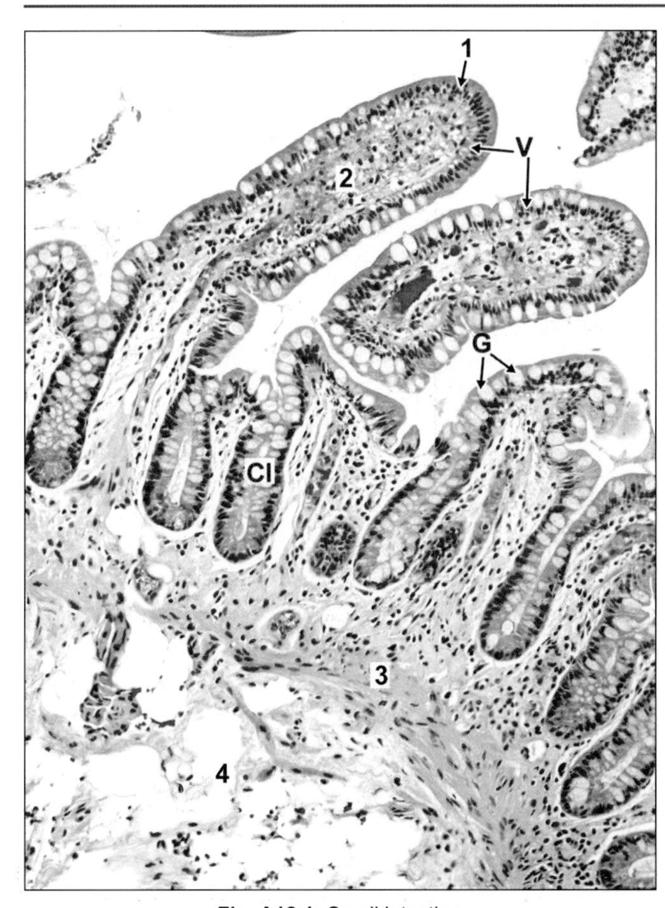

Fig. A18.1: Small intestine

The structure of jejunum should be regarded as typical for small intestine.

Key

1. Lining epithelium
2. Lamina propria
3. Muscularis mucosa

(1–3: Mucosa)

4. Submucosa

V. Villi

Cl. Crypts of Lieberkuhn

G. Goblet cells

Duodenum

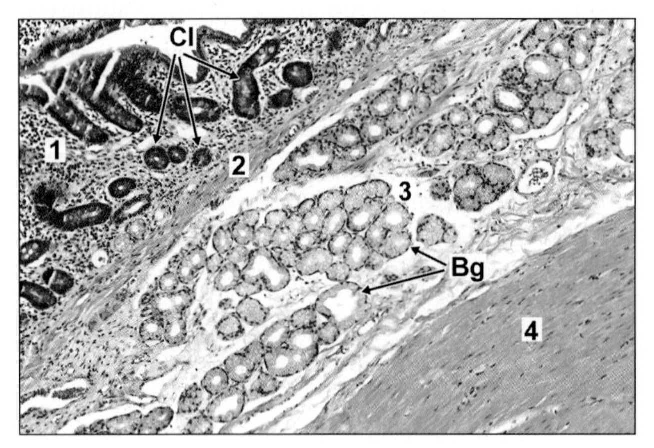

Fig. A18.2: Duodenum

- The general structure of the duodenum is the same as that described for the small intestine, except that the submucosa is packed with mucous secreting glands of Brunner.

Key

1. Lamina propria
2. Muscularis mucosa
3. Submucosa
4. Muscularis externa

Bg. Brunner's gland present in submucosa

Cl. Crypts of Lieberkuhn present in lamina propria

Note: Intestinal crypts lie 'above' the muscularis mucosae while the glands of Brunner lie 'below' it. The presence of the glands of Brunner is a distinctive feature of the duodenum.

Ileum

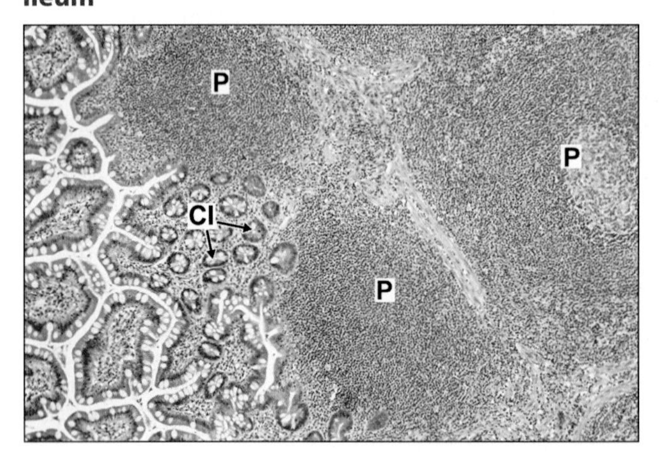

Fig. A18.3: Ileum

The general structure of the ileum is similar to that of the jejunum except for

- The entire thickness of the lamina propria is infiltrated with lymphocytes amongst which typical lymphatic follicles can be seen which extend into the submucosa. These lymphatic follicles are called as Peyer's patches.
- In the region overlying the Peyer's patch villi may be rudimentary or absent.

Key

P. Peyer's patches

Cl. Crypts of Lieberkuhn

Presence of Peyer's patches is the most distinguishing feature of ileum.

Large Intestine

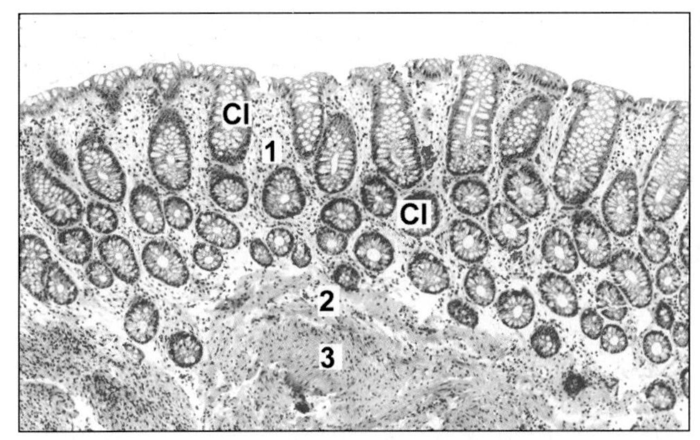

Fig. A19.1: Large intestine

- The most important feature to note in large intestine is the absence of villi.
- The lamina propria shows numerous uniformly arranged crypts of Lieberkuhn in the entire thickness.
- The surface of the mucosa, and the crypts, are lined by columnar cells amongst which there are numerous goblet cells.
- The muscularis mucosae, submucosa and circular muscle coat are similar to those in the small intestine. However, the longitudinal muscle coat is gathered into three thick bands called taenia coli.
- The longitudinal muscle is thin in the intervals between the taenia.

Note: The mucosa is cut obliquely so that the deeper parts of the crypts appear circular.

Key

Cl. Crypts of Lieberkuhn 1. Lamina propria 2. Muscularis mucosa 3. Submucosa

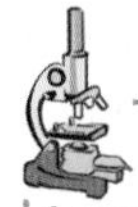

A section of the large intestine is easily distinguished from that of the small intestine because of the absence of villi; and from the stomach because of the presence of goblet cells (which are absent in the stomach).

GALL BLADDER

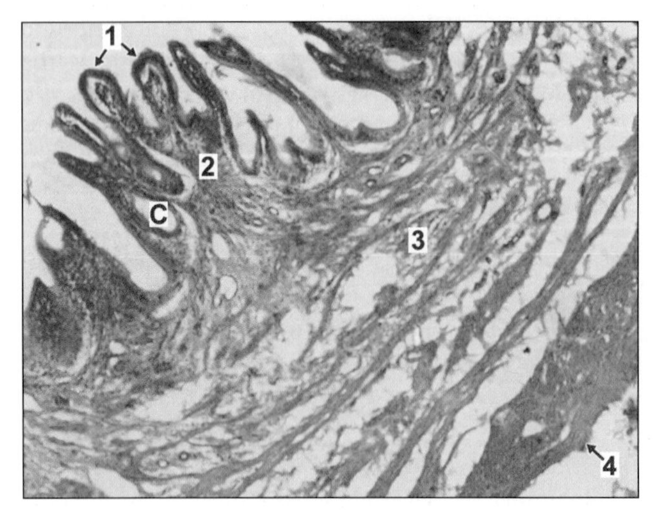

Fig. A20.1: Gall Bladder

It is characterised by

- The mucous membrane is lined by tall columnar cells with brush border. The mucosa is highly folded and some of the folds might look like villi.
- Crypts may be found in lamina propria.
- Submucosa is absent.
- The muscle coat is poorly developed there being numerous connective tissue fibres amongst the muscle fibres. This is called fibromuscular coat.
- A serous covering lined by flattened mesothelium is present.

Key

1. Mucous membrane lined by tall columnar cells with brush border
2. Lamina propria
3. Fibromuscular coat
4. Serosa

C. Crypt in lamina propria

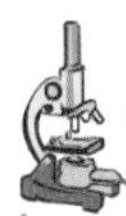

Gall Bladder can be differentiated from small intestine by:

- Absence of villi
- Absence of goblet cells
- Absence of submucosa
- Absence of proper muscularis externa

URINARY BLADDER

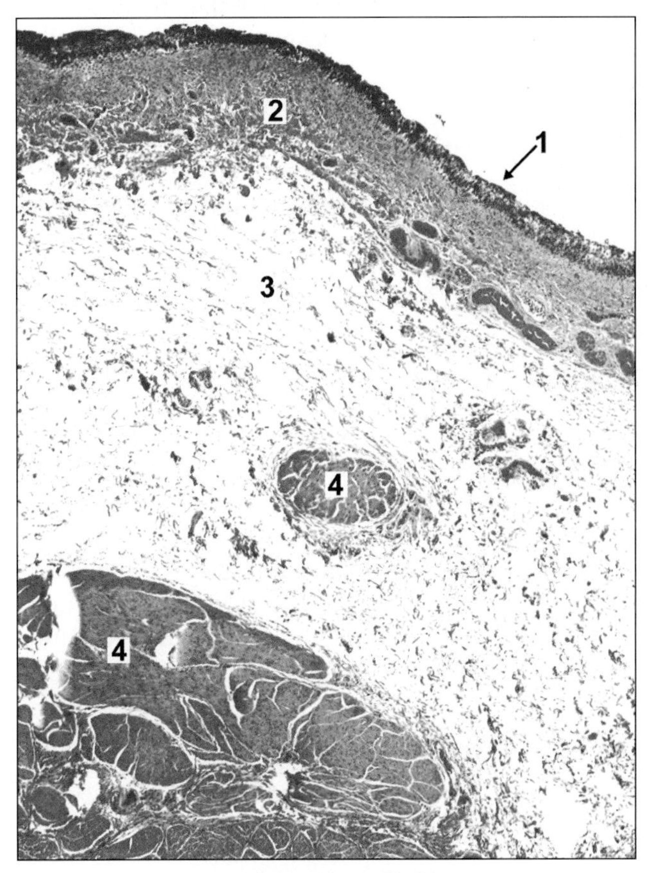

Fig. A21.1: Urinary Bladder

- The organ is easily recognised because the mucous membrane is lined by transitional epithelium (Observe that the nuclei are distributed in the entire thickness of the epithelium and the appearance is uniform).
- The epithelium rests on a layer of lamina propria.
- The muscle layer is thick. It has inner and outer longitudinal layers between which there is a layer of circular or oblique fibres. The distinct muscle layers may not be distinguishable.
- The outer surface is lined in parts by peritoneum (not seen in slide).

Key

1. Transitional epithelium
2. Lamina propria
3. Interstitial connective tissue
4. Smooth muscle bundles

UTERUS

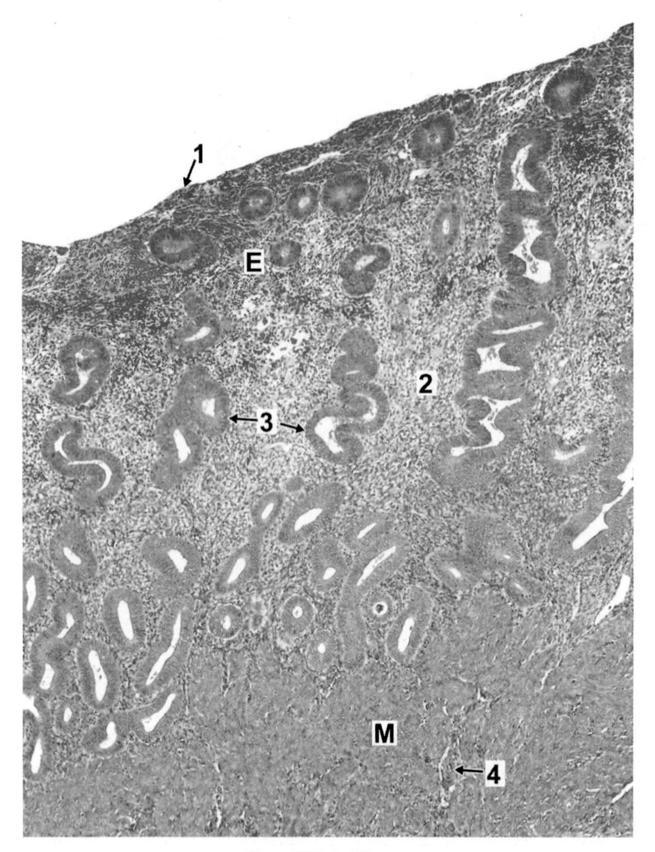

Fig. A23.1: Uterus

- The wall of the uterus consists of a mucous membrane (called the endometrium) and a very thick layer of smooth muscle (the myometrium). The thickness of the muscle layer helps to identify the uterus easily.
- The endometrium has a lining of columnar epithelium that rests on a stroma of connective tissue. Blood vessels are present in the lower portion of endometrium.
- Numerous tubular uterine glands dip into the stroma.
- The appearance of the endometrium varies considerably depending upon the phase of the menstrual cycle.

Key

1. Columnar epithelium
2. Connective tissue
3. Uterine glands
4. Blood vessels

E. Endometrium

M. Myometrium

You might confuse this appearance with that of the large intestine. To distinguish between the two remember that in the uterine epithelium there are no goblet cells. The uterine glands are not as closely packed as the crypts of the large intestine. There is no muscularis mucosae and submucosa. The mucosa (endometrium) rests directly on the muscle layer. The muscle fibres run irregularly and clear circular and longitudinal layers of muscle cannot be made out.

TRACHEA

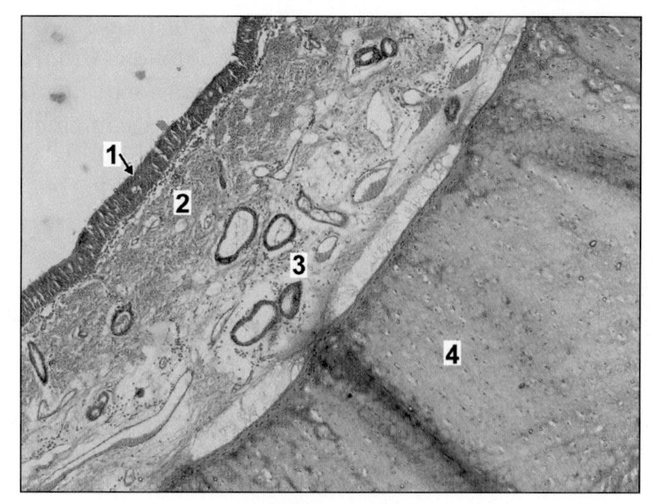

Fig. A24.1: Trachea

- The wall of the trachea is made up of layers but these are quite different from other tissues described above.
- Mucosa is formed by pseudostratified ciliated columnar epithelium with goblet cells and the underlying lamina propria.
- Submucosa is found deep to mucosa and is made up of loose connective tissue containing mucous and serous glands, blood vessels and ducts of the glands.
- The next layer is made up of hyaline cartilage. Chondrocytes increase in size from periphery to centre. They may appear as isogenous groups surrounded by darkly stained territorial matrix.
- External to the cartilage is the outer covering of collagen fibres called adventitia (not seen in slide).

Key

1. Pseudostratified ciliated columnar epithelium
2. Lamina propria
3. Submucosa
4. Hyaline cartilage

EYEBALL

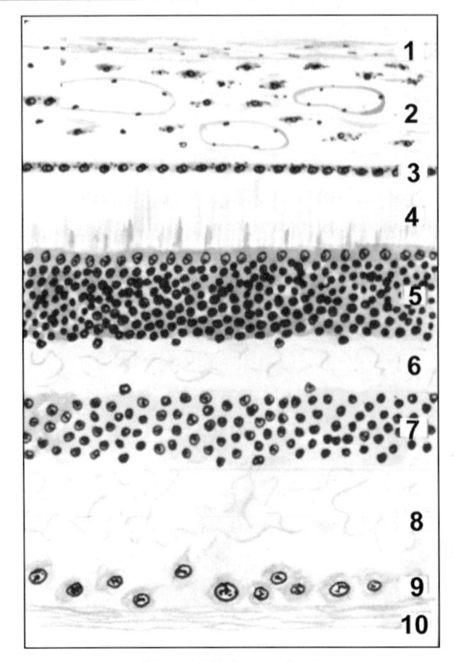

Fig. A25.1: Eyeball

The wall of the eyeball is made up of several layers as follows (from without inwards):

1. Sclera, made up of collagen fibres.
2. Choroid, containing blood vessels and pigment cells.
   The remaining layers are subdivisions of the retina.
3. Pigment cell layer.
4. Layer of rods and cones.
5. Outer nuclear layer.
6. Outer plexiform layer.
7. Inner nuclear layer.
8. Inner plexiform layer.
9. Layer of ganglion cells.
10. Layer of optic nerve fibres.

The appearance is not likely to be confused with any other tissue.

SOME ORGANS CONSISTING PREDOMINANTLY OF ACINI OR ALVEOLI

Acini are rounded structures found in glands. Each acinus appears as a rounded or oval mass. It is made up of cells but their boundaries are often not distinct. Each cell is more or less triangular, its base lying near the periphery of the acinus and its apex near the centre of the acinus. Each acinus has a small lumen that is often not seen. Acini are of two types. Some of them are dark staining (and usually bluish) and are termed serous acini. Others are very light staining. These are mucous acini. In the mammary gland, the secretory unit is termed alveolus, not acinus. Please do not confuse the alveoli of mammary gland with the alveoli of lung!

MIXED SALIVARY GLAND (SUBMANDIBULAR GLAND)

- The submandibular gland is characterised by the presence of both serous and mucous acini. In some cases serous cells are present in relation to mucous acini forming demilunes (or crescents).

Key

1. Serous acini
2. Mucous acinus
3. Mucous acinus with serous demilune
4. Duct

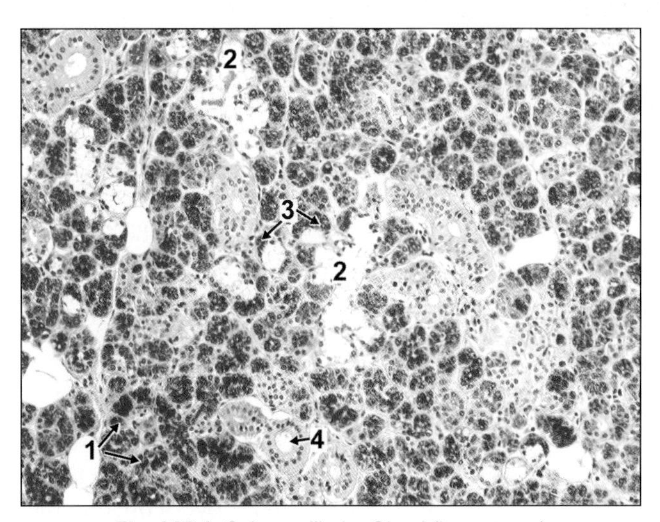

Fig. A26.1: Submandibular Gland (Low power)

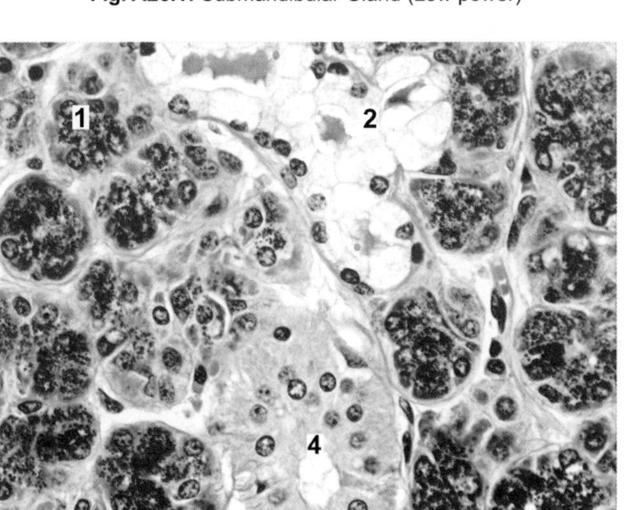

Fig. A26.2: Submandibular Gland (High power)

PANCREAS

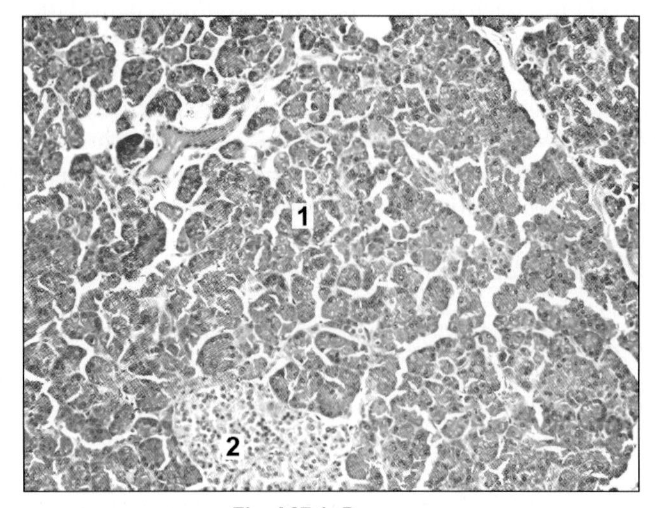

Fig. A27.1: Pancreas

Key

1. Serous acini
2. Pancreatic islets

- This gland is made up of serous acini.
- The cells forming the acini of the pancreas are highly basophilic (bluish staining). The lumen of the acinus is seldom seen.
- Amongst the acini some ducts are present.
- Some acini may show pale staining 'centro-acinar cells'.
- At some places the acini are separated by areas where aggregations of cells is quite different from those of the acini. These aggregations form the pancreatic islets: these islets have an endocrine function.

MAMMARY GLAND

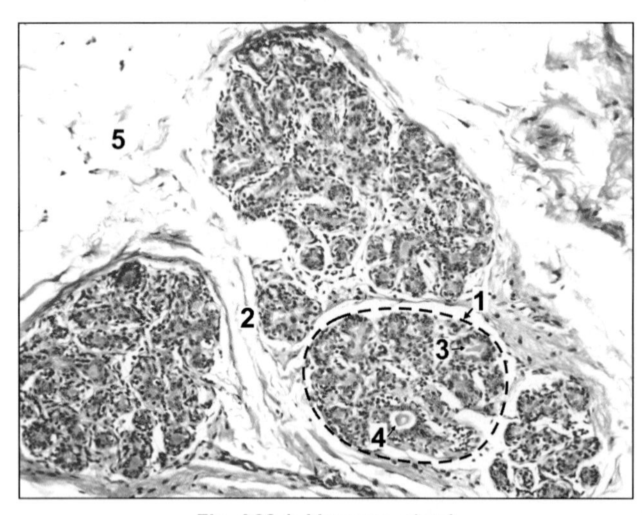

Fig. A28.1: Mammary gland

Key

1. Lobule
2. Connective tissue
3. Alveoli
4. Duct
5. Adipose tissue

- Mammary gland consists of lobules of glandular tissue separated by considerable quantity of connective tissue and fat.
- Nonlactating mammary glands contain more connective tissue and less glandular tissue.
- The glandular elements or alveoli are distinctly tubular. They are lined by cuboidal epithelium and have a large lumen so that they look like ducts. Some of them may be in form of solid cords of cells.
- Extensive branching of duct system seen.

SOME ORGANS SHOWING MULTIPLE TUBULAR ELEMENTS

KIDNEY

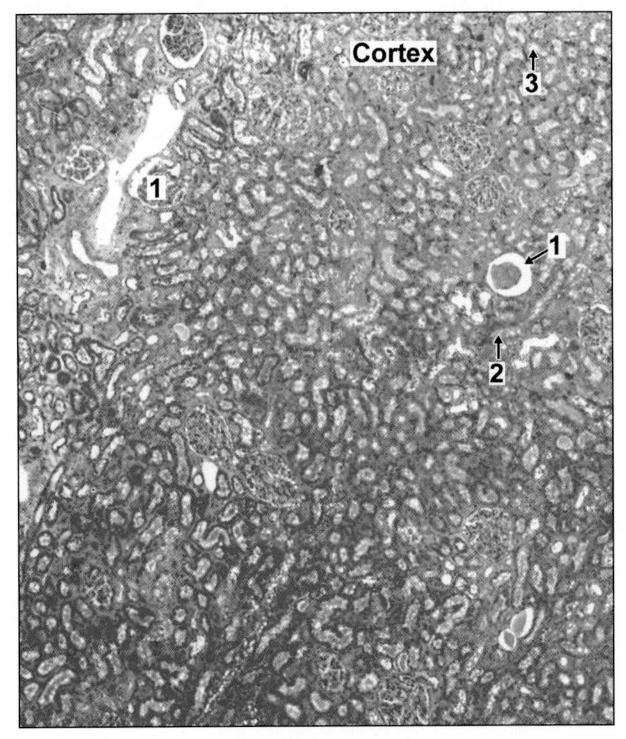

Fig. A27.1: Kidney

When we look at a section of the kidney we see that most of the area is filled with a very large number of tubules. These are of various shapes and have different types of epithelial lining. This fact by itself suggests that the tissue is the kidney.

- The kidney is covered by a capsule.
- Deep to the capsule there is the cortex.
- Deep to the cortex there is the medulla of the kidney.
- In the cortex circular structures called renal corpuscles are present surrounding which there are tubules.
- The lumen of proximal convoluted tubules is small and indistinct. It is lined by cuboidal epithelium with brush border. The distal convoluted tubule has a simple cuboidal epithelium and presents a distinct lumen.
- The medulla shows cut sections of collecting ducts and loop of Henle. Collecting ducts are lined by simple cuboidal epithelium and loops of Henle are lined by simple squamous epithelium.
- Cut sections of blood vessels are seen both in the cortex and medulla.

Key

1. Renal corpuscle
2. Proximal convoluted tubules
3. Distal convoluted tubules

EPIDIDYMIS

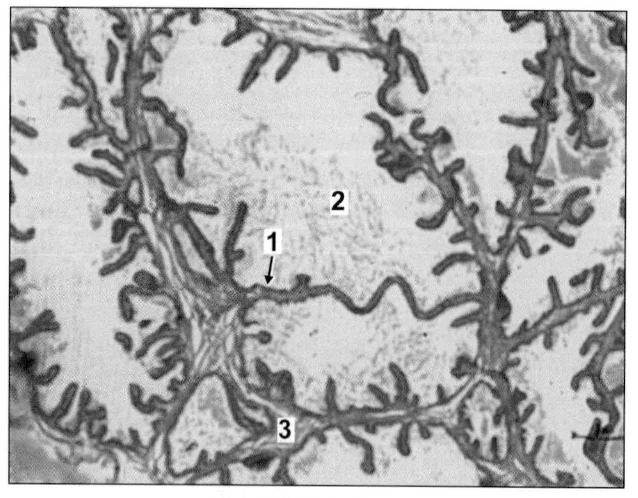

Fig. A30.1: Epididymis

- The body of the epididymis is a long convoluted duct.
- It shows cut sections of tubules lined by pseudostratified columnar epithelium in which there are tall columnar cells and shorter basal cells that do not reach the lumen. The columnar cells bear stereocilia.
- Smooth muscles are present around each tubule.
- Clumps of spermatozoa are present in the lumen of the tubule.

Key

1. Pseudostratified columnar epithelium with stereocilia
2. Spermatozoa clumps
3. Smooth muscle fibres

SEMINAL VESICLE

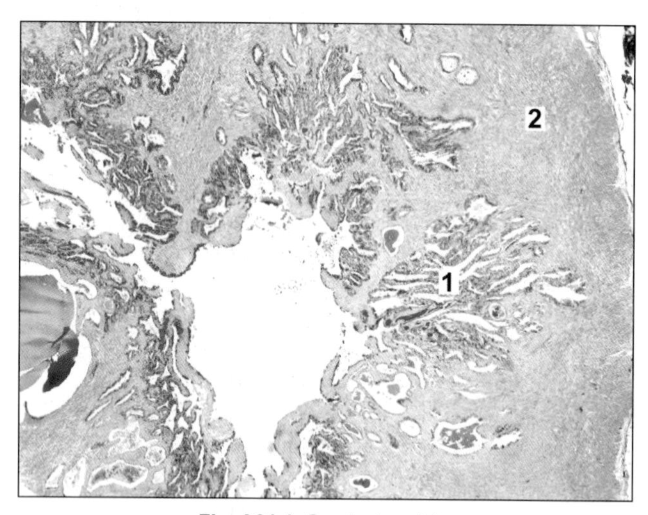

Fig. A31.1: Seminal vesicle

- The seminal vesicle shows highly convoluted tubule and irregular lumen.
- It has an outer covering of connective tissue, a thin layer of smooth muscle and an inner mucosa.
- The mucosal lining is thrown into numerous folds that branch and anastomose to form a network.
- The lining epithelium is usually simple columnar or pseudostratified.

Key

1. Folds of mucosa
2. Muscle wall

SOME ORGANS THAT ARE SEEN IN THE FORM OF ROUNDED ELEMENTS THAT ARE NOT CLEARLY TUBULAR

TESTIS

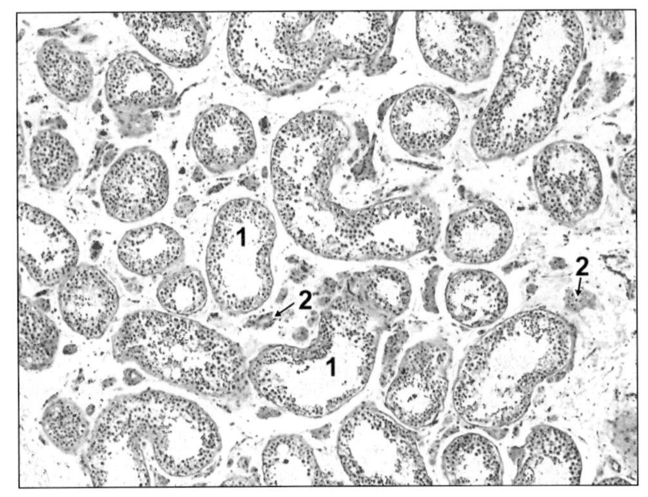

Fig. A32.1: Testis

- The testis has an outer fibrous layer, the tunica albuginea deep to which are seen a number of seminiferous tubules cut in various directions.
- The tubules are separated by connective tissue, containing blood vessels and groups of interstitial cells.
- Each seminiferous tubule is lined by several layers of germinal cells which will eventually form the spermatozoa.

Key

1. Seminiferous tubule
2. Interstitial cells of Leydig

PROSTATE

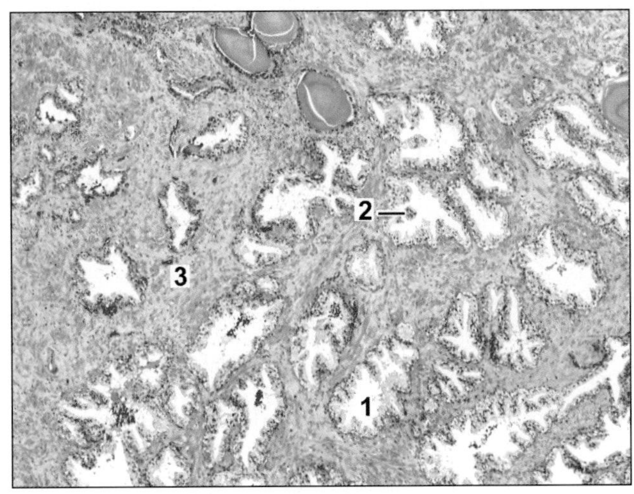

Fig. A33.1: Prostate

- The prostate consists of glandular tissue embedded in prominent fibromuscular tissue.
- The glandular tissue is in the form of follicles with serrated edges. They are lined by columnar epithelium.
- The lumen may contain amyloid bodies called corpora amylacea (pink stained).
- The follicles are separated by broad bands of fibromuscular stroma.

Key

1. Follicles lined by columnar epithelium
2. Corpora amylacea
3. Fibromuscular tissue

The two features that help to recognise a section of the prostate are:
- The irregular (serrated) outlines of secretory elements and
- Their wide separation by fibromuscular tissue.

THYROID GLAND

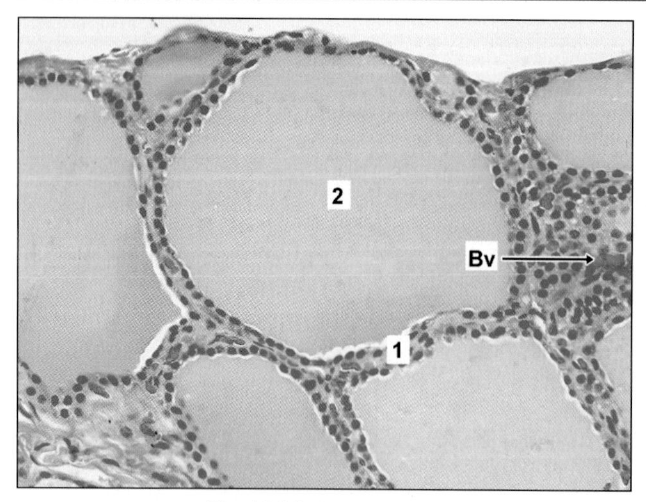

Fig. A34.1: Thyroid gland

- The thyroid gland is made up of follicles lined by cuboidal epithelium.
- The follicles contain pink staining colloid. In the intervals between the follicles, there is some connective tissue.
- Parafollicular cells are present in relation to the follicles and also as groups in the connective tissue.

Key

Bv. Blood vessel
1. Follicles lined by cuboidal epithelium
2. Pink stained colloidal material

SOME TISSUES THAT APPEAR AS COLLECTIONS OF CELLS

LIVER

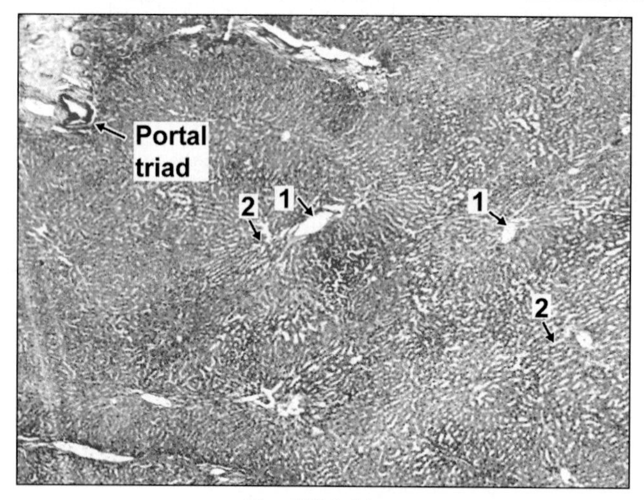

Fig. A35.1: Liver

- Many hexagonal areas called hepatic lobules are present. The lobules are partially separated by connective tissue.
- Each lobule has a small round space in the centre. This is the central vein.
- A number of broad irregular cords seem to pass from this vein to the periphery of the lobule.
- These cords are made up of polygonal liver cells (hepatocytes).
- Along the periphery of the lobules there are angular intervals filled by connective tissue. Each such area contains a branch of the portal vein, a branch of the hepatic arter, and an interlobular bile duct. These three constitute a portal triad. The identification of hepatic lobules and of portal triads is enough to recognize liver tissue.

Key

1. Central vein
2. Radiating cords of hepatocytes

SUPRARENAL GLAND

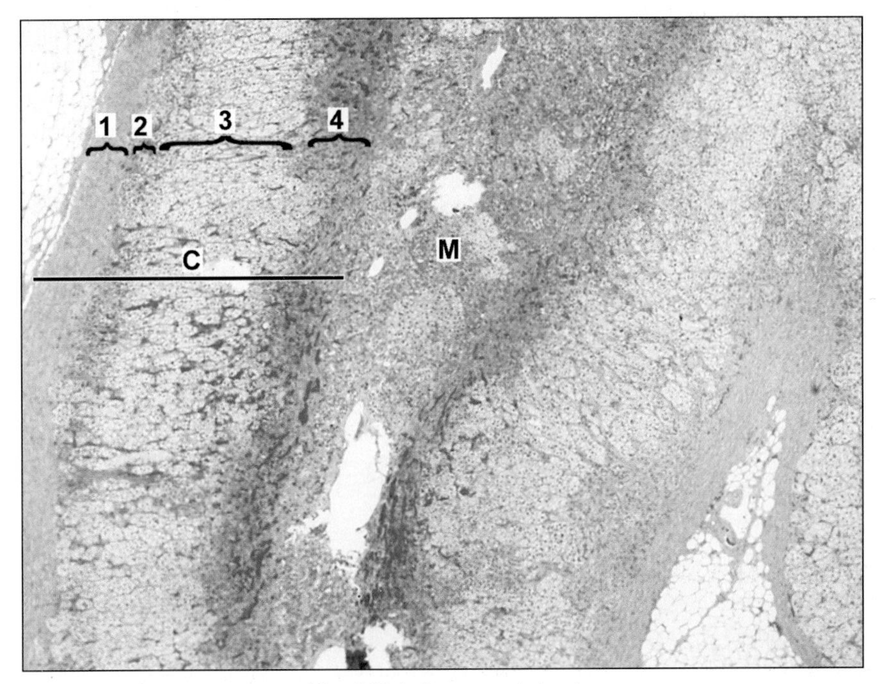

Fig. A36.1: Suprarenal gland

The suprarenal gland is made up a large number of cells arranged in layers.

- It consists of an outer cortex and an inner medulla.
- The cortex is divisible into three zones.
    - The zona glomerulosa is most superficial. Here the cells are arranged in the form of inverted U-shaped structures or acinus-like groups.
    - In the zona fasciculata the cells are arranged in straight columns (typically two cell thick). Sinusoids intervene between the columns. The cells of this zone appear pale.
    - The zona reticularis is made up of cords of cells that branch and form a network.
- The medulla is made up of groups of cells separated by wide sinusoids. Some sympathetic neurons are also present.

Key

1. Capsule
2. Zona glomerulosa
3. Zona fasciculata
4. Zona reticularis

C. Cortex
M. Medulla

HYPOPHYSIS CEREBRI

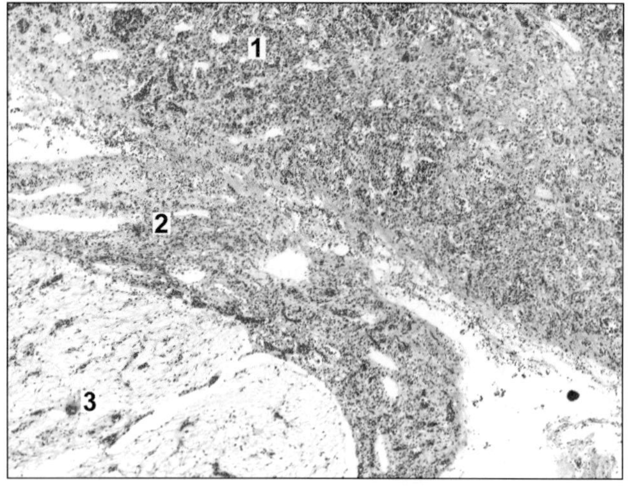

Fig. A37.1: Hypophysis cerebri

- The hypophysis cerebri consists of three main parts.
  - The pars anterior is cellular. It consists of groups or cords of cells with numerous sinusoids between them.
  - The pars intermedia is variable in structure.
  - The pars posterior consists of fibres, and is lightly stained

Key

1. Pars anterior
2. Pars intermedia
3. Pars posterior

PARATHYROID GLANDS

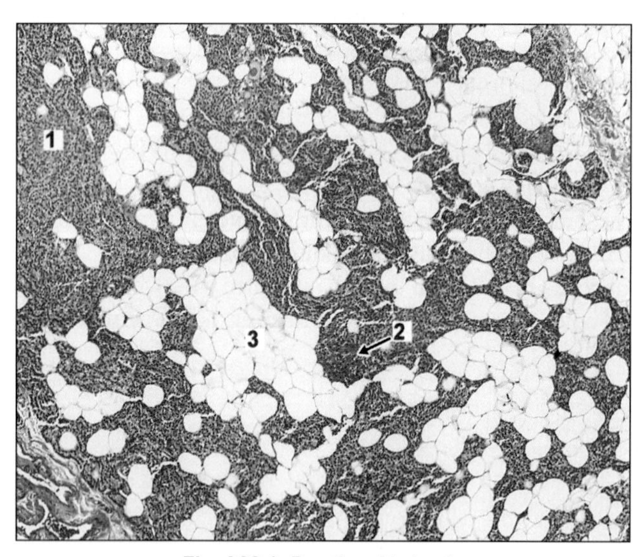

Fig. A38.1: Parathyroid glands

- These glands are made up of masses of cells with numerous capillaries in between.
- Most of the cells (of which only nuclei are seen) are the chief cells (small basophilic cells).
- Oxyphilic cells and adipose cells are also seen. The number of oxyphilic cells is very less as compared to chief cells.

Key

1. Chief cells
2. Oxyphillic cells
3. Adipose cells

CEREBRAL CORTEX

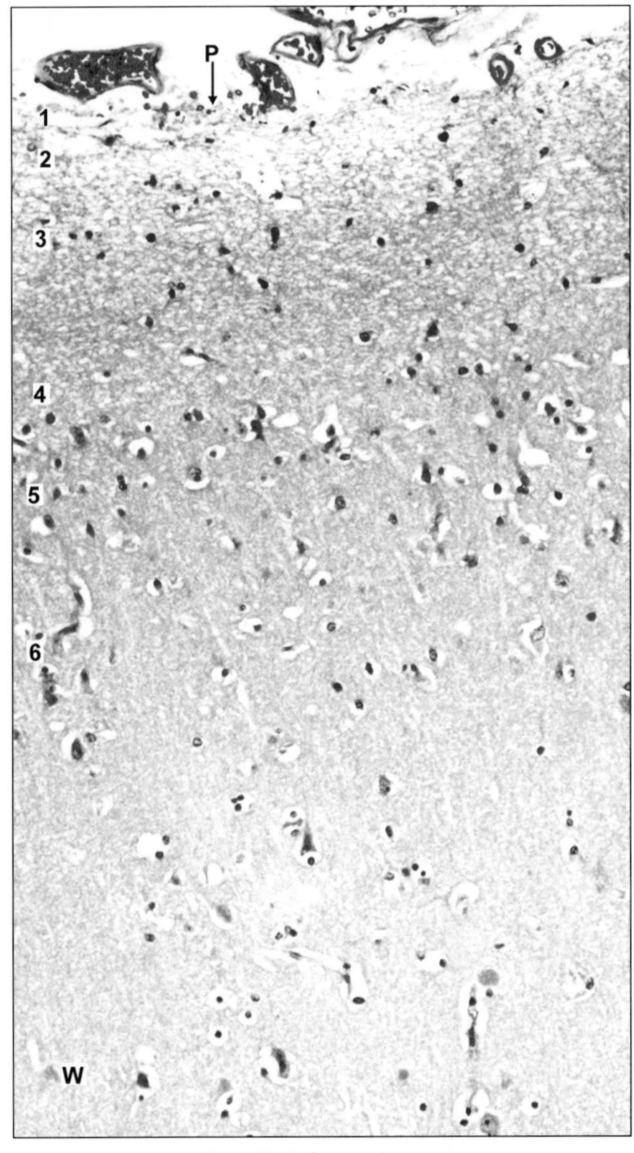

Fig. A39.1: Cerebral cortex

- The cerebral cortex consists of many cells having varied shapes. Nerve fibres, myelinated and unmyelinated, are also present. Blood vessels are seen. The nerve fibres are arranged in several layers as follows (from without inwards).
  - Molecular layer consisting mostly of nerve fibres.
  - External granular layer with densely packed nuclei.
  - Pyramidal cell layer with large triangular cells.
  - Internal granular layer.
  - Ganglionic layer with large cells.
  - Multiform layer with cells of varied shapes.

Key

1. Molecular layer
2. External granular
3. Pyramidal cell layer
4. Internal granular layer
5. Ganglionic layer
6. Multiform layer

P. Piamater
W. White matter

SENSORY GANGLIA

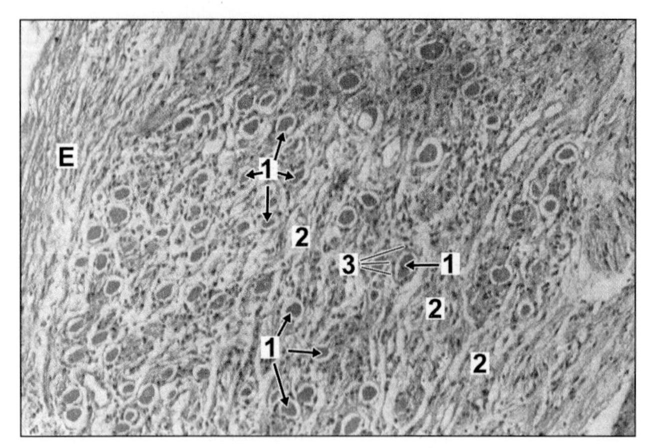

Fig. A40.1: Sensory ganglia

- Ganglia are of two types, sensory and autonomic.
- In sensory ganglion large pseudounipolar neurons are arranged in groups that are separated by bundles of nerve fibres.
- Each neuron has a vesicular nucleus with a prominent nucleolus. The neuron is surrounded by a ring of satellite cells.

Key

1. Pseudounipolar neurons
2. Nerve fibres
3. Satellite cells

E–Epineurium

AUTONOMIC GANGLIA

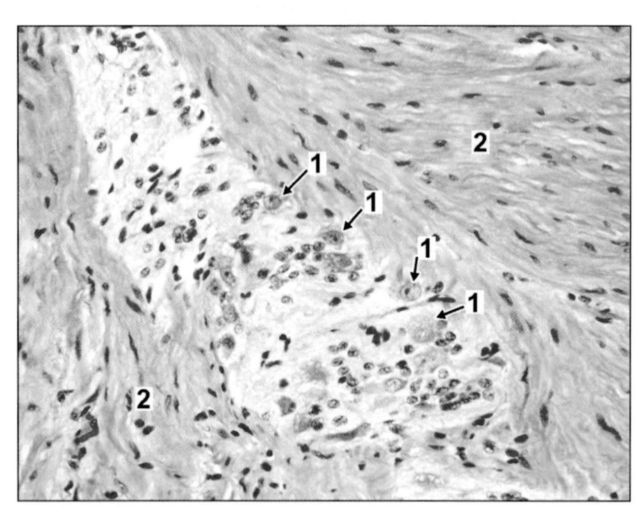

Fig. A41.1: Autonomic ganglia

- Autonomic ganglion consists of multipolar neurons which are not arranged in groups, but are scattered amongst nerve fibres.
- Satellite cells are present, but are less prominent than in sensory ganglia.

Key

1. Multipolar neurons
2. Nerve fibres

MISCELLANEOUS STRUCTURES THAT DO NOT FIT IN ANY OF THE GROUPS

LUNG

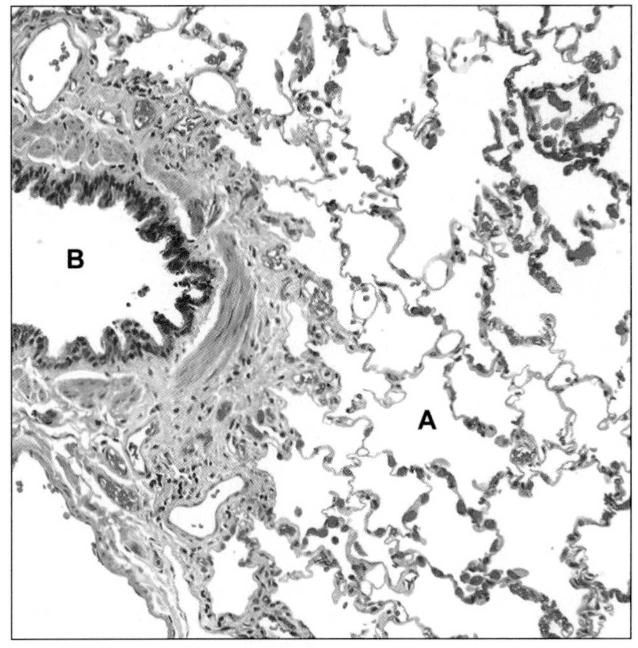

Fig. A42.1: Lung

- The lung substance is made up of numerous alveoli.
- Alveoli are thin-walled spaces lined by simple squamous epithelium.
- The structure of large bronchus is similar to that of the trachea. Smooth muscle, cartilage and glands are present in its wall; and it is lined by pseudostratified ciliated columnar epithelium with goblet cells.
- Small bronchioles are lined by a simple columnar epithelium, and have a wall of smooth muscle. There is no cartilage in their walls.
- Bronchioles subdivide and when their diameters are approximately 1 mm or less, they are called terminal bronchioles.
- Arteries are seen near the bronchioles.
- Respiratory bronchiole and alveolar duct are also present.
- The lung surface is covered by pleura. It consists of a lining of mesothelium resting on a layer of connective tissue.
- This slide shows a medium-sized bronchiole surrounded by alveoli.

Key

A. Alveoli
B. Bronchiole

OVARY

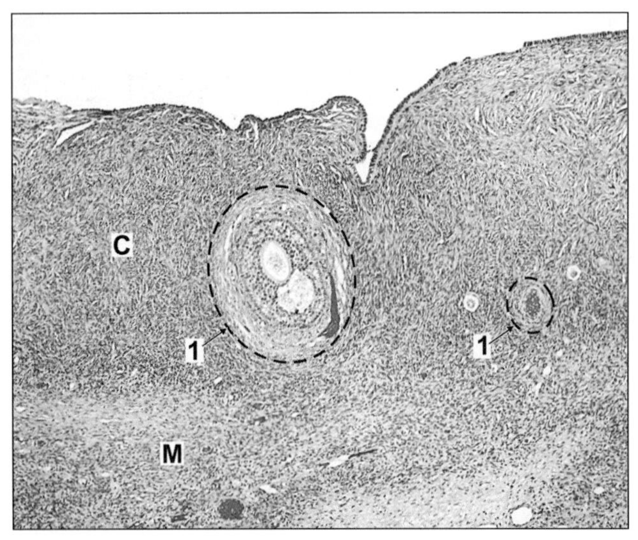

Fig. A43.1: Ovary

Key

1. Ovarian follicles
C. Cortex
M. Medulla

- The surface is covered by a cuboidal epithelium. Deep to the epithelium there is a layer of connective tissue that constitutes the tunica albuginea.
- The substance of the ovary has an outer cortex in which follicles of various sizes are present; and an inner medulla consisting of connective tissue containing numerous blood vessels.
- Just deep to the tunica albuginea many primordial follicles each of which contains a developing ovum surrounded by flattened follicular cells are present. Large follicles have a follicular cavity surrounded by several layers of follicular cells, membrana granulosa.
- The cells surrounding the ovum constitute the cumulus oophoricus.
- The follicle is surrounded by a condensation of connective tissue which forms a capsule for it.
- The capsule consists of an inner cellular part (the theca interna), and an outer fibrous part (the theca externa). The theca interna and externa are collectively called theca folliculi.
- The follicle is surrounded by a stroma made up of reticular fibres and fusiform cells.

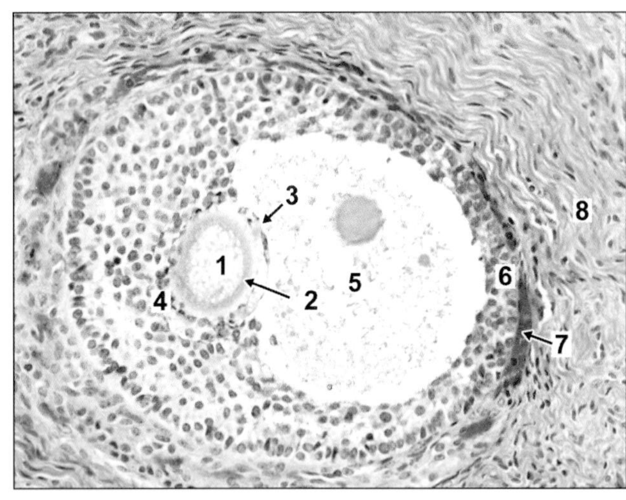

Fig. A43.2: Graafian follicle

Key

1. Maturing oocyte
2. Zona pellucida
3. Cumulus oophoricus
4. Discus proligerus
5. Antrum folliculi
6. Membrana granulosa
7. Capsule
8. Stroma

SPINAL CORD

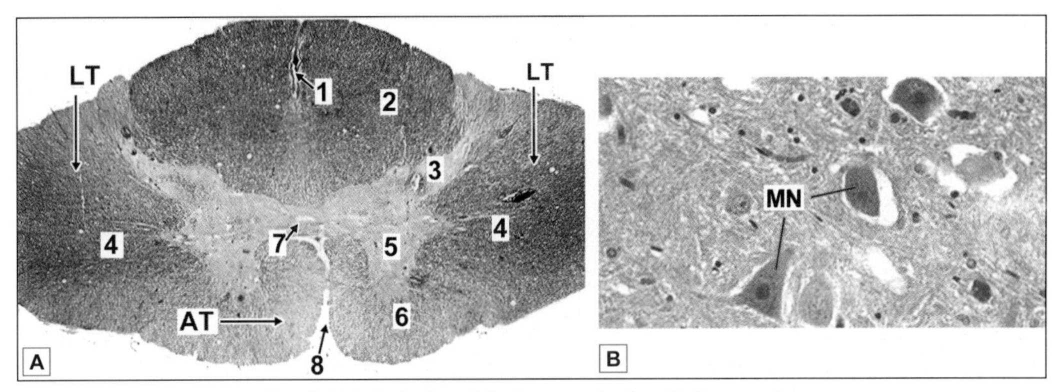

Fig. A44.1: Spinal cord **A.** Panoramic view **B.** Grey matter

- The spinal cord has a characteristic oval shape. It is made up of white matter (made up mainly of myelinated fibres), and grey matter (containing neurons and unmyelinated fibres). The grey matter lies towards the centre and is surrounded all round by white matter. The grey matter consists of a centrally placed mass and projections (horns) that pass forwards and backwards.

Note: The stain used for the slide is Luxol Fast Blue.

Key

1. Posterior median septum
2. Posterior white column
3. Posterior grey column
4. Lateral white column
5. Anterior grey column
6. Anterior white column
7. Central canal lying in grey commissure. The fibres in front of the grey commissure form the anterior white commissure
8. Anterior median sulcus.

AT. Anterior motor tracts
LT. Lateral motor tracts
MN. Multipolar neurons in grey matter

CEREBELLUM

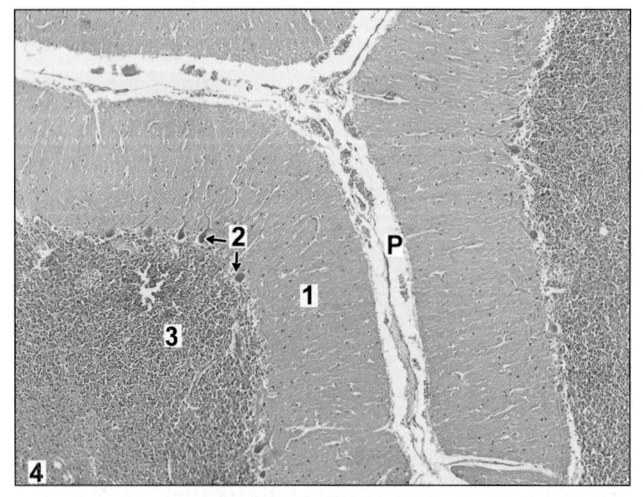

Fig. A45.1: Cerebellum

- The cerebellum contains leaf like folia. The core of each folium is formed by pink staining fibres of the white matter.
- The layers overlying the white matter form the cerebellar cortex which consists of (from without inwards):
  1. Molecular layer
  2. Purkinje cells
  3. Granule cell layer
- The cortex is covered by piamater.

Key

1. Molecular layer
2. Purkinje cells
3. Granule cell layer
4. White matter

P. Piamater

PERIPHERAL NERVE

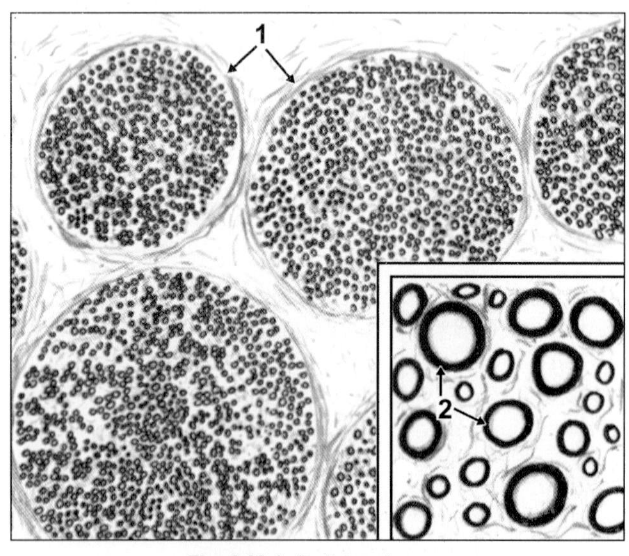

Fig. A46.1: Peripheral nerve

- The nerve has been fixed in osmic acid that stains the myelin sheaths (around nerve fibres) black. The myelin sheaths, therefore, appear as black rings.
- The nerve fibres are arranged in bundles that are held by connective tissue (called perineurium).
- The connective tissue around individual nerve fibres is called endoneurium.

Key

1. Perineurium
2. Endoneurium

Chapter 1

Light Microscopy and Tissue Preparation

—*Sabita Mishra*

A light microscope is an optical device that uses visible light for illumination and lenses to magnify a specimen or tissue section for detailed visualisation. Simple microscopes utilise a single lens while compound microscopes have a number of lenses in combination. In light microscopy the tissue is visualized against a bright background so it also referred to as bright field microscopy and is best suited to view stained tissue sections. In contrast a dark field microscopy is where unstained specimens, e.g. living cells, are observed. The light enters from the periphery and scattered light enters the objective lens showing a bright specimen against a dark background. All routine histological techniques involve the bright field microscopy.

Antonie Philips van Leeuwenhoek in 1673 invented the concept of using combination of convex lenses to magnify small structures and visualise them. His invention of this primitive microscope detected small structures like bacteria, yeast and blood cells. The invention of microscope led to the discovery and description of cell by Robert Hooke, an English scientist in 1655. Modern microscopy started with the invention of achromatic lens by Lister in 1829.

COMPONENTS OF A LIGHT MICROSCOPE (FIG. 1.1)

A light microscope has optical parts and non-optical parts.

Optical Parts

The functioning of the microscope is based on the optics of the lenses.

Illuminating Device

Most of the advanced microscopes come with a built-in illuminator using low voltage bulbs for transmitted light. The brightness of light can be adjusted. Older monocular microscopes used mirrors to reflect light from an external source.

- ***Condenser:*** Collects and focuses light from the light source onto the specimen being viewed. The condenser is close to the stage and has an ***aperture (iris diaphragm)*** that controls the amount of the light coming up through the condenser.

Aperture closed	Aperture open
Light comes in centre	Image is brighter
Contrast high	Contrast low

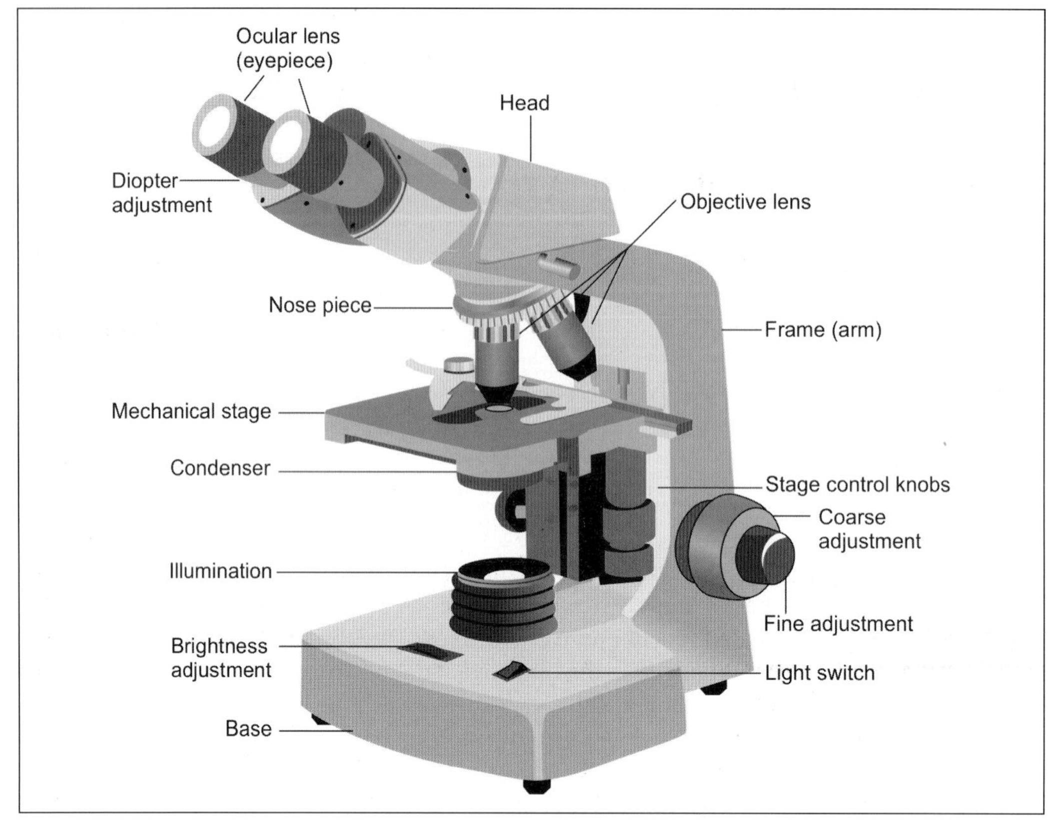

Fig. 1.1: Binocular light microscope

- ***Objective lenses:*** These lenses are attached to the nosepiece of the microscope. The objective lens is responsible for magnifying the specimen/section to be visualized. The type and quality of an objective lens influences the performance of a microscope. A standard microscope has three, four, or five objective lenses that range in power from 4x to 100x. The objective lens collects maximum amount of light from the object to form a high quality ***magnified real image.***
- ***Eye piece:*** The viewer looks through the eyepiece to observe the magnified image. Eye pieces may be monocular, binocular or combined photo binocular (trinocular). It is the final stage of the optical path of the microscope and produces a ***magnified virtual image*** which is seen by the eye. The eye piece has a power of 10X. In a binocular microscope diopter adjustments are present to adjust the focusing of the eye piece

Non-Optical Components of Microscope

Body Tube (Head)

A cylindrical tube that connects the eyepiece to the objective lenses. Standard length 160 mm.

- ***Arm:*** The arm connects the body tube to the base of the microscope.
- ***Coarse adjustment:*** Mechanical knobs that bring the specimen into general focus.

- ***Fine adjustment:*** Fine tunes the focus and increases the detail of the specimen. Individual user has to adjust according to his/her own vision to observe the tissue section.
- ***Nosepiece:*** A rotating turret that houses the objective lenses. The viewer rotates the nosepiece to select different objective lenses.
- ***Stage:*** The flat platform where the slide is placed and lies perpendicular to the optical pathway. It has ***Stage clips*** that hold the slide in place. ***Stage Control Knobs*** move the stage left and right or up and down. The stage also has a vernier caliper attached to it so that the viewer can come back to any reference point by the help of the caliper. An aperture in the middle of the stage allows light from the illuminator to reach the specimen.
- ***Base:*** The base supports the microscope. The illuminator with its power switch is located on the base.

Specimen or slide: The specimen is the object being examined. Most specimens are mounted on slides, flat rectangles of thin glass. Stained tissue sections are mounted on a glass slide with a coverslip placed over it. This allows the slide to be easily inserted or removed from the microscope. It also allows the specimen to be labelled, transported, and stored without any damage. The slide is placed on the stage for viewing.

PRINCIPLES OF A CONVENTIONAL BRIGHT FIELD MICROSCOPE

The word compound refers to the fact that two lenses, the objective lens and the eyepiece (or ocular), work together to produce the final magnified image that is projected onto the eye of the observer

Magnification

Magnification (M final) is calculated by the formula
M final = M **(objctive)** × **M(eyepiece)**.

Resolution

Given sufficient light an unaided eye can distinguish 2 points lying 0.2 mm apart. This distance is called resolution of a normal eye. By assembling a combination of lenses the distance can be increased and the eye can visualise objects closer than 0.2 mm.

Resolution of a microscope is dependent on

a. wave length of light 400-800 λ
b. Numerical apertur**e (NA):** collecting power of the objective and condenser lens
NA= n × sin u
Where n = refractive index of media between cover slip and objective lens
sin u = Angle between optical axis of lens and outer most ray (r/h)
NA = n × r/h (Fig. 1.2)
Refractive index of air =1
Refractive index of oil immersion 1.51
Resolution of microscope **(r)** = **0.6λ/NA**
Resolution of light microscope is 0.2 μ

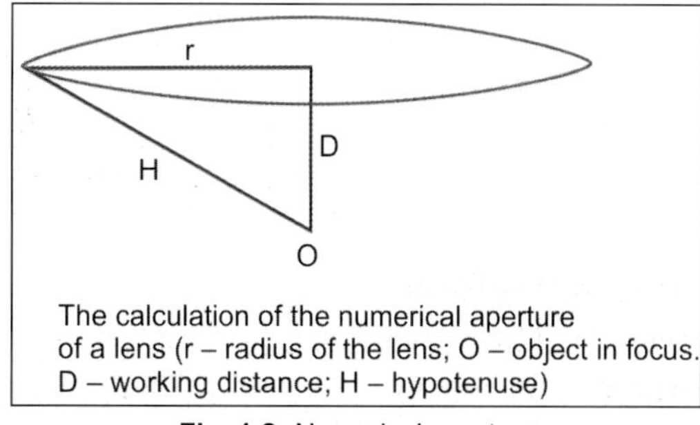

The calculation of the numerical aperture of a lens (r – radius of the lens; O – object in focus. D – working distance; H – hypotenuse)

Fig. 1.2: Numerical aperture

Working of a Light Microscope (Flow chart 1.1, Fig. 1.3)

To view a section /specimen under a light microscope:
Light from the light source enters the condenser, passes through specimen and is magnified by the objective lens. The real magnified image formed by the objective lens is further magnified by the eyepiece. Thus the viewer observes a magnified virtual image.

Flowchart 1.1: Working of a light microscope

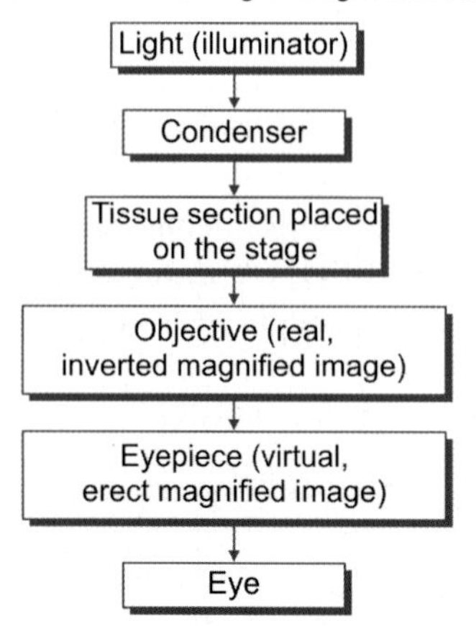

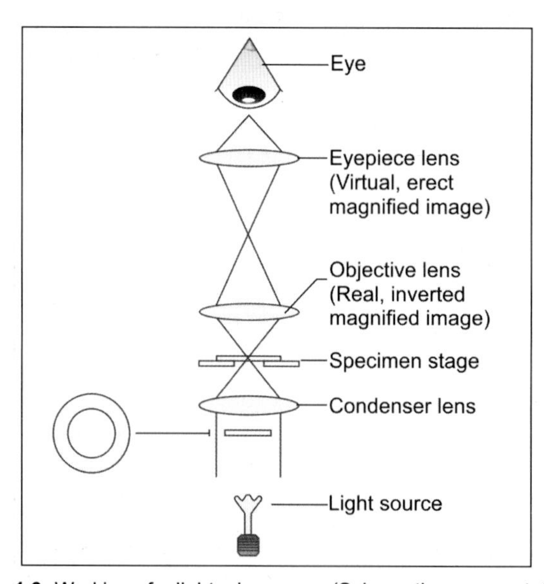

Fig. 1.3: Working of a light microscope (Schematic representation)

Axial Aberrations

When light passes through the lens it suffers a number of aberrations which result in image degradation. The optical parts are the condenser, objective and eyepiece. The best (and most expensive) lenses have the least aberrations.

Commonly seen aberrations are

- ***Chromatic aberration:*** Production of a coloured spectrum of white light. Different lenses corrected for chromatic aberrations are listed in Table 1.1.

Table 1.1: Different lenses corrected for chromatic aberrations

Achromatic	Fluorite	Apochromatic
Corrected for red and blue focused at same length	Spread of secondary spectrum reduced	Fully corrected for three colours
Green shorter focus Violet longer focus	Type of correction though similar to achromatic	In combination with fluorite causes elimination of all colours

- ***Spherical aberration:*** A defect of single lenses due to their curved surface. Light passing through the periphery of the lens is refracted to a greater extent than through the central part. This is corrected by using a compound lens.

Illumination

Critical Illumination

This is used with simple equipment and a separate light source. Light source is focused in the same plane as the object, when the object is in focus.

Kohler Illumination

High intensity microscopes have a small light source that is insufficient to fill the whole field with light and are usually supplied by an auxiliary lens and iris which increases the apparent light source. With Kohler illumination auxiliary lens of the lamp focuses the enlarged image on to the iris diaphragm of the sub-stage condenser. The resolving power of critical and Kohler are similar, but Kohler illumination provides an evenly illuminated view and displaces critical illumination.

PRACTICAL TIPS IN USING A BRIGHT FIELD MICROSCOPE

- Mount the specimen with the coverslip facing up on the stage
- Optimise the lighting
- Adjust the condenser
- Focus, locate, and centre the specimen
- Adjust eyepiece separation and focus
- Select an objective lens for viewing
- Move up the magnification in steps

TYPES OF MICROSCOPES

- ***Dark-field microscopy:*** The specimen is illuminated from the side and only scattered light enters the objective lens which results in bright objects against dark background. Images produced by dark-field microscopy are low resolution and details cannot be seen. Dark-field microscopy is especially useful for visualisation of small particles, such as bacteria.
- ***Phase contrast microscopy*** and ***differential-interference-contrast*** allow objects that differ slightly in refractive index or thickness to be distinguished within unstained or living cells.

- ***Fluorescence microscopy:*** A fluorochrome is excited with ultraviolet light and the resulting visible fluorescence is viewed. This produces a bright image in a dark background. There are some natural fluorescence substances which fluoresce when ultraviolet light falls on them, called primary fluorescent substances. Certain fluorescent dyes when added to the tissue lead to a secondary fluorescence which are visualised by a fluorescent microscope.
- ***Confocal microscopy:*** The confocal scanning optical microscope is designed to illuminate an object in a serial fashion, point by point, where a small beam of light (from a LASER) is scanned across the object rapidly in an X-Y raster pattern. The images are digitized and stored.
- ***Electron microscopy:*** The property of accelerated electrons in vacuum to behave like light and travel in a straight line has been exploited in the invention of the electron microscope. Instead of glass lenses here one uses electromagnetic lenses. The wave length of the electrons in vacuum is 10,0000 times less than light. The resolving power of an electron microscope is 0.2 nm. In transmission electron microscopy the beam of electron passes through the tissue which is a thin section less than 100 nm. To prepare such thin section one uses an ultramicrotome and instead of steel blades, the sections are cut with laboratory prepared fresh glass knives. The sections are picked up in grids and are stained with uranyl acetate and lead citrate before viewing. Transmission electron microscope is used for ultrastructural studies. In scanning electron microscope the beam of electrons are reflected back from the surface, thus giving us the surface view.

TISSUE PROCESSING

To visualise the microstructure of any tissue under a light microscope, the specimen has to go through a thorough protocol of tissue fixation, tissue processing, sectioning and staining.

STEPS INVOLVED IN TISSUE PREPARATION

- ***Tissue collection:*** Commonly tissue is obtained from, autopsy, surgical procedures, experimental animals (rabbit, rats, mice, etc.) either perfused or decapitated (Guideline of ethics are always observed in any experimental study).
- ***Fixation:*** The primary objective of fixation is that stained section of any tissue must maintain clear and consistent morphological features to almost that what was existing during life.
- ***Effects of fixation:*** It coagulates the tissue proteins and constituents, thus minimising their, loss during tissue processing, hardens the tissue and makes it insensitive to hypotonic or hypertonic solution.

  Commonly used fixative is formaldehyde and glutaraldehyde

  Formaldehyde is a cross-linking fixative which acts by creating a covalent bond between proteins in the tissue. Formaldehyde is a gas and is soluble in water to an extent of 40% by weight. 10% methanol is added to it as a stabiliser. Paraformaldehyde is a polymer of formaldehyde available as a white crystalline powder.
- ***Tissue processing:*** Principle of tissue processing involves replacement of all extracellular water from the tissue and replacing it with a medium that provides sufficient rigidity to enable sectioning without any damage or distortion to the tissue.

STEPS IN TISSUE PROCESSING

- ***Dehydration:*** Removal of water by a dehydrating agent. Commonly used dehydrating agent is alcohol in ***descending grades*** (e.g. 100%, 90%, 70%, 50%, 30%).
- ***Clearing:*** Making the tissue clear by removing the dehydrating agent e.g. xyline, chloroform
- ***Infiltration:*** Permeating the tissue with a support media.
- ***Embedding:*** Paraffin wax is routinely used as an embedding media. It has a melting point of 45-55 degree. Other embedding media used are celloidin and resins.
  The tissue is embedded and orientated in the media and forms a solid block at room temperature. The tissue blocks are ready for sectioning.
- ***Sectioning:*** A rotary microtome is used to cut sections of 5-7 μ thick for routine histology. The sections are cut and picked up on clean glass slides under a water bath. They are dried before staining.
- ***Staining: Haematoxylin and eosin*** is routinely used for all teaching slides. Morphological identification becomes easier. Haematoxylin is a basic dye and stains the nucleus blue while eosin is an acidic dye and stains the cytoplasm pink. Once the sections are stained they are mounted and are ready for viewing under a light microscope.

Chapter 2

Cell Structure

INTRODUCTION

Cell is the fundamental structural and functional unit of the body. A cell is bounded by a ***cell membrane*** (or ***plasma membrane***) within which is enclosed a complex material called ***protoplasm***. The protoplasm consists of a central more dense part called the ***nucleus***; and an outer less dense part called the ***cytoplasm***. The nucleus is separated from the cytoplasm by a nuclear membrane. The cytoplasm has a fluid base (matrix) which is referred to as the ***cytosol*** or ***hyaloplasm***. The cytosol contains a number of ***organelles*** which have distinctive structure and functions. Many of them are in the form of membranes that enclose spaces. These spaces are collectively referred to as the ***vacuoplasm***.

THE CELL MEMBRANE

The membrane separating the cytoplasm of the cell from surrounding structures is called the ***cell membrane*** or the ***plasma membrane***.

It has highly selective permeability properties so that the entry and exit of compounds are regulated. The cellular metabolism is in turn influenced and probably regulated by the membrane. Thus the membrane is metabolically very active.

Basic Membrane Structure

When suitable preparations are examined by electron microscope (EM), the average cell membrane is seen to be about 7.5 nm thick. It consists of two densely stained layers separated by a lighter zone; thus creating a trilaminar appearance.

Membranes are mainly made up of lipids, proteins and small amounts of carbohydrates. The contents of these compounds vary according to the nature of the membrane.

Lipids in Cell Membrane

The trilaminar structure of membranes is produced by the arrangement of lipid molecules (predominantly phospholipids) that constitute the basic framework of the membrane.

Each phospholipid molecule consists of an enlarged head in which the phosphate portion is located; and of two thin tails (Fig. 2.1). The head end is called the ***polar end*** while the tail end is the ***non-polar end***. The head end is soluble in water and is said to be ***hydrophilic***. The tail end is insoluble and is said to be ***hydrophobic***.

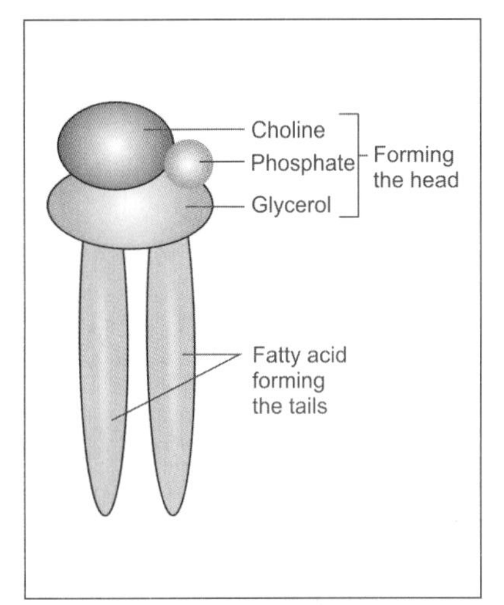

Fig. 2.1: Structure of a phospholipid molecule (phosphatidyl choline) seen in a cell membrane (Schematic representation)

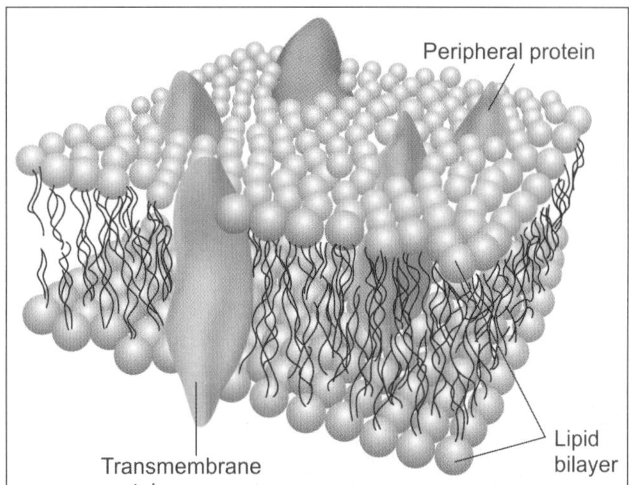

Fig. 2.2 : Fluid mosaic model of membrane (Schematic representation)

When such molecules are suspended in an aqueous medium, they arrange themselves so that the hydrophilic ends are in contact with the medium; but the hydrophobic ends are not. They do so by forming a bi-layer which forms the basis of ***fluid mosaic model of membrane*** (Singer and Nicolson 1972) (Fig. 2.2).

The dark staining parts of the membrane (seen by EM) are formed by the heads of the molecules, while the light staining intermediate zone is occupied by the tails, thus giving the membrane its trilaminar appearance.

Because of the manner of its formation, the membrane is to be regarded as a fluid structure that can readily reform when its continuity is disturbed. For the same reasons, proteins present within the membrane can move freely within the membrane.

Added Information

- As stated above phospholipids are the main constituents of cell membranes. They are of various types including phosphatidylcholine, sphingomyelin, phosphatidylserine and phosphatidyl-ethanolamine.
- Cholesterol provides stability to the membrane.
- Glycolipids are present only over the outer surface of cell membranes. One glycolipid is galactocerebroside, which is an important constituent of myelin. Another category of glycolipids seen are ganglionosides.

Proteins in Cell Membrane

The proteins are present in the form of irregularly rounded masses. Most of them are embedded within the thickness of the membrane and partly project on one of its surfaces (either outer or inner). However, some proteins occupy the entire thickness of the membrane and may project out of both its surfaces (Fig. 2.2). These are called ***transmembrane proteins*** (Fig. 2.3).

The proteins of the membrane are of great significance as follows:

- Membrane proteins help to maintain the structural integrity of the cell by giving attachment to cytoskeletal filaments. They also help to provide adhesion between cells and extracellular materials.
- Some proteins play a vital role in transport across the membrane and act as pumps. Ions get attached to the protein on one surface and move with the protein to the other surface.
- Some proteins are so shaped that they form passive channels through which substances can diffuse through the membrane. However, these channels can be closed by a change in the shape of the protein.
- Other proteins act as receptors for specific hormones or neurotransmitters.
- Some proteins act as enzymes.

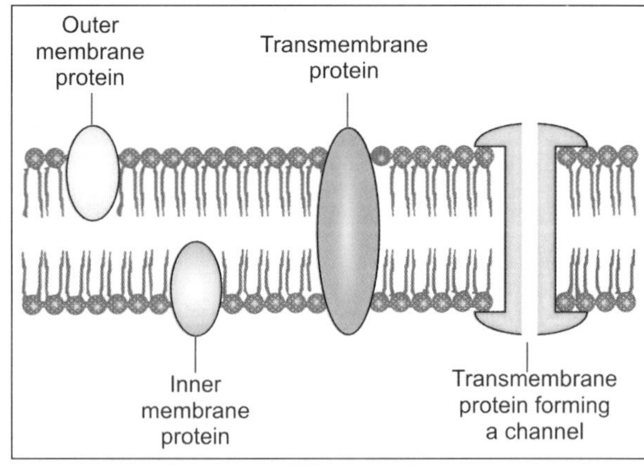

Fig. 2.3: Some varieties of membrane proteins (Schematic representation)

Carbohydrates of Cell Membranes

In addition to the phospholipids and proteins, carbohydrates are present at the surface of the membrane. They are attached either to the proteins (forming glycoproteins) or to the lipids (forming glycolipids) (Fig. 2.4). The carbohydrate layer is specially well developed on the external surface of the plasma membrane forming the cell boundary. This layer is referred to as the cell coat or ***glycocalyx***.

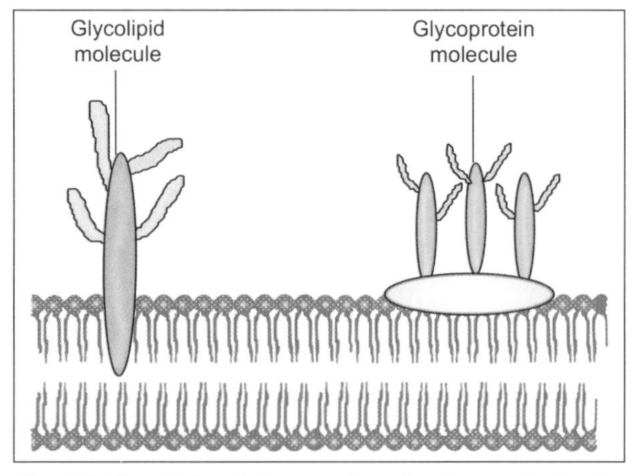

Fig. 2.4: Glycolipid and glycoprotein molecules attached to the outer aspect of cell membrane (Schematic representation)

Added Information

The glycocalyx is made up of the carbohydrate portions or glycoproteins and glycolipids present in the cell membrane. Some functions attributed to the glycocalyx are as follows:

- Special adhesion molecules present in the layer enable the cell to adhere to specific types of cells, or to specific extracellular molecules.
- The layer contains antigens. These include major histocompatibility complexes (MHC). In erythrocytes, the glycocalyx contains blood group antigens.
- Most molecules in the glycocalyx are negatively charged causing adjoining cells to repel one another. This force of repulsion maintains the 20 nm interval between cells. However, some molecules that are positively charged adhere to negatively charged molecules of adjoining cells, holding the cells together at these sites.

Functions of Cell Membrane

The cell membrane is of great importance in regulating the following activities:

- The membrane maintains the shape of the cell.
- It controls the passage of all substances into or out of the cell.
- The cell membrane forms a sensory surface. This function is most developed in nerve and muscle cells. The plasma membranes of such cells are normally polarised—the external surface bears a positive charge and the internal surface bears a negative charge, the potential difference being as much as 100 mv. When suitably stimulated, there is a selective passage of sodium and potassium ions across the membrane reversing the charge. This is called ***depolarisation***. It results in contraction in the case of muscle, or in generation of a nerve impulse in the case of neurons.
- The surface of the cell membrane bears ***receptors*** that may be specific for particular molecules (e.g., hormones or enzymes). Stimulation of such receptors (e.g., by the specific hormone) can produce profound effects on the activity of the cell. Receptors also play an important role in absorption of specific molecules into the cell as described below.
- Cell membranes may show a high degree of specialisation in some cells. For example, the membranes of rod and cone cells (present in the retina) bear proteins that are sensitive to light.

Role of Cell Membrane in Transport of Material into or out of the Cell

It has already been discussed that some molecules can enter cells by passing through passive channels in the cell membrane. Large molecules enter the cell by the process of ***endocytosis*** (Fig. 2.5). In this process the molecule invaginates a part of the cell membrane, which first surrounds the molecule, and then separates (from the rest of the cell membrane) to form an ***endocytic vesicle***. This vesicle can move through the cytosol to other parts of the cell.

The term ***pinocytosis*** is applied to a process similar to endocytosis when the vesicles (then called ***pinocytotic vesicles***) formed are used for absorption of fluids (or other small molecules) into the cell.

Some cells use the process of endocytosis to engulf foreign matter (e.g., bacteria). The process is then referred to as ***phagocytosis***.

Molecules produced within the cytoplasm (e.g., secretions) may be enclosed in membranes to form vesicles that approach the cell membrane and fuse with its internal surface. The vesicle then ruptures releasing the molecule to the exterior. The vesicles in question are called ***exocytic vesicles***, and the process is called ***exocytosis*** or ***reverse pinocytosis*** (Fig. 2.6).

In a Nutshell:

- Cell membrane controls the passage of substance in and out of the cell.
- Small molecules pass through passive channels.
- Large molecules enter the cell by the process of endocytosis.
- Absorption of fluids is by pinocytosis.
- If the engulfed material is a foreign body (e.g., bacteria), the term phagocytosis is used.
- When the vesicles release the molecule to the exterior, the process is then referred to as exocytosis or reverse pinocytosis.

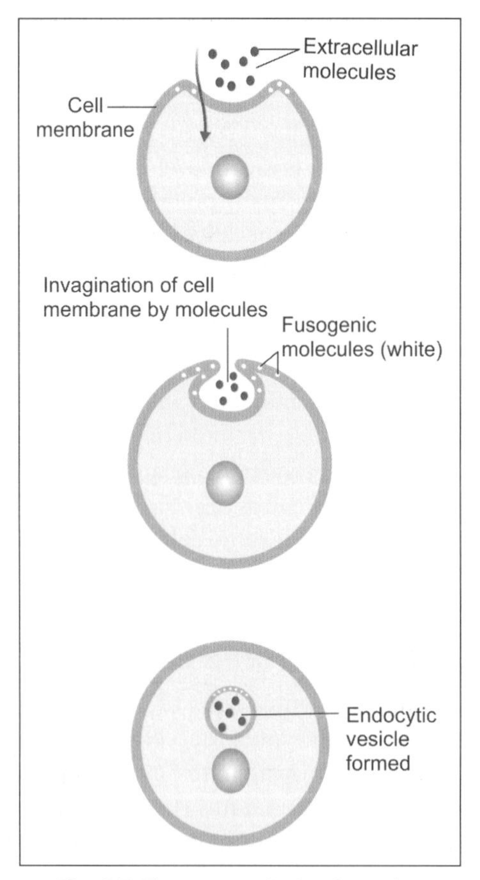

Fig. 2.5: Three stages in the absorption of extracellular molecules by endocytosis (Schematic representation)

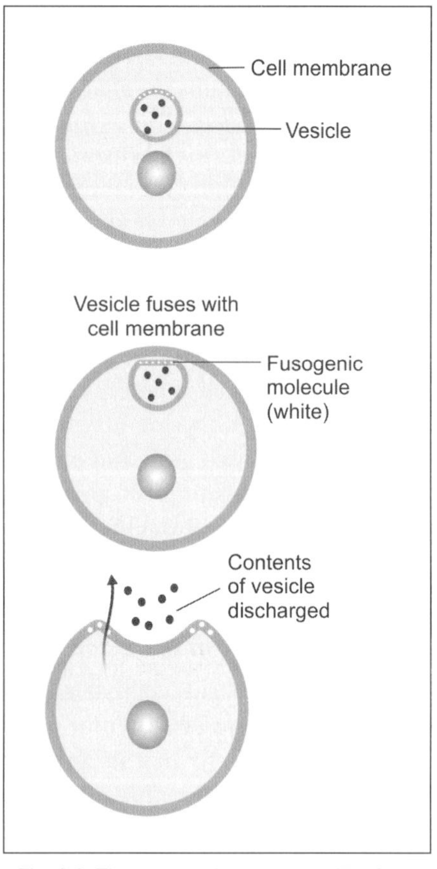

Fig. 2.6: Three stages in exocytosis. The fusogenic proteins facilitate adhesion of the vesicle to the cell membrane (Schematic representation)

Added Information

- As endocytic vesicles are derived from cell membrane, and as exocytic vesicles fuse with the later, there is a constant transfer of membrane material between the surface of the cell and vesicles within the cell.
- Areas of cell membrane which give origin to endocytic vesicles are marked by the presence of ***fusogenic proteins*** that aid the formation of endocytic vesicles. Fusogenic proteins also help in exocytosis by facilitating fusion of membrane surrounding vesicles with the cell membrane.
- When viewed by EM, areas of receptor mediated endocytosis are seen as depressed areas called ***coated pits*** (Fig. 2.7). The membrane lining the floor of the pits is thickened because of the presence of a protein called ***clathrin***. This protein forms a scaffolding around the developing vesicle

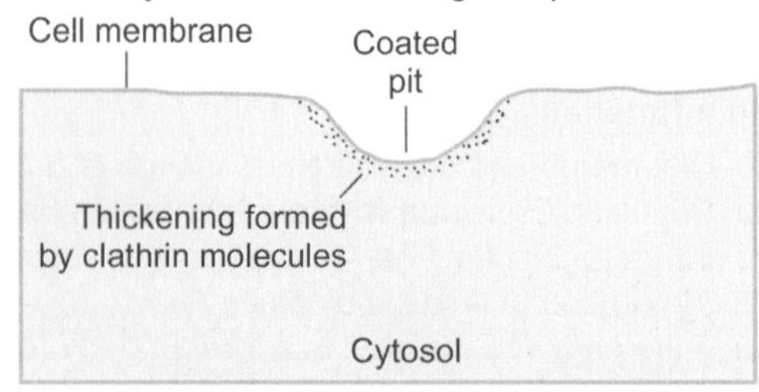

Fig. 2.7: Coated pit as seen by electron microscope in cell membrane (Schematic representation)

contd...

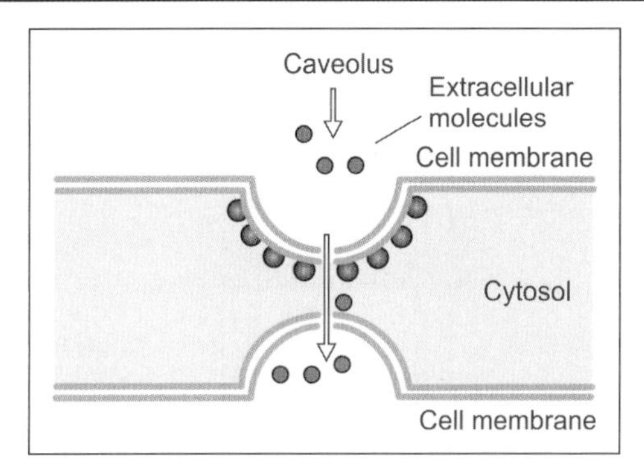

Fig. 2.8: How extracellular molecules can pass through the entire thickness of a cell (transcytosis). Caveolae are involved (Schematic representation)

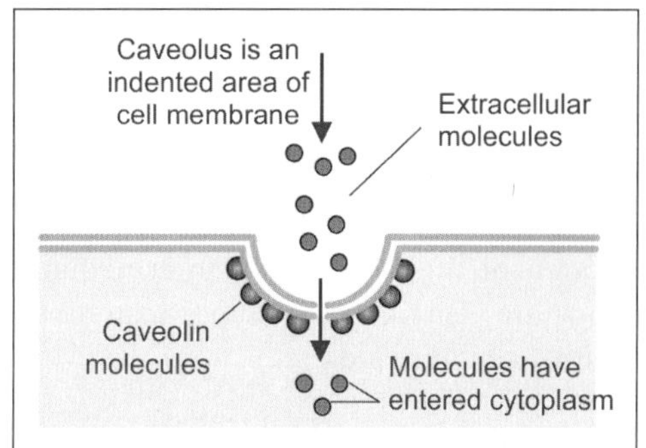

Fig. 2.9: How extracellular molecules enter the cytosol through caveolae. Endocytic vesicles are not formed. The process is called potocytosis (Schematic representation)

and facilitates its separation from the cell membrane. Thereafter, the clathrin molecules detach from the surface of the vesicle and return to the cell membrane.

- The term ***transcytosis refers*** to a process where material is transferred right through the thickness of a cell. The process is seen mainly in flat cells (e.g., endothelium). The transport takes place through invaginations of cell membrane called ***caveolae***. A protein caveolin is associated with caveolae (Fig. 2.8). Caveolae differ from coated pits in that they are not transformed into vesicles. Caveolae also play a role in transport of extracellular molecules to the cytosol (without formation of vesicles) (Fig. 2.9).

CONTACTS BETWEEN ADJOINING CELLS

In tissues in which cells are closely packed, the cell membranes of adjoining cells are separated over most of their extent by a narrow space (about 20 nm). This contact is sufficient to bind cells loosely together, and also allows some degree of movement of individual cells.

In some regions the cell membranes of adjoining cells come into more intimate contact. These areas can be classified as follows:

Classification of Cell Contacts

Unspecialised Contacts

These are contacts that do not show any specialised features on EM examination. At such sites adjoining cell membranes are held together by some glycoprotein molecules, present in the cell membrane, called as ***cell adhesion molecules*** (CAMs). These molecules occupy the entire thickness of the cell membrane (i.e., they are transmembrane proteins). At its cytosolic end, each CAM is in contact with ***an intermediate protein*** (or ***link protein***) (that appears to hold the CAM in place). Fibrous elements of the cytoskeleton are attached to this intermediate protein (and thus indirectly to CAMs). The other end of the CAM juts into the 20 nm intercellular space, and comes in contact with a similar molecule from the opposite cell membrane. In this way a path is established through which forces can be transmitted from the cytoskeleton of one cell to another (Fig. 2.10).

CAMs and intermediate proteins are of various types. Contacts between cells can be classified on the basis of the type of CAMs proteins present. The adhesion of some CAMs is dependent on the presence of calcium ions; while some others are not dependent on them (Table 2.1). Intermediate proteins are also of various types (catenins, vinculin, α-actinin, etc.).

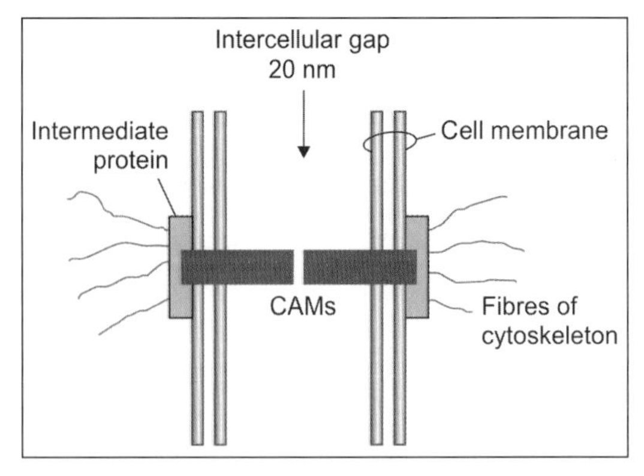

Fig. 2.10: Basic structure of an unspecialised contact between two cells (Schematic representation)

Specialised Junctional Structures

These junctions can be recognised by EM. The basic mode of intercellular contact in them is similar to that described above and involves CAMs, intermediate proteins and cytoskeletal elements.

Types of Specialised Junctions (Fig. 2.11)

- ***Anchoring junctions*** or ***adhesive junctions*** bind cells together mechanically to their neighbouring cells. They can be of the following types:
  - ***Adhesive spots*** (also called ***desmosomes*** or ***maculae adherens***).
  - ***Adhesive belts*** or ***zonula adherens.***
  - ***Adhesive strips*** or ***fascia adherens.***

Table 2.1: Types of cell adhesion molecules (CAM)

Type of CAM	Subtypes	Present in
Calcium dependent	Cadherins (of various types)	Most cells including epithelia
	Selectins	Migrating cells, e.g., Leucocytes
	Integrins	Between cells and intercellular substances. About 20 types of integrins, each attaching to a special extracellular molecule
Calcium independent	Neural cell adhesion molecule (NCAM)	Nerve cells
	Intercellular adhesion molecule (ICAM)	Leucocytes

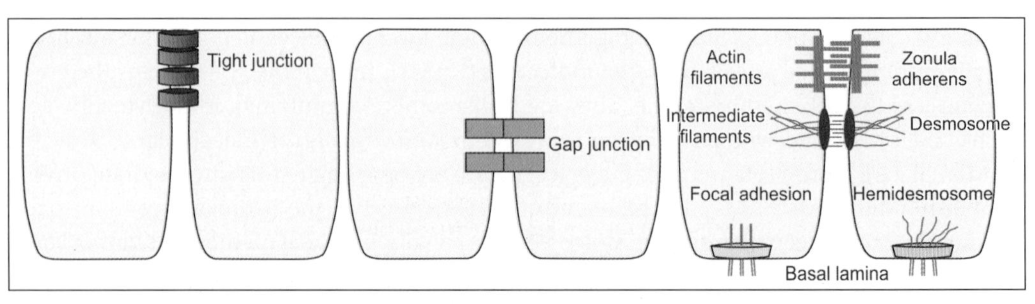

Fig. 2.11: Various types of specialised junctions (Schematic representation)

Modified anchoring junctions attach cells to extracellular material. Such junctions are seen as ***hemidesmosomes*** or as ***focal spots***.

- ***Occluding junctions*** (***zonula occludens*** or ***tight junctions***). Apart from holding cells together, these junctions form barriers to movement of material through intervals between cells.
- ***Communicating junctions*** (or ***gap junctions***). Such junctions allow direct transport of some substances from cell to cell.

Anchoring Junctions

Adhesion Spots (Desmosomes, Macula Adherens)

These are the most common type of junctions between adjoining cells. Desmosomes are present where strong anchorage between cells is needed, e.g., between cells of the epidermis.

As seen by EM, a desmosome is a small circumscribed area of attachment (Fig. 2.12). At the site of a desmosome, the plasma membrane (of each cell) is thickened because of the presence of a dense layer of proteins on its inner surface (i.e., the surface towards the cytoplasm).

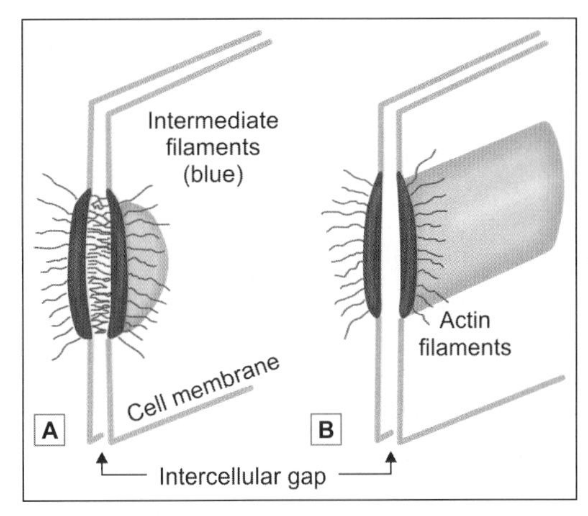

Fig. 2.12: A. Electron microscope appearance of a desmosome; **B.** Electron microscope appearance of zonula adherens (Schematic representation)

The thickened areas of the two sides are separated by a gap of 25 nm. The region of the gap is rich in glycoproteins. The thickened areas of the two membranes are held together by fibrils that appear to pass from one membrane to the other across the gap. We now know that the fibrils seen in the intercellular space represent CAMs (Fig. 2.13).

The thickened area (or plaque) seen on the cytosolic aspect of the cell membrane is produced by the presence of intermediate (link) proteins. Cytoskeletal filaments attached to the thickened area are intermediate filaments. CAMs seen in desmosomes are integrins (desmogleins I, II). The link proteins are desmoplakins.

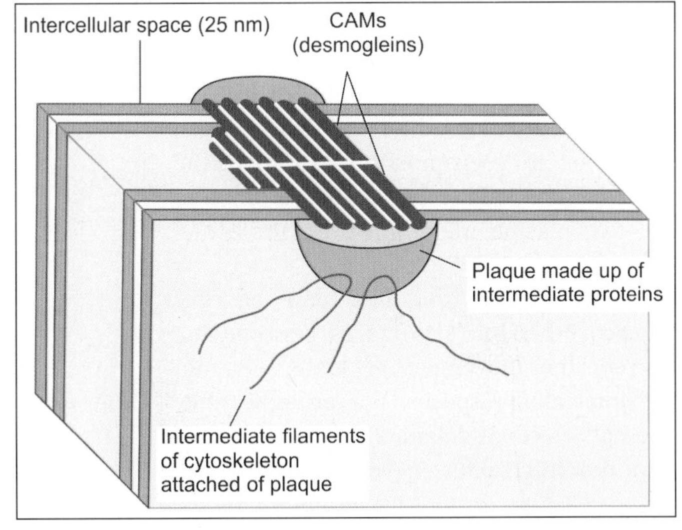

Fig. 2.13: Detailed structure of a desmosome (in the epidermis) (Schematic representation)

Adhesive Belts (Zonula Adherens)

In some situations, most typically near the apices of epithelial cells, we see a kind of junction called the ***zonula adherens*** or ***adhesive belt*** (Fig. 2.12B). This is similar to a desmosome in being marked by thickenings of the two plasma membranes, to the cytoplasmic aspects of which fibrils are attached. However, the junction differs from a desmosome as follows:

- Instead of being a small circumscribed area of attachment, the junction is in the form of a continuous band passing all around the apical part of the epithelial cell.
- The gap between the thickenings of the plasma membranes of the two cells is not traversed by filaments.

The CAMs present are cadherins. In epithelial cells zona adherens are located immediately deep to occluding junctions (Fig. 2.16).

Adhesive Strips (Fascia Adherens)

These are similar to adhesive belts. They differ from the latter in that the areas of attachment are in the form of short strips (and do not go all round the cell). These are seen in relation to smooth muscle, intercalated discs of cardiac muscle, and in junctions between glial cells and nerves.

Hemidesmosomes

These are similar to desmosomes, but the thickening of cell membrane is seen only on one side. At such junctions, the 'external' ends of CAMs are attached to extracellular structures. Hemidesmosomes are common where basal epidermal cells lie against connective tissue.

The cytoskeletal elements attached to intermediate proteins are keratin filaments (as against intermediate filaments in desmosomes). As in desmosomes, the CAMs are integrins.

Focal Spots

These are also called ***focal adhesion plaques***, or ***focal contacts***. They represent areas of local adhesion of a cell to extracellular matrix. Such junctions are of a transient nature (e.g., between a leucocyte and a vessel wall). Such contacts may send signals to the cell and initiate cytoskeletal formation.

The CAMs in focal spots are integrins. The intermediate proteins (that bind integrins to actin filaments) are α-actinin, vinculin and talin.

Occluding Junctions (Zonula Occludens)

Like the zonula adherens, the zonula occludens are seen most typically near the apices of epithelial cells. At such a junction the two plasma membranes are in actual contact (Fig. 2.14A).

These junctions not only bind the cells to each other but also act as barriers that prevent the movement of molecules into the intercellular spaces. For example, they prevent intestinal contents from permeating into the intercellular spaces between the lining cells by virtue of occluding junctions. Zonulae occludens are, therefore, also called ***tight junctions*** (Fig. 2.14A).

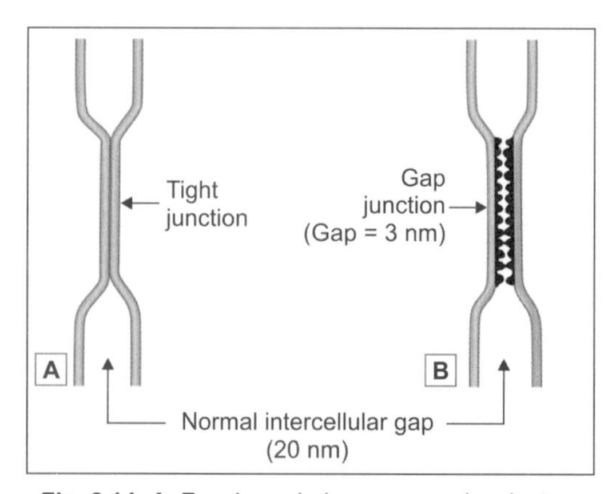

Fig. 2.14: A. Zonula occludens as seen by electron microscope; **B.** Gap junction as seen by electron microscope (Schematic representation)

Added Information

Recent studies have provided a clearer view of the structure of tight junctions (Fig. 2.15). Adjoining cell membranes are united by CAMs that are arranged in the form of a network that 'stitches' the two membranes together.

Other functions attributed to occluding junctions are as follows:

- These junctions separate areas of cell membrane that are specialised for absorption or secretion (and lie on the luminal side of the cell) from the rest of the cell membrane.
- Areas of cell membrane performing such functions bear specialised proteins. Occluding junctions prevent lateral migration of such proteins.
- In cells involved in active transport against a concentration gradient, occluding junctions prevent back diffusion of transported substances.

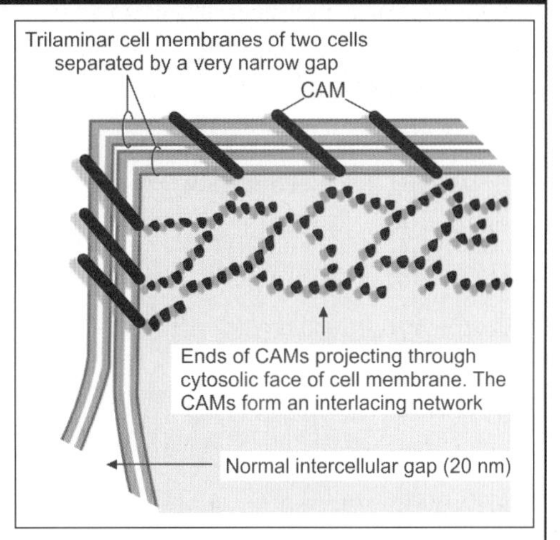

Fig. 2.15: Detailed structure of part of an occluding junction (Schematic representation)

Apart from epithelial cells, zonulae occludens are also present between endothelial cells.

In some situations occlusion of the gaps between the adjoining cells may be incomplete and the junction may allow slow diffusion of molecules across it. These are referred to as ***leaky tight junctions***.

Junctional Complex

Near the apices of epithelial cells the three types of junctions described above, namely zonula occludens, zonula adherens and macula adherens are often seen arranged in that order (Fig. 2.16). They collectively form a junctional complex. In some complexes, the zonula occludens may be replaced by a leaky tight junction, or a gap junction.

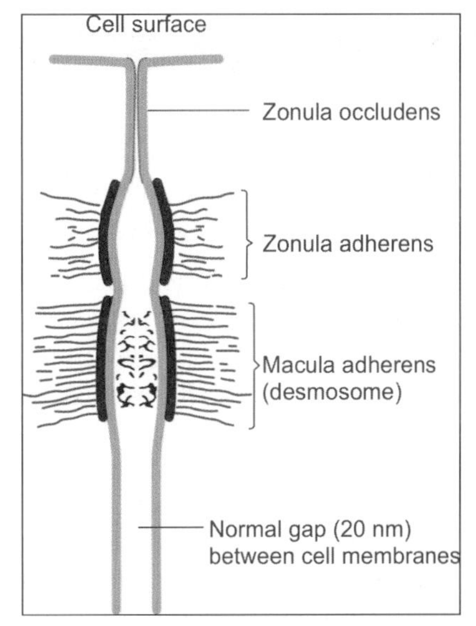

Fig. 2.16: A junctional complex (schematic representations)

Communicating Junctions (Gap Junctions)

At these junctions, the plasma membranes are not in actual contact (as in a tight junction), but lie very close to each other; the gap being reduced (from the normal 20 nm to 3 nm) (Fig. 2.14B). In transmission electronmicrographs this gap is seen to contain bead-like structures (Fig. 2.17A). A minute canaliculus passing through each 'bead' connects the cytoplasm of the two cells thus allowing the free passage of some substances (sodium, potassium, calcium, metabolites) from one cell to the other. Gap junctions are, therefore, also called ***maculae communicantes***. They are widely distributed in the body.

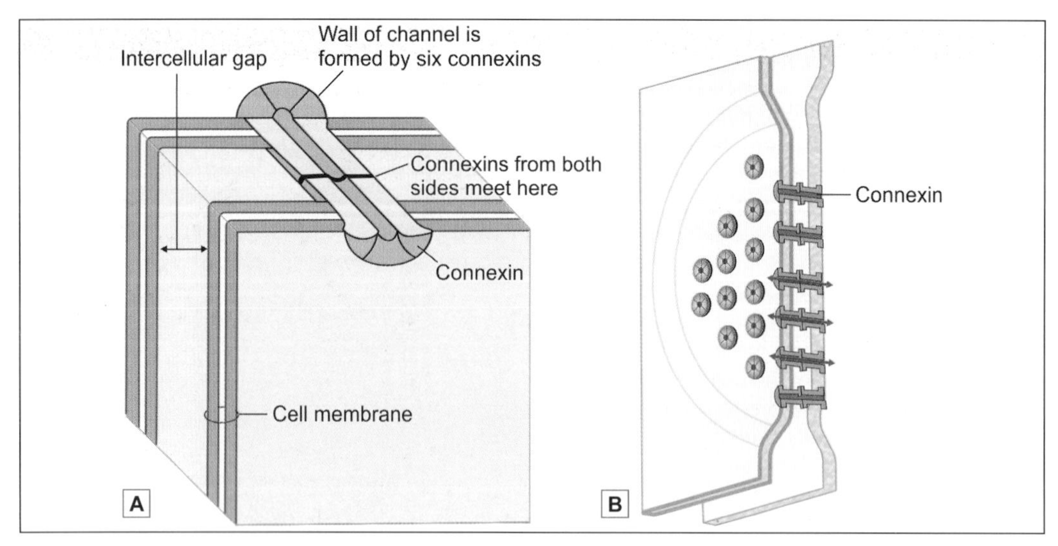

Fig. 2.17: A communicating junction (gap junction). **A.** As seen under electron microscope; **B.** To show the constitution of one channel of a communicating junction (Schematic representation)

The wall of each channel is made up of six protein elements (called ***nexins***, or ***connexons***). The 'inner' ends of these elements are attached to the cytosolic side of the cell membrane while the 'outer' ends project into the gap between the two cell membranes (Fig. 2.17B). Here they come in contact with (and align perfectly with) similar connexins projecting into the space from the cell membrane of the opposite cell to complete the channel.

Added Information

- Changes in pH or in calcium ion concentration can close the channels of gap junctions. By allowing passing of ions they lower transcellular electrical resistance. Gap junctions form electrical synapses between some neurons.
- The number of channels present in a gap junction can vary considerably.

CELL ORGANELLES

The cytoplasm of a typical cell contains various structures that are referred to as organelles. They include the endoplasmic reticulum (ER), ribosomes, mitochondria, the Golgi complex and various types of vesicles (Fig. 2.18).

The cell organelles can be membrane bound or without membrane (Table 2.2). The cytosol also contains a cytoskeleton made up of microtubules, microfilaments and intermediate filaments. Centrioles are closely connected with microtubules.

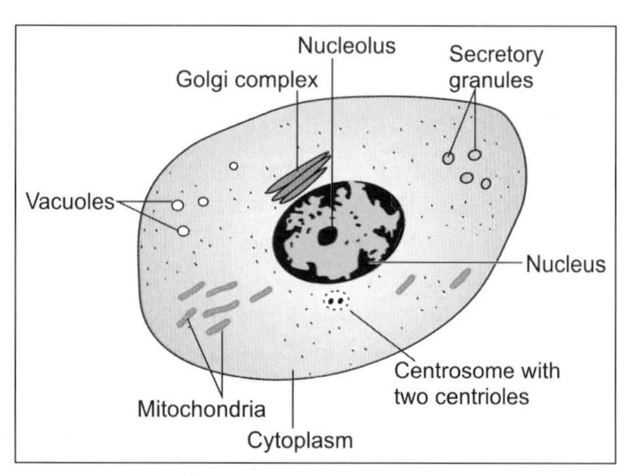

Fig. 2.18: Some features of a cell that can be seen with a light microscope (Schematic representation)

Table 2.2: Cell organelles

Membrane bound cell organelles	Non-membranous cell organelles
• Endoplasmic reticulum (ER) • Golgi complex • Mitochondria • Membrane bound vesicles including - Phagosomes - Lysosomes - Peroxisomes - Exocytic vesicles	• Cytoskeleton including - Microfilaments - Microtubules - Intermediate filaments • Ribosomes

Endoplasmic Reticulum

It is a network of interconnecting membranes enclosing channels or cisternae, that are continuous from outer nuclear envelope to outer plasma membrane (Fig. 2.19).

Because of the presence of the endoplasmic reticulum (ER) the cytoplasm is divided into two components, one within the channels and one outside them. The cytoplasm within the channels is called the ***vacuoplasm***, and that outside the channels is the ***hyaloplasm*** or ***cytosol***.

The proteins, glycoproteins and lipoproteins are synthesised in the ER. ER is very prominent in cells actively synthesising proteins, e.g., immunoglobulin secreting plasma cells.

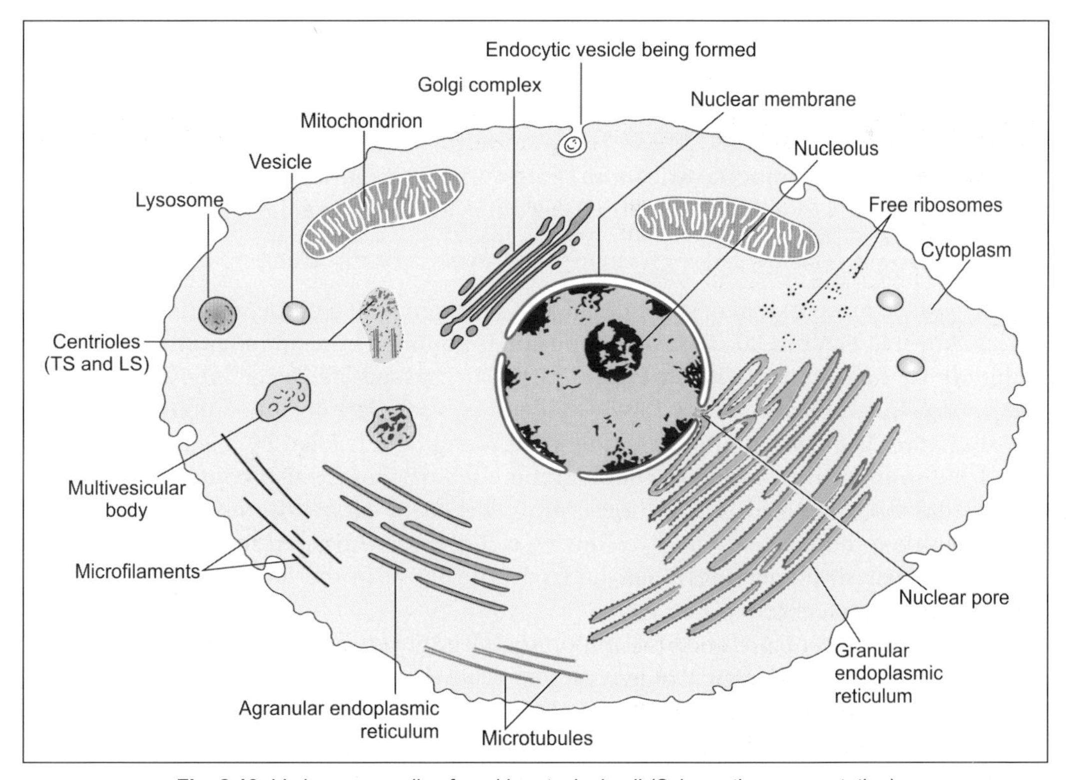

Fig. 2.19: Various organelles found in a typical cell (Schematic representation)

According to the EM appearance, the ER is generally classified into ***rough*** (granular) and ***smooth*** (agranular) varieties (Fig. 2.19). The rough appearance of ***rough ER*** is due to ribosomes attached to cytoplasmic side of membrane which play an important role in protein synthesis. It is a prominent feature of cells synthesising proteins. The lumen of rough ER is continuous with the perinuclear space (between the inner and outer nuclear membranes). It is also continuous with the lumen of smooth ER.

Smooth ER is responsible for further processing of proteins synthesised in rough ER. It is also responsible for synthesis of lipids, specially that of membrane phospholipids (necessary for membrane formation). Most cells have very little smooth ER. It is a prominent feature of cells processing lipids.

Products synthesised by the ER are stored in the channels within the reticulum. Ribosomes and enzymes are present on the 'outer' surfaces of the membranes of the reticulum.

Ribosomes

As discussed, ribosomes are present in relation to rough endoplasmic reticulum. They may also lie free in the cytoplasm. They may be present singly in which case they are called ***monosomes***; or in groups which are referred to as ***polyribosomes*** (or ***polysomes***). Each ribosome consists of proteins and RNA (ribonucleic acid) and is about 15 nm in diameter. The ribosome is made up of two subunits small (40S) and large (60S) classified on the basis of their sedimentation rates. Ribosomes play an essential role in protein synthesis.

Mitochondria

Mitochondria can be seen with the light microscope in specially stained preparations (Fig. 2.18). They are so called because they appear either as granules or as rods (*mitos* = granule; *chondrium* = rod).

The number of mitochondria varies from cell to cell being greatest in cells with high metabolic activity (e.g., in secretory cells and sperms). Erythrocytes do not contain mitochondria.

Mitochondria vary in size, most of them being 0.5–2 μm in length. Mitochondria are large in cells with a high oxidative metabolism.

A schematic presentation of some details of the structure of a mitochondrion (as seen by EM) is shown in Figure 2.20. The mitochondrion is bounded by a smooth ***outer membrane*** within which there is an ***inner membrane***; the two being separated by an ***intermembranous space***. The inner membrane is highly folded on itself forming incomplete partitions called ***cristae***. The space bounded by the inner membrane is filled by a granular material called the ***matrix***. This matrix contains numerous enzymes. It also contains some RNA and dioxyribonucleic acid (DNA). These are believed to carry information that enables mitochondria to duplicate themselves during cell division. An interesting fact, discovered recently, is that all mitochondria are derived from those in the fertilised ovum and are entirely of maternal origin.

Mitochondria are of great functional importance. It is the ***power house of the cell***. It contains many enzymes including some that play an important part in Kreb's cycle [tricarboxylic acid (TCA) cycle]. Adenosine triphosphate (ATP) and guanosine triphosphate (GTP) are formed in mitochondria from where they pass to other parts of the cell and provide energy for various cellular functions.

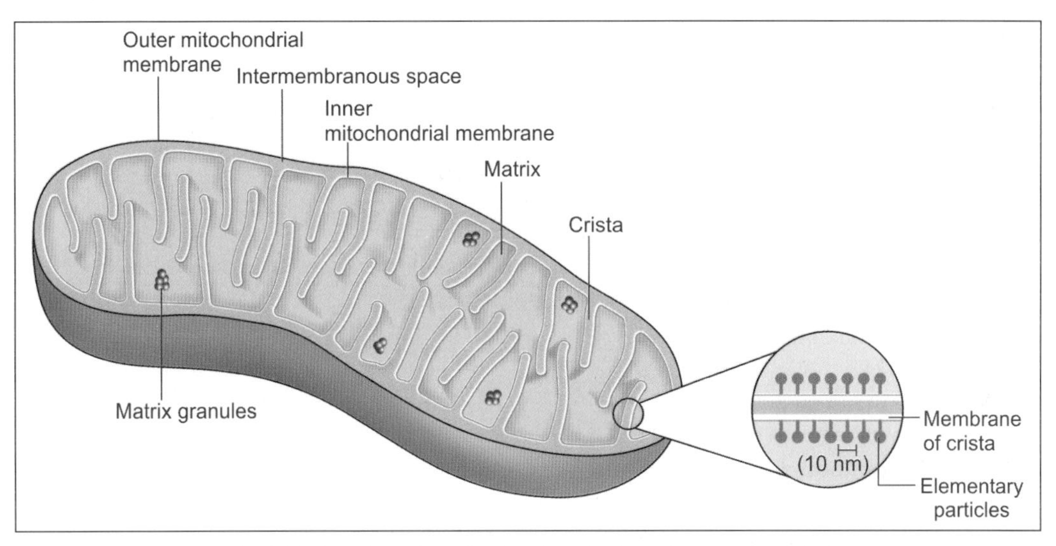

Fig. 2.20: Structure of a mitochondrion (Schematic representation)

Added Information

The enzymes of the TCA cycle are located in the matrix of mitochondria, while enzymes associated with the respiratory chain and ATP production are present on the inner mitochondrial membrane. Enzymes for conversion of ADP to ATP are located in the intermembranous space. Enzymes for lipid synthesis and fatty acid metabolism are located in the outer membrane.

Clinical Correlation

Mitochondrial Abnormalities

Mitochondrial DNA can be abnormal. This interferes with mitochondrial and cell functions, resulting in disorders referred to as **mitochondrial cytopathy syndromes**. The features (which differ in intensity from patient to patient) includes muscle weakness, degenerative lesions in the brain and high levels of lactic acid. The condition can be diagnosed by EM examination of muscle biopsies. The mitochondria show characteristic para-crystalline inclusions.

Golgi Complex

The Golgi complex (Golgi apparatus, or merely Golgi) was known to microscopists long before the advent of electron microscopy. In light microscopic preparations suitably treated with silver salts, the Golgi complex can be seen as a small structure of irregular shape, usually present near the nucleus (Fig. 2.18).

When examined with the EM, the complex is seen to be made up of network of flattened smooth membranes similar to those of smooth ER (Fig. 2.21). The membranes form the walls of a number of flattened sacs that are stacked over one another. Towards their margins the sacs are continuous with small rounded vesicles. The cisternae of the Golgi complex form an independent system. Their lumen is not in communication with that of ER. Material from ER reaches the Golgi complex through vesicles.

From a functional point of view the Golgi complex is divisible into three regions (Fig. 2.22). The region nearest the nucleus is the ***cis face*** (or ***cis Golgi***). The opposite face (nearest the cell membrane) is the ***trans face*** (also referred to as ***trans Golgi***). The intermediate part (between the cis face and the trans face) is the ***medial Golgi***.

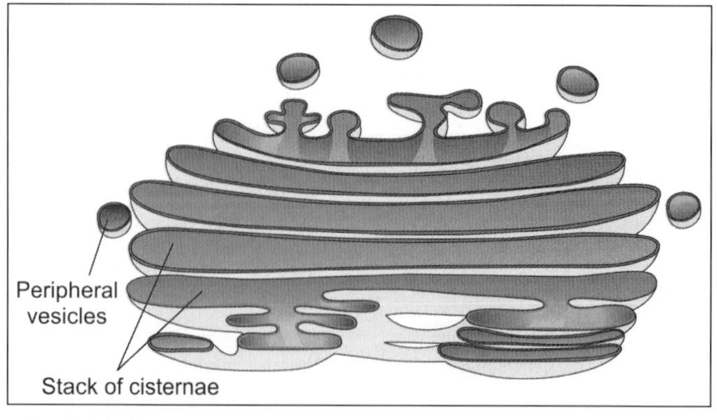

Fig. 2.21: Structure of the Golgi complex (Schematic representation)

Material synthesised in rough ER travels through the ER lumen into smooth ER. Vesicles budding off from smooth ER transport this material to the cis face of the Golgi complex. Some proteins are phosphorylated here. From the cis face all these materials pass into the medial Golgi. Here sugar residues are added to proteins to form protein-carbohydrate complexes (glycoproteins).

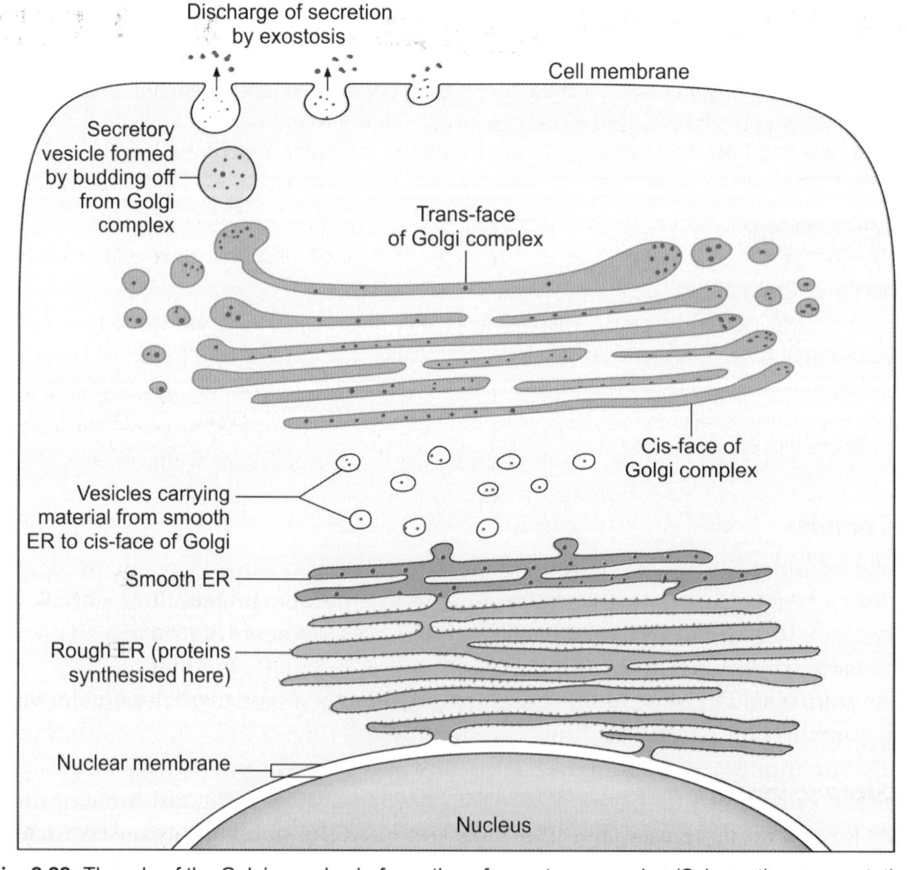

Fig. 2.22: The role of the Golgi complex in formation of secretory vacuoles (Schematic representation)

Finally, all material passes to the trans face, which performs the function of protein sorting, packaging and secretion. The membranes of the Golgi complex contain appropriate enzymes for the functions performed by them. As proteins pass through successive sacs of Golgi they undergo a process of purification.

Membrane Bound Vesicles

The cytoplasm of a cell may contain several types of vesicles. The contents of any such vesicle are separated from the rest of the cytoplasm by a membrane which forms the wall of the vesicle.

Vesicles are formed by budding off from existing areas of membrane. Some vesicles serve to store material. Others transport material into or out of the cell, or from one part of a cell to another. Vesicles also allow exchange of membrane between different parts of the cell.

Details of the appearances of various types of vesicles will not be considered here. However, the student must be familiar with their terminology given below.

Phagosomes

Solid 'foreign' materials, including bacteria, may be engulfed by a cell by the process of ***phagocytosis***. In this process, the material is surrounded by a part of the cell membrane. This part of the cell membrane then separates from the rest of the plasma membrane and forms a free floating vesicle within the cytoplasm. Such membrane bound vesicles, containing solid ingested material are called ***phagosomes*** (also see lysosomes).

Pinocytotic Vesicles

Some fluid may also be taken into the cytoplasm by a process similar to phagocytosis. In the case of fluids, the process is called ***pinocytosis*** and the vesicles formed are called ***pinocytotic vesicles***.

Exocytic Vesicles

Just as material from outside the cell can be brought into the cytoplasm by phagocytosis or pinocytosis, materials from different parts of the cell can be transported to the outside by vesicles. Such vesicles are called ***exocytic vesicles***, and the process of discharge of cell products in this way is referred to as ***exocytosis*** (or ***reverse pinocytosis***).

Secretory Granules

The cytoplasm of secretory cells frequently contains what are called ***secretory granules***. These can be seen with the light microscope. With the EM, each 'granule' is seen to be a membrane bound vesicle containing secretion. The appearance, size and staining reactions of these secretory granules differ depending on the type of secretion. These vesicles are derived from the Golgi complex.

Other Storage Vesicles

Materials such as lipids, or carbohydrates, may also be stored within the cytoplasm in the form of membrane bound vesicles.

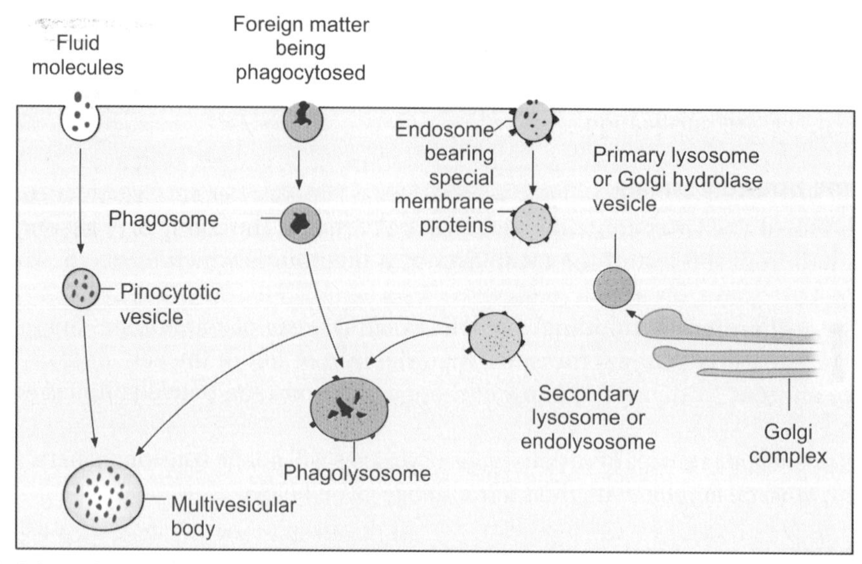

Fig. 2.23: Schematic representation to show how lysosomes, phagolysosomes and multivesicular bodies are formed

Lysosomes

These membrane bound vesicles contain enzymes that can destroy unwanted material present within a cell. Such material may have been taken into the cell from outside (e.g., bacteria); or may represent organelles that are no longer of use to the cell. The enzymes present in lysosomes include (amongst others) proteases, lipases, carbohydrases and acid phosphatase, (As many as 40 different lysosomal enzymes have been identified).

Lysosomes belong to what has been described as the ***acid vesicle system***. The vesicles of this system are covered by membrane which contains H^+ATPase. This membrane acts as a H^+ pump creating a highly acid environment within the vesicle (up to pH 5). The stages in the formation of a lysosome are as follows:

- Acid hydrolase enzymes synthesised in ER reach the Golgi complex where they are packed into vesicles (Fig. 2.23). The enzymes in these vesicles are inactive because of the lack of an acid medium (These are called ***primary lysosomes*** or ***Golgi hydrolase vesicles***) (Flow chart 2.1).
- These vesicles fuse with other vesicles derived from cell membrane (***endosomes***). These endosomes possess the membrane proteins necessary for producing an acid medium. The

Flow chart 2.1: Types of lysosomes

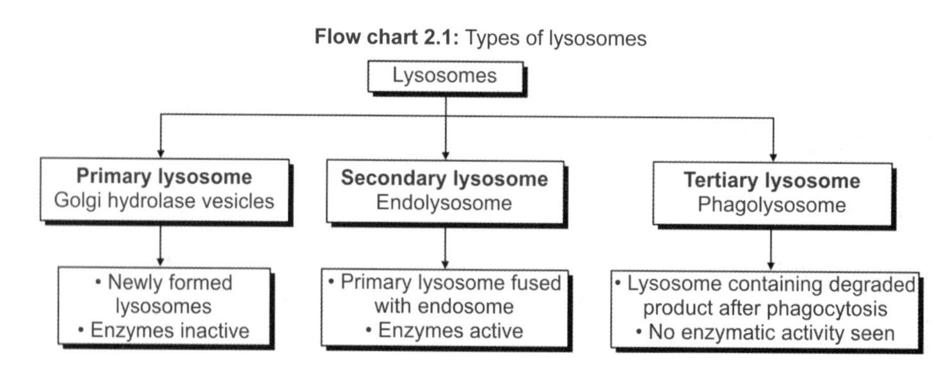

product formed by fusion of the two vesicles is an ***endolysosome*** (or ***secondary lysosome***) (Flow chart 2.1).

- H^+ ions are pumped into the vesicle to create an acid environment. This activates the enzymes and a mature lysosome is formed.

Lysosomes help in 'digesting' the material within phagosomes as follows. A lysosome, containing appropriate enzymes, fuses with the phagosome so that the enzymes of the former can act on the material within the phagosome. These bodies consisting of fused phagosomes and lysosomes are referred to as ***phagolysosomes*** (Fig. 2.23).

In a similar manner lysosomes may also fuse with pinocytotic vesicles. The structures formed by such fusion often appear to have numerous small vesicles within them and are, therefore, called ***multivesicular bodies*** (Fig. 2.23).

After the material in phagosomes or pinocytotic vesicles has been 'digested' by lysosomes, some waste material may be left. Some of it is thrown out of the cell by exocytosis. However, some material may remain within the cell in the form of membrane bound ***residual bodies***.

Lysosomal enzymes play an important role in the destruction of bacteria phagocytosed by the cell. Lysosomal enzymes may also be discharged out of the cell and may influence adjoining structures.

Note: Lysosomes are present in all cells except mature erythrocytes. They are a prominent feature in neutrophil leucocytes.

Clinical Correlation

Genetic defects can lead to absence of specific acid hydrolases that are normally present in lysosomes. As a result some molecules cannot be degraded, and accumulate in lysosomes. Examples of such disorders are ***lysosomal glycogen storage disease***, in which there is abnormal accumulation of glycogen, and ***Tay-Sach's disease***, in which lipids accumulate in lysosomes and lead to neuronal degeneration resulting in seizures, muscle rigidity and blindness.

Peroxisomes

These are similar to lysosomes in that they are membrane bound vesicles containing enzymes. The enzymes in most of them react with other substances to form hydrogen peroxide, which is used to detoxify various substances by oxidising them. The enzymes are involved in oxidation of very long chain fatty acids. Hydrogen peroxide resulting from the reactions is bactericidal. Some peroxisomes contain the enzyme catalase which converts the toxic hydrogen peroxide to water, thus preventing the latter from accumulating in the cell.

Peroxisomes are most prominent in cells of the liver and in cells of renal tubules.

Pathologial Correlation

- **Adrenoleukodystrophy** (Brown-Schilder's disease) characterised by progressive degeneration of liver, kidney and brain is a rare autosomal recessive condition due to insufficient oxidation of very long chain fatty acids by peroxisomes.
- **Zellweger syndrome** is characterised by formation of empty peroxisomes or peroxisomal ghosts inside the cells due to a defect in transportation of proteins into the peroxisomes.
- **Primary hyperoxaluria** is due to the defective peroxisomal metabolism of glyoxyalate derived from glycine.

THE CYTOSKELETON

The cytoplasm is permeated by a number of fibrillar elements that collectively form a supporting network. This network is called the ***cytoskeleton***. Apart from maintaining cellular architecture the cytoskeleton facilitates cell motility (e.g., by forming cilia), and helps to divide the cytosol into functionally discrete areas. It also facilitates transport of some constituents through the cytosol, and plays a role in anchoring cells to each other.

The elements that constitute the cytoskeleton consist of the following:

- Microfilaments
- Microtubules
- Intermediate filaments.

Microfilaments

These are about 5 nm in diameter. They are made up of the protein ***actin***. Individual molecules of actin are globular (***G-actin***). These join together (polymerise) to form long chains called ***F-actin, actin filaments*** or ***microfilaments***.

Actin filaments form a meshwork just subjacent to the cell membrane. This meshwork is called the ***cell cortex*** (The filaments forming the meshwork are held together by a protein called ***filamin***). The cell cortex helps to maintain the shape of the cell. The meshwork of the cell cortex is labile. The filaments can separate (under the influence of actin severing proteins), and can reform in a different orientation. That is how the shape of a cell is altered.

Microvilli contain bundles of actin filaments, and that is how they are maintained. Filaments also extend into other protrusions from the cell surface.

Microtubules

Microtubules are about 25 nm in diameter (Fig. 2.24). The basic constituent of microtubules is the protein tubulin (composed of subunits α and β). Chains of tubulin form protofilaments. The wall of a microtubule is made up of thirteen protofilaments that run longitudinally (Fig. 2.24). The tubulin protofilaments are stabilised by microtubule associated proteins (MAPs).

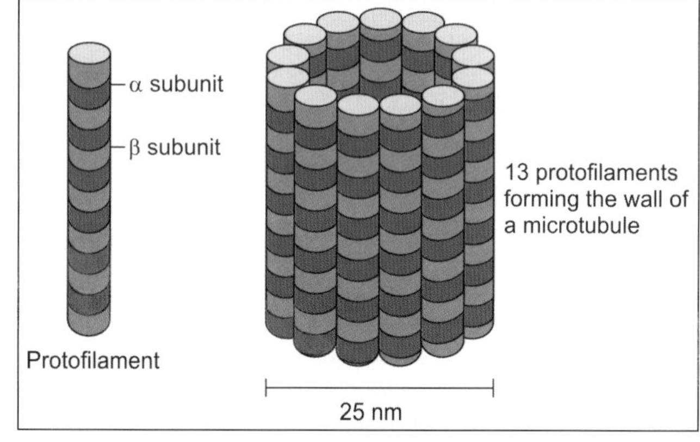

Fig. 2.24: Schematic representation to show how a microtubule is constituted

Microtubules are formed in centrioles (see below) which constitute a ***microtubule organising centre***.

The roles played by microtubules are as follows:

- As part of the cytoskeleton, they provide stability to the cell. They prevent tubules of ER from collapsing.
- Microtubules facilitate transport within the cell. Some proteins (dynein, kinesin) present in membranes of vesicles, and in organelles, attach these to microtubules, and facilitate movement along the tubules. Such transport is specially important in transport along axons.

- In dividing cells microtubules form the mitotic spindle.
- Cilia are made up of microtubules (held together by other proteins).

Intermediate Filaments

These are so called as their diameter (10 nm) is intermediate between that of microfilaments (5 nm) and of microtubules (25 nm). The proteins constituting these filaments vary in different types of cells.

They include ***cytokeratin*** (in epithelial cells), ***neurofilament protein*** (in neurons), ***desmin*** (in muscle), ***glial fibrillary acidic protein*** (in astrocytes); ***Laminin*** (in the nuclear lamina of cells), and ***vimentin*** (in many types of cells).

The roles played by intermediate filaments are as follows:

- Intermediate filaments link cells together. They do so as they are attached to transmembrane proteins at desmosomes. The filaments also facilitate cell attachment to extracellular elements at hemidesmosomes.
- In the epithelium of the skin the filaments undergo modification to form keratin. They also form the main constituent of hair and nails.
- The neurofilaments of neurons are intermediate filaments. Neurofibrils help to maintain the cylindrical shape of axons.
- The nuclear lamina consists of intermediate filaments.

Table 2.3 summarises the characteristics of the three types of cytoskeleton filaments.

Table 2.3: Characteristics of Three Types of Cytoskeletal Filaments

	Microfilaments	Intermediate filaments	Microtubules
Shape	Double-stranded linear helical arrangement	Rope like fibres	Long non-branching chains
Diameter	5 nm	10 nm	25 nm
Basic protein subunit	Monomer of G-actin (protein)	Various intermediate filament proteins including cytokeratin, neurofilament protein, desmin, laminin and vimentin	Dimers of α- and β-tubulin
Location in cell	• Forms a network just subjacent to cell membrane • Core of microvilli • Contractile elements of muscles	• Extend across cytoplasm connecting desmosomes and hemidesmosomes • The nuclear lamina • In skin epithelium (in the form of keratin)	• Mitotic spindle • Core of cilia
Major functions	Provide essential components to contractile elements of muscle cells (sarcomeres)	Provide mechanical strength and link cells together	• Provide network for movement of organelles within cell • Facilitate transport of organelles within the cell • Provide movement for cilia

Clinical Correlation

- **Kartagener syndrome:** It is characterised by defect in the organisation of microtubules that can immobilise the cilia of respiratory epithelium resulting in the inability of the respiratory system to clear accumulated secretions.
- **Alzheimer's disease:** Defect in the proper assembly of intermediate filaments leads to various diseases. The changes in the neurofilaments within the brain lead to Alzheimer's disease. The disease is characterised by accumulation of tangles inside the neurons. The intracellular deposition of masses of filaments may cause disruption of cytoskeleton architecture and subsequent neuronal cell death leading to dementia (memory loss).

Centrioles

All cells capable of division (and even some which do not divide) contain a pair of structures called ***centrioles***. With the light microscope, the two centrioles are seen as dots embedded in a region of dense cytoplasm which is called the ***centrosome***. With the EM, the centrioles are seen to be short cylinders that lie at right angles to each other. When we examine a transverse section across a centriole (by EM) it is seen to consist essentially of a series of microtubules arranged in a circle. There are nine groups of tubules, each group consisting of three tubules (triplets) (Fig. 2.25).

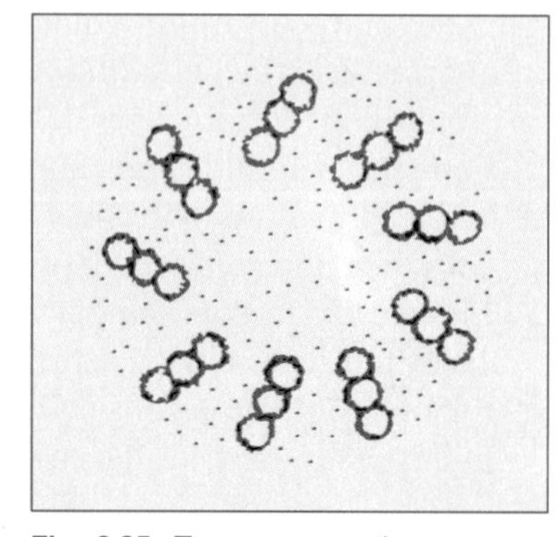

Fig. 2.25: Transverse section across a centriole (near its base). Note nine groups of tubules, each group having three microtubules (Schematic representation)

Centrioles play an important role in the formation of various cellular structures that are made up of microtubules. These include the mitotic spindles of dividing cells, cilia, flagella, and some projections of specialised cells (e.g., the axial filaments of spermatozoa). It is of interest to note that cilia, flagella and the tails of spermatozoa, all have the 9 + 2 configuration of microtubules that are seen in a centriole.

THE NUCLEUS

The nucleus constitute the central, more dense, part of the cell. It is usually rounded or ellipsoid. Occasionally it may be elongated, indented or lobed. It is the largest cell organelle measuring 4–10 µm in diameter. All cells in the body contain nucleus, except mature red blood cells (RBCs) in circulation.

Nuclear Components

- Chromatin
- Nucleolus
- Nuclear membrane
- Nucleoplasm.

Chromatin

In usual class-room slides stained with haematoxylin and eosin, the nucleus stains dark purple or blue while the cytoplasm is usually stained pink. In some cells the nuclei are relatively large and light staining. Such nuclei appear to be made up of a delicate network of fibres—the material making up the fibres of the network is called ***chromatin*** (because of its affinity for dyes). At some places the chromatin is seen in the form of irregular dark masses that are called ***heterochromatin***. At other places the network is loose and stains lightly—the chromatin of such areas is referred to as ***euchromatin***. Nuclei which are large and in which relatively large areas of euchromatin can be seen are referred to as ***open-face nuclei.*** Nuclei that are made up mainly of heterochromatin are referred to as ***closed-face nuclei*** (Fig. 2.26).

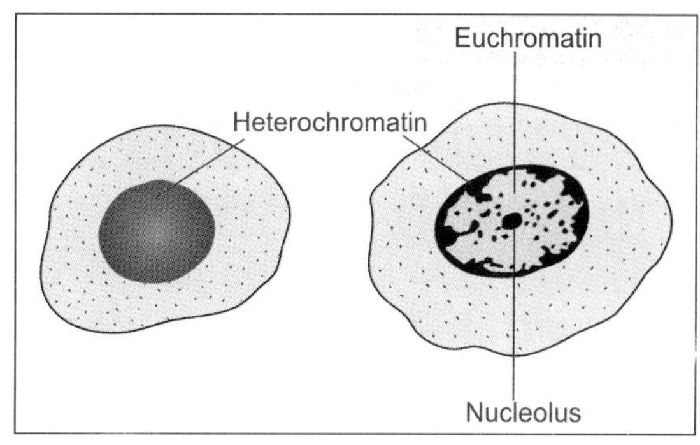

Fig. 2.26: Comparison of a heterochromatic nucleus (left) and a euchromatic nucleus (right) (Schematic representation)

Nature and Significance of Chromatin

In recent years, there has been a considerable advance in our knowledge of the structure and significance of chromatin. It is made up of a substance called DNA; and of proteins. Most of the proteins in chromatin are ***histones***. Some non-histone proteins are also present.

Each eukaryotic cell contains genetic information encoded in DNA structure. The length of the DNA molecule is 100,000 times longer than nucleus. Therefore, the DNA must be highly folded and tightly packed in the cell nucleus, which is accomplished by the formation of chromatin.

Filaments of DNA form coils around histone complexes. The structure formed by a histone complex and the DNA fibre coiled around it is called a ***nucleosome***. Nucleosomes are attached to one another forming long chains (Fig. 2.27). These chains are coiled on themselves (in a helical manner) to form filaments 30 nm in diameter. These filaments constitute chromatin.

Filaments of chromatin are again coiled on themselves (***supercoiling***), and this coiling is repeated several times. Each coiling produces a thicker filament. In this way a filament of DNA that is originally 50 mm long can be reduced to a chromosome only 5 μm in length. (A little calculation will show that this represents a reduction in length of 10,000 times).

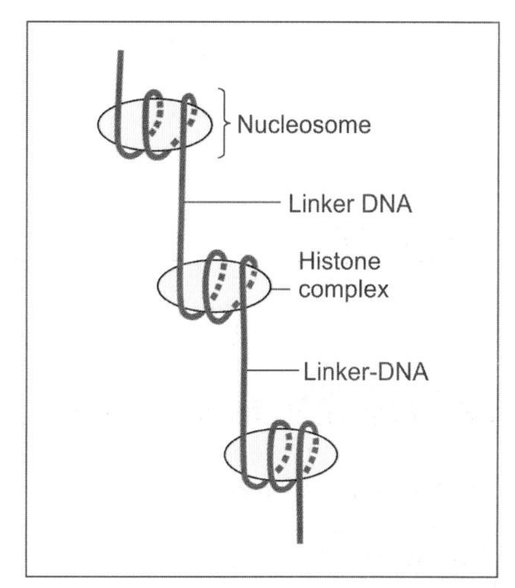

Fig. 2.27: Structure of a chromatin fibre. The DNA fibril makes two turns around a complex formed by histones to form a nucleosome. Nucleosomes give the chromatin fibre the appearance of a beaded string. The portion of the DNA fibre between the nucleosomes is called linker-DNA (Schematic representation)

Heterochromatin represents areas where chromatin fibres are tightly coiled on themselves forming 'solid' masses. In contrast, euchromatin represents areas where coiling is not so marked. During cell division, the entire chromatin within the nucleus becomes very tightly coiled and takes on the appearance of a number of short, thick, rod-like structures called ***chromosomes***. Chromosomes are made up of DNA and proteins. Proteins stabilise the structure of chromosomes.

Added Information

Some details of the formation of a histone complex are shown in Figure 2.28. Five types of histones are recognised. These are H1, H2A, H2B, H3 and H4. Two molecules each of H2A, H2B, H3 and H4 join to form a granular mass, the ***nucleosome core***. The DNA filament is wound ***twice*** around this core, the whole complex forming a nucleosome. The length of the DNA filament in one nucleosome contains 146 nucleotide pairs. One nucleosome is connected to the next by a short length of ***linker DNA***. Linker DNA is made up of about 50 nucleotide pairs.

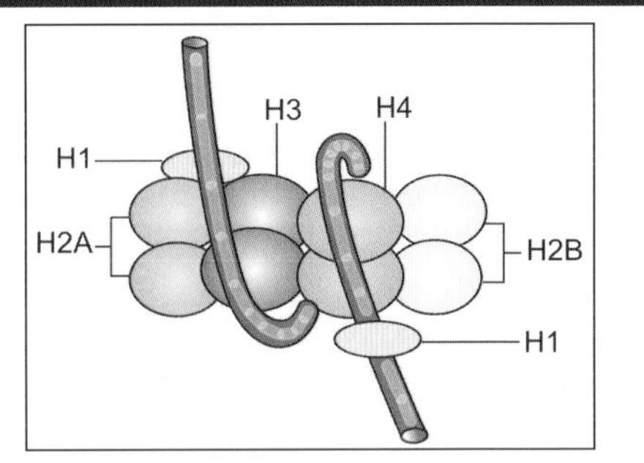

Fig. 2.28: Detailed composition of a histone complex forming the nucleosome core (Schematic representation)

Nucleoli

In addition to the masses of heterochromatin (which are irregular in outline), the nucleus shows one or more rounded, dark staining bodies called ***nucleoli***.

These are spherical and about 1–3 µm in diameter. They stain intensely both with haematoxylin and eosin, the latter giving them a slight reddish tinge. In ordinary preparations, they can be distinguished from heterochromatin by their rounded shape (In contrast masses of heterochromatin are very irregular). Nucleoli are larger and more distinct in cells that are metabolically active.

Using histochemical procedures that distinguish between DNA and RNA, it is seen that the nucleoli have a high RNA content.

Nucleoli are site where ribosomal RNA is synthesised. The templates for this synthesis are located on the related chromosomes. Ribosomal RNA is at first in the form of long fibres that constitute the fibrous zone of nucleoli. It is then broken up into smaller pieces (ribosomal subunits) that constitute the granular zone. Finally, this RNA leaves the nucleolus, passes through a nuclear pore, and enters the cytoplasm where it takes part in protein synthesis.

Added Information

With the EM nucleoli are seen to have a central filamentous zone (***pars filamentosa***) and an outer granular zone (***pars granulosa***) both of which are embedded in an amorphous material (***pars amorphosa***) (Fig. 2.29).

Nucleoli are formed in relationship to the secondary constrictions of specific chromosomes (discussed later). These regions are considered to be nucleolar organising centres. Parts of the chromosomes located within nucleoli constitute the ***pars chromosoma*** of nucleoli.

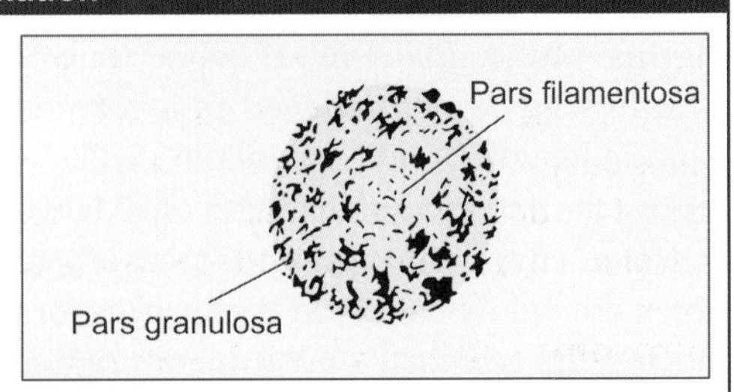

Fig. 2.29: Electron microscopic structure of a nucleolus

Nucleoplasm

Besides chromatin and nucleolus, the nucleus also contains various small granules, fibres and vesicles (of obscure function). The spaces between the various constituents of the nucleus are filled by a base called the ***nucleoplasm***.

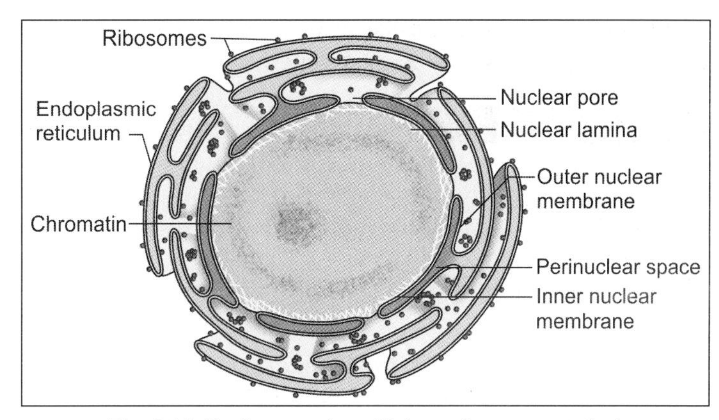

Fig. 2.30: Nuclear envelope (Schematic representation)

Nuclear Membrane

With the EM the nucleus is seen to be surrounded by a double-layered ***nuclear membrane*** or ***nuclear envelope***. The outer nuclear membrane is continuous with endoplasmic reticulum. The space between the inner and outer membranes is the ***perinuclear space***. This is continuous with the lumen of rough endoplasmic reticulum. The inner layer of the nuclear membrane provides attachment to the ends of chromosomes.

Deep to the inner membrane there is a layer containing proteins and a network of filaments this layer is called the ***nuclear lamina***. Specific proteins present in the inner nuclear membrane give attachment to filamentous proteins of the nuclear lamina. These proteins (called ***Laminins***) form a scaffolding that maintains the spherical shape of the nucleus.

At several points the inner and outer layers of the nuclear membrane fuse leaving gaps called ***nuclear pores***. Each pore is surrounded by dense protein arranged in the form of eight complexes. These proteins and the pore together form the ***pore complex***.

Nuclear pores represent sites at which substances can pass from the nucleus to the cytoplasm and vice versa (Fig. 2.30). The nuclear pore is about 80 nm across. It is partly covered by a diaphragm that allows passage only to particles less than 9 nm in diameter. A typical nucleus has 3,000–4,000 pores.

It is believed that pore complexes actively transport some proteins into the nucleus, and ribosomes out of the nucleus.

CHROMOSOMES

Haploid and Diploid Chromosomes

During cell division, the chromatin network in the nucleus becomes condensed into a number of thread-like or rod-like structures called ***chromosomes***. The number of chromosomes in each cell is fixed for a given species, and in man it is 46. This is referred to as the ***diploid number*** (diploid = double). However, in spermatozoa and in ova the number is only half the diploid number, i.e., 23—this is called the ***haploid number*** (haploid = half).

Autosomes and Sex Chromosomes

The 46 chromosomes in each cell can again be divided into 44 ***autosomes*** and two ***sex chromosomes***. The sex chromosomes may be of two kinds, X or Y. In a man, there are 44 autosomes, one X chromosome and one Y chromosome; while in a woman, there are 44 autosomes and two X chromosomes in each cell. When we study the 44 autosomes we find that they really consist of 22 pairs, the two chromosomes forming a pair being exactly alike (***homologous chromosomes***). In a woman, the two X chromosomes form another such pair; but in a man this pair is represented by one X and one Y chromosome. One chromosome of each pair is obtained (by each individual) from the mother, and one from the father.

As the two sex chromosomes of a female are similar the female sex is described as ***homogametic***; in contrast the male sex is ***heterogametic***.

Structure of Fully Formed Chromosomes

Each chromosome consists of two parallel rod-like elements that are called ***chromatids*** (Fig. 2.31). The two chromatids are joined to each other at a narrow area that is light staining and is called the ***centromere*** (or ***kinetochore***). In this region the chromatin of each chromatid is most highly coiled and, therefore, appears to be thinnest. The chromatids appear to be 'constricted' here and this region is called the ***primary constriction***.

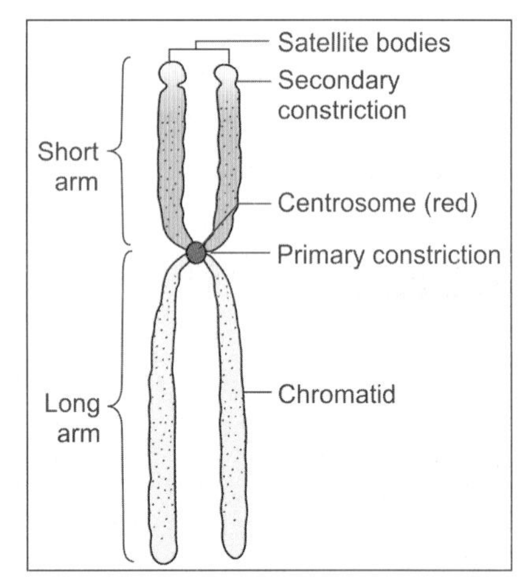

Fig. 2.31: A typical chromosome. Note that this chromosome is submetacentric (Schematic representation).

Typically the centromere is not midway between the two ends of the chromatids, but somewhat towards one end. As a result each chromatid can be said to have a ***long arm*** and a ***short arm***. Such chromosomes are described as being ***submetacentric*** (when the two arms are only slightly different in length); or as ***acrocentric*** (when the difference is marked) (Fig. 2.32). In some chromosomes, the two arms are of equal length such chromosomes are described as ***metacentric***. Finally, in some chromosomes the centromere may lie at one end, such a chromosome is described as ***telocentric***.

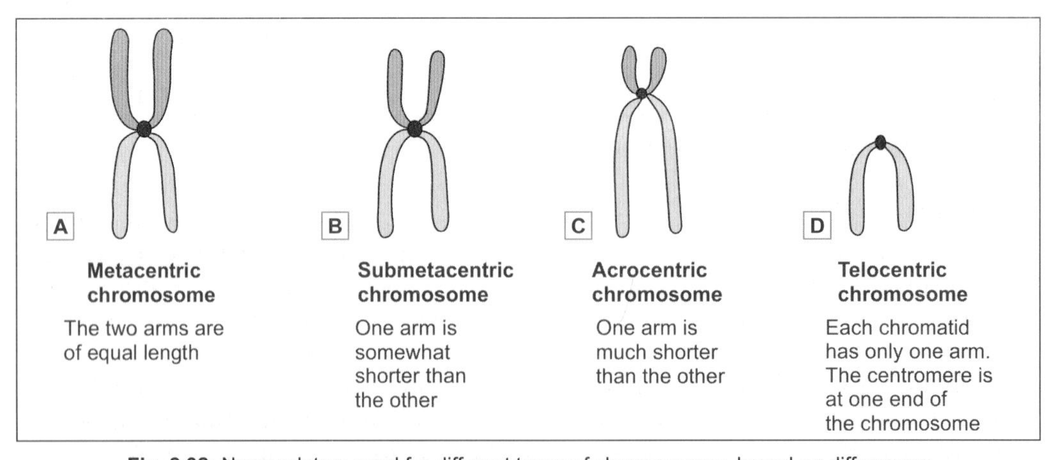

Fig. 2.32: Nomenclature used for different types of chromosomes, based on differences in lengths of the two arms of each chromatid (Schematic representation)

Differences in the total length of chromosomes, and in the position of the centromere are important factors in distinguishing individual chromosomes from each other. Additional help in identification is obtained by the presence in some chromosomes of ***secondary constrictions***. Such constrictions lie near one end of the chromatid. The part of the chromatid 'distal' to the constriction may appear to be a rounded body almost separate from the rest of the chromatid such regions are called ***satellite bodies*** (Secondary constrictions are concerned with the formation of nucleoli and are, therefore, called ***nucleolar organising centres***). Considerable help in identification of individual chromosomes is also obtained by the use of special staining procedures by which each chromatid can be seen to consist of a number of dark and light staining transverse bands.

Chromosomes are distinguishable only during mitosis. In the interphase (between successive mitoses) the chromosomes elongate and assume the form of long threads. These threads are called ***chromonemata*** (singular = ***chromonema***).

Pathological Correlation

Karyotyping

Using the criteria described above, it is now possible to identify each chromosome individually and to map out the chromosomes of an individual. This procedure is called ***karyotyping***. For this purpose a sample of blood from the individual is put into a suitable medium in which lymphocytes can multiply. After a few hours a drug (colchicine, colcemid) that arrests cell division at a stage when chromosomes are most distinct, is added to the medium. The dividing cells are then treated with hypotonic saline so that they swell up. This facilitates the proper spreading out of chromosomes. A suspension containing the dividing cells is spread out on a slide and suitably stained. Cells in which the chromosomes are well spread out (without overlap) are photographed. The photographs are cut-up and the chromosomes arranged in proper sequence. In this way a map of chromosomes is obtained, and abnormalities in their number or form can be identified. In many cases specific chromosomal abnormalities can be correlated with specific diseases.

(For details of chromosomal abnormalities see the author's HUMAN EMBRYOLOGY).

In recent years greater accuracy in karyotyping has been achieved by use of several different banding techniques, and by use of computerised analysis.

It has been estimated that the total DNA content of a cell (in all chromosomes put together) is represented by about 6×10^9 nucleotide pairs. Of these 2.5×10^8 are present in chromosome 1 (which is the largest chromosome). The Y-chromosome (which is the smallest chromosome) contains 5×10^7 nucleotide pairs.

In the region of the centromere, the DNA molecule is specialised for attachment to the spindle. This region is surrounded by proteins that form a mass. This mass is the kinetochore. The ends of each DNA molecule are also specialised. They are called telomeres.

Significance of Chromosomes

Each cell of the body contains within itself a store of information that has been inherited from precursor cells. This information (which is necessary for the proper functioning of the cell) is stored in chromatin. Each chromosome bears on itself a very large number of functional segments that are called ***genes***. Genes represent 'units' of stored information which guide the performance of particular cellular functions, which may in turn lead to the development of particular features of an individual or of a species. Recent researches have told us a great deal about the way in which chromosomes, and genes store and use information.

The nature and functions of a cell depend on the proteins synthesised by it. Proteins are the most important constituents of our body. They make up the greater part of each cell and of intercellular substances. Enzymes, hormones, and antibodies are also proteins.

It is, therefore, not surprising that one cell differs from another because of the differences in the proteins that constitute it. Individuals and species also owe their distinctive characters to their proteins. We now know that chromosomes control the development and functioning of cells by determining what type of proteins will be synthesised within them.

Chromosomes are made up predominantly of a nucleic acid called ***DNA***, and all information is stored in molecules of this substance. When the need arises this information is used to direct the activities of the cell by synthesising appropriate proteins. To understand how this becomes possible we must consider the structure of DNA in some detail.

Basic Structure of DNA

DNA in a chromosome is in the form of very fine fibres. If we look at one such fibre, it has the appearance as shown in Figure 2.33. It is seen that each fibre consists of two strands that are twisted spirally to form what is called a ***double helix***. The two strands are linked to each other at regular intervals (Note the dimensions shown in Fig. 2.33).

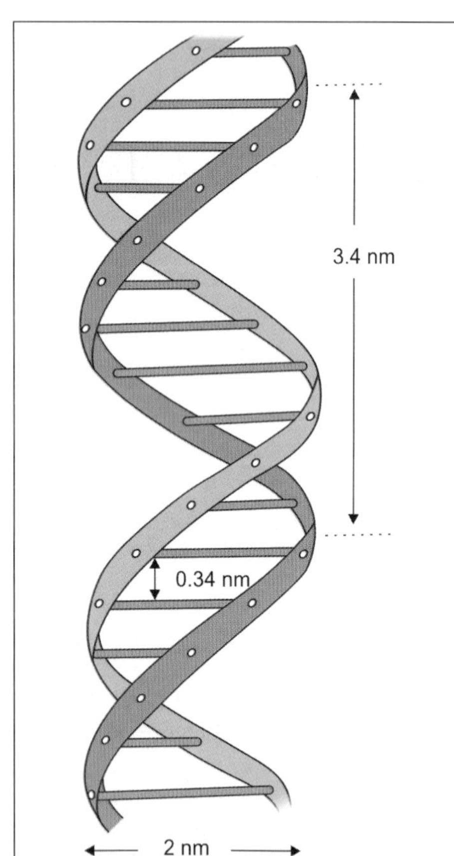

Fig. 2.33: Part of a DNA molecule arranged in the form of a double helix (Schematic representation)

Each strand of the DNA fibre consists of a chain of ***nucleotides***. Each nucleotide consists of a sugar deoxyribose, a molecule of phosphate and a base (Fig. 2.34). The phosphate of one nucleotide is linked to the sugar of the next nucleotide (Fig. 2.35). The base that is attached to the sugar molecule may be ***adenine, guanine, cytosine*** or ***thymine***. The two strands of a DNA fibre are joined together by the linkage of a base on one strand with a base on the opposite strand (Fig. 2.36).

This linkage is peculiar in that adenine on one strand is always linked to thymine on the other strand, while cytosine is always linked to guanine. Thus the two strands are complementary and the arrangement of bases on one strand can be predicted from the other.

The order in which these four bases are arranged along the length of a strand of DNA

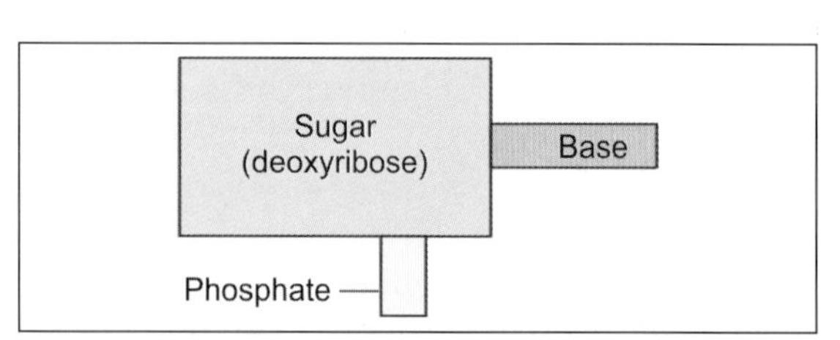

Fig. 2.34: Composition of a nucleotide. The base may be adenine, cytosine, guanine or thymine (Schematic representation)

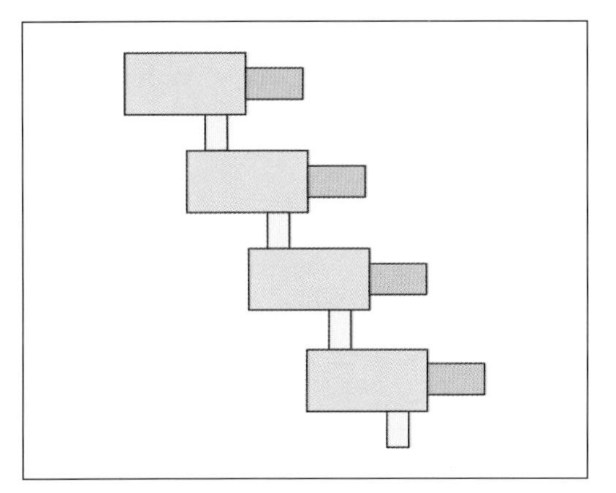

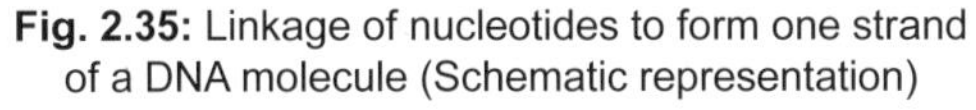

Fig. 2.35: Linkage of nucleotides to form one strand of a DNA molecule (Schematic representation)

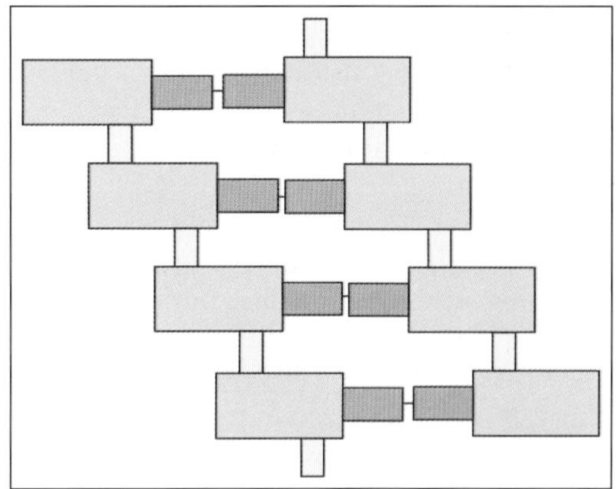

Fig. 2.36: Linkage of two chains of nucleotides to form part of a DNA molecule (Schematic representation)

determines the nature of the protein that can be synthesised under its influence. Every protein is made up of a series of amino acids; the nature of the protein depending upon the amino acids present, and the sequence in which they are arranged. Amino acids may be obtained from food or may be synthesised within the cell. Under the influence of DNA these amino acids are linked together in a particular sequence to form proteins.

Added Information

In the preceding paragraphs the structure of DNA has been described in the simplest possible terms. We will now consider some details:

- The structure of the sugar deoxyribose is shown in Figure 2.37. Note that there are five carbon atoms; and also note how they are numbered.
- Next observe, in Figure 2.38, that C-3 of one sugar molecule is linked to C-5 of the next molecule through a phosphate linkage (P). It follows that each strand of DNA has a 5' end and a 3' end.
- Next observe that although the two chains forming DNA are similar they are arranged in opposite directions. In Figure 2.38, the 5' end of the left chain, and the 3' end of the right chain lie at the upper end of the figure. The two chains of nucleotides are, therefore, said to be ***antiparallel***.
- The C-1 carbon of deoxyribose give attachment to a base. This base is attached to a base of the opposite chain as already described.
- The reason why adenine on one strand is always linked to thymine on the other strand is that the structure of these two molecules is complementary and hydrogen bonds are easily formed between them. The same is true for cytosine and guanine.

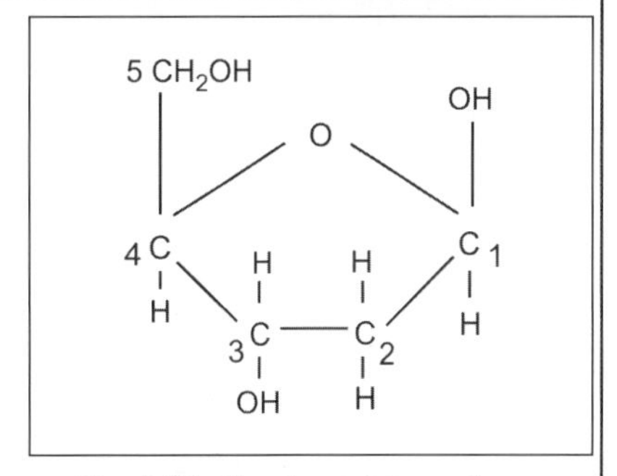

Fig. 2.37: Structure of deoxyribose. Note the numbering of carbon atoms (Schematic representation)

Ribonucleic Acid

In addition to DNA, cells contain another important nucleic acid called ***ribonucleic acid*** or ***RNA***. The structure of a molecule of RNA corresponds fairly closely to that of one strand of a DNA molecule, with the following important differences:

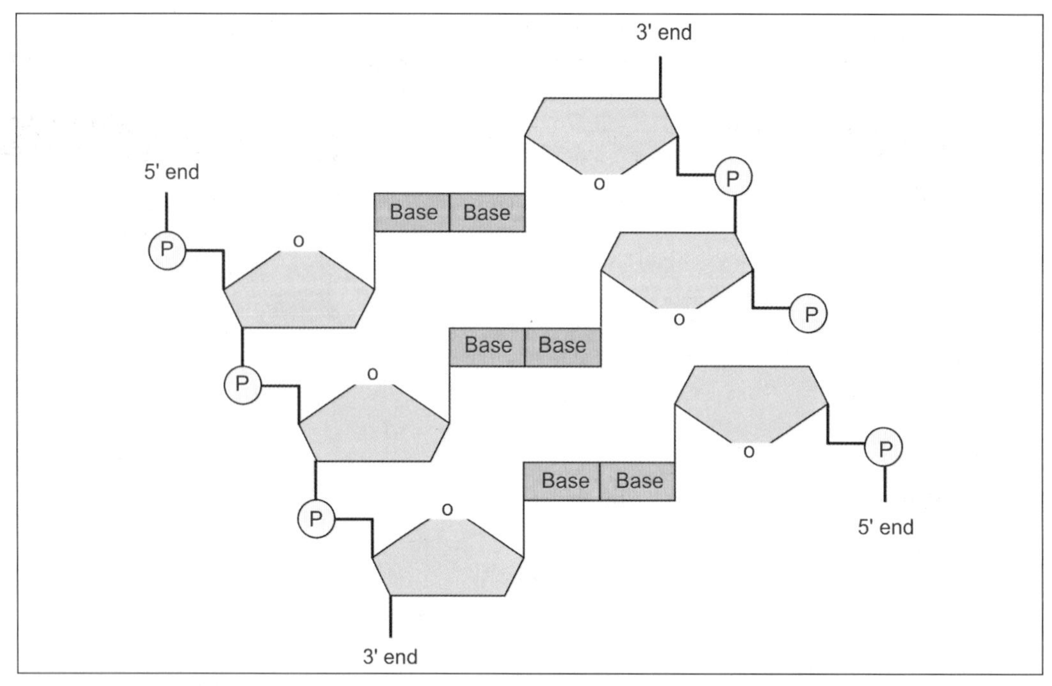

Fig. 2.38: Nucleotides are linked to form a chain of DNA.
The asymmetric placing of bonds gives a helical shape to the chain (Schematic representation)

- RNA contains the sugar ribose instead of deoxyribose.
- Instead of the base thymine it contains uracil.

RNA is present both in the nucleus and in the cytoplasm of a cell. It is present in three main forms namely ***messenger RNA*** (***mRNA***), ***transfer RNA*** (***tRNA***) and ***ribosomal RNA (rRNA)***. Messenger RNA acts as an intermediary between the DNA of the chromosome and the amino acids present in the cytoplasm and plays a vital role in the synthesis of proteins from amino acids.

Some forms of RNA are confined to nuclei. The small nuclear RNAs (SnRNA) are concerned with RNA splicing.

Synthesis of Protein

A protein is made up of amino acids that are linked together in a definite sequence. This sequence is determined by the order in which the bases are arranged in a strand of DNA. Each amino acid is represented in the DNA molecule by a sequence of three bases (***triplet code***). It has been mentioned earlier that there are four bases in all in DNA, namely adenine, cytosine, thymine and guanine. These are like letters in a word. They can be arranged in various combinations so that as many as sixty four 'code words' can be formed from these four bases. There are only about twenty amino acids that have to be coded for so that each amino acid has more than one code. The code for a complete polypeptide chain is formed when the codes for its constituent amino acids are arranged in proper sequence. That part of the DNA molecule that bears the code for a complete polypeptide chain constitutes a ***structural gene*** or ***cistron***.

At this stage it must be emphasised that a chromosome is very long and thread-like. Only short lengths of the fibre are involved in protein synthesis at a particular time.

The main steps in the synthesis of a protein may now be summarised as follows (Fig. 2.39):

- The two strands of a DNA fibre separate from each other (over the area bearing a particular cistron) so that the ends of the bases that were linked to the opposite strand are now free.
- A molecule of mRNA is synthesised using one DNA strand as a guide (or ***template***), in such a way that one guanine base is formed opposite each cytosine base of the DNA strand, cytosine is formed opposite guanine, adenine is formed opposite thymine, and uracil is formed opposite adenine. In this way the code for the sequence in which amino acids are to be linked is passed on from DNA of the chromosome to mRNA. This process is called ***transcription*** (Transcription takes place under the influence of the enzyme RNA polymerase).

  That part of the mRNA strand that bears the code for one amino acid is called a ***codon***.
- This molecule of mRNA now separates from the DNA strand and moves from the nucleus to the cytoplasm (passing through a nuclear pore).
- In the cytoplasm, the mRNA becomes attached to a ribosome.
- As mentioned earlier, the cytoplasm also contains another form of RNA called tRNA. In fact there are about twenty different types of tRNA each corresponding to one amino acid. On one side tRNA becomes attached to an amino acid. On the other side it bears a code of three bases (***anticodon***) that are complementary to the bases coding for its amino acid on mRNA. Under the influence of the ribosome several units of tRNA, along with their amino acids, become arranged along side the strand of mRNA in the sequence determined by the code on mRNA. This process is called ***translation***.

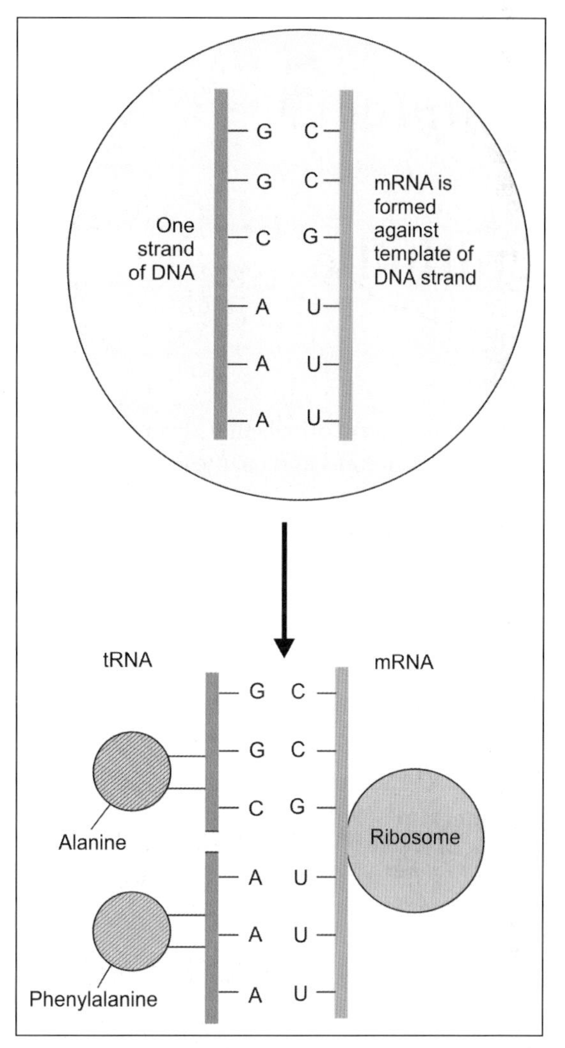

Fig. 2.39: Proteins are synthesised under the influence of DNA. The process is actually more complex as explained in the text (Schematic representation)

- The amino acids now become linked to each other to form a polypeptide chain. The amino acids are linked up exactly in the order in which their codes are arranged on mRNA, which in turn is based on the code on the DNA molecule. Chains of amino acids formed in this way constitute polypeptide chains. Proteins are formed by union of polypeptide chains.

The flow of information from DNA to RNA and finally to protein has been described as the ***'central dogma of molecular biology'***.

Chapter 3

Epithelia

One or more layers of cells that cover the outer surface (of the body) or line the luminal surface of tubular structures and cavities of the body are called epithelia (singular = epithelium).

CHARACTERISTIC FEATURES OF EPITHELIAL TISSUE

- Very cellular with little intercellular space (20 nm)
- Usually avascular
- Cells rest on a basement membrane
- Cells show polarity
- Cells may display surface modifications

FUNCTIONS

- Protection
- Absorption
- Secretion
- Exchange

CLASSIFICATION OF EPITHELIA (FLOW CHART 3.1)

An epithelium may consist of only one layer of cells when it is called a ***unilayered*** or ***simple*** epithelium. Alternatively, it may be ***multilayered*** (***stratified***) or it can be ***pseudostratified***.

- ***Unilayered (simple) epithelia:*** Single layer of cells resting on a basement membrane. It may be further classified according to the shape of the cells constituting them.
  - When the cells are flattened, their height being very little as compared to their width. Such an epithelium is called as a ***squamous epithelium***.
  - When the height and width of the cells of the epithelium are more or less equal (i.e. they look like squares in section) it is described as a ***cuboidal epithelium***.
  - When the height of the cells of the epithelium is distinctly greater than their width, it is described as a ***columnar epithelium***.
- ***Pseudostratified columnar epithelia:*** In true sense this is a simple epithelium as each cell rests on the basement membrane. This epithelium gives an appearance of a multilayered epithelium due to unequal height and shape of cells.
- ***Multilayered (stratified) epithelia:*** Epithelia which consist of multiple layers with the basal layer resting on the basement membrane. The epithelium is named according to the shape of cells of the most superficial layer.

Flow chart 3.1: Classification of epithelia

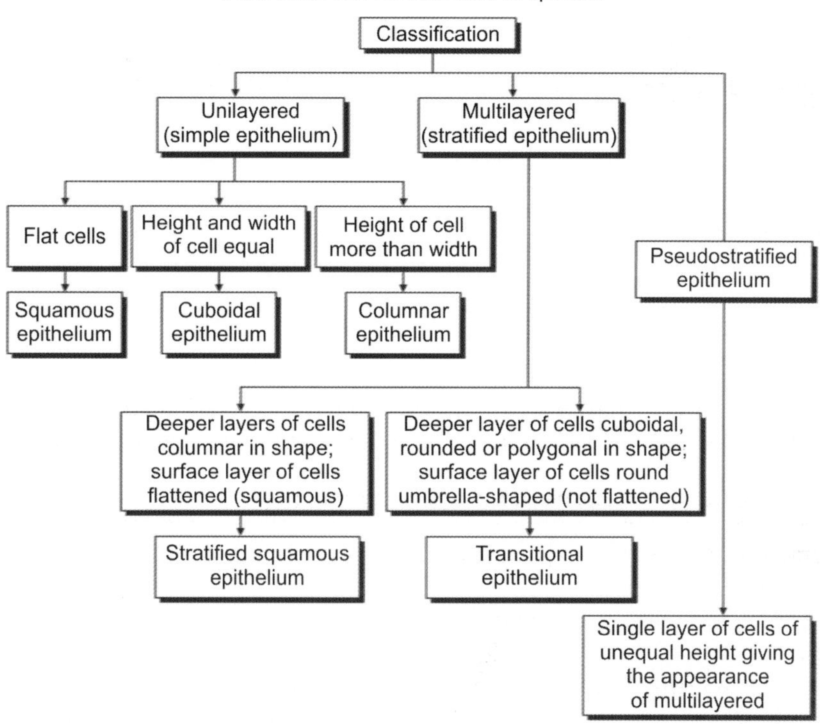

- ***Stratified squamous***: The deeper layers are columnar, but in proceeding towards the surface of the epithelium the cells become increasingly flattened (or squamous). It may be noted that all cells in this kind of epithelium are not squamous.
- ***Stratified cuboidal***: The surface cells are cuboidal in shape.
- ***Stratified columnar***: The surface cells are columnar in shape.

- ***Transitional epithelium:*** In this type of multilayered epithelium all layers are made up of cuboidal, polygonal or round cells. The cells towards the surface of the epithelium are round. As transitional epithelium is confined to the urinary tract, it is also called ***urothelium***.

SIMPLE EPITHELIUM

Squamous Epithelium

Description

In surface view, the cells have polygonal outlines that interlock with those of adjoining cells (Plate 3.1D).

In a section, the cells appear flattened their height being much less as compared to their width (Fig. 3.1).

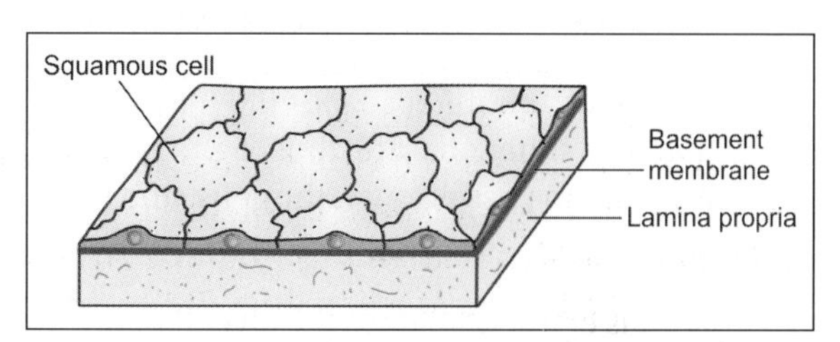

Fig. 3.1: Simple squamous epithelium (schematic representation)

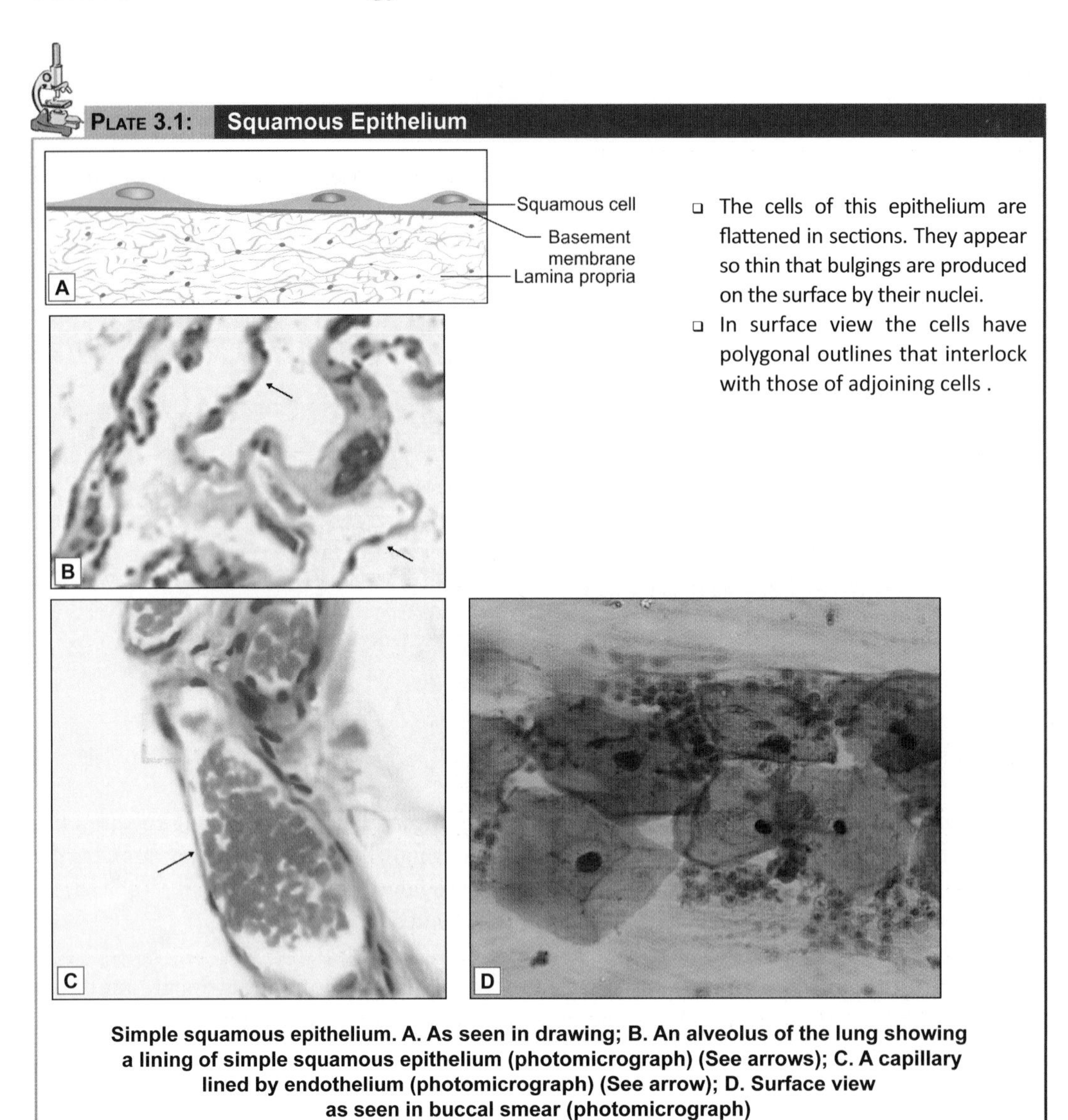

- The cells of this epithelium are flattened in sections. They appear so thin that bulgings are produced on the surface by their nuclei.
- In surface view the cells have polygonal outlines that interlock with those of adjoining cells .

Simple squamous epithelium. A. As seen in drawing; B. An alveolus of the lung showing a lining of simple squamous epithelium (photomicrograph) (See arrows); C. A capillary lined by endothelium (photomicrograph) (See arrow); D. Surface view as seen in buccal smear (photomicrograph)

The cytoplasm of cells forms only a thin layer. The nuclei produce bulgings of the cell surface (Plate 3.1).

With the electron microscope (EM) the junctions between cells are marked by occluding junctions. The junctions are thus tightly sealed and any substance passing through the epithelium has to pass through the cells, and not between them.

Location

- Squamous epithelium lines the alveoli of the lungs.
- It lines the free surface of the serous pericardium, the pleura, and the peritoneum; here it is called ***mesothelium***.

PLATE 3.2: Cuboidal Epithelium

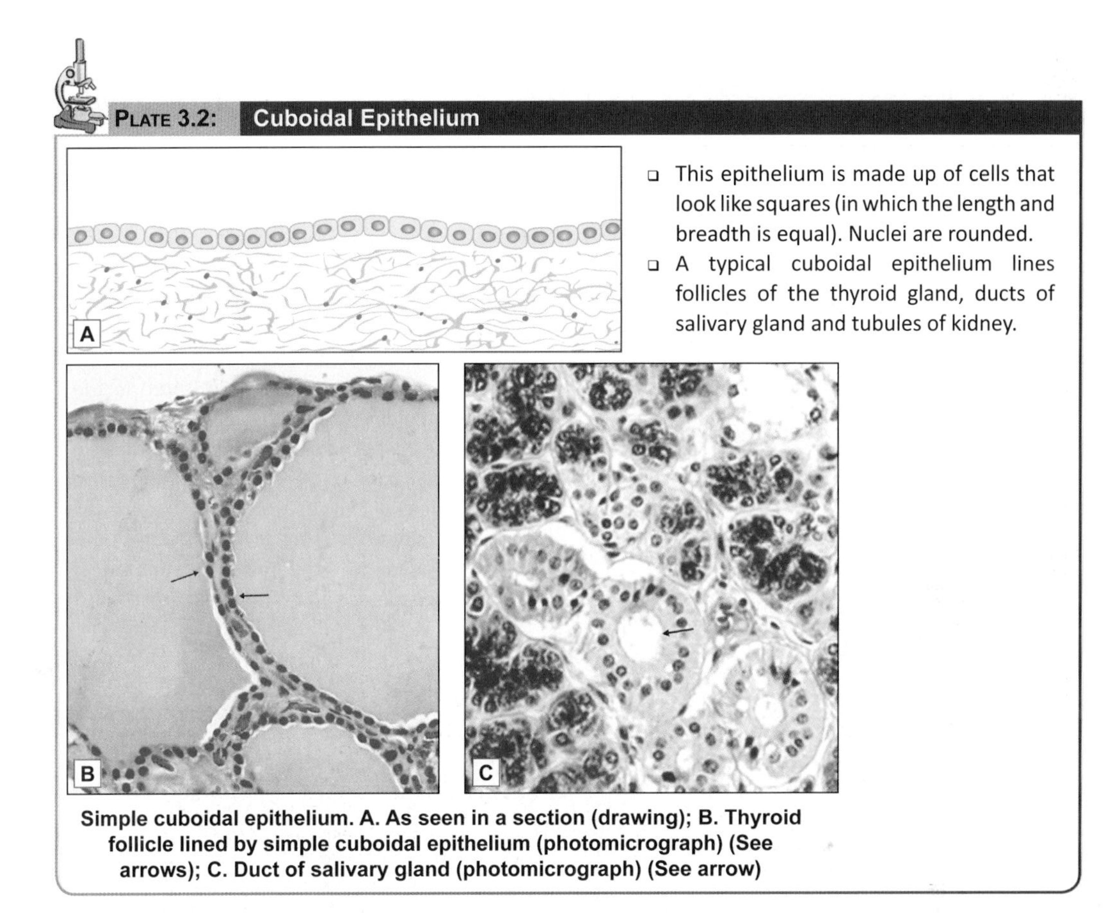

- This epithelium is made up of cells that look like squares (in which the length and breadth is equal). Nuclei are rounded.
- A typical cuboidal epithelium lines follicles of the thyroid gland, ducts of salivary gland and tubules of kidney.

Simple cuboidal epithelium. A. As seen in a section (drawing); B. Thyroid follicle lined by simple cuboidal epithelium (photomicrograph) (See arrows); C. Duct of salivary gland (photomicrograph) (See arrow)

- It lines the inside of the heart, where it is called ***endocardium***; and of blood vessels and lymphatics, where it is called ***endothelium***.
- Squamous epithelium is also found lining some parts of the renal tubules, and in some parts of the internal ear.

Function

It helps in rapid transport of substances, diffusion of gases and filtration of fluids.

Cuboidal Epithelium

Description

In cuboidal epithelium, the height of the cells is about the same as their width. The nuclei are usually rounded (Plate 3.2).

In sectional view cells appear cuboidal in shape. When viewed from surface, cells are hexagonal in shape (Fig. 3.2).

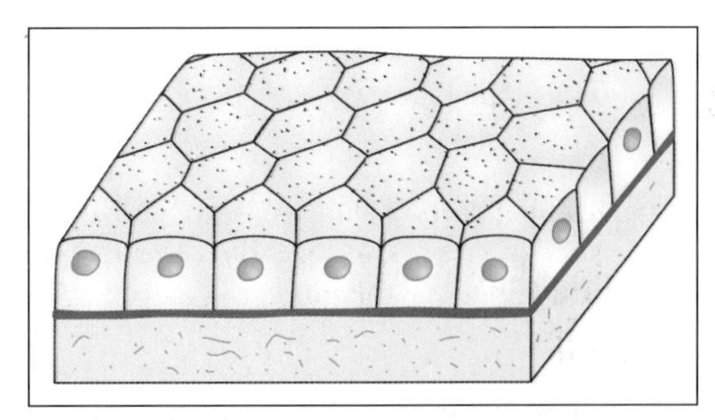

Fig. 3.2: Simple cuboidal epithelium. Note that the cells appear cuboidal in section and hexagonal in surface view (schematic representation).

Plate 3.3: Simple Columnar Epithelium

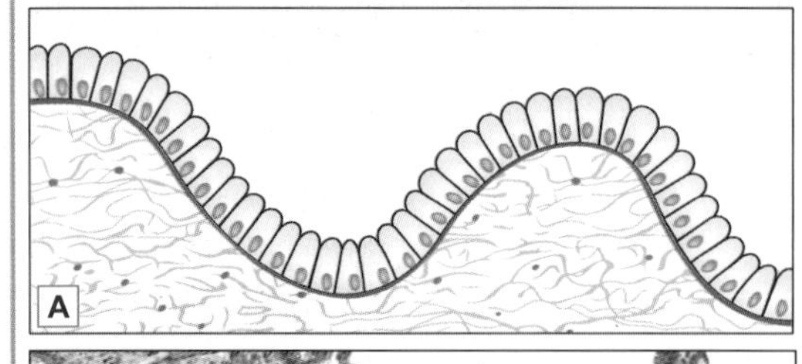

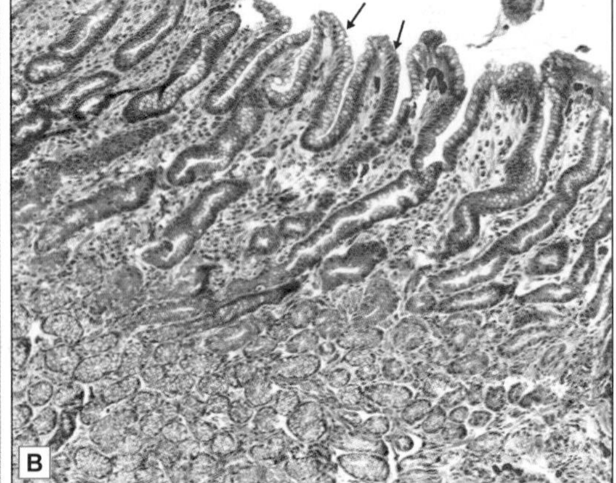

Simple columnar epithelium A. As seen in a section (drawing); B. Lining the mucosa of the stomach (photomicrograph)

- The height of the cells in this epithelium is much greater than their width. The nuclei are oval being elongated in the same direction as the cells. They lie near the bases of the cells. Because of this we see a zone of clear cytoplasm above the nuclei
- A simple columnar epithelium (non-ciliated) lines the mucous membrane of the stomach and of the large intestine.

Location

- A typical cuboidal epithelium may be seen in the follicles of the thyroid gland, in the ducts of many glands, and on the surface of the ovary.
- Other sites are the choroid plexuses, the inner surface of the lens, and the pigment cell layer of the retina.
- A cuboidal epithelium with a prominent brush border is seen in the proximal convoluted tubules of the kidneys.

Function

It is mainly concerned with secretory and absorptive functions.

Columnar Epithelium

Description

Cells of the epithelium are much taller compared to their width. Nuclei are elongated and located in the lower half of the cells. All nuclei are placed at the same level in neighbouring cells (Fig. 3.3 and Plate 3.3).

In vertical section the cells of this epithelium are rectangular. On surface view (or in transverse section) the cells are polygonal (Fig. 3.3).

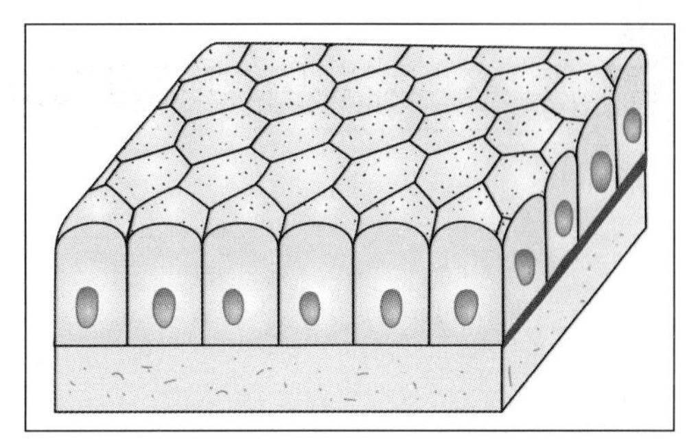

Fig. 3.3: Simple columnar epithelium. Note the basally placed oval nuclei. The cells appear hexagonal in surface view (Schematic representation)

Columnar epithelium can be further classified according to the nature of the free surfaces of the cells as follows.

- In some situations, the cell surface has no particular specialisation, this is ***simple columnar epithelium.***
- In some situations, the cell surface bears cilia. This is ***ciliated columnar epithelium*** (Fig. 3.4A).
- In other situations the surface is covered with microvilli (Fig. 3.4B). Although the microvilli are visible only with the EM, with the light microscope the region of the microvilli is seen as a ***striated border*** (when the microvilli are arranged regularly) or as a ***brush border*** (when the microvilli are irregularly placed).

Location

- Simple columnar epithelium is present over the mucous membrane of the stomach and the large intestine.
- Columnar epithelium with a striated border is seen most typically in the small intestine, and with a brush border in the gallbladder (Plate 3.4).
- Ciliated columnar epithelium lines most of the respiratory tract, the uterus, and the uterine tubes. It is also seen in the efferent ductules of the testis, parts of the middle ear and auditory tube; and in the ependyma lining the central canal of the spinal cord and the ventricles of the brain.

Function

- Some columnar cells have a secretory function. The apical parts of their cytoplasm contain secretory vacuoles. Secretory columnar cells are scattered in the mucosa of the stomach and intestines.
- In the intestines many of them secrete mucous which accumulates in the apical part of the cell making it very light staining. These cells acquire a characteristic shape and are called ***goblet cells.***
- Some columnar cells secrete enzymes.
- In the respiratory tract the cilia move mucous accumulating in the bronchi (and containing

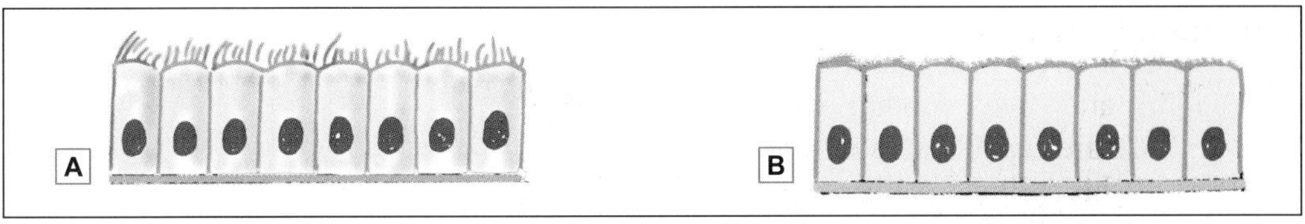

Fig. 3.4: A. Columnar epithelium showing cilia; **B.** Columnar epithelium showing a striated border made up of microvilli (Schematic representation)

PLATE 3.4: **Columnar Epithelium showing Striated Border**

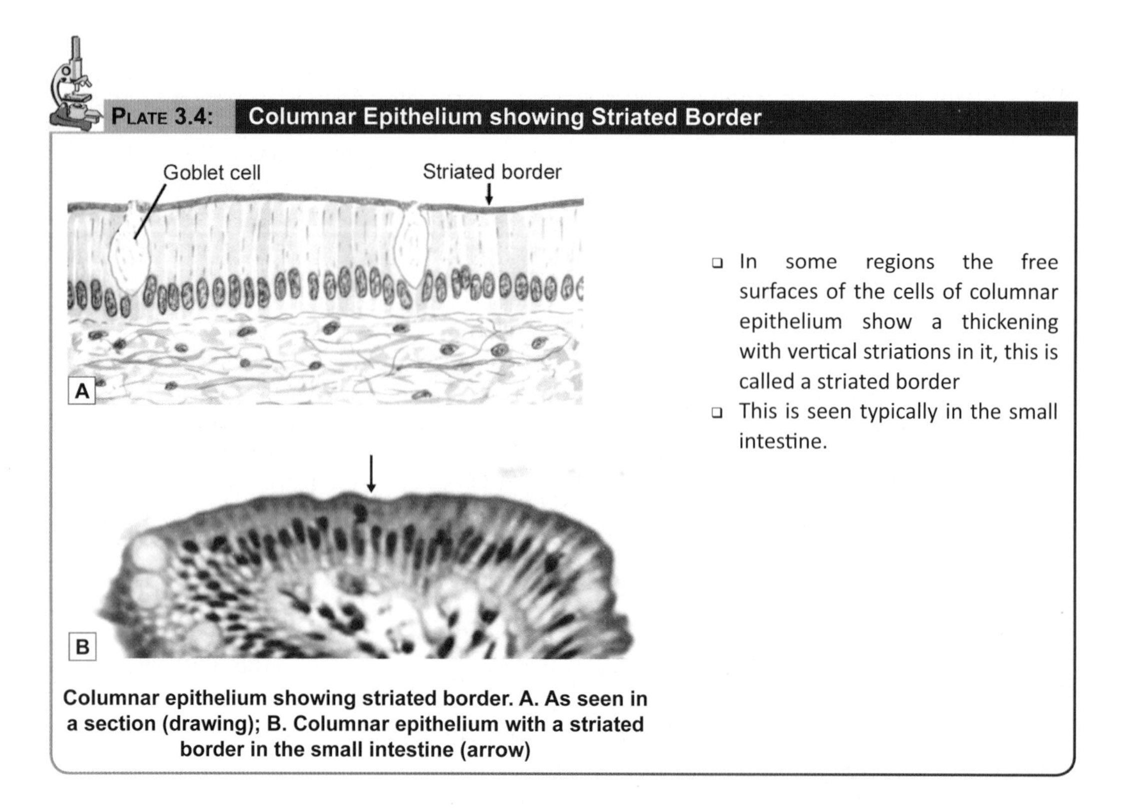

- In some regions the free surfaces of the cells of columnar epithelium show a thickening with vertical striations in it, this is called a striated border
- This is seen typically in the small intestine.

Columnar epithelium showing striated border. A. As seen in a section (drawing); B. Columnar epithelium with a striated border in the small intestine (arrow)

trapped dust particles) towards the larynx and pharynx. When excessive this mucous is brought out as sputum during coughing. In the uterine tubes the movements of the cilia help in the passage of ova towards the uterus.

- Microvilli increase the surface area for absorption.

PSEUDOSTRATIFIED EPITHELIUM

Description

It is not a true stratified epithelium but appears to be stratified (Fig. 3.5 and Plate 3.5). Normally, in columnar epithelium the nuclei lie in a row, towards the basal part of the cells. Sometimes, however, the nuclei appear to be arranged in two or more layers giving the impression that the epithelium is more than one cell thick. The cells are attached to the basement membrane but are of different heights, some cells are short and basal, while others are tall and columnar.

The epithelium may bear cilia (ciliated epithelium) and may contain goblet cells. The cilia are capable of movement. At some sites this epithelium may display stereocilia.

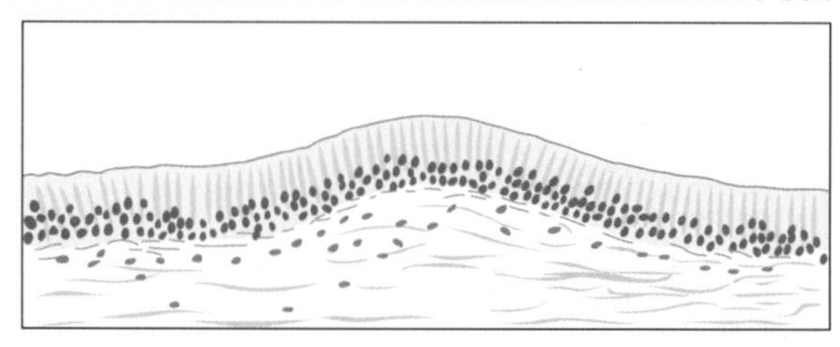

Fig. 3.5: Pseudostratified columnar epithelium as seen in a section.

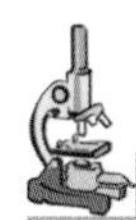

PLATE 3.5: Pseudostratified Ciliated Columnar Epithelium

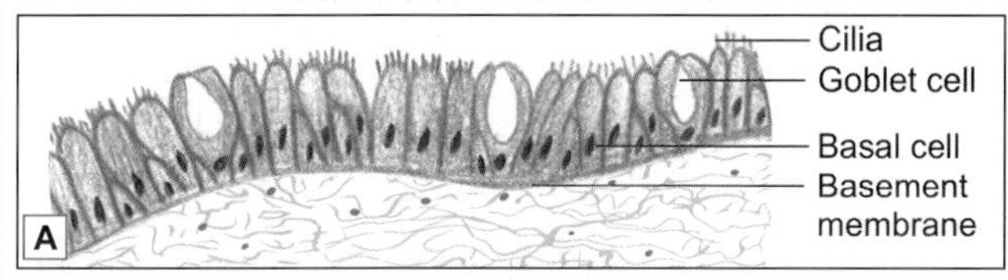

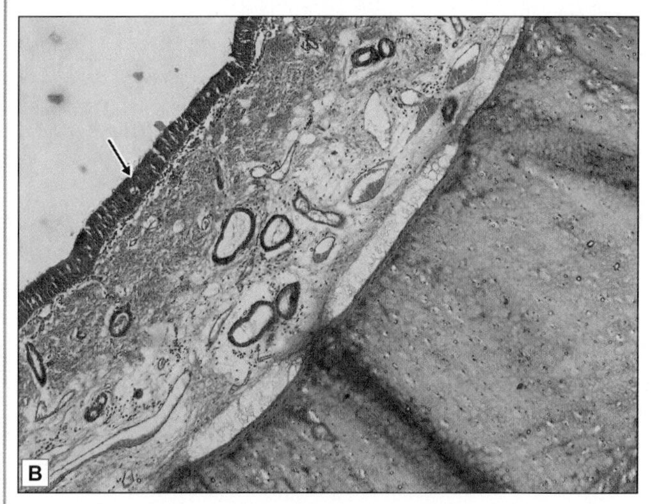

Pseudostratified ciliated columnar epithelium. A. As seen in a section (drawing); B. Pseudostratified ciliated columnar epithelium in trachea (photomicrograph) (see arrow)

- It is not a true stratified epithelium but appears to be stratified. Normally, in columnar epithelium the nuclei lie in a row, towards the basal part of the cells. Sometimes, however, the nuclei appear to be arranged in two or more layers giving the impression that the epithelium is more than one cell thick
- In some situations, pseudostratified columnar epithelium bears hair-like projections called cilia
- Pseudostratified ciliated columnar epithelium is seen in trachea and in large bronchi.

Function

The tall columnar cells are secretory in nature, while the short, basal cells are stem cells which constantly replace the tall cells. The cilia help in clearance of the mucous. The stereocilia help in absorption.

Location

- Non-ciliated pseudostratified columnar epithelium is found in some parts of the auditory tube, the ductus deferens, and the male urethra (membranous and penile parts).
- Ciliated pseudostratified columnar epithelium is seen in the trachea and in large bronchi.
- Pseudostratified columnar epithelium with stereocilia (long microvilli) is seen in epididymis.

STRATIFIED EPITHELIUM

Stratified Squamous Epithelium

Description

This type of epithelium is made up of several layers of cells (Fig. 3.6).

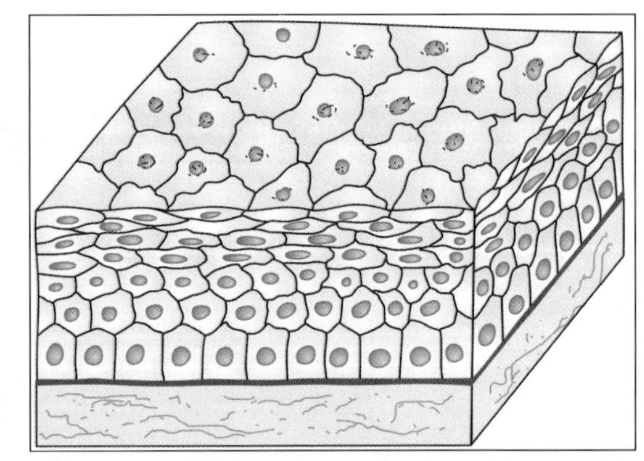

Fig. 3.6: Stratified squamous epithelium. There is a basal layer of columnar cells that rests on the basement membrane. Overlying the columnar cells of this layer there are a few layers of polygonal cells or rounded cells. Still more superficially, the cells undergo progressive flattening, becoming squamous (Schematic representation)

The cells of the deepest (or basal) layer rest on the basement membrane, they are usually columnar in shape. Lying over the columnar cells there are polyhedral or cuboidal cells. As we pass towards the surface of the epithelium these cells become progressively more flat, so that the most superficial cells consist of flattened squamous cells.

Stratified squamous epithelium can be divided into two types—***non-keratinised*** and ***keratinised***.

- ***Non-keratinised stratified squamous epithelium:*** In situations where the surface of the squamous epithelium remains moist, the most superficial cells are living and nuclei can be seen in them. This kind of epithelium is described as non-keratinised stratified squamous epithelium (Fig. 3.7A and Plate 3.6).
- ***Keratinised stratified squamous epithelium:*** At places where the epithelial surface is dry (as in the skin) the most superficial cells die and lose their nuclei. These cells contain a substance called ***keratin***, which forms a non-living covering over the epithelium. This kind of epithelium constitutes keratinised stratified squamous epithelium (Fig. 3.7B and Plate 3.6).
- ***Stratified squamous epithelium (both keratinised and non-keratinised)*** is found over those surfaces of the body that are subject to friction. As a result of friction the most superficial layers are constantly being removed and are replaced by proliferation of cells from the basal layer. This layer, therefore, shows frequent mitoses.

Location

- Keratinised stratified squamous epithelium covers the skin of whole of the body and forms the epidermis.

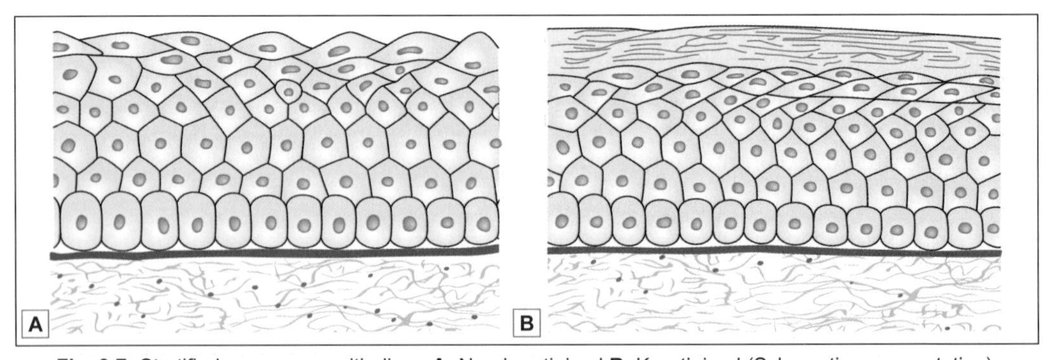

Fig. 3.7: Stratified squamous epithelium. A. Non-keratinised B. Keratinised (Schematic representation).

PLATE 3.6: Stratified Squamous Epithelium

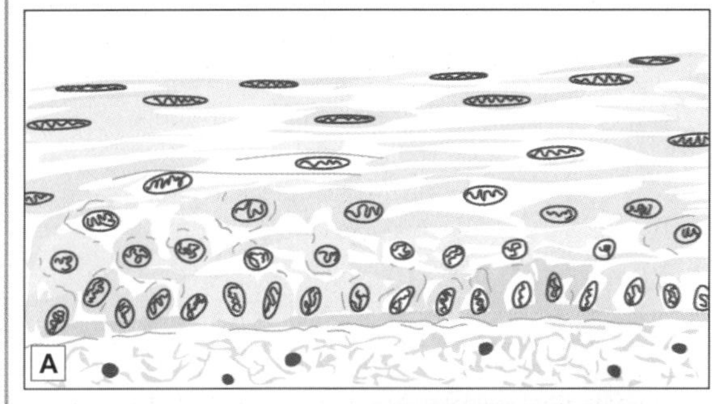

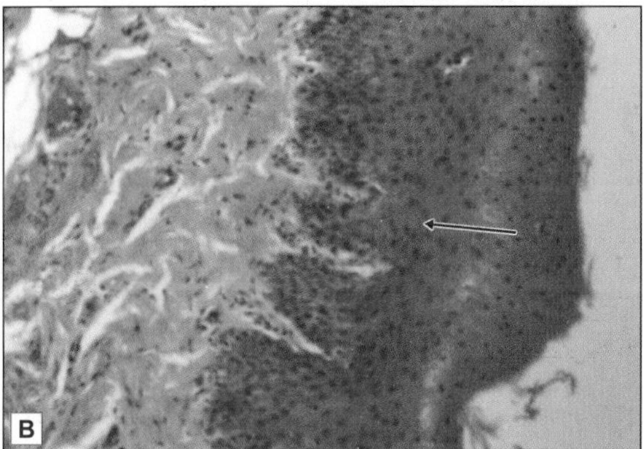

Non-keratinised stratified squamous epithelium A. As seen in a section (drawing); B. As seen in oesophagus (photomicrograph)

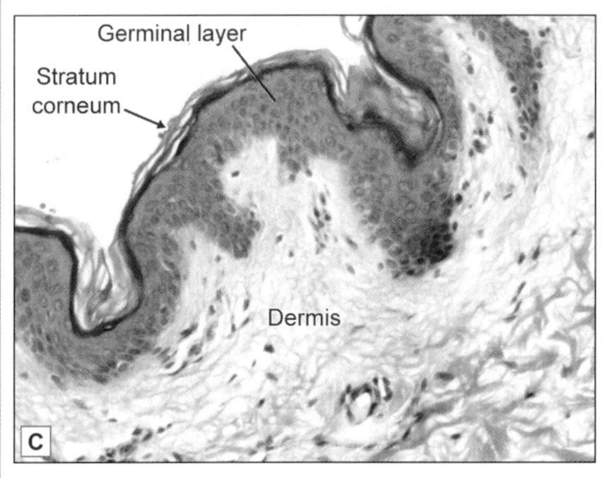

Keratinised stratified squamous epithelium as seen in skin (photomicrograph)

Plate 3.6A and B show non-keratinised stratified squamous epithelium:

- Although this is called stratified squamous epithelium, only the most superficial cells are squamous (flattened)
- The cells in the deepest (or basal) layer are columnar. In the middle layers they are polyhedral, while the more superficial layers show increasing degrees of flattening
- The nuclei are oval in the basal layer, rounded in the middle layer, and transversely elongated in the superficial layers
- The surface layer shows squamous cells with flattened nuclei
- This kind of epithelium is seen lining some internal organs like the oesophagus or the vagina.

Plate 3.6C shows keratinised stratified squamous epithelium

- Here the deeper layers are covered by additional layers that represent stages in the conversion of cells into non-living fibres. This process is called keratinisation (or cornification)
- The surface layer is made up of keratin which appears as fibres. No cellular outline or nuclei can be seen
- It is seen typically in epidermis of the skin.

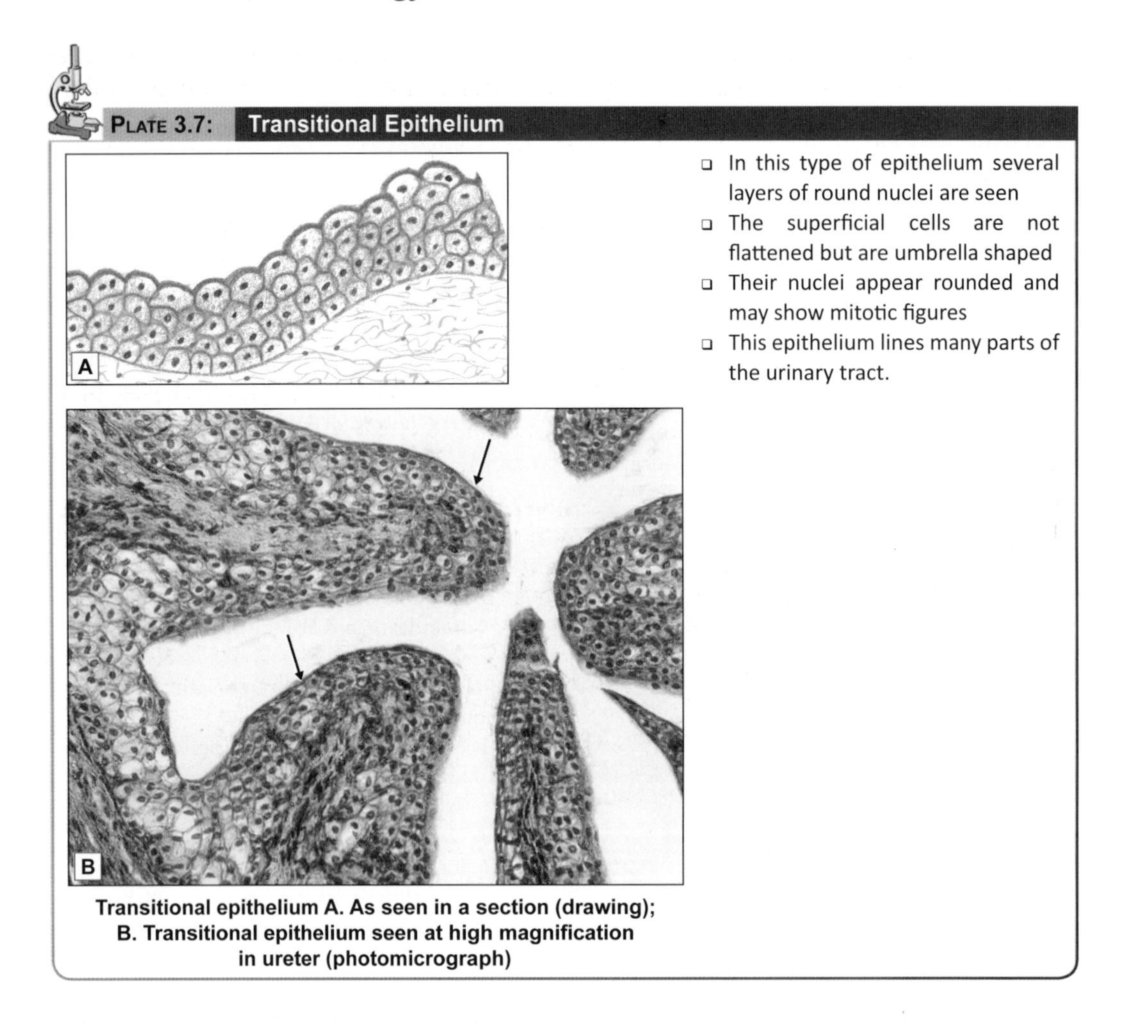

Plate 3.7: Transitional Epithelium

- In this type of epithelium several layers of round nuclei are seen
- The superficial cells are not flattened but are umbrella shaped
- Their nuclei appear rounded and may show mitotic figures
- This epithelium lines many parts of the urinary tract.

Transitional epithelium A. As seen in a section (drawing); B. Transitional epithelium seen at high magnification in ureter (photomicrograph)

- Non-keratinised stratified squamous epithelium covers wet surfaces exposed to wear and tear. It is seen lining the mouth, the tongue, the oro- and laryngopharynx, the oesophagus, the vagina and the cornea. Under pathological conditions the epithelium in any of these situations may become keratinised.

Function

- It is protective in nature.
- Keratin prevents dehydration of underlying tissue.

Transitional Epithelium

Description

This is a multilayered epithelium and is 4 to 6 cells thick. It differs from stratified squamous epithelium in that the cells at the surface are not squamous. The deepest cells are columnar or cuboidal. The middle layers are made up of polyhedral or pear-shaped cells. The cells of the surface layer are large and often shaped like an umbrella (Plate 3.7).

In the urinary bladder, it is seen that cells of transitional epithelium can be stretched considerably without losing their integrity. When stretched it appears to be thinner and the cells become flattened.

Location

Transitional epithelium is found in the renal pelvis and calyces, the ureter, the urinary bladder, and part of the urethra. Because of this distribution it is also called ***urothelium***.

Function

At the surface of the epithelium the plasma membranes are unusual. Embedded in the lipid layer of the membranes there are special glycoproteins. It is believed that these glycoproteins make the membrane impervious and resistant to the toxic effects of substances present in urine, and thus afford protection to adjacent tissues.

Added Information

- With the electron microscope the cells of transitional epithelium are seen to be firmly united to one another by numerous desmosomes. Because of these connections the cells retain their relative position when the epithelium is stretched or relaxed.
- The cells in the basal layer of transitional epithelium show occasional mitoses, but these are much less frequent than those in stratified squamous epithelium, as there is normally little erosion of the surface. Many cells of the superficial (luminal) layers of the epithelium may contain two nuclei. In some cells the nucleus is single, but contains multiples of the normal number of chromosomes (i.e. it may be polyploid).
- According to some workers, all cells of transitional epithelium reach the basal lamina through thin processes. Even though the cells are stratified, they retain a contact with the basement membrane. Hence, this is a transition from unilayered to multilayered epithelium.

Stratified Columnar or Cuboidal Epithelium

Description

This epithelium consists of two or more layers of columnar or cuboidal cells (Figs 3.8 and 3.9).

Location

Stratified cuboidal and columnar epithelium is seen in large ducts of exocrine glands like sweat glands, pancreas, and salivary glands.

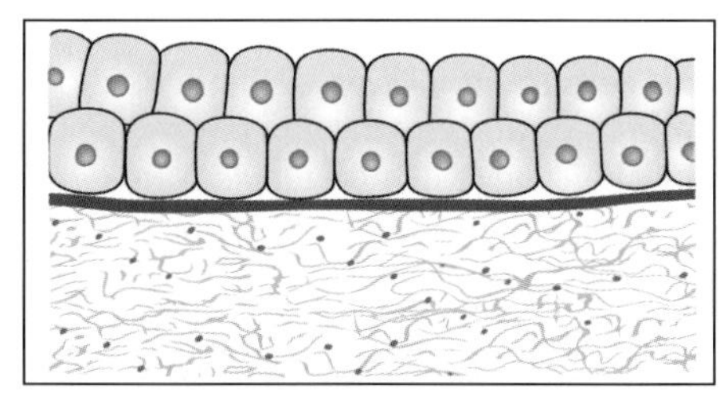

Fig. 3.8: Stratified cuboidal epithelium (Schematic representation)

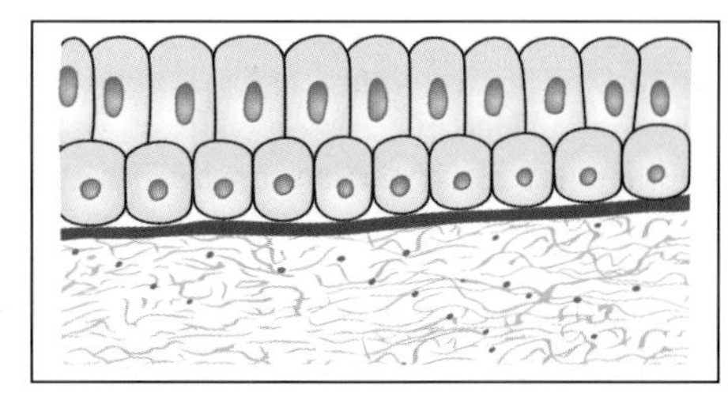

Fig. 3.9: Stratified columnar epithelium (Schematic representation)

Function

Like all stratified epithelia it is protective in function and it also helps in conducting the secretion of the glands.

Pathological Correlation

Tumours arising from epithelia

A tumour (or neoplasm) can arise from any tissue if there is uncontrolled growth of cells. Such a tumour may be benign, when it remains localised; or may be malignant. A malignant growth invades surrounding tissues. Cells of the tumour can spread to distant sites (through lymphatics or through the bloodstream) and can start growing there producing secondaries or metastases.

A malignant tumour arising from an epithelium is called a carcinoma. If it arises from a squamous epithelium it is a squamous cell carcinoma; and if it arises from glandular epithelium it is called an adenoma.

Quite commonly cells in tumours resemble those of the tissue from which they are derived, and this is useful in pathological diagnosis. However, in metastases of fast growing tumours the cells may not show the characteristics of the tissue of origin (undifferentiated tumour), and it may be difficult to find out the location of the primary growth. In such cases diagnosis can be aided by localisation of proteins that are present only in epithelia. As mentioned above this can be done by using immunohistochemical techniques.

Added Information

- The shape of epithelial cells is related to the amount of contained cytoplasm and organelles. These in turn are related to metabolic activity. Squamous cells are least active. Columnar cells contain abundant mitochondria and endoplasmic reticulum and are highly active.
- Laterally, epithelial cells are in contact with other epithelial cells. The contact between adjoining cells is generally an intimate one because of the presence of desmosomes, zonulae adherens, and zonulae occludens. The intimate contact ensures that materials passing through the epithelium have to pass through the cells, rather than between them.
- We have seen that cilia are present on the free surfaces of some epithelial cells. The surface area of an epithelial cell may be greatly increased by the presence of microvilli, or of basolateral folds.
- Some epithelial cells contain pigment. Such cells are present in the skin, the retina and the iris.
- Epithelia are generally devoid of blood vessels. Their cells obtain nutrition by diffusion from blood vessels in underlying tissues. In contrast, delicate nerve fibres frequently penetrate into the intervals between epithelial cells.
- Epithelia have considerable capacity for repair after damage. They grow rapidly after injury, to repair the defect.
- It should be remembered that epithelial cells that look alike (on superficial examination) could have very different functions. For example, cuboidal cells lining follicles of the thyroid gland have very little in common with cuboidal cells covering the surface of the ovary.
- Epithelial cells in which transport of ions is an important function (e.g. renal tubules) are marked by the presence of basolateral folds, and the presence of large numbers of mitochondria, which provide adenosine triphosphate (ATP) for ion transport. Tight junctions between the cells prevent passive diffusion of ions.
- Epithelial cells contain some proteins not present in non-epithelial cells. These include cytokeratin (present in intermediate filaments). Such proteins can be localised using immunohistochemical techniques.

BASEMENT MEMBRANE

- Epithelial cells rest on a thin basement membrane.
- In multi-layered epithelia, the deepest cells lie on this membrane.
- A distinct basement membrane cannot be seen in haematoxylin and eosin (H & E) preparations, but can be well demonstrated using the periodic acid Schiff (PAS) method. The latter stains the glycoproteins present in the membrane.
- Under the EM a basement membrane is seen to have a ***basal lamina*** (nearest the epithelial cells) and a ***reticular lamina*** or ***fibroreticular lamina*** (consisting of reticular tissue and merging into surrounding connective tissue). The basal lamina is divisible into the ***lamina densa*** containing fibrils; and the ***lamina lucida*** which appears to be transparent. The lamina lucida lies against the cell membranes of epithelial cells.

Functions of Basement Membrane

- It provides adhesion on one side to epithelial cells (or parenchyma); and on the other side to connective tissue (mainly collagen fibres).
- It acts as a barrier to the diffusion of molecules. The barrier function varies with location (because of variations in pore size). Large proteins are prevented from passing out of blood vessels, but (in the lung) diffusion of gases is allowed.
- Recent work suggests that basement membranes may play a role in cell organisation, as molecules within the membrane interact with receptors on cell surfaces. Substances present in the membrane may influence morphogenesis of cells to which they are attached.
- The membranes may influence the regeneration of peripheral nerves after injury, and may play a role in re-establishment of neuromuscular junctions.

PROJECTIONS FROM THE CELL SURFACE

Many epithelial cells show projections from the cell surface. The various types of projections are as follows:
- Cilia
- Microvilli
- Stereocilia

Cilia

These can be seen with the light microscope, as minute hair-like projections from the free surfaces of some epithelial cells.

In the living animal cilia can be seen to be motile.

Structure

The detailed structure of cilia can only be made out by electron microscopy.

The free part of each cilium is called the ***shaft***. The region of attachment of the shaft to the cell surface is called the ***base*** (also called the ***basal body***, ***basal granule***, or ***kinetosome***). The free end of the shaft tapers to a tip. Each cilium is 10 µm in length and 0.25 µm in diameter.

It consists of (1) an outer covering that is formed by an extension of the cell membrane; and (2) an inner core (***axoneme***), that is formed by microtubules arranged in a definite manner (Fig. 3.11).

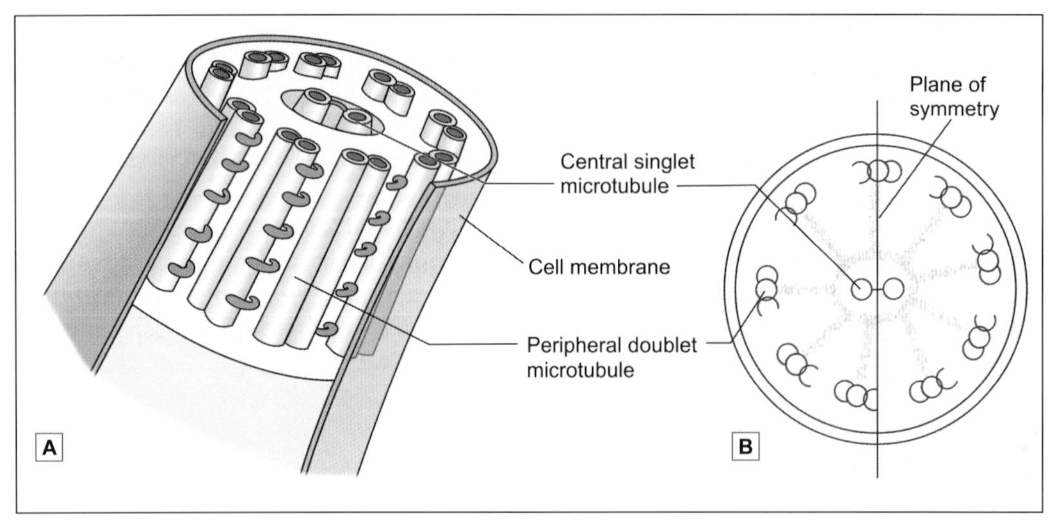

Fig. 3.10: Cilium. **A.** Ultrastructure **B.** Internal struture (Schematic representation)

It has a striking similarity to the structure of a centriole. There is a central pair of tubules that is surrounded by nine pairs of tubules. The outer tubules are connected to the inner pair by radial structures (which are like the spokes of a wheel). Other projections pass outwards from the outer tubules (Fig. 3.11).

As the tubules of the shaft are traced towards the tip of the cilium, it is seen that one tubule of each outer pair ends short of the tip so that near the tip each outer pair is represented by one tubule only. Just near the tip, only the central pair of tubules is present.

At the base of the cilium one additional tubule is added to each outer pair so that here the nine outer groups of tubules have three tubules each, exactly as in the centriole.

Microtubules in cilia are bound with proteins (dynein and nexin). Nexin holds the microtubules together. Dynein molecules are responsible for bending of tubules, and thereby for movements of cilia (Fig. 3.11).

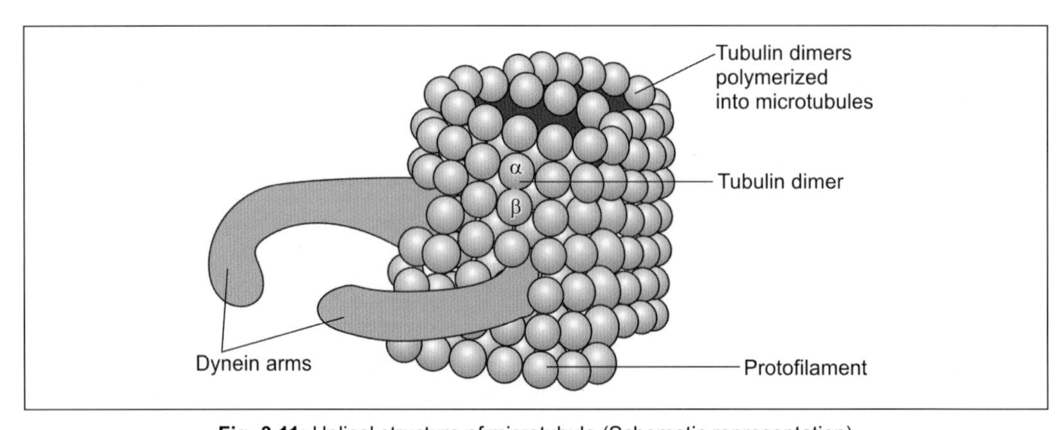

Fig. 3.11: Helical structure of microtubule (Schematic representation)

Functional Significance

The cilia lining an epithelial surface move in coordination with one another the total effect being that like a wave. As a result fluid, mucous, or small solid objects lying on the epithelium can be caused to move in a specific direction.

- Movements of cilia lining the respiratory epithelium help to move secretions in the trachea and bronchi towards the pharynx.
- Ciliary action helps in the movement of ova through the uterine tube, and of spermatozoa through the male genital tract.
- In some situations there are cilia-like structures that perform a sensory function. They may be non-motile, but can be bent by external influences. Such 'cilia' present on the cells in the olfactory mucosa of the nose are called ***olfactory cilia***, they are receptors for smell. Similar structures called ***kinocilia*** are present in some parts of the internal ear.

Pathological Correlation

Abnormalities of cilia

Cilia can be abnormal in persons with genetic defects that interfere with synthesis of ciliary proteins. This leads to the **immotile cilia syndrome**. As secretions are not removed from respiratory passages the patient has repeated and severe chest infections. Women affected by the syndrome may be sterile as movement of ova along the uterine tube is affected. Ciliary proteins are present in the tails of spermatozoa, and an affected male may be sterile because of interference with the motility of spermatozoa.

Ciliary action is also necessary for normal development of tissues in embryonic life. Migration of cells during embryogenesis is dependent on ciliary action, and if the cilia are not motile various congenital abnormalities can result.

Note: ***Flagella:*** These are somewhat larger processes having the same basic structure as cilia. In the human body the best example of a flagellum is the tail of the spermatozoon. The movements of flagella are different from those of cilia. In a flagellum, movement starts at its base. The segment nearest the base bends in one direction. This is followed by bending of succeeding segments in opposite directions, so that a wave-like motion passes down the flagellum.

Microvilli

Structure

Microvilli are finger-like projections from the cell surface (Fig. 3.4B) that can be seen by electron microscope (EM) measuring 1–2 mm in length and about 75–90 μm in diameter. Microvilli are covered by a polysaccharide surface coat called as ***glycocalyx***.

Each microvillus consists of an outer covering of plasma membrane and a cytoplasmic core in which there are numerous microfilaments (actin filaments). The filaments are continuous with actin filaments of the cell cortex.

Numerous enzymes and glycoproteins, concerned with absorption have been located in microvilli.

With the light microscope the free borders of epithelial cells lining the small intestine appear to be thickened, the thickening has striations perpendicular to the surface. This ***striated border*** of light microscopy has been shown by EM to be made up of long microvilli arranged parallel to one another.

In some cells the microvilli are not arranged so regularly. With the light microscope the microvilli of such cells give the appearance of a ***brush border***.
See Table 3.1 for differences between cilia and microvilli.

Functional Significance

Microvilli are non-motile processes which greatly increase the surface area of the cell and are, therefore, seen most typically at sites of active absorption, e.g. the intestine, and the proximal convoluted tubules of the kidney.

Stereocilia

Stereocilia are very long, thick microvilli measuring about 5–10 μm in length. They are non-motile.

Stereocilia are seen on receptor cells in the internal ear, and on the epithelium of the epididymis. They increase the cells surface area for absorption in epididymis and probably function in signal generation in hair cells of internal ear.

Table 3.1: Differences between cilia and microvilli

Cilia	Microvilli
10 μm in length and 0.25 μm in diameter	1–2 mm in length and about 75–90 μm in diameter
Motile	Non-motile
Contains 9+2 pattern of microtubules	Contains numerous microfilaments (actin filaments)
Concerned with movement of ova through uterine tube and movement of secretions in trachea and bronchi towards pharynx	Concerned with absorptive functions
Seen over lining epithelium of respiratory tract and uterine tube	Seen over intestinal epithelium and proximal convulated tubule of kidney

Note: ***Basal folds:*** In some cells the cell membrane over the basal or lateral aspect of the cell shows deep folds (basolateral folds). Like microvilli, basolateral folds are an adaptation to increase cell surface area. Basal folds are seen in renal tubular cells, and in cells lining the ducts of some glands. Lateral folds are seen in absorptive cells lining the gut.

Chapter 4

Glands

Some epithelial cells may be specialised to perform a secretory function. Such cells, present singly or in groups, constitute glands.

CLASSIFICATION OF GLANDS

Glands may be unicellular or multicellular.

- ***Unicellular:*** Unicellular glands are interspersed amongst other (non-secretory) epithelial cells. They can be found in the epithelium lining the intestines.
- ***Multicellular:*** Most glands are ***multicellular***. Such glands develop as diverticulae from epithelial surfaces. The 'distal' parts of the diverticulae develop into secretory elements, while the 'proximal' parts form ducts through which secretions reach the epithelial surface, e.g., lacrimal gland, parotid gland, etc.

Glands that pour their secretions on to an epithelial surface, directly or through ducts are called ***exocrine glands*** (or ***externally secreting glands***). Some glands lose all contact with the epithelial surface from which they develop and they pour their secretions into blood. Such glands are called ***endocrine glands***, ***internally secreting glands***, or ***duct-less glands***.

CLASSIFICATION OF EXOCRINE GLANDS (FLOW CHART 4.1)

Exocrine glands can be further classified on the basis of:

- ***Branching of ducts:***
  - ***Simple:*** When all the secretory cells of an exocrine gland discharge into one duct, the gland is said to be a ***simple gland***, e.g., gastric glands, sweat glands, etc.
  - ***Compound:*** Sometimes there are a number of groups of secretory cells, each group discharging into its own duct. These ducts unite to form larger ducts that ultimately drain on to an epithelial surface. Such a gland is said to be a ***compound gland***, e.g., parotid gland, pancreas, etc.
- ***Shape of the secretory unit:*** Both in simple and in compound glands the secretory cells may be arranged in various ways:
  - ***Tubular glands:*** Glands with secretory unit tubular in shape. The tube may be straight, coiled or branched, e.g., gastric glands.
  - ***Acinar glands:*** Glands with secretory unit round or oval in shape, e.g., salivary glands.
  - ***Alveolar glands:*** Glands with secretory unit flask-shaped. However, it may be noted that the terms acini and alveoli are often used as if they were synonymous. Glands in which the secretory elements are greatly distended are called ***saccular glands***.

Flow chart 4.1: Classification of glands

- Glands
  - Without ducts → Endocrine
  - With ducts → Exocrine
    - Unicellular → Goblet cell
    - Multicellular
      - On the basis of number of ducts that drain the gland
        - Simple
        - Compound
      - On the basis of shape of the secretory unit
        - Tubular
        - Acinar
        - Alveolar
      - On the basis of nature of secretory product
        - Serous
        - Mucous
        - Mixed
      - On the basis of secretory mechanism
        - Merocrine
        - Apocrine
        - Holocrine

Note: Combinations of the above may be present in a single gland. From what has been said above it will be seen that an exocrine gland may be:

- Unicellular
- Simple tubular
- Simple alveolar (or acinar)
- Compound tubular
- Compound alveolar
- Compound tubulo-alveolar (or racemose)

Some further subdivisions of these are shown in Fig. 4.1.

- ***Nature of their secretions: mucous glands*** and ***serous glands***.
  - ***Mucous glands:*** Cell of mucous acini are tall with flat nuclei at their bases. The lumen of these acini is larger than the serous acini. In mucous glands the secretion contains mucopolysaccharides. The secretion collects in the apical parts of the cells. As a result nuclei are pushed to the base of the cell, and may be flattened. In classroom slides stained with haematoxylin and eosin, the secretion within mucous cells remains unstained so that they have an 'empty' look. However, the stored secretion can be brightly stained using a special procedure called the ***periodic acid Schiff (PAS)*** method. Unicellular cells secreting mucus are numerous in the intestines, they are called ***goblet cells*** because of their peculiar shape (Fig. 4.2).
  - ***Serous glands:*** Cells of serous acini are triangular in shape with a rounded nucleus. Their nuclei are centrally placed. The secretions of serous glands are protein in nature. The cytoplasm of these cells is granular and often stains bluish with haematoxylin and eosin. The lumen of their acini is small. A comparison of mucous and serous acini has been given in Table 4.1.

Note: Some glands contain both serous and mucous elements (mixed).

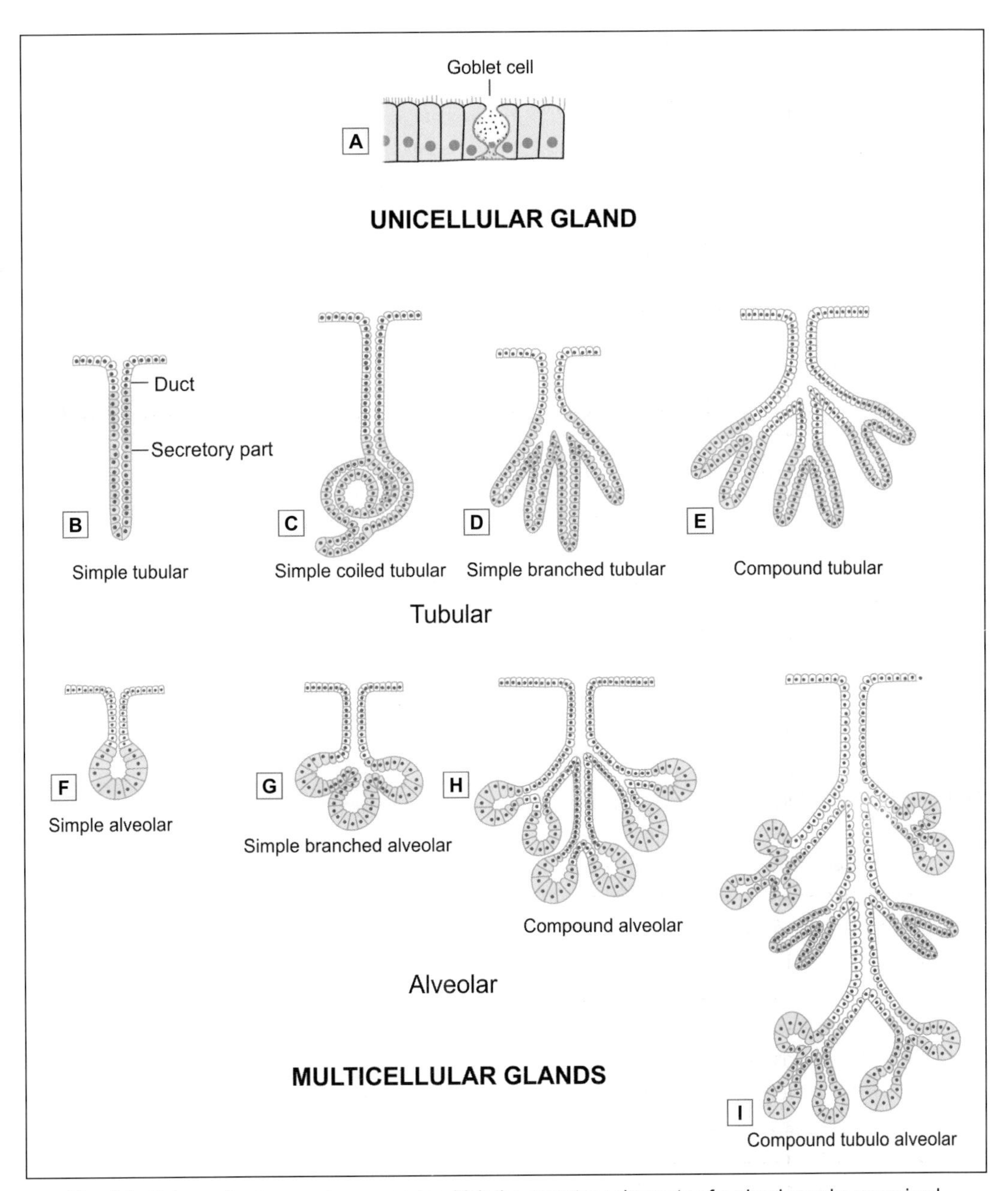

Fig. 4.1: Scheme to show various ways in which the secretory elements of a gland may be organised. (A) unicellular gland; (B to G) multicellular glands with a single duct are simple glands, (H and I) multicellular glands with branching duct system are compound glands

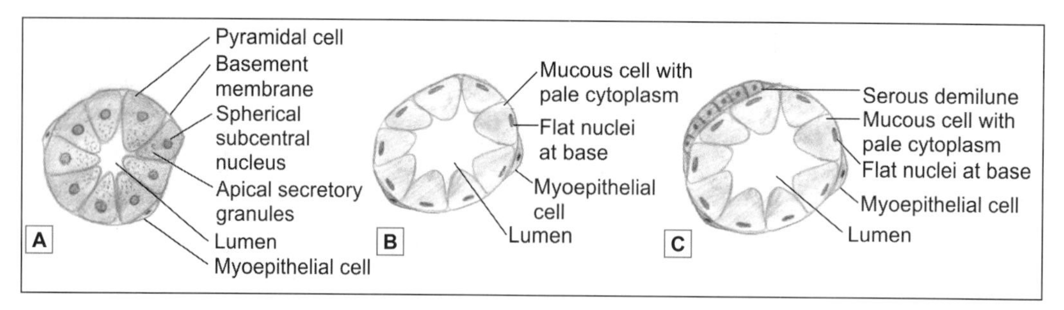

Fig. 4.2: Types of acini. **A.** Serous; **B.** Mucous; **C.** Mucous acini with serous demilune (Schematic representation)

Table 4.1: Comparison between serous and mucous acini

Serous acini	Mucous acini
Triangular cells with rounded nucleus at the base. Cell boundaries are indistinct	Tall cells with flat nucleus at the base. Cell boundaries are distinct
Contain zymogen granules	Contain mucoid material
Darkly stained with H & E (because of the presence of zymogen granules, the colour varies from pink to dark purple)	Lightly stained and appear empty with H & E
Thin watery secretion	Thick mucoid secretion
Example: Parotid gland	Example: Sublingual gland

Added Information

Epithelia in secretory portions of glands show specialisations of structure depending upon the nature of secretion as follows:

- Cells that are protein secreting (e.g., hormone producing cells) have a well developed rough endoplasmic reticulum (ER), and a supranuclear Golgi complex. Secretory granules often fill the apical portions of the cells. The staining characters of the granules differ in cells producing different secretions (the cells being described as acidophil, basophil, etc).
- Mucin secreting cells have a well developed rough ER (where the protein component of mucin is synthesised) and a very well developed Golgi complex (where proteins are glycosylated).
- Steroid producing cells are characterised by the presence of extensive smooth ER and prominent mitochondria.

- ***The manner in which their secretions are poured out of the cells:***
  - ***Merocrine:*** In most exocrine glands secretions are thrown out of the cells by a process of exocytosis, the cell remaining intact, this manner of secretion is described as ***merocrine***, e.g., goblet cell (sometimes also called ***eccrine*** or ***epicrine***) (Fig. 4.3A).
  - ***Apocrine:*** In some glands the apical parts of the cells are shed off to discharge the secretion, this manner of secretion is described as ***apocrine***. An example of apocrine secretion is seen in some atypical sweat glands and in mammary glands (Fig. 4.3B).
  - ***Holocrine:*** In some glands, the entire cell disintegrates while discharging its secretion. This manner of discharging secretion is described as ***holocrine***, and is seen typically in sebaceous glands (Fig. 4.3C).

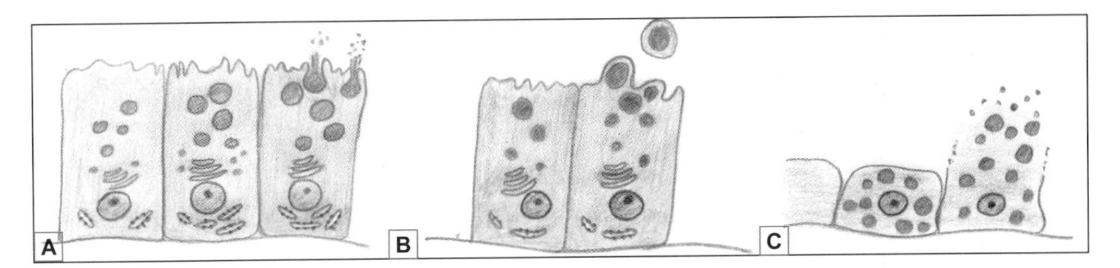

Fig. 4.3: Types of glands based on the manner in which their secretions are poured out of the cells. **A.** Merocrine; **B.** Apocrine; **C.** Holocrine (Schematic representation)

STRUCTURAL ORGANISATION

Exocrine Glands

All exocrine glands have basically the same structural organisation consisting of three components—parenchyma, stroma and duct system.

- ***Parenchyma:*** The secretory cells of a gland constitute its ***parenchyma***.
- ***Stroma:*** The connective tissue in which the parenchyma lies is called the ***stroma***. The glandular tissue is often divisible into lobules separated by connective tissue septa. Aggregations of lobules may form distinct lobes. The connective tissue covering the entire gland forms a ***capsule*** for it. Blood vessels and nerves pass along connective tissue septa to reach the secretory elements. Their activity is under nervous or hormonal control (Fig. 4.4).
- ***Duct system:*** The ducts convey the secretory product of the gland. When a gland is divided into lobes the ducts draining it may be ***intralobular*** (lying within a lobule), ***interlobular*** (lying in the intervals between lobules), or ***interlobar*** (lying between adjacent lobes), in increasing order of size (Fig. 4.4).

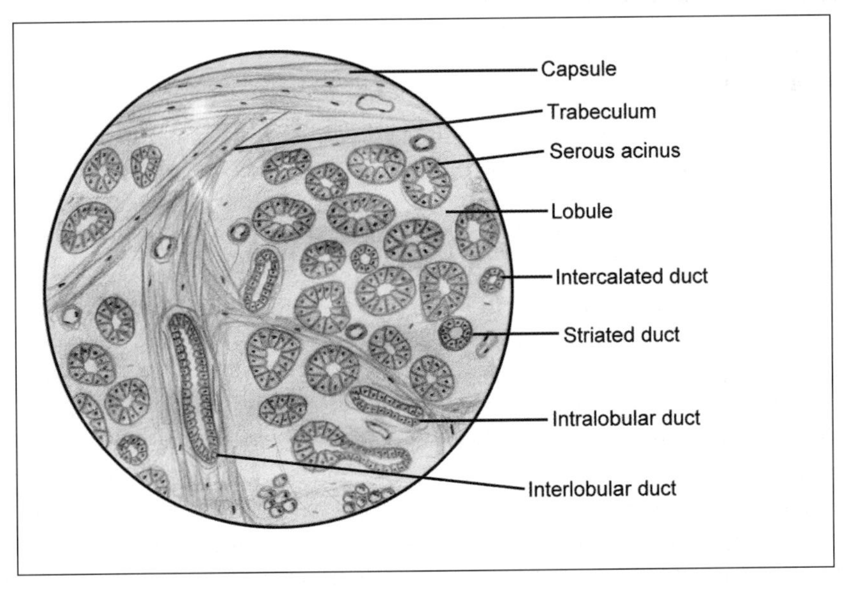

Fig. 4.4: Structural organisation of an exocrine gland (Schematic representation)

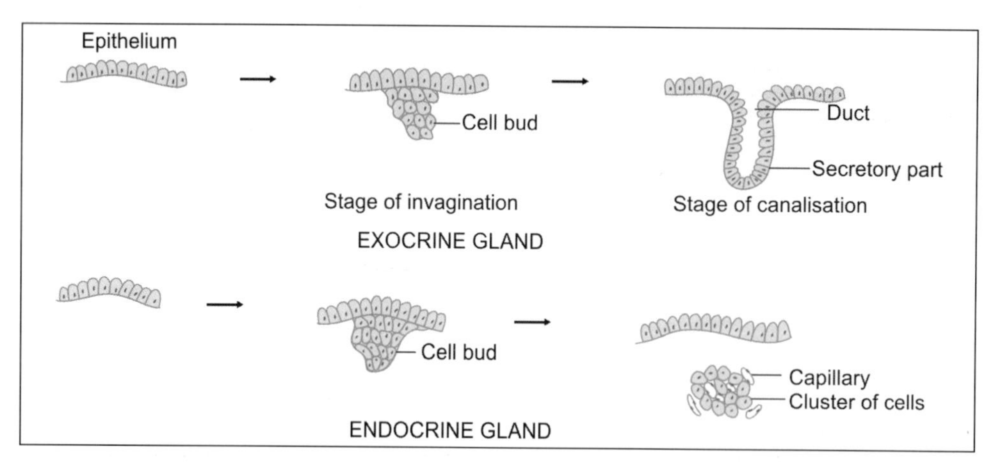

Fig. 4.5: Development of glands (Schematic representation)

Endocrine Glands

Endocrine glands are usually arranged in cords or in clumps that are intimately related to a rich network of blood capillaries or of sinusoids. In some cases (for example the thyroid gland) the cells may form rounded follicles.

Endocrine cells and their blood vessels are supported by delicate connective tissue, and are usually surrounded by a capsule.

Clinical Correlation

Neoplasms can arise from the epithelium lining a gland. A benign growth arising in a gland is an ***adenoma***; and a malignant growth is an ***adenocarcinoma***.

DEVELOPMENT OF GLANDS

Glands, both exocrine and endocrine develop as diverticula of the epithelium. As shown in the Fig. 4.5, the ***exocrine*** develops as a solid bud from the epithelium into the underlying connective tissue. Soon it elongates, undergoes canalisation and displays a secretory and conducting portion. The conducting part forms the duct and is continuous with the epithelium. Hence, an endocrine gland discharges its secretions through a duct.

Endocrine gland also develops in a similar manner to exocrine gland but with further development it breaks the continuity with overlying epithelium. It appears as a clump of cells. Soon these groups of cells get surrounded by blood vessels into which they pour their secretions.

Note: In this chapter we have considered the general features of glands. Further details of the structure of exocrine and endocrine glands will be considered while studying individual glands.

Chapter 5

General Connective Tissue

INTRODUCTION

The term ***connective tissue*** is applied to a tissue that fills the interstices between more specialised elements.

Connective tissue serves to hold together, and to support, different elements within an organ. For this reason, such connective tissue is to be found in almost every part of the body. It is conspicuous in some regions and scanty in others. This kind of connective tissue is referred to as ***general connective tissue*** to distinguish it from more specialised connective tissues that we will consider separately. It is also called ***fibro-collagenous tissue***.

Components of Connective Tissue (Fig. 5.1 and Flow chart 5.1)

- Fibres
- Cells
- Ground substance

Many tissues and organs of the body are made up mainly of aggregations of closely packed cells, e.g., epithelia, and solid organs like the liver. In contrast, cells are relatively few in connective tissue, and are widely separated by a prominent ***intercellular substance***. The intercellular substance is in the form of a ***ground substance*** within which there are numerous ***fibres*** (Fig. 5.2). Connective tissue can assume various forms depending upon the nature of the ground substance, and of the type of fibres and cells present.

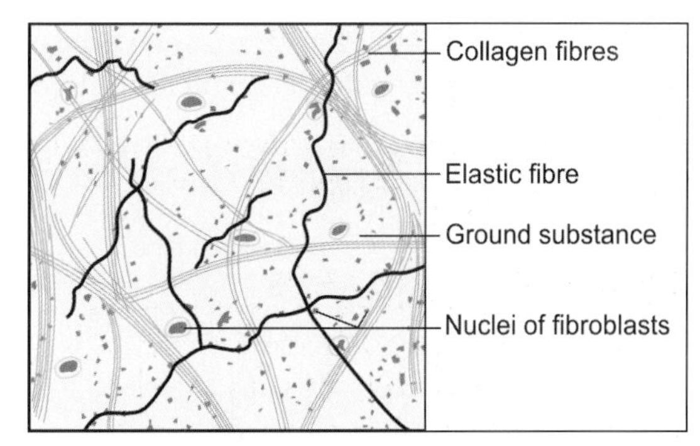

Fig. 5.1: Stretch preparation of omentum showing loose areolar tissue (Schematic representation)

FIBRES OF CONNECTIVE TISSUE

The most conspicuous components of connective tissue are the fibres within it. These are of three main types: collagen, reticular and elastic.

Collagen fibres are most numerous. They can be classified into various types. ***Reticular fibres*** were once described as a distinct variety of fibres, but they are now regarded as one variety of collagen fibre.

Flow chart 5.1: Basic components of connective tissue

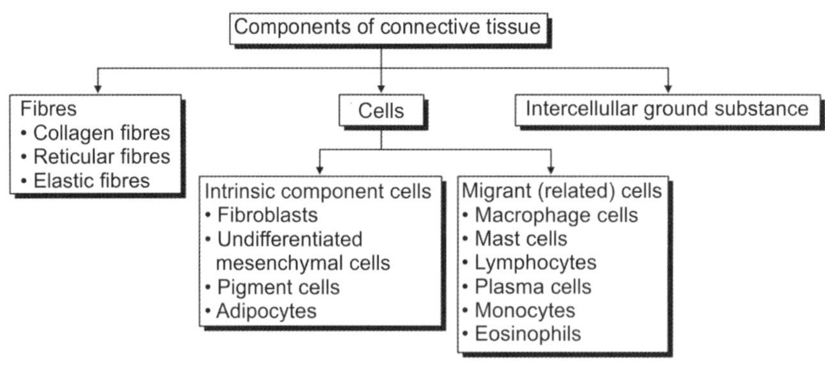

Collagen Fibres

Collagen fibres are most abundant type of connective tissue fibres. With the light microscope collagen fibres are seen in bundles (Figs 5.2 and 5.3A). The bundles may be straight or wavy depending upon how much they are stretched. The bundles are made up of collections of individual collagen fibres which are 1–12 μm in diameter.

The bundles often branch, or anastomose with adjacent bundles, but the individual fibres do not branch.

With the EM each collagen fibre is seen to be made of fibrils that are 20–200 nm in diameter. Each fibril consists of a number of microfibrils (3.5 nm in diameter). At high magnifications of

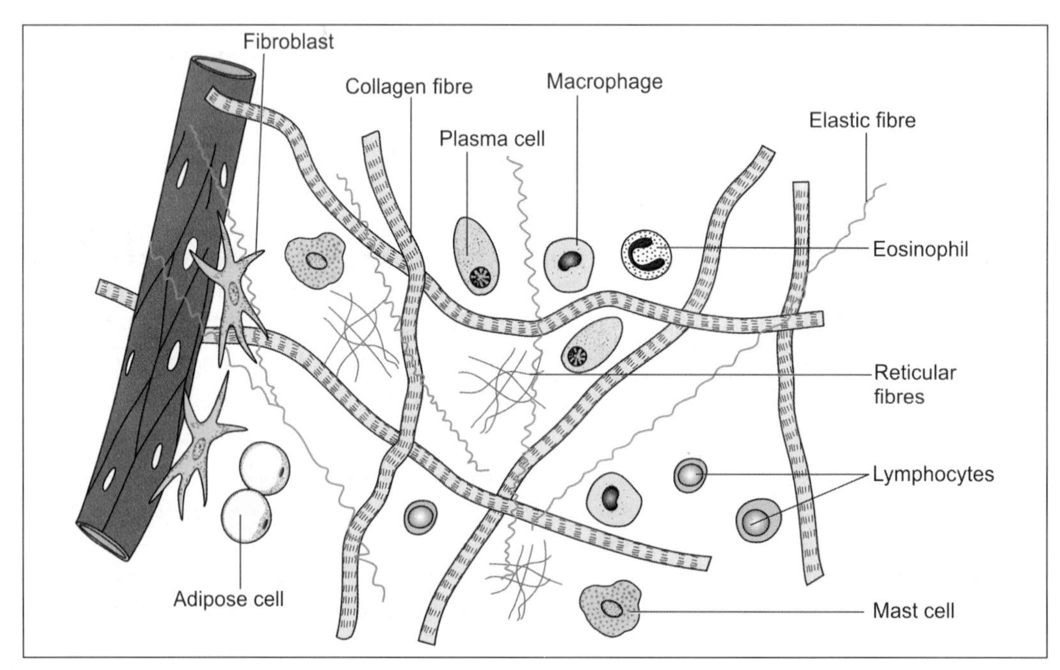

Fig. 5.2: Components of loose connective tissue (Schematic representation)

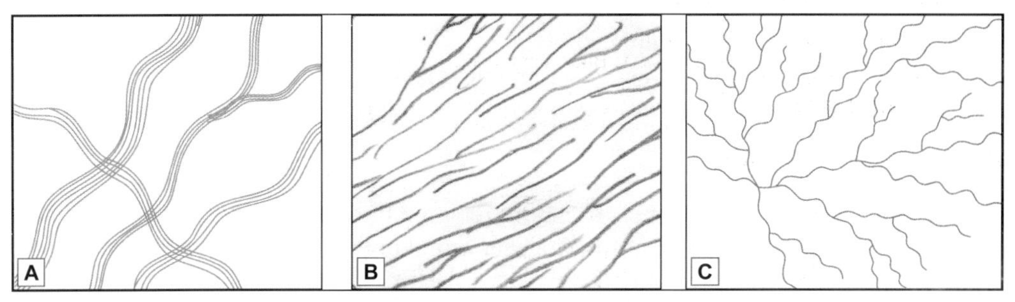

Fig. 5.3: Fibres of connective tissue. **A.** Collagen fibres; **B.** Reticular fibres; **C.** Elastic fibres (Schematic representation)

the EM each fibril shows characteristic cross striations (or periods) after every 67 nm interval (in unfixed tissue).

Staining Characters

Bundles of collagen fibres appear white with the unaided eye. In sections stained with haematoxylin and eosin, collagen fibres are stained light pink. With special methods they assume different colours depending upon the dye used. Two commonly used methods are ***Masson's trichrome*** with which the fibres stain blue and the ***Van Gieson*** method with which they stain red. After silver impregnation the fibres are stained brown.

Chemical Nature

Collagen fibres are so called because they are made up mainly of a protein called ***collagen***. Collagen is made up of molecules of ***tropocollagen***.

Microfibrils of collagen are chains of tropocollagen molecules. Each molecule of tropocollagen is 300 nm in length. Within a fibre, the molecules of tropocollagen are arranged in a regular overlapping pattern which is responsible for producing the cross striated appearance of the fibres.

Each molecule of tropocollagen is made up of three polypeptide chains. The chains are arranged in the form of a triple helix. The polypeptide chains are referred to as procollagen.

Added Information

Each procollagen chain consists of a long chain of amino acids that are arranged in groups of three (triplets). Each triplet contains the amino acid glycine. The other two amino acids in each triplet are variable. Most commonly these are hydroxyproline and hydroxylysine. Variations in the amino acid pattern give rise to several types of collagen as described below.

Physical Properties

Collagen fibres can resist considerable tensile forces (i.e., stretching) without significant increase in their length. At the same time they are pliable and can bend easily.

When polarised light is thrown on the fibres the light is split into two beams that are refracted in different directions. This is called ***birefringence***: it is an indication of the fact that each collagen fibre is made up of finer fibrils.

Collagen fibres swell and become soft when treated with a weak acid or alkali. They are destroyed by strong acids (This fact is sometimes made use of to soften collagen fibres to facilitate preparation of anatomical specimens). On boiling, collagen is converted into gelatin.

Production of Collagen Fibres

Tropocollagen is synthesised by fibroblast and released into extracellular space, where they get polymerised to form collagen fibres. Collagen is not only synthesised by fibroblast but by other cells also, e.g., Chondroblasts in cartilage, osteoblast in bone and smooth muscle in blood vessel.

Added Information

The mechanism of the production of collagen fibres by fibroblasts has been extensively studied. Amino acids necessary for synthesis of fibres are taken into the cell. Under the influence of ribosomes located on rough endoplasmic reticulum, the amino acids are bonded together to form polypeptide chains (α-chains) (Flow chart 5.2). A procollagen molecule is formed by joining together of three such chains. Molecules of procollagen are transported to the exterior of the cell where they are acted upon by enzymes (released by the fibroblast) to form tropocollagen. Collagen fibres are formed by aggregation of tropocollagen molecules. Vitamin C and oxygen are necessary for collagen formation, and wound repair may be interfered with if either of these is deficient. In this connection it may be noted that fibroblasts are themselves highly resistant to damaging influences and are not easily destroyed.

There are observations that indicate that the orientation of collagen fibres depends on the stresses imposed on the tissue. If fibroblasts growing in tissue culture are subjected to tension in a particular direction, the cells exposed to the tension multiply faster than others; they orientate themselves along the line of stress; and lay down fibres in that direction. It follows that in the embryo collagen fibres would tend to be laid down wherever they are required to resist a tensile force. In this way tendons, ligaments, etc., will tend to develop wherever they are called for (This cannot, of course, be a complete explanation. Genetic factors must play a prominent role in development of these structures).

Flow chart 5.2: Scheme to show how collagen fibres are synthesised

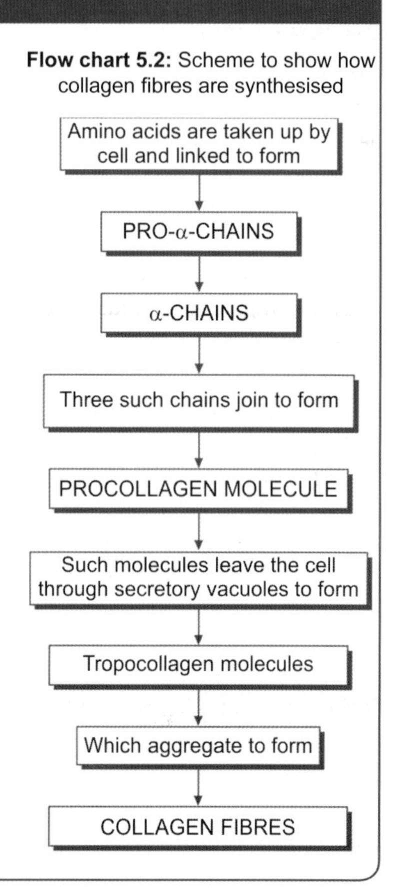

Varieties of Collagen and their Distribution

Several types of collagen (more than 25) are recognised depending upon the diameter of fibres, the prominence of cross striations, and other features.

- ***Type I***: These are collagen fibres of classical description having the properties described above. They are found in connective tissue, tendons, ligaments, fasciae, aponeuroses, etc. They are also present in the dermis of the skin, and in meninges. They form the fibrous basis of bone and of fibrocartilage. Type I fibres are of large diameter (about 250 nm) and have prominent cross striations.

- ***Type II***: These are of two subtypes. The larger of these are about 100 nm in diameter, while the narrower fibres are about 20 nm in diameter. In type II collagen striations are less prominent than in type I. Type II collagen fibres form the fibrous basis of hyaline cartilage. Fine type II fibres are also present in the vitreous body.
- ***Type III***: These form the reticular fibres described below.
- ***Type IV:*** This type of collagen consists of short filaments that form sheets. It is present in the basal laminae of basement membranes. It is also seen in the lens capsule.
- ***Type V:*** This type of collagen is found in blood vessels and foetal membranes.

Reticular Fibres

These fibres are a variety of collagen fibres and are composed of collagen type III. They show periodicity (striations) of 67 nm. They differ from typical (Type I) collagen fibres as follows:

- They are much finer and have uneven thickness.
- They form a network (or reticulum) by branching, and by anastomosing with each other. They do not run in bundles (Fig. 5.3B).
- They can be stained specifically by silver impregnation, which renders them black. They can thus be easily distinguished from type I collagen fibres which are stained brown. Because of their affinity for silver salts reticular fibres are sometimes called ***argentophil fibres*** (Fig. 5.4).
- Reticular fibres contain more carbohydrates than Type I fibres (which is probably the reason why they are argentophil).

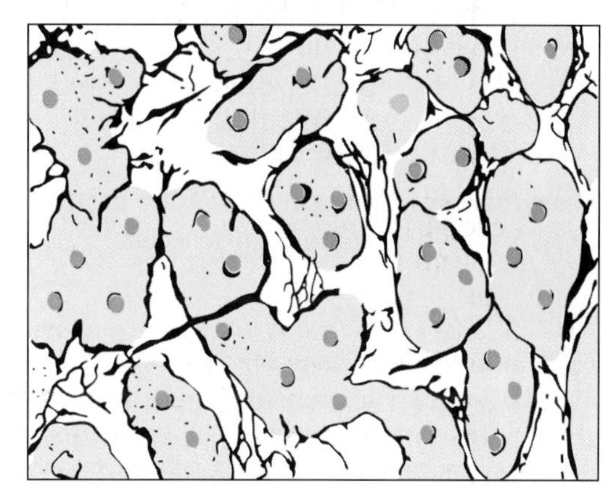

Fig. 5.4: Reticular fibres (black) forming a network in the liver. The white spaces represent sinusoids (Schematic representation)

Reticular fibres provide a supporting network in lymphoid organs like the spleen, lymph nodes and bone marrow; most glands, including the liver (Fig. 5.4); and the kidneys. Reticular fibres form an essential component of all basement membranes. They are also found in relation to smooth muscle and nerve fibres. Reticular fibres are synthesised by fibroblasts and reticular cells (special variety of fibroblasts).

Elastic Fibres

Elastic fibres are much fewer than those of collagen. They run singly (not in bundles), branch and anastomose with other fibres (Fig. 5.3C).

Elastic fibres are thinner than those of collagen (0.1–0.2 μm) (Fig. 5.1). In some situations elastic fibres are thick (e.g., in the ligamenta flava). In other situations (as in walls of large arteries) they form fenestrated membranes.

With the EM each elastic fibre is seen to have a central amorphous core and an outer layer of fibrils (Fig. 5.5). The outer fibrils are made up of a glycoprotein called ***fibrillin***. Periodic striations are not present in elastic fibres.

Staining Characters

Elastic fibres do not stain with the usual stains for collagen. They can be demonstrated by staining with orcein, with aldehyde fuchsin, and by Verhoeff's method.

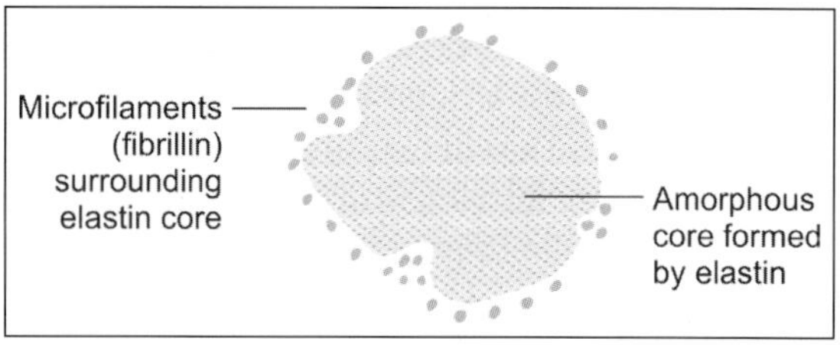

Fig. 5.5: Electron microscopy appearance of an elastic fibre as seen in transverse section (Schematic representations)

Chemical Nature

Elastic fibres are composed mainly of a protein called ***elastin*** that forms their central amorphous core. Elastin is made up of smaller units called ***tropoelastin***. Elastin contains a high quantity of the amino acids valine and alanine. Another amino acid called desmosine is found exclusively in elastic tissue. The outer fibrils of elastic fibres are composed of the glycoprotein fibrillin.

Production of Elastic Fibres

Elastic fibres of connective tissue are produced by fibroblasts. In some situations elastic tissue can be formed by smooth muscle cells.

Physical Properties

As their name implies elastic fibres can be stretched (like a rubber band) and return to their original length when tension is released. They are highly refractile and are, therefore, seen as shining lines in unstained preparations. Relaxed elastic fibres do not show birefringence, but when stretched the fibres become highly birefringent.

Unlike collagen, elastic fibres are not affected by weak acids or alkalies, or by boiling. However, they are digested by the enzyme elastase.

CELLS OF CONNECTIVE TISSUE

Various types of cells are present in connective tissue. These can be classed into two distinct categories (Flow chart 5.1).

- ***Cells that are intrinsic components of connective tissue:***

In typical connective tissue the most important cells are ***fibroblasts***. Others present are ***undifferentiated mesenchymal cells, pigment cells***, and ***fat cells***. Other varieties of cells are present in more specialised forms of connective tissues.

- ***Cells that belong to the immune system and are identical or closely related with certain cells present in blood and in lymphoid tissues***

These include ***macrophage cells*** (or ***histiocytes***), ***mast cells***, ***lymphocytes, plasma cells, monocytes*** and ***eosinophils***.

Fibroblasts

These are the most numerous cells of connective tissue. They are called fibroblasts because they are concerned with the production of collagen fibres. They also produce reticular and elastic fibres. Where associated with reticular fibres they are usually called ***reticular cells***.

Fibroblasts are present in close relationship to collagen fibres. They are 'fixed' cells, i.e., they are not mobile. In tissue sections these cells appear to be spindle shaped, and the nucleus

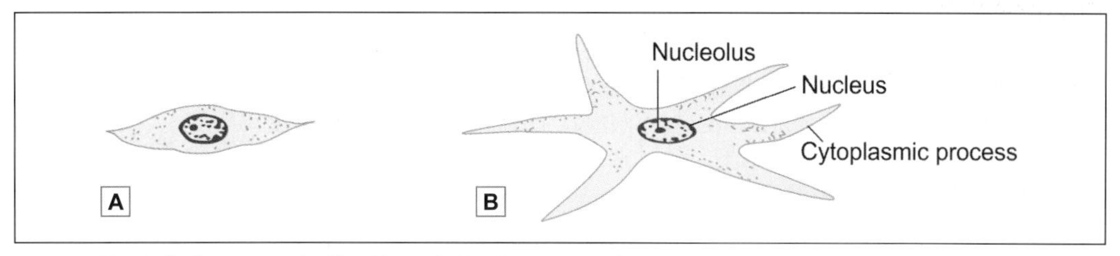

Fig. 5.6: Structure of a fibroblast. **A.** Profile view; **B.** Structure view (Schematic representation)

appears to be flattened. When seen from the surface the cells show branching processes (Fig. 5.6). The nucleus is large, euchromatic, and has prominent nucleoli.

The amount of cytoplasm and of organelles varies depending upon activity. In inactive fibroblasts that are the cytoplasm is scanty, organelles are few, and the nucleus may become heterochromatic. Inactive fibroblasts are often called ***fibrocytes***. In contrast to fibrocytes active fibroblasts have abundant cytoplasm (characteristic of cells actively engaged in protein synthesis). The endoplasmic reticulum, the Golgi complex and mitochondria become much more conspicuous.

Fibroblasts become very active when there is need to lay down collagen fibres. This occurs, for example, in wound repair. When the need arises fibroblasts can give rise, by division, to more fibroblasts. They are, however, regarded to be specialised cells and cannot convert themselves to other cell types.

Myofibroblasts

Under EM some cells resembling fibroblasts have been shown to contain actin and myosin arranged as in smooth muscle, and are contractile. They have been designated as myofibroblasts. In tissue repair such cells probably help in retraction and shrinkage of scar tissue.

These cells are pleuripotent and develop into new cells when stimulated. They are found along the periphery of blood vessels and hence, also called as ***adventitial cells***.

Undifferentiated Mesenchymal Cells

Embryonic connective tissue is called ***mesenchyme***. It is made up of stellate small cells with slender branching processes that join to form a fine network called as ***undifferentiated mesenchymal cells*** (Fig. 5.7).

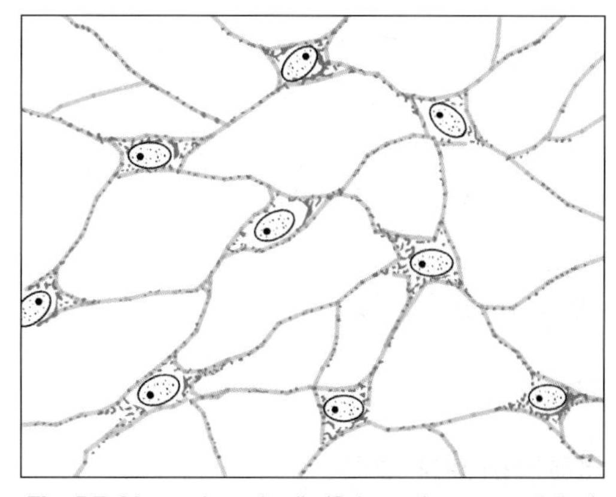

Fig. 5.7: Mesenchymal cells (Schematic representation)

Note: It is from such a tissue that the various elements of mature connective tissue are derived. As more specialised types of cells (e.g., fibroblasts) are formed they lose the ability to transform themselves into other types. At one time it was believed that fibroblasts were relatively undifferentiated cells, and that when the need arose they could transform themselves into other types. However, it is now believed that mature fibroblasts are not able to do so. It is also believed that some undifferentiated mesenchymal cells persist as such and these are the cells from which other types can be formed when required.

Pigment Cells

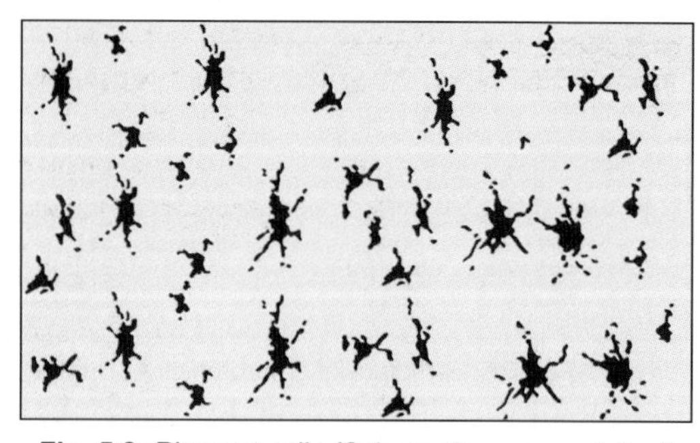

Fig. 5.8: Pigment cells (Schematic representation)

Pigment cells are easily distinguished as they contain brown pigment (melanin) in their cytoplasm (Fig. 5.8). They are most abundant in connective tissue of the skin, and in the choroid and iris of the eyeball.

Along with pigment containing epithelial cells they give the skin, the iris, and the choroid their dark colour. Variations in the number of pigment cells, and in the amount of pigment in them accounts for differences in skin colour of different races, and in different individuals.

Of the many cells that contain pigment in their cytoplasm only a few are actually capable of synthesising melanin. Such cells, called ***melanocytes***, are of neural crest origin.

The remaining cells are those that have engulfed pigment released by other cells. Typically, such cells are star shaped (stellate) with long branching processes. In contrast to melanocytes such cells are called ***chromatophores*** or ***melanophores***. They are probably modified fibroblasts.

Pigment cells prevent light from reaching other cells. The importance of this function in relation to the eyeball is obvious. Pigment cells in the skin protect deeper tissues from the effects of light (specially ultraviolet light). The darker skin of races living in tropical climates is an obvious adaptation for this purpose.

Fat Cells (Adipocytes)

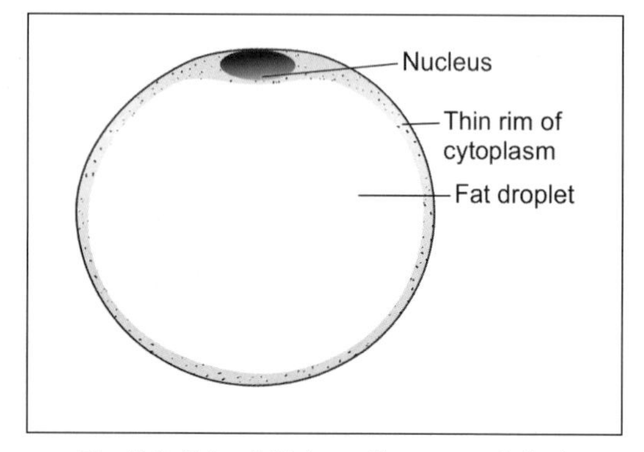

Fig. 5.9: Fat cell (Schematic representation)

Although some amount of fat (lipids) may be present in the cytoplasm of many cells, including fibroblasts, some cells store fat in large amounts and become distended with it (Fig. 5.9). These are called ***fat cells, adipocytes***, or ***lipocytes***. Aggregations of fat cells constitute ***adipose tissue.***

Each fat cell contains a large droplet of fat that almost fills it (Fig. 5.9). As a result the cell becomes rounded (when several fat cells are closely packed they become polygonal because of mutual pressure).

The cytoplasm of the cell forms a thin layer just deep to the plasma membrane.

The nucleus is pushed against the plasma membrane and is flattened resembling a signet ring. Adipocytes are incapable of division.

Macrophage Cells

Macrophage cells of connective tissue are part of a large series of cells present in the body that have similar functions. These collectively form the ***mononuclear phagocyte system***.

Macrophage cells of connective tissue are also called ***histiocytes*** or ***clasmatocytes*** (Fig. 5.10). They have the ability to phagocytose (eat up) unwanted material. Such material is usually organic: it includes bacteria invading the tissue, and damaged tissues. Macrophages also phagocytose inorganic particles injected into the body (e.g., India ink).

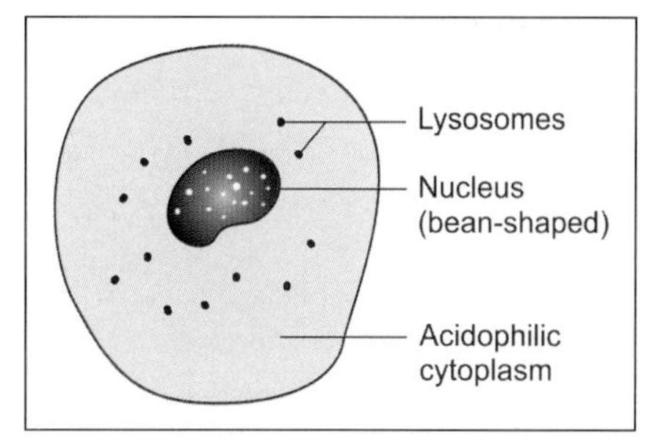

Fig. 5.10: Macrophage cell (histiocyte) (Schematic representation)

In ordinary preparations of tissue it is difficult to distinguish macrophages from other cells. However, if an animal is injected with India ink (or trypan blue, or lithium carmine) particles of it are taken up into the cytoplasm of macrophages, thus making them easy to recognise.

Macrophages are usually described as 'fixed' when they are attached to fibres; or as motile (or 'free'). Fixed macrophages resemble fibroblasts in appearance, but free macrophages are rounded. However, all macrophages are capable of becoming mobile when suitably stimulated. The nuclei of macrophages are smaller, and stain more intensely than those of fibroblasts. They are often kidney shaped (Fig. 5.10). With the EM the cytoplasm is seen to contain numerous lysosomes that help in 'digesting' material phagocytosed by the macrophage. Sometimes macrophages may fuse together to form multinucleated ***giant cells***.

Apart from direct phagocytic activity, macrophages play an important role in immunological mechanisms.

Mast Cells

These are small round or oval cells (***mastocytes***, or ***histaminocytes***) (Fig. 5.11). The nucleus is small and centrally placed. Irregular microvilli (filopodia) are present on the cell surface.

The distinguishing feature of these cells is the presence of numerous granules in the cytoplasm. The granules can be demonstrated with the PAS stain. They also stain with dyes like toluidine blue or alcian blue: with them the nuclei stain blue, but the granules stain purple to red (When components of a cell or tissue stain in a colour different from that of the dye used, the staining is said to be ***metachromatic***). On the basis of the staining reactions the granules are known to contain acid mucopolysaccharides.

With the EM the 'granules' are seen to be vesicles, each of which is surrounded by a membrane (In other words they are membrane bound vesicles).

The granules contain various substances. The most important of these is histamine. Release of histamine is associated with the production of allergic reactions when a tissue is exposed to an antigen to which it is sensitive (because of previous exposure).

Apart from histamine mast cells may contain various enzymes, and factors that attract eosinophils or neutrophils.

Mast cells differ considerably in size and in number from species to species, and at different sites in the same animal.

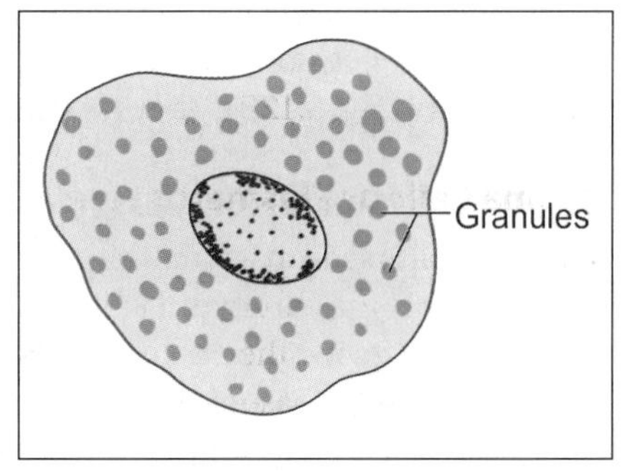

Fig. 5.11: Mast cell (Schematic representation)

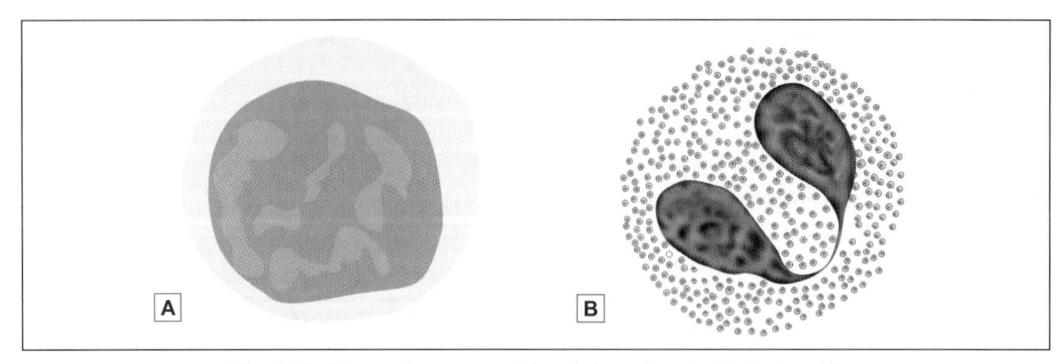

Fig. 5.12: Schematic representation **A.** Lymphocyte; **B.** Eosinophil

They are most frequently seen around blood vessels and nerves. Mast cells are probably related in their origin to basophils of blood. They may represent modified basophil cells.

Lymphocytes (Fig. 5.12A)

Lymphocytes represent one variety of leukocytes (white blood cells) present in blood. Large aggregations of lymphocytes are present in lymphoid tissues. They reach connective tissue from these sources, and are specially numerous when the tissue undergoes inflammation.

Lymphocytes play an important role in defence of the body against invasion by bacteria and other organisms. They have the ability to recognise substances that are foreign to the host body; and to destroy these invaders by producing antibodies against them.

The lymphocytes are derived from stem cells present in bone marrow. They are of two types. ***B-lymphocytes*** pass through blood to reach other tissues directly. Some B-lymphocytes mature into ***plasma cells*** described below. The second type of lymphocytes, called ***T-lymphocytes***, travel (through blood) from bone marrow to the thymus. After undergoing a process of maturation in this organ they again enter the bloodstream to reach other tissues. Both B-lymphocytes and T-lymphocytes can be seen in connective tissue.

Other Leukocytes

Apart from lymphocytes two other types of leukocytes may be seen in connective tissue. ***Monocytes*** are closely related in function to macrophages. ***Eosinophils*** (so called because of the presence of eosinophilic granules in the cytoplasm) are found in the connective tissue of many organs (Fig. 5.12B). They increase in number in allergic disorders.

Plasma Cells or Plasmatocytes

A plasma cell is seen to be small and round basophilic cytoplasm (Fig. 5.13). It can be recognised by the fact that the chromatin in its nucleus forms four or five clumps near the periphery (of the nucleus) thus giving the nucleus a resemblance to a cartwheel.

With the EM the basophilia is seen to be due to the fact that the cytoplasm is filled with rough endoplasmic reticulum, except for a small region near the nucleus where a well developed Golgi complex is located.

Both these features are indicative of the fact that plasma cells are engaged in considerable synthetic activity. They produce antibodies that may be discharged locally; may enter the circulation; or may be stored within the cell itself in the form of inclusions called ***Russell's bodies***.

Very few plasma cells can be seen in normal connective tissue. Their number increases in the presence of certain types of inflammation. It is believed that plasma cells represent B-lymphocytes that have matured and have lost their power of further division.

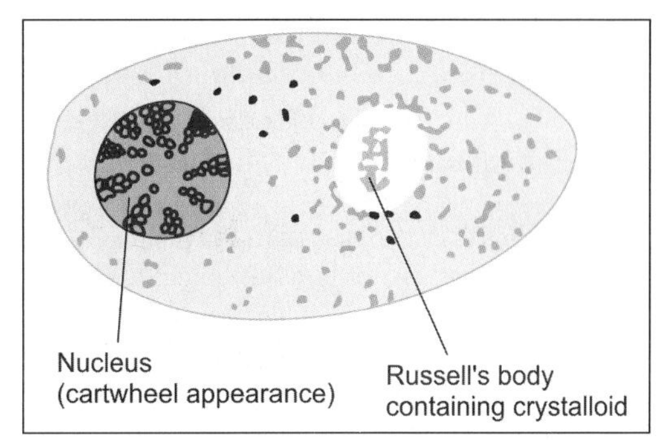

Fig. 5.13: Plasma cell (Schematic representation)

INTERCELLULAR GROUND SUBSTANCE OF CONNECTIVE TISSUE

The interfibrillar transparent homogenous viscous solution is called ***ground substance or matrix***. It fills the space between cells and fibres.

If a small quantity of fresh areolar tissue is spread out on a slide, and is examined under a microscope, the spaces between the fibre bundles appear to be empty. If such a preparation is treated with silver nitrate the spaces are seen to be filled with a brown staining material.

By the use of a technique called freeze drying, tissue can be prepared for sectioning without the use of extraneous chemicals. In areolar tissue prepared in this way the ground substance stains metachromatically with toluidine blue. It can also be stained with the PAS method. These early observations tell us that the ground substance is rich in protein-carbohydrate complexes or ***proteoglycans***.

Various types of proteoglycans are known. Each of them is a complex formed by protein and long chained polysaccharides called ***glucosaminoglycans***. The glycosaminoglycans, and the tissues in which each type is present are given in Table 5.1.

Table 5.1: Glycosaminoglycans present in various tissues

Tissue	Chondroitin sulphate	Dermatan sulphate	Heparan sulphate	Heparin	Keratan sulphate	Hyaluronic acid
Typical connective tissue	+	–	–	–	–	+
Cartilage	+	–	–	–	+	+
Bone	+	–	–	–	–	
Skin	+	+		+	–	+
Basement membrane	–	–	+	–	–	–
Others	–	Blood vessels and heart	Lung arteries	Mast cells Lungs Liver	Cornea Intervertebral discs	Synovial fluid

In addition to proteoglycans the ground substance also contains ***structural glycoproteins***. Their main function is to facilitate adhesion between various elements of connective tissue. Intercellular ground substance is synthesised by fibroblasts. Osteoblasts, chondroblasts, and even smooth muscle cells can also produce ground substance.

Added Information

With the exception of hyaluronic acid, all other glycosaminoglycans listed in the Table 5.1 have the following features.

- They are linked with protein (to form proteoglycans).
- They carry sulphate groups (SO_3^-) and carboxyl groups (COO^-) which give them a strong negative charge.
- The proteoglycans formed by them are in the form of long chains that do not fold. Because of this they occupy a large space (or ***domain***), and hold a large amount of water. They also hold Na^+ ions.

  Retained water and proteoglycans form a gel that gives a certain degree of stiffness to connective tissue; and helps it to resist compressive forces.
- Because of the arrangement of molecules within it, ground substance acts like a sieve. The size of the pores of the sieve can be altered (by change in orientation of molecules, and by change in the charges on them). In this way ground substance forms a selective barrier. This barrier function is specially important in basement membranes. In the kidney this barrier prevents large protein molecules from passing (from blood) into urine. However, exchange of gases is permitted in the lungs.

DIFFERENT FORMS OF CONNECTIVE TISSUE

Loose Areolar Connective Tissue

It consists of loosely arranged collagen fibres that appear to enclose large spaces and abundant ground substance. Spaces are also called ***areolae***, and such tissue is also referred to as ***areolar tissue.*** It gets distorted easily; hence it allows the tissue to move freely, e.g., Endomysium, subperiosteal tissue, lamina propria.

Dense Collagenous Connective Tissue (Dense Fibrous Connective Tissue)

In loose areolar tissue the fibre bundles are loosely arranged with wide spaces in between them. In many situations the fibre bundles are much more conspicuous, and form a dense mass. This kind of tissue is referred to as ***fibrous tissue***. It appears white in colour and hence also called as ***white fibrous tissue***. There are two kinds of dense connective tissue, regular and irregular according to the orientation of their fibrous components.

In some situations the bundles of collagen are arranged parallel to one another in a very orderly manner. This kind of tissue is called ***regular fibrous tissue*** (or regular connective tissue), and the best example of it is to be seen in tendons (Plate 5.1). Many ligaments are also made up of similar tissue, but in them there may be different layers in which fibres run in somewhat different directions. A similar arrangement is also seen in sheets of deep fascia, intermuscular septa, aponeuroses, the central tendon of the diaphragm, the fibrous pericardium and dura mater.

In other situations the collagen bundles do not show such a regular arrangement, but interlace in various directions forming ***dense irregular tissue*** (Plate 5.2). Such tissue is found

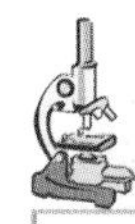

PLATE 5.1: Dense Regular Collagenous Connective Tissue (Tendon)

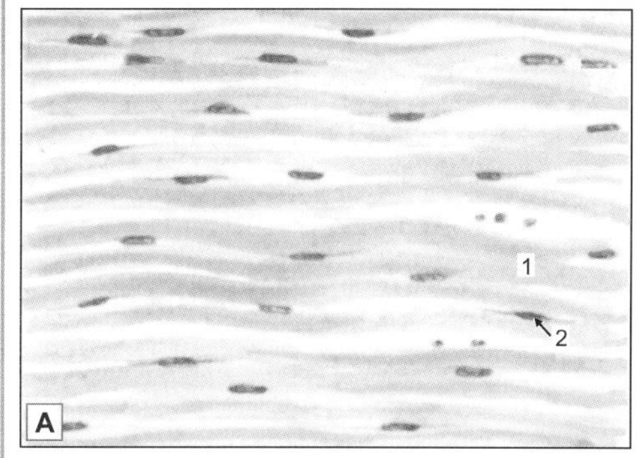

It is characterised by:

- Presence of collagen fibres, (or fibre bundles) arranged in orderly fashion parallel to each other
- Nuclei of some cells (mainly fibroblasts) are seen between the bundles of collagen. They are elongated (elliptical)
- Ground substance is less in amount.

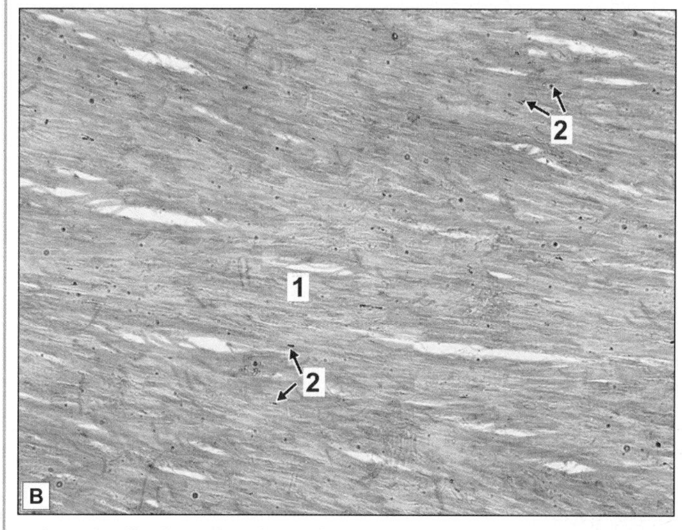

Key

1. Collagen fibres
2. Nuclei of fibroblast

Longitudinal section through a tendon A. As seen in drawing; B. Photomicrograph

in the dermis, connective tissue sheaths of muscles and nerves, capsules of glands, the sclera, the periosteum, and the adventitia of blood vessels.

Tendons

Tendons are composed of collagen fibres that run parallel to each other. The fibres are arranged in the form of bundles (Fig. 5.14). These bundles are united by areolar tissue, which contains numerous fibroblasts. In longitudinal sections

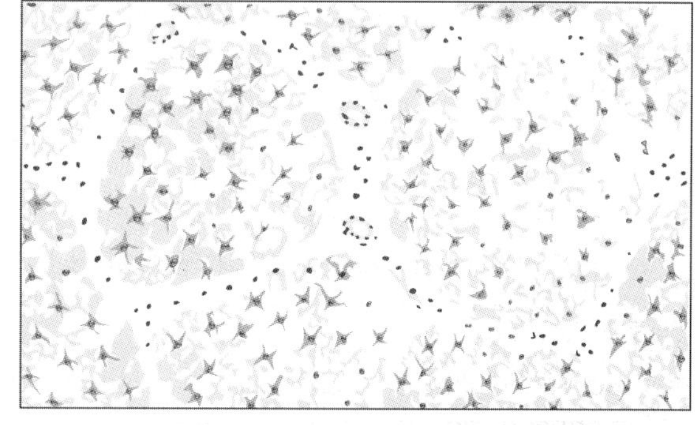

Fig. 5.14: Transverse section through tendon (Schematic representation)

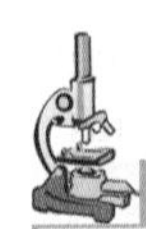

PLATE 5.2: Dense Irregular Connective Tissue (Dermis of Skin)

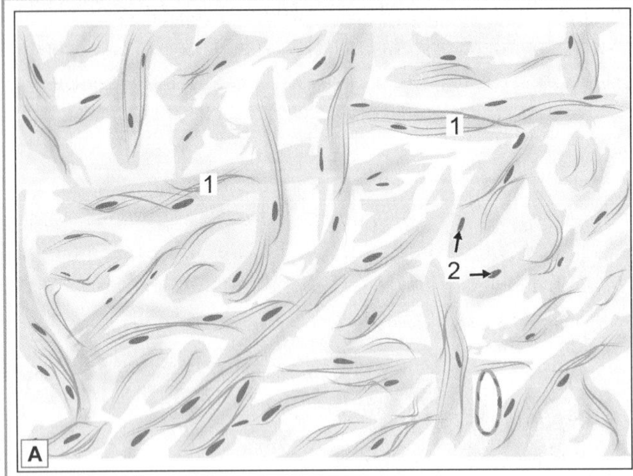

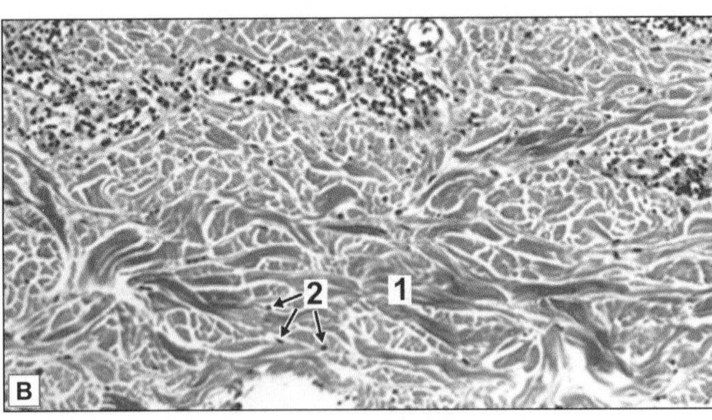

Dense irregular connective tissue as seen in dermis of skin.
A. Drawing; B. Photomicrograph

It is characterised by:

- Irregularly arranged bundles of collagen fibres that stain pink. In stretch preparation they are seen in wavy bundles. Other fibres present (elastic, reticular) can be seen only with special stains
- Few cells (fibroblast) and less ground substance.

Key

1. Collagen fibres
2. Nuclei of fibroblast

through a tendon (Plate 5.1A) the fibroblasts, and their nuclei, are seen to be elongated. In transverse sections, the fibroblasts are stellate (Fig. 5.14). Tendons serve to concentrate the pull of a muscle on a relatively small area of bone. By curving around bony pulleys, or under retinacula, they allow alterations in the direction of pull. Tendons also allow the muscle mass to be placed at a convenient distance away from its site of action (Imagine what would happen if there were no tendons in the fingers!).

Connective Tissue with Special Properties

Elastic Tissue

It is specialised dense connective tissue formed by elastic fibres. In contrast to fibrous tissue, which appears white elastic tissue is yellow in colour. Some ligaments are made up of elastic tissue. These include the ligamentum nuchae (on the back of the neck); and the ligamenta flava (which connect the laminae of adjoining vertebrae). The vocal ligaments (of the larynx) are also made up of elastic fibres. Elastic fibres are numerous in membranes that are required to stretch periodically. For example the deeper layer of superficial fascia covering the anterior abdominal wall has a high proportion of elastic fibres to allow for distension of the abdomen.

Elastic fibres may fuse with each other to form sheets (usually fenestrated). Such sheets form the main support for the walls of large arteries (e.g., the aorta) . In smaller arteries they form the internal elastic lamina.

Reticular Tissue

It is loose connective tissue made up of reticular fibres. In many situations (e.g., lymph nodes, glands) these fibres form supporting networks for the cells (Fig. 5.4). In some situations (bone marrow, spleen, lymph nodes) the reticular network is closely associated with ***reticular cells***. Most of these cells are fibroblasts, but some may be macrophages.

Mucoid Tissue

In contrast to all the connective tissues the most conspicuous component of mucoid tissue is a jelly-like ground substance rich in hyaluronic acid. Scattered through this ground substance there are star-shaped fibroblasts, some delicate collagen fibres and some rounded cells (Fig. 5.15). This kind of tissue is found in the umbilical cord. The vitreous of the eyeball is a similar tissue.

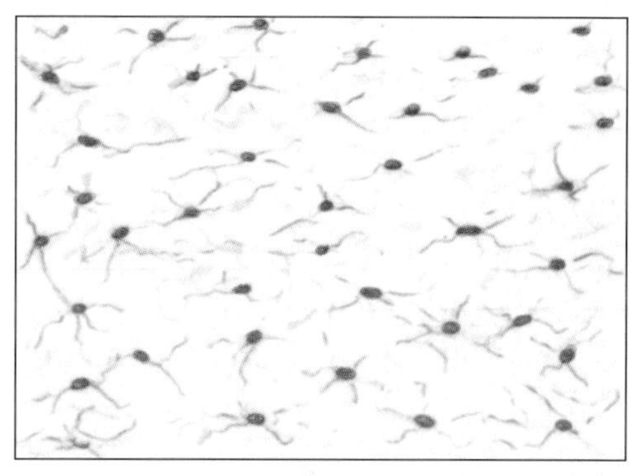

Fig. 5.15: Mucoid tissue (Schematic representation)

Adipose Tissue

Adipose tissue is basically an aggregation of fat cells also called adipocytes. It is found subcutaneously throughout the body except over the eyelid, auricle, penis and scrotum.

Distribution

- It is present in the superficial fascia over most of the body. This subcutaneous layer of fat is called the ***panniculus adiposus***. It is responsible for giving a smooth contour to the skin. However, fat is not present in the superficial fascia of the eyelids, the scrotum and the penis. The distribution of subcutaneous fat in different parts of the body is also different in the male and female and is responsible (to a great extent) for the differences in body contours in the two sexes. In women it forms a thicker and more even layer: this is responsible for the soft contours of the female body. Subcutaneous fat is not present in animals that have a thick coat of fur.
- Adipose tissue fills several hollow spaces in the body. These include the orbits, the axillae and the ischiorectal fossae. In the adult much of the space in marrow cavities of long bones is filled by fat in the form of yellow bone marrow. Much fat is also present in synovial folds of many joints filling spaces that would otherwise have been empty during certain phases of movement.
- Fat is present around many abdominal organs, specially the kidneys (***perinephric fat***).
- Considerable amounts of fat may be stored in the greater omentum, and in other peritoneal folds.

PLATE 5.3: Adipose Tissue

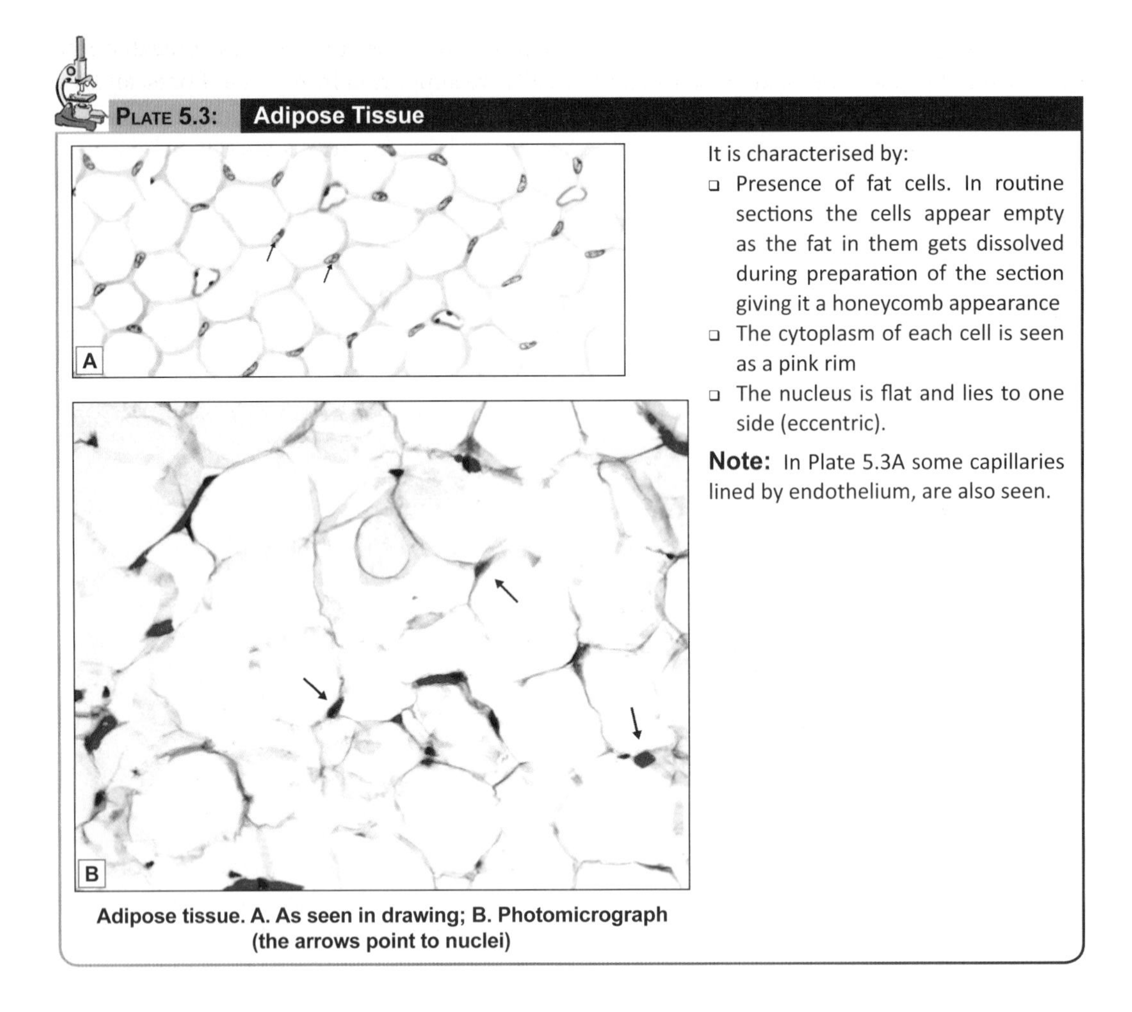

Adipose tissue. A. As seen in drawing; B. Photomicrograph (the arrows point to nuclei)

It is characterised by:

- Presence of fat cells. In routine sections the cells appear empty as the fat in them gets dissolved during preparation of the section giving it a honeycomb appearance
- The cytoplasm of each cell is seen as a pink rim
- The nucleus is flat and lies to one side (eccentric).

Note: In Plate 5.3A some capillaries lined by endothelium, are also seen.

Structure

Fat cells can be seen easily by spreading out a small piece of fresh omentum taken from an animal, on a slide. They are best seen in regions where the layer of fat is thin. The fat content can be brightly stained by using certain dyes (Sudan III, Sudan IV) (Fig. 5.16).

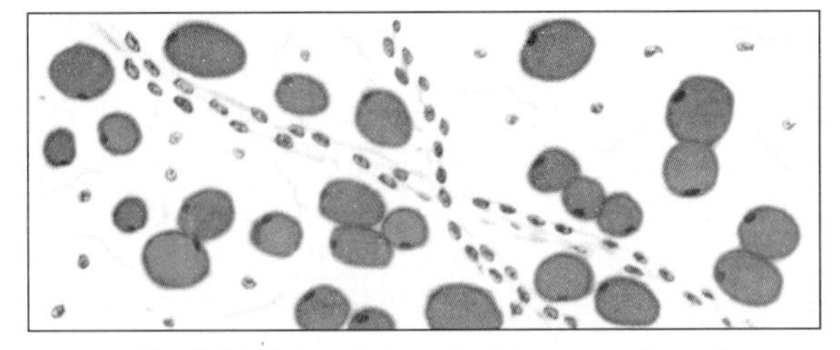

Fig. 5.16: Fat cells in a stretch preparation of omentum stained with a specific stain for fat (Sudan IV) (Schematic representation)

During the preparation of usual class room slides, the tissues have to be treated with fat solvents (like xylene or benzene) which dissolve out the fat, so that in such preparations fat cells look like rounded empty spaces (Plate 5.3). The fat content of the cells can be preserved by cutting sections after freezing the tissue (frozen sections): in this process the tissue is not exposed to fat solvents.

Fat cells may be scattered singly in some situations, but they are usually aggregated into groups that form lobules of adipose tissue. The cells are supported by reticular fibres, and the lobules are held together by areolar tissue.

Adipose tissue is richly supplied with blood, and is rich in enzyme systems.

Functions

- It acts as a store house of nutrition, fat being deposited when available in excess; and being removed when deficient in the diet.
- In many situations fat performs a mechanical function. The fat around the kidneys keeps them in position. If there is a sudden depletion of this fat the kidneys may become mobile (***floating kidney***). The fat around the eyeball performs an important supporting function and allows the eyeball to move smoothly. In the palms and soles, and over the buttocks fat has a cushioning effect protecting underlying tissues from pressure. In such areas adipose tissue may contain many elastic fibres.

  It has been observed that in situations in which the presence of fat serves an important mechanical function, this fat is the last to be depleted in prolonged starvation.
- The subcutaneous fat has been regarded as an insulation against heat loss, and would certainly perform this function if the layer of adipose tissue is thick. This may be one reason why girls (who have a thicker layer of subcutaneous fat) feel less cold than boys at the same temperature. The whale (a warm blooded mammal) can survive in very cold water because it has a very thick layer of subcutaneous fat.
- Some workers feel that adipose tissue contributes to warmth, not so much by acting as an insulator, but by serving as a heat generator. The heat generated can be rapidly passed on to neighbouring tissues because of the rich blood supply of adipose tissue.

Production

Some earlier workers regarded fat cells to be merely fibroblasts that had accumulated fat in their cytoplasm. They believed that after the fat was discharged the fat cell reverted to a fibroblast. However, most authorities now believe that fat cells are derived from specific cells (***lipoblasts***) arising during development from undifferentiated mesenchymal cells: they regard adipose tissue to be a specialised tissue. Some observations that point in this direction are as follows:

- When an animal puts on fat it is because of an increase in the size of fat cells rather than an increase in their number.
- Mature adipose tissue does not appear to have any capacity for regeneration. If a pad of fat is partially excised compensatory hypertrophy cannot be observed in the remaining part.
- In man the fat in fats cells is in the form of triglyceride: it consists mainly of oleic acid, and of smaller amounts of lineolic and palmitic acids. The composition of fat differs from species to species and is influenced by diet. The esterification of triglyceride results in the liberation of large amounts of heat.

Types

Adipose tissue is of following two types:

i. Yellow (white) or unilocular adipose tissue (adult type)

ii. Brown or multilocular adipose tissue (embryonic type).

The yellow adipose tissue is the adult type which has been described in detail above.

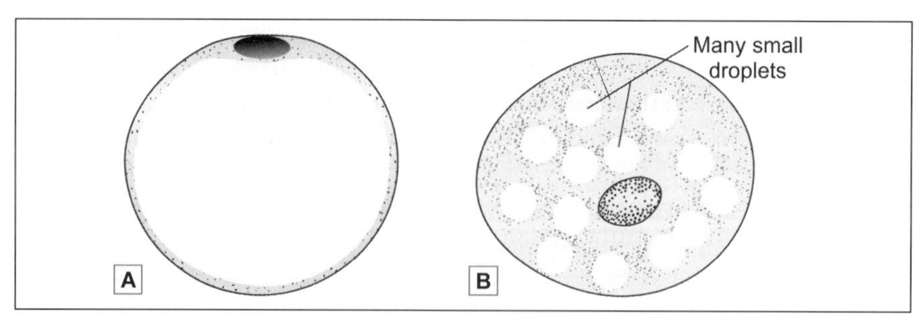

Fig. 5.17: A. Unilocular adipose tissue (white adipose tissue cell); **B.** Multilocular adipose tissue (brown adipose tissue cell) (Schematic representation)

Brown Adipose Tissue

In some parts of the body adipose tissue has a brownish colour (in distinction to the yellowish colour of ordinary fat). The cells in this type of tissue differ from those in ordinary adipose tissue as follows:

- They are smaller than in typical adipose tissue.
- The fat in the cytoplasm occurs in the form of several small droplets. Hence brown fat is also called ***multilocular adipose tissue*** (while the typical variety is described as ***unilocular adipose tissue***) (Fig. 5.17).
- The cytoplasm and nucleus of the cell are not pushed to the periphery. The cytoplasm contains numerous mitochondria (which are few in typical fat cells).

Brown adipose tissue is abundant in the newborn, but most of it is lost during childhood. Brown fat is also abundant in hibernating animals in whom it serves mainly as a heat generator when the animal comes out of hibernation.

Other Connective Tissues

Bone and cartilage are regarded as forms of connective tissue as the cells in them are widely separated by intercellular substance. The firmness of cartilage, and the hardness of bone, is because of the nature of the ground substance in them. Blood is also included amongst connective tissues as the cells are widely dispersed in a fluid intercellular substance, the plasma.

SUMMARY OF THE FUNCTIONS OF CONNECTIVE TISSUE

Mechanical Functions

- In the form of loose connective tissue, it holds together structures like skin, muscles, blood vessels, etc. It binds together various layers of hollow viscera. In the form of areolar tissue and reticular tissue it forms a framework that supports the cellular elements of various organs like the spleen, lymph nodes, and glands, and provides capsules for them.
- The looseness of areolar tissue facilitates movement between structures connected by it. The looseness of superficial fascia enables the movement of skin over deep fascia. In hollow organs this allows for mobility and stretching.
- In the form of deep fascia connective tissue provides a tight covering for deeper structures (specially in the limbs and neck) and helps to maintain the shape of these regions.
- In the form of ligaments it holds bone ends together at joints.

- In the form of deep fascia, intermuscular septa and aponeuroses, connective tissue provides attachment for the origins and insertions of many muscles.
- In the form of tendons it transmits the pull of muscles to their insertion.
- Thickened areas of deep fascia form retinacula that hold tendons in place at the wrist and ankle.
- Both areolar tissue and fascial membranes provide planes along which blood vessels, lymphatics, and nerves travel. The superficial fascia provides passage to vessels and nerves going to the skin, and supports them.
- In the form of dura mater it provides support to the brain and spinal cord.

Other Functions

- In the form of adipose tissue it provides a store of nutrition. In cold weather the fat provides insulation and helps to generate heat.
- Because of the presence of cells of the immune system (macrophages and plasma cells), connective tissue helps the body to fight against invading foreign substances (including bacteria) by destroying them, or by producing antibodies against them.
- Because of the presence of fibroblasts connective tissue helps in laying down collagen fibres necessary for wound repair.
- By virtue of the presence of undifferentiated mesenchymal cells connective tissue can help in regeneration of tissues (e.g., cartilage and bone) by providing cells from which specialised cells can be formed.
- Deep fascia plays a very important role in facilitating venous return from the limbs (specially the lower limbs). When muscles of the limb contract, increase in their thickness is limited by the deep fascia. As a result, veins deep to the fascia are pressed upon. Because of the presence of valves in the veins, this pressure causes blood to flow towards the heart. In this way deep fascia enables muscles to act as pumps that push venous blood towards the heart.

Clinical Correlation

Some Diseases of Connective Tissue

Mutations in genes that are responsible for production of collagen can lead to a number of diseases. The main feature of these diseases is that there is reduced strength in the tissues concerned. Collagen plays an important role in giving strength to bone. When collagen is not properly formed bones are weak and break easily. This condition is called ***osteogenesis imperfecta***.

In other collagen diseases the skin may become abnormally extensible, and joints may be lax (because of improperly formed ligaments ***(Ehlers-Danlos syndrome)***.

Mutations in genes coding for fibrillin can result in abnormalites in organs where elastic fibres play an important role. For example, there may be subluxation of the lens (due to weakness of the suspensory ligament). The tunica media of the aorta may be weak and this can lead to rupture of the vessel. It appears that fibrillin has something to do with the control of bone growth, and when fibrillin is deficient the person becomes abnormally tall. The features mentioned above constitute ***Marfan's syndrome***.

Chapter 6

Cartilage

Cartilage is a tissue that forms the 'skeletal' basis of some parts of the body, e.g., the auricle of the ear, or the lower part of the nose. Feeling these parts readily demonstrates that while cartilage is sufficiently firm to maintain its form, it is not rigid like bone. It can be bent, returning to its original form when the bending force is removed.

Cartilage is considered to be a modified connective tissue. It resembles ordinary connective tissue in that the cells in it are widely separated by a considerable amount of intercellular material or ***matrix***. The latter consists of a homogeneous ***ground substance*** within which fibres are embedded (Some authorities use the term matrix as an equivalent of ground substance, while others include embedded fibres under the term). Cartilage differs from typical connective tissue mainly in the nature of the ground substance. This is firm and gives cartilage its characteristic consistency. Three main types of cartilage can be recognised depending on the number and variety of fibres in the matrix. These are ***hyaline cartilage***, ***fibrocartilage*** and ***elastic cartilage***.

As a rule the free surfaces of hyaline and elastic cartilage are covered by a fibrous membrane called the ***perichondrium***, but fibrocartilage is not.

Before we consider the features of individual varieties of cartilage we shall examine some features of cartilage cells and of the matrix.

GENERAL FEATURES OF CARTILAGE

- Cartilage is derived (embryologically) from mesenchyme. Some mesenchymal cells differentiate into cartilage forming cells or ***chondroblasts***. Chondroblasts produce the intercellular matrix as well as the collagen fibres that form the intercellular substance of cartilage. Chondroblasts that become imprisoned within this matrix become chondrocytes. Some mesenchymal cells that surround the developing cartilage form the ***perichondrium***. Apart from collagen fibres and fibroblasts, the perichondrium contains cells that are capable of transforming themselves into cartilage cells when required.
- Cartilage has very limited ability for regeneration (after destruction by injury or disease). Defects in cartilage are usually filled in by fibrous tissue.
- During fetal life, cartilage is much more widely distributed than in the adult. The greater part of the skeleton is cartilaginous in early fetal life. The ends of most long bones are cartilaginous at the time of birth, and are gradually replaced by bone. The replacement is completed only

after full growth of the individual (i.e., by about 18 years of age). Replacement of cartilage by bone is called ***ossification***. Ossification of cartilage has to be carefully distinguished from ***calcification***, in which the matrix hardens because of the deposition in it of calcium salts, but true bone is not formed.

- Cartilage is usually described as an avascular tissue. However, the presence of ***cartilage canals***, through which blood vessels may enter cartilage, is well documented. Each canal contains a small artery surrounded by numerous venules and capillaries. Cartilage cells receive their nutrition by diffusion from vessels in the perichondrium or in cartilage canals. Cartilage canals may also play a role in the ossification of cartilage by carrying bone forming cells into it.
- ***Growth of cartilage:*** Newly formed cartilage grows by multiplication of cells throughout its substance. This kind of growth is called ***interstitial growth***. Interstitial growth is possible only as long as the matrix is sufficiently pliable to allow movement of cells through it. As cartilage matures the matrix hardens and the cartilage cells can no longer move widely apart; in other words interstitial growth is no longer possible. At this stage, when a cartilage cell divides the daughter cells remain close together forming cell nests. Further growth of cartilage now takes place only by addition of new cartilage over the surface of existing cartilage; this kind of growth is called ***appositional growth***. It is possible because of the presence of cartilage forming cells in the deeper layers of the perichondrium.

COMPONENTS OF CARTILAGE

Like ordinary connective tissue, cartilage is made up of:

- Cells—chondrocytes
- Ground substance/matrix
- Fibres—collagen fibres.

Cartilage Cells

The cells of cartilage are called ***chondrocytes***. They lie in spaces (or ***lacunae***) present in the matrix. At first the cells are small and show the features of metabolically active cells. The nucleus is euchromatic. Mitochondria, endoplasmic reticulum (ER) and Golgi complex are prominent. Some authorities use the term ***chondroblasts*** for these cells (This term is used mainly for embryonic cartilage producing cells). As the cartilage cells mature they enlarge, often reaching a diameter of 40 µm or more. The nuclei become heterochromatic and organelles become less prominent. The cytoplasm of chondrocytes may also contain glycogen and lipids.

Ground Substance

The ground substance of cartilage is made up of complex molecules containing proteins and carbohydrates (proteoglycans).

The carbohydrates are chemically ***glycosaminoglycans*** (GAG). They include chondroitin sulphate, keratin sulphate and hyaluronic acid. The core protein is ***aggrecan***. The proteoglycan molecules are tightly bound. Along with the water content, these molecules form a firm gel that gives cartilage its firm consistency.

Collagen Fibres of Cartilage

The collagen fibres present in cartilage are (as a rule) chemically distinct from those in most other tissues. They are described as type II collagen. However, fibrocartilage and the perichondrium, contain the normal type I collagen.

TYPES OF CARTILAGE

Hyaline Cartilage

Hyaline cartilage is so called because it is transparent (*hyalos* = glass).

Its intercellular substance appears to be homogeneous, but using special techniques it can be shown that many collagen fibres are present in the matrix.

The fibres are arranged so that they resist tensional forces. Hyaline cartilage has been compared to a tyre. The ground substance (corresponding to the rubber of the tyre) resists compressive forces, while the collagen fibres (corresponding to the treads of the tyre) resist tensional forces.

In haematoxylin and eosin stained preparations, the matrix is stained blue, i.e., it is basophilic. However, the matrix just under the perichondrium is acidophilic (Plate 6.1).

Towards the centre of a mass of hyaline cartilage the chondrocytes are large and are usually present in groups (of two or more). The groups are formed by division of a single parent cell. The cells tend to remain together as the dense matrix prevents their separation. Groups of cartilage cells are called ***cell-nests*** (or ***isogenous cell groups***). Towards the periphery of the cartilage, the cells are small and elongated in a direction parallel to the surface. Just under the perichondrium the cells become indistinguishable from fibroblasts.

Immediately around lacunae housing individual chondrocytes, and around cell nests the matrix stains deeper than elsewhere giving the appearance of a capsule. This deep staining matrix is newly formed and is called the ***territorial matrix*** or ***lacunar capsule***. In contrast the pale staining matrix separating cell nests is the ***interstitial matrix***.

Calcification of hyaline cartilage is often seen in old people. The costal cartilages or the large cartilages of the larynx are commonly affected. In contrast to hyaline cartilage, elastic cartilage and fibrocartilage do not undergo calcification. Although articular cartilage is a variety of hyaline cartilage, it does not undergo calcification or ossification.

Pathological Correlation

With advancing age, the collagen fibres of hyaline cartilage (which are normally difficult to see) become much more prominent, so that in some places hyaline cartilage is converted to fibrocartilage. It has been said that the transformation of hyaline cartilage to fibrocartilage is one of the earliest signs of ageing in the body.

Distribution of Hyaline Cartilage

- ***Costal cartilages:*** These are bars of hyaline cartilage that connect the ventral ends of the ribs to the sternum, or to adjoining costal cartilages. They show the typical structure of hyaline cartilage described above. The cellularity of costal cartilage decreases with age.
- ***Articular cartilage:*** The articular surfaces of most synovial joints are lined by hyaline cartilage. These articular cartilages provide the bone ends with smooth surfaces between which there is very little friction. They also act as shock absorbers. Articular cartilages

Plate 6.1: Hyaline Cartilage

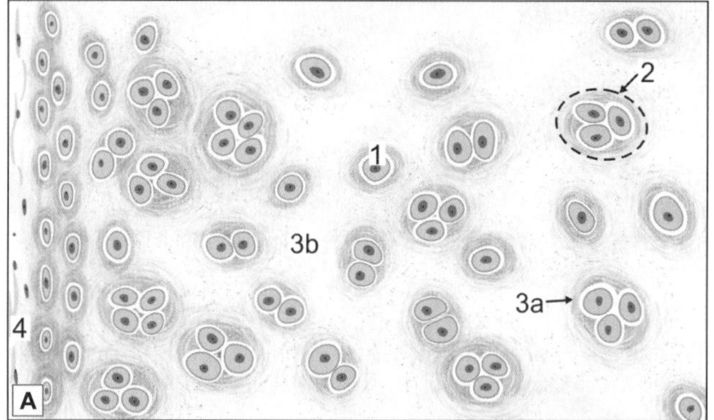

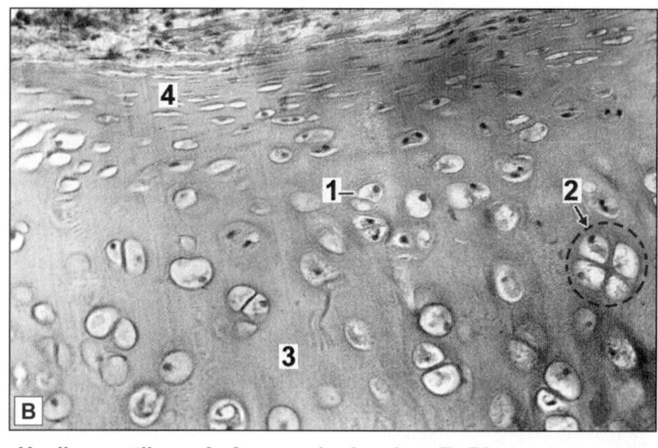

Hyaline cartilage. A. As seen in drawing; B. Photomicrograph

- Hyaline cartilage is characterized by isogenous groups of chondrocytes called as cell nest
- ***Chondrocytes*** are surrounded by a homo-geneous basophilic matrix which separates the cells widely
- Chondrocytes increase in size from periphery to centre
- Near the surface of the cartilage the cells are flattened and merge with the cells of the overlying connective tissue. This connective tissue forms the ***perichondrium***
- Perichondrium displays an outer fibrous and inner cellular layer.

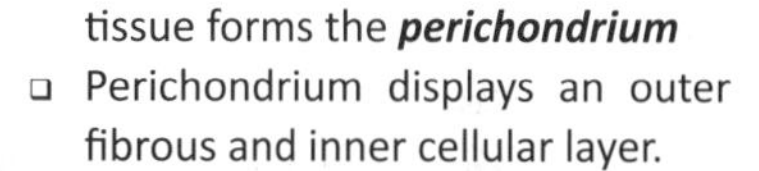

Key

1. Chondrocyte
2. Cell nest
3. Basophilic matrix
   a. Territorial matrix
   b. Interterritorial matrix
4. Perichondrium

are not covered by perichondrium. Their surface is kept moist by synovial fluid that also provides nutrition to them.

- ***Other sites where hyaline cartilage is found:***
  - The skeletal framework of the larynx is formed by a number of cartilages. Of these the thyroid cartilage, the cricoid cartilage and the arytenoid cartilage are composed of hyaline cartilage.
  - The walls of the trachea and large bronchi contain incomplete rings of cartilage. Smaller bronchi have pieces of cartilage of irregular shape in their walls.
  - Parts of the nasal septum and the lateral wall of the nose are made up of pieces of hyaline cartilage.
  - In growing children long bones consist of a bony ***diaphysis*** (corresponding to the shaft) and of one or more bony ***epiphyses*** (corresponding to bone ends or projections). Each epiphysis is connected to the diaphysis by a plate of hyaline cartilage called the ***epiphyseal plate***. This plate is essential for bone growth.

Elastic Cartilage

Elastic cartilage (or yellow fibrocartilage) is similar in many ways to hyaline cartilage.

The main difference between hyaline cartilage and elastic cartilage is that instead of collagen fibres, the matrix contains numerous elastic fibres that form a network (Plate 6.2). The fibres are difficult to see in haematoxylin and eosin stained sections, but they can be clearly visualised if special methods for staining elastic fibres are used (Fig. 6.1). The surface of elastic cartilage is covered by perichondrium.

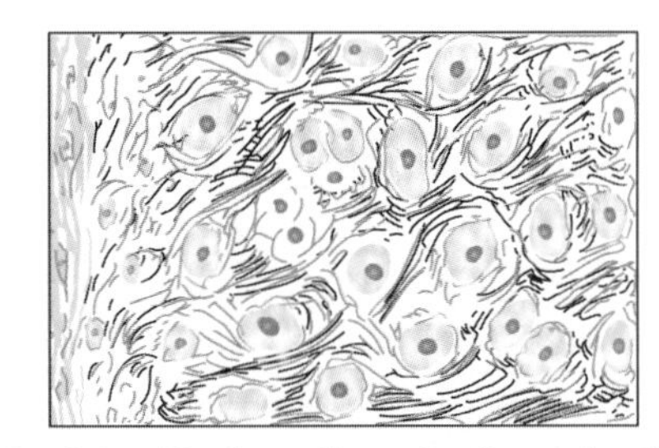

Fig. 6.1: Elastic cartilage. Section stained by Verhoeff's method in which elastic fibres are stained bluish black (Schematic representation)

Elastic cartilage possesses greater flexibility than hyaline cartilage and readily recovers its shape after being deformed.

PLATE 6.2: Elastic Cartilage

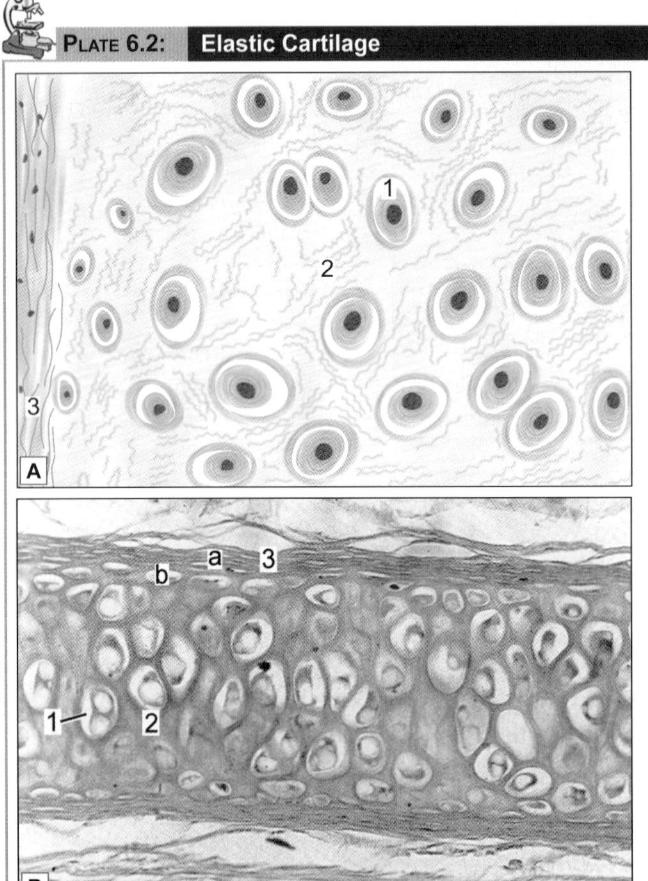

- Elastic cartilage is characterised by presence of chondrocytes within lacuna surrounded by bundles of elastic fibres
- Perichondrium is present showing an outer fibrous and inner cellular layer.

Key

1. Lacuna with chondrocytes
2. Cartilage matrix with elastic fibres
3. Perichondrium
   a. Outer fibrous layer
   b. Inner cellular layer

Elastic cartilage. A. As seen in drawing; B. Photomicrograph

Distribution of Elastic Cartilage

- It forms the 'skeletal' basis of the ***auricle*** (or pinna) and of the ***lateral part of the external acoustic meatus***.
- The wall of the ***medial part of the auditory tube*** is made of elastic cartilage.
- The ***epiglottis*** and two small ***laryngeal cartilages*** (corniculate and cuneiform) consist of elastic cartilage. The apical part of the arytenoid cartilage contains elastic fibres but the major portion of it is hyaline.

Note that all the sites mentioned above are concerned either with the production or reception of sound.

Fibrocartilage

On superficial examination this type of cartilage (also called ***white fibrocartilage***) looks very much like dense fibrous tissue (Plate 6.3). However, in sections it is seen to be cartilage because it contains typical cartilage cells surrounded by capsules. The matrix is pervaded by numerous collagen bundles amongst which there are some fibroblasts. The fibres merge with those of surrounding connective tissue.

PLATE 6.3: Fibrocartilage

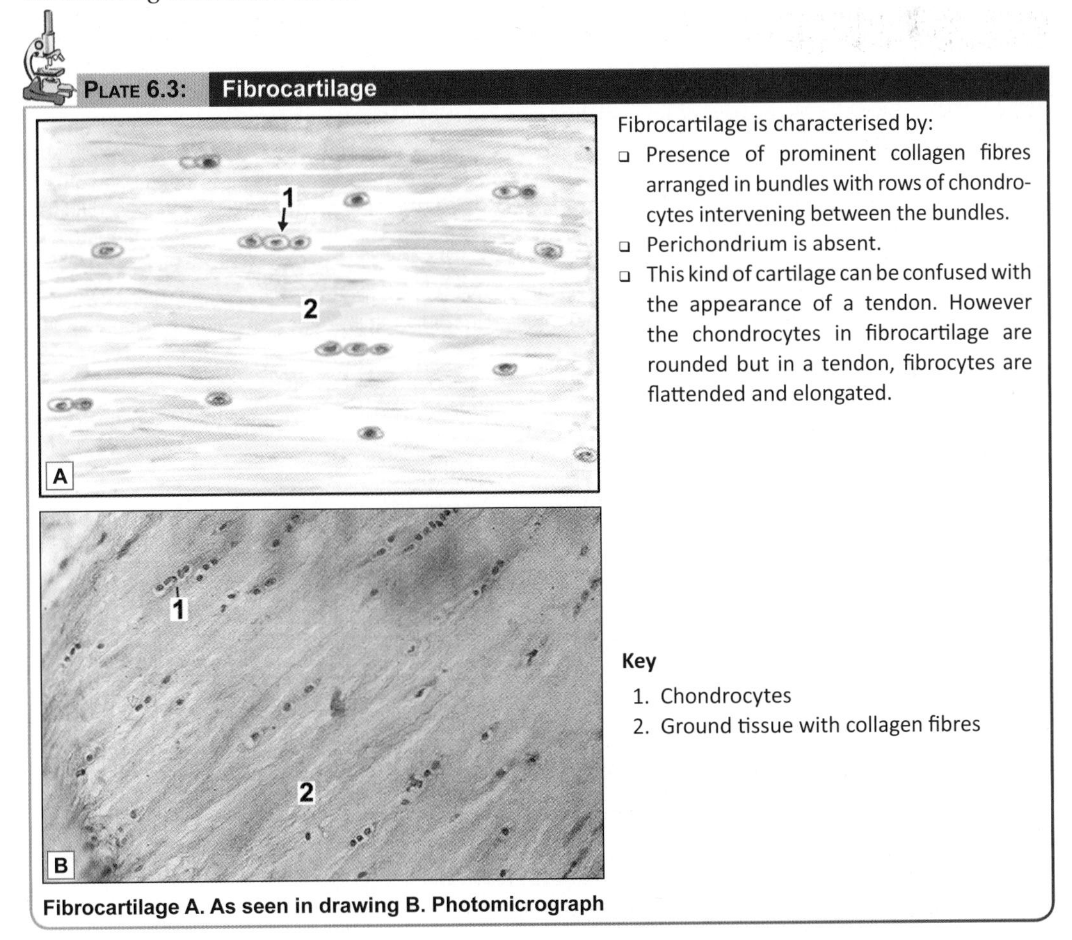

Fibrocartilage is characterised by:

- Presence of prominent collagen fibres arranged in bundles with rows of chondrocytes intervening between the bundles.
- Perichondrium is absent.
- This kind of cartilage can be confused with the appearance of a tendon. However the chondrocytes in fibrocartilage are rounded but in a tendon, fibrocytes are flattended and elongated.

Key

1. Chondrocytes
2. Ground tissue with collagen fibres

Fibrocartilage A. As seen in drawing B. Photomicrograph

There is no perichondrium over the cartilage. This kind of cartilage has great tensile strength combined with considerable elasticity. The collagen in fibrocartilage is different from that in hyaline cartilage in that it is type I collagen (identical to that in connective tissue), and not type II. The fibrocartilage in contrast to hyaline cartilage does not undergo calcification.

Distribution of Fibrocartilage

- Fibrocartilage is most conspicuous in secondary cartilaginous joints or ***symphyses***. These include the joints between bodies of vertebrae (where the cartilage forms intervertebral discs); the pubic symphysis; and the manubriosternal joint.
- In some synovial joints the joint cavity is partially or completely subdivided by an ***articular disc***. These discs are made up of fibrocartilage (Examples are discs of the temporomandibular and sternoclavicular joints, and menisci of the knee joint).
- The ***glenoidal labrum*** of the shoulder joint and the ***acetabular labrum*** of the hip joint are made of fibrocartilage.
- In some situations where tendons run in deep grooves on bone, the grooves are lined by fibrocartilage. Fibrocartilage is often present where tendons are inserted into bone.

Pathological Correlation

Cartilage-forming (Chondroblastic) Tumours

The tumours which are composed of frank cartilage or derived from cartilage-forming cells are included in this group. This group comprises benign lesions like osteocartilaginous exostoses (osteochondromas), enchondroma, chondroblastoma and chondromyxoid fibroma, and a malignant counterpart, chondrosarcoma.

- ***Osteocartilaginous Exostoses or Osteochondromas:*** These are the commonest of benign cartilage-forming lesions. Exostoses arise from metaphyses of long bones as exophytic lesions, most commonly lower femur and upper tibia and upper humerus.
- ***Enchondroma:*** Enchondroma is the term used for the benign cartilage-forming tumour that develops centrally withn the interior of the affected bone, while chondroma refers to the peripheral development of lesion.
- ***Chondroblastoma:*** Chondroblastoma is a relatively rare benign tumour arising from the epiphysis of long bones adjacent to the epiphyseal cartilage plate. Most commonly affected bones are upper tibia and lower femur and upper humerus.
- ***Chondrosarcoma:*** Chondrosarcoma is a malignant tumour of chondroblasts.

Chapter 7

Bone

Bone is a rigid form of connective tissue in which the extracelluar matrix is impregnated with inorganic salts, mainly calcium phosphate and carbonate, that provide hardness.

GENERAL FEATURES

If we examine a longitudinal section across a bone (such as the humerus), we see that the wall of the shaft is tubular and encloses a large ***marrow cavity***. The wall of the tube is made up of a hard dense material that appears, on naked eye examination, to have a uniform smooth texture with no obvious spaces in it. This kind of bone is called ***compact bone***.

When we examine the end of long bone and diploë of flat bones we find that the marrow cavity does not extend into them. They are filled by a meshwork of tiny rods or plates of bone and contain numerous spaces, the whole appearance resembling that of a sponge. This kind of bone is called ***spongy*** or ***cancellous bone*** (*cancel = cavity*). The spongy bone at the bone ends is covered by a thin layer of compact bone, thus providing the bone ends with smooth surfaces (Fig. 7.1). Small bits of spongy bone are also present over the wall of the marrow cavity.

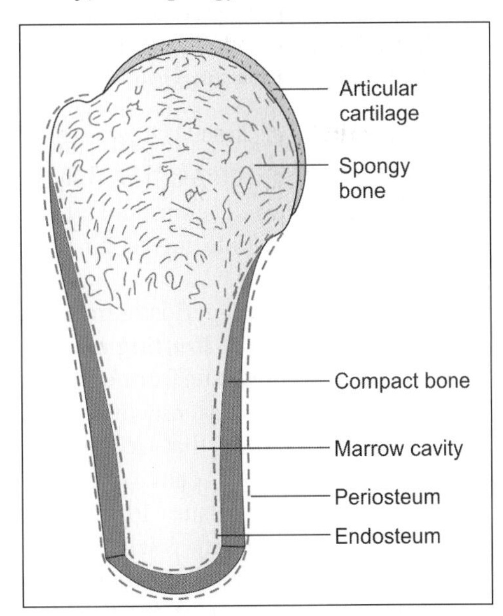

Fig. 7.1: Some features of bone structure as seen in a longitudinal section through one end of a long bone (Schematic representation)

Where the bone ends take part in forming joints they are covered by a layer of articular cartilage. With the exception of the areas covered by articular cartilage, the entire outer surface of bone is covered by a membrane called the ***periosteum***. The wall of the marrow cavity is lined by a membrane called the ***endosteum***.

The marrow cavity and the spaces of spongy bone (present at the bone ends) are filled by a highly vascular tissue called ***bone marrow***. At the bone ends, the marrow is red in colour. Apart from blood vessels this ***red marrow*** contains numerous masses of blood forming cells (***haemopoietic tissue***). In the shaft of the bone of an adult, the marrow is yellow. This ***yellow marrow*** is made up predominantly of fat cells.

Some islands of haemopoietic tissue may be seen here also. In bones of a fetus or of a young child, the entire bone marrow is red. The marrow in the shaft is gradually replaced by yellow marrow with increasing age.

THE PERIOSTEUM

The external surface of any bone is, as a rule, covered by a membrane called periosteum (Fig. 7.1).

Note: The only parts of the bone surface devoid of periosteum are those that are covered with articular cartilage.

The periosteum consists of two layers, outer and inner. The outer layer is a fibrous membrane. The inner layer is cellular. In young bones, the inner layer contains numerous osteoblasts, and is called the ***osteogenetic layer*** (This layer is sometimes described as being distinct from periosteum). In the periosteum covering, the bones of an adult osteoblasts are not conspicuous, but osteoprogenitor cells present here can form osteoblasts when need arises e.g., in the event of a fracture. Periosteum is richly supplied with blood. Many vessels from the periosteum enter the bone and help to supply it.

Functions of Periosteum

- The periosteum provides a medium through which muscles, tendons and ligaments are attached to bone. In situations where very firm attachment of a tendon to bone is necessary, the fibres of the tendon continue into the outer layers of bone as the ***perforating fibres of Sharpey*** (Fig. 7.2). The parts of the fibres that lie within the bone are ossified; they have been compared to 'nails' that keep the lamellae in place.
- Because of the blood vessels passing from periosteum into bone, the periosteum performs a nutritive function.
- Because of the presence of osteoprogenitor cells in its deeper layer, the periosteum can form bone when required. This role is very important during development. It is also important in later life for repair of bone after fracture.
- The fibrous layer of periosteum is sometimes described as a ***limiting membrane*** that prevents bone tissue from 'spilling out' into neighbouring tissues. This is based on the observation that if periosteum is torn, osteogenic cells may extend into surrounding tissue forming bony projections (***exostoses***). Such projections are frequently seen on the bones of old persons. The concept of the periosteum as a limiting membrane helps to explain how ridges and tubercles are formed on

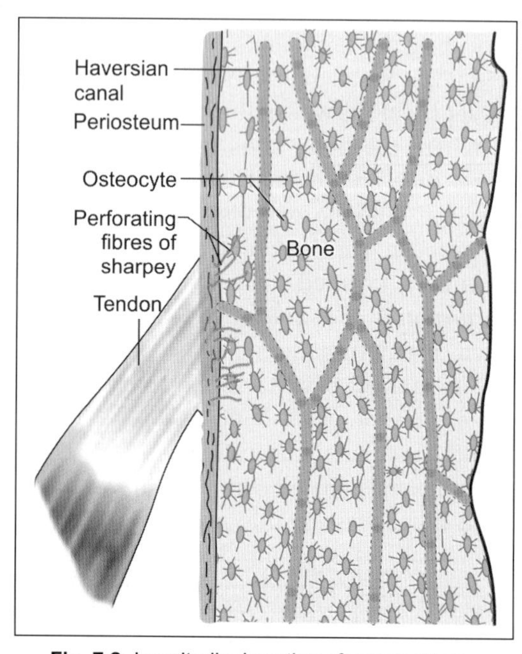

Fig. 7.2: Longitudinal section of compact bone

the surface of a bone. At sites where a tendon pulls upon periosteum, the latter tends to be lifted off from bone. The 'gap' is filled by proliferation of bone leading to the formation of a tubercle (Such views are, however, hypothetical).

ELEMENTS COMPRISING BONE TISSUE

Like cartilage, bone is a modified connective tissue. It consists of bone cells or ***osteocytes*** (Fig. 7.3) that are widely separated from one another by a considerable amount of intercellular substance. The latter consists of a homogeneous ground substance or matrix in which collagen fibres and mineral salts (mainly calcium and phosphorus) are deposited.

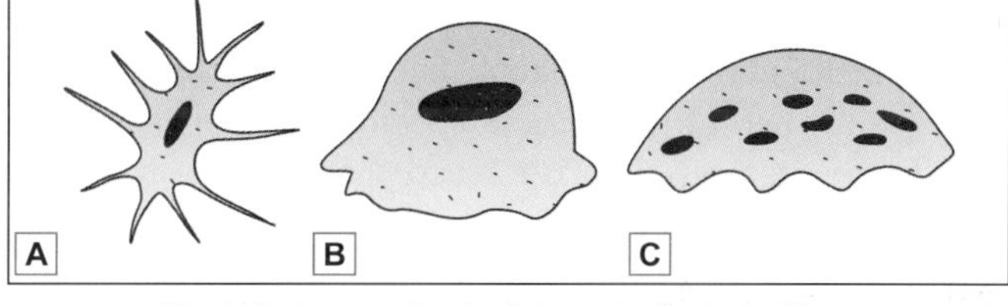

Fig. 7.3: Bone cells. **A.** Osteocyte; **B.** Osteoblast; **C.** Osteoclast (Schematic representation)

In addition to mature bone cells (osteocytes) two additional types of cells are seen in developing bone. These are bone producing cells or ***osteoblasts***, and bone removing cells or ***osteoclasts*** (Figs 7.3 and 7.4). Other cells present include ***osteoprogenitor cells*** from which osteoblasts and osteocytes are derived; cells lining the surfaces of bone; cells belonging to periosteum; and cells of blood vessels and nerves which invade bone from outside.

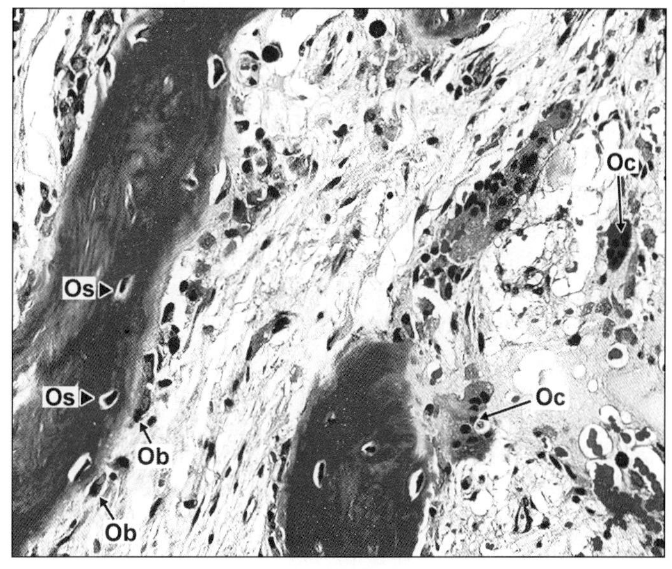

Fig. 7.4: Photomicrograph of bone cells showing osteoblast rimming the suface of the bone; multinucleated osteoclast and osteocyte Ob-osteoblast (short arrow); Oc-osteoclast (long arrow); Os-osteocyte (arrow head)

Osteoprogenitor cells

These are stem cells of mesenchymal origin that can proliferate and convert themselves into osteoblasts whenever there is need for bone formation. They resemble fibroblasts in appearance. In the fetus such cells are numerous at sites where bone formation is to take place. In the adult, osteoprogenitor cells are present over bone surfaces (on both the periosteal and endosteal aspects).

Osteoblasts

These are bone forming cells derived from osteoprogenitor cells. They are found lining growing surfaces of bone, sometimes giving an epithelium-like appearance. However, on closer examination it is seen that the cells are of varied shapes (oval, triangular, cuboidal, etc.) and that there are numerous gaps between adjacent cells.

The nucleus of an osteoblast is ovoid and euchromatic. The cytoplasm is basophilic because of the presence of abundant rough endoplasmic reticulum. This, and the presence of a well developed Golgi complex, signifies that the cell is engaged in considerable synthetic activity. Numerous slender cytoplasmic processes radiate from each cell and come into contact with similar processes of neighbouring cells.

Osteoblasts are responsible for laying down the organic matrix of bone including the collagen fibres. They are also responsible for calcification of the matrix. Alkaline phosphatase present in the cell membranes of osteoblasts plays an important role in this function. Osteoblasts are believed to shed off ***matrix vesicles*** that possibly serve as points around which formation of hydroxyapatite crystals takes place. Osteoblasts may indirectly influence the resorption of bone by inhibiting or stimulating the activity of osteoclasts.

Pathological Correlation

- A benign tumour arising from osteoblasts is called an **osteoma**.
- A malignant tumour arising from the same cells is called an **osteosarcoma**. Osteosarcomas are most commonly seen in bones adjoining the knee joint. They can spread to distant sites in the body through the blood stream.

Osteocytes

These are the cells of mature bone. They lie in the lacunae of bone, and represent osteoblasts that have become 'imprisoned' in the matrix during bone formation. Delicate cytoplasmic processes arising from osteocytes establish contacts with other osteocytes and with bone lining cells present on the surface of bone.

In contrast to osteoblasts, osteocytes have eosinophilic or lightly basophilic cytoplasm. This is to be correlated with

- The fact that these cells have negligible secretory activity
- The presence of only a small amount of endoplasmic reticulum in the cytoplasm.

Osteocytes are present in greatest numbers in young bone, the number gradually decreasing with age.

Functions

The functions attributed to osteocytes are:

- They probably maintain the integrity of the lacunae and canaliculi, and thus keep open the channels for diffusion of nutrition through bone
- They play a role in removal or deposition of matrix and of calcium when required.

Osteoclasts

These are bone removing cells. They are found in relation to surfaces where bone removal is taking place (Bone removal is essential for maintaining the proper shape of growing bone). At such locations the cells occupy pits called ***resorption bays*** or ***lacunae of Howship***. Osteoclasts are very large cells (20 to 100 μm or even more in diameter). They have numerous nuclei: up to 20 or more. The cytoplasm shows numerous mitochondria and lysosomes containing acid phosphatase. At sites of bone resorption the surface of an osteoclast shows many folds that are described as a ***ruffled membrane.*** Removal of bone by osteoclasts involves demineralisation and removal of matrix. Bone removal can be stimulated by factors secreted by osteoblasts, by macrophages, or by lymphocytes. It is also stimulated by the parathyroid hormone.

Recent studies have shown that osteoclasts are derived from monocytes of blood. It is not certain whether osteoclasts are formed by fusion of several monocytes, or by repeated division of the nucleus, without division of cytoplasm.

Bone Lining Cells

These cells form a continuous epithelium-like layer on bony surfaces where active bone deposition or removal is not taking place. The cells are flattened. They are present on the periosteal surface as well as the endosteal surface. They also line spaces and canals within bone. It is possible that these cells can change to osteoblasts when bone formation is called for (In other words many of them are osteoprogenitor cells).

Organic and Inorganic Constituents of Bone Matrix

The ground substance (or matrix) of bone consists of an organic matrix in which mineral salts are deposited.

The Organic Matrix

This consists of a ground substance in which collagen fibres are embedded. The ***ground substance*** consists of glycosaminoglycans, proteoglycans and water. Two special glycoproteins ***osteonectin*** and ***osteocalcin*** are present in large quantity. They bind readily to calcium ions and, therefore, play a role in mineralisation of bone. Various other substances including chondroitin sulphates, phospholipids and phosphoproteins are also present.

The ***collagen fibres*** are similar to those in connective tissue (Type I collagen) (They are sometimes referred to as ***osteoid collagen***). The fibres are usually arranged in layers, the fibres within a layer running parallel to one another. Collagen fibres of bone are synthesised by osteoblasts.

The matrix of bone shows greater density than elsewhere immediately around the lacunae, forming capsules around them, similar to those around chondrocytes in cartilage. The term ***osteoid*** is applied to the mixture of organic ground substance and collagen fibres (before it is mineralised).

The Inorganic Ions

The ions present are predominantly calcium and phosphorus (or phosphate). Magnesium, carbonate, hydroxyl, chloride, fluoride, citrate, sodium and potassium are also present in significant amounts. Most of the calcium, phosphate and hydroxyl ions are in the form of needle-shaped crystals that are given the name ***hydroxyapatite*** ($Ca_{10}[PO_4]_6[OH]_2$). Hydroxyapatite crystals lie parallel to collagen fibres and contribute to the lamellar appearance of bone. Some amorphous calcium phosphate is also present.

About 65% of the dry weight of bone is accounted for by inorganic salts, and 35% by organic ground substance and collagen fibres. (Note that these percentages are for dry weight of bone. In living bone about 20% of its weight is made up by water). About 85% of the total salts present in bone are in the form of calcium phosphate; and about 10% in the form of calcium carbonate. Ninety seven percent of total calcium in the body is located in bone.

The calcium salts present in bone are not 'fixed'. There is considerable interchange between calcium stored in bone and that in circulation. When calcium level in blood rises calcium is deposited in bone; and when the level of calcium in blood falls calcium is withdrawn from bone to bring blood levels back to normal. These exchanges take place under the influence of hormones (parathormone produced by the parathyroid glands, and calcitonin produced by the thyroid gland).

Pathological Correlation

- The nature of mineral salts of bone can be altered under certain conditions. If the content of fluoride ions in drinking water is high, the fluoride content of bone increases considerably. This can lead to drastic alterations in bone, including the outgrowth of numerous abnormal projections (exostoses); narrowing of foramina leading to compression of nerves, or even of the spinal cord; and abnormalities in histological structure. The condition called fluorosis is seen in many parts of India.
- Normal calcium and phosphorus are easily substituted by radioactive calcium (Ca45) or radioactive phosphorus (P32) if the latter are ingested. Calcium can also be replaced by radioactive strontium, radium or lead. Presence of radioactive substances in bone can lead to the formation of tumours and to leukaemia. All these changes occur in individuals who are exposed to radioactivity from any source: in particular from a nuclear explosion.
- Radioisotopes of calcium and phosphorus can also serve a useful purpose when used with care. If these substances are administered to an animal they get deposited wherever new bone is being formed. Areas of such deposits can be localised, in sections, by a process called autoradiography. In this way radioactive isotopes can be used for study of the patterns of bone growth.

Added Information

Role of Inorganic Salts in Providing Strength to Bone

Mineral salts can be removed from bone by treating it with weak acids. Chelating agents (e.g., ethylene diamine tetraacetic acid or EDTA) are also used. The process, called decalcification, is necessary for preparing sections of bone. When mineral salts are removed by decalcification the tissue becomes soft and pliable: a long bone like the fibula, or a rib, can even be tied into a knot. This shows that the rigidity of bone is mainly due to the presence of mineral salts in it.

Conversely, the organic substances present in bone can be destroyed by heat (as in burning). The form of the bone remains intact, but the bone becomes very brittle and breaks easily. It follows that the organic matter contributes substantially to the strength of bone. Resistance to tensile forces is mainly due to the presence of collagen fibres.

TYPES OF BONE

- **Based on histology**
  - Compact bone
  - Spongy bone
- **Based on maturity**
  - Mature/ lamellar bone
  - Immature/woven bone
- **Based on manner of development**
  - Cartilage bone
  - Membrane bone.

Lamellar Bone

When we examine the structure of any bone of an adult, we find that it is made up of layers or ***lamellae*** (Fig. 7.5). This kind of bone is called ***lamellar bone***. Each lamellus is a thin plate of bone consisting of collagen fibres and mineral salts that are deposited in a gelatinous ground substance. Even the smallest piece of bone is made up of several lamellae placed over one another. Between adjoining lamellae we see small flattened spaces or ***lacunae***.

Each lacuna contains one osteocyte. Spreading out from each lacuna there are fine canals or ***canaliculi*** that communicate with those from other lacunae (Fig. 7.6). The canaliculi are occupied by delicate cytoplasmic processes of osteocytes.

The lamellar appearance (Fig. 7.5) of bone depends mainly on the arrangement of collagen fibres. The fibres of one lamellus run parallel to each other; but those of adjoining lamellae run at varying angles to each other. The ground substance of a lamellus is continuous with that of adjoining lamellae.

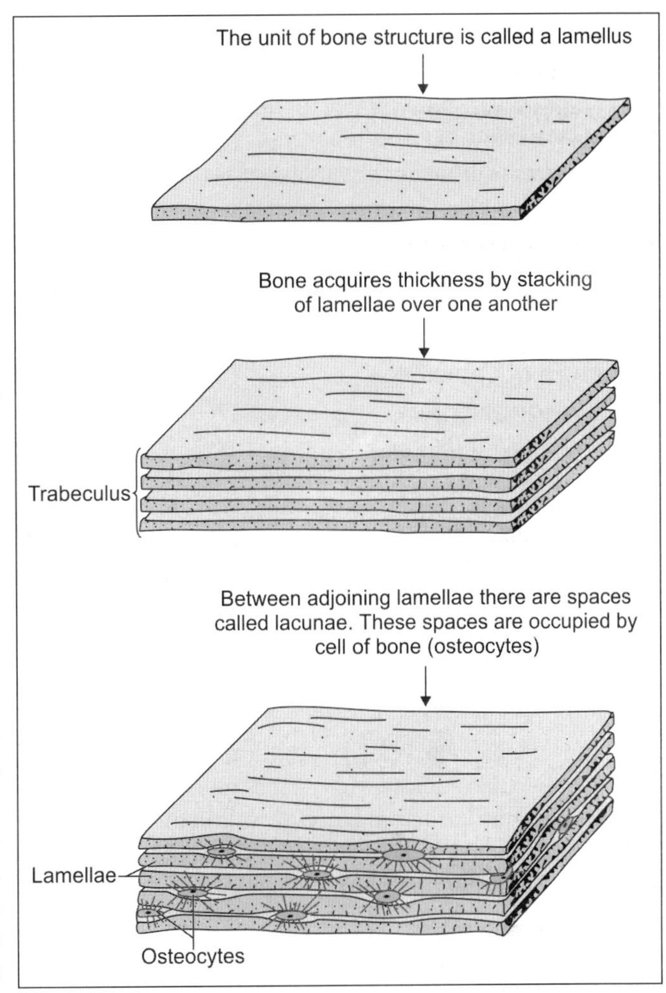

Fig. 7.5: How lamellae constitute bone (Schematic representation)

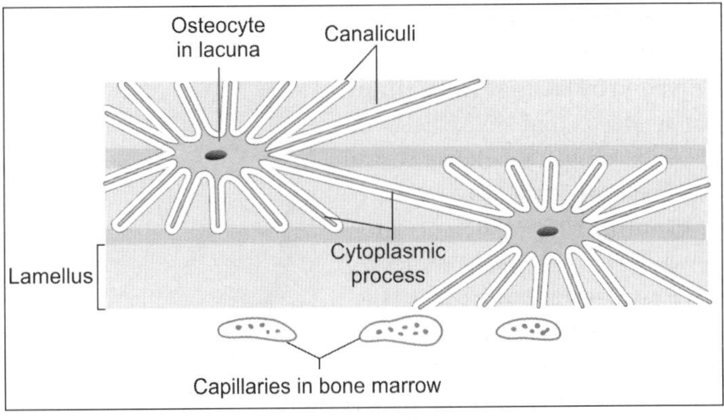

Fig. 7.6: Relationship of osteocytes to bone lamellae (Schematic representation)

Woven Bone

In contrast to mature bone, newly formed bone does not have a lamellar structure. The collagen fibres are present in bundles that appear to run randomly in different directions, interlacing with each other. Because of the interlacing of fibre bundles this kind of bone is called ***woven bone***. All newly formed bone is woven bone. It is later replaced by lamellar bone.

Clinical Correlation

Paget's disease of bone or osteitis deformans

Abnormal persistence of woven bone is a feature of Paget's disease. It was first described by Sir James Paget in 1877. Paget's disease of bone is an osteolytic and osteosclerotic bone disease of uncertain etiology involving one (monostotic) or more bones (polyostotic). The condition affects predominantly males over the age of 50 years. The bones are weak and there may be deformities.

Structure of Cancellous Bone (Spongy Bone)

Cancellous bone is made up of a meshwork of bony plates or rods called ***trabeculae***. Each trabeculus is made up of a number of lamellae (described above) between which there are lacunae containing osteocytes. Canaliculi, containing the processes of osteocytes, radiate from the lacunae (Fig. 7.7).

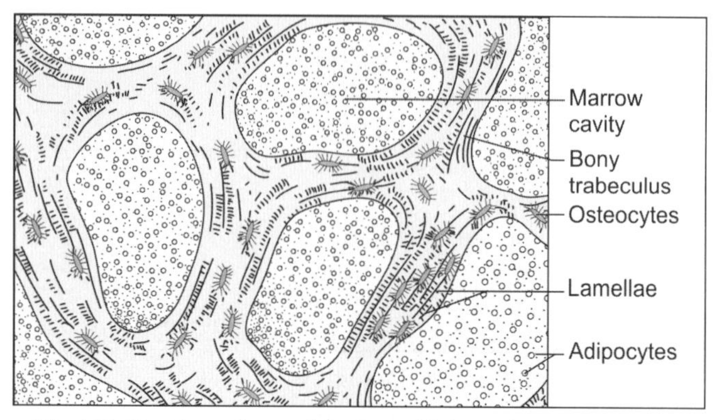

Fig. 7.7: Structure of cancellous bone (Schematic representation)

The trabeculae enclose wide spaces that are filled in by bone marrow (Fig. 7.7 and Plate 7.1). They receive nutrition from blood vessels in the bone marrow.

The trabeculae are covered externally by vascular endosteum containing osteoblasts, osteoclasts and osteoprogenitor cells.

Structure of Compact Bone

When we examine a section of compact bone we find that this type of bone is also made up of concentric lamellae, and is pervaded by lacunae (containing osteocytes), and by canaliculi (Fig. 7.8). Most of the lamellae are arranged in the form of concentric rings that surround a narrow ***Haversian canal*** present at the centre of each ring. The Haversian canal is occupied by blood vessels, nerve fibres, and some cells. One Haversian canal and the lamellae around it constitute a ***Haversian system*** or ***osteon*** (Figs 7.8 and 7.9).

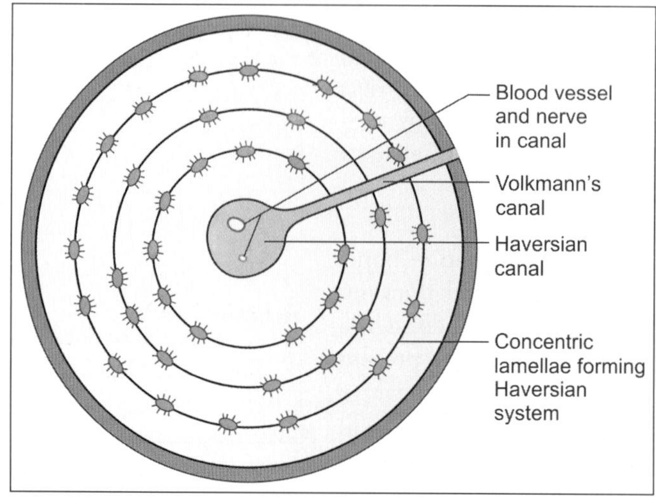

Fig. 7.8: Haversian system in compact bone (Schematic representation)

PLATE 7.1: Cancellous Bone

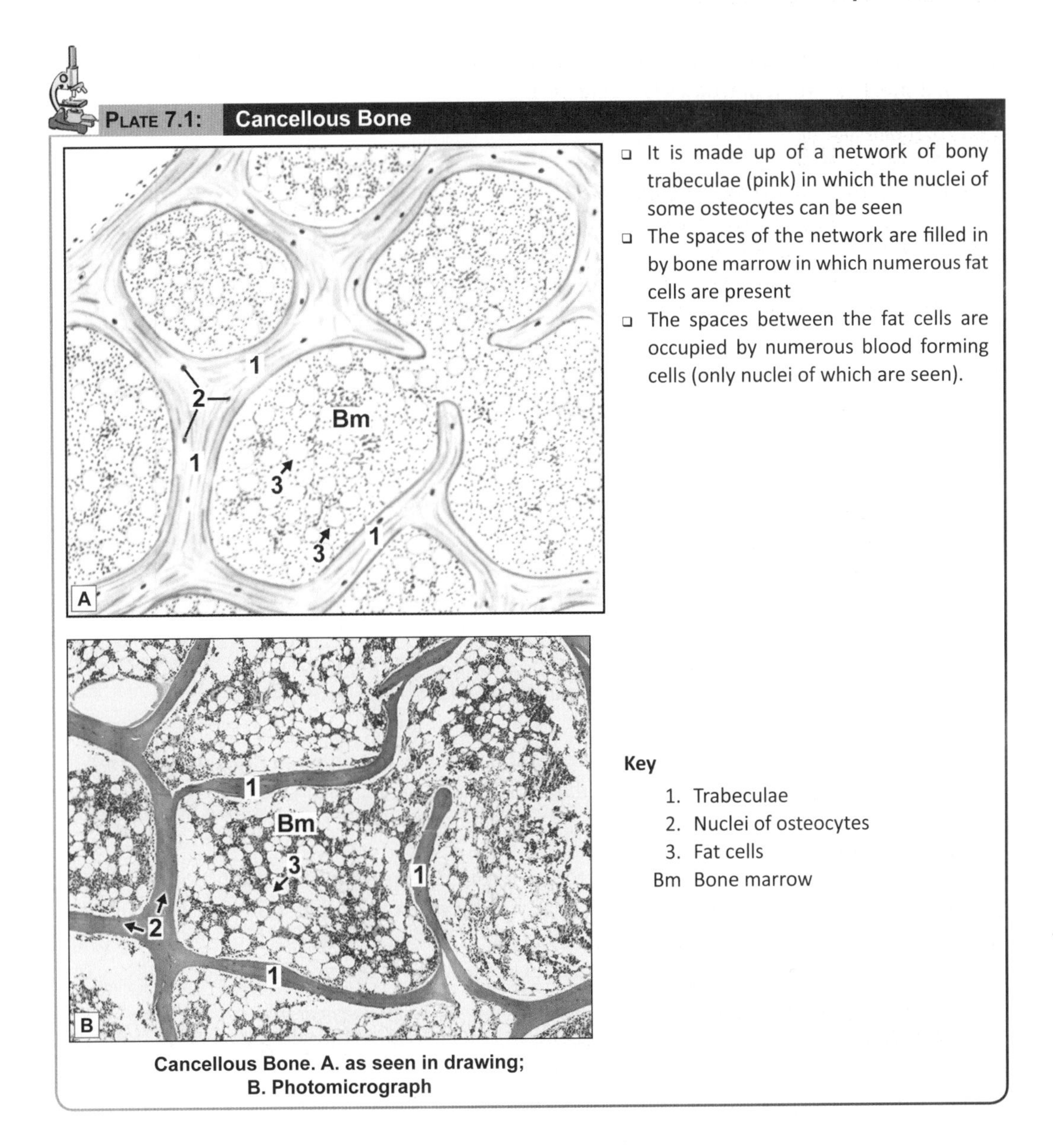

- It is made up of a network of bony trabeculae (pink) in which the nuclei of some osteocytes can be seen
- The spaces of the network are filled in by bone marrow in which numerous fat cells are present
- The spaces between the fat cells are occupied by numerous blood forming cells (only nuclei of which are seen).

Key

1. Trabeculae
2. Nuclei of osteocytes
3. Fat cells

Bm Bone marrow

Cancellous Bone. A. as seen in drawing; B. Photomicrograph

Compact bone consists of several such osteons. Between adjoining osteons there are angular intervals that are occupied by ***interstitial lamellae*** (Fig. 7.9 and Plate 7.2). These lamellae are remnants of osteons, the greater parts of which have been destroyed (as explained later). Near the surface of compact bone the lamellae are arranged parallel to the surface: these are called ***circumferential lamellae*** (Plate 7.2).

When we examine longitudinal sections through compact bone (Fig. 7.9) we find that the Haversian canals (and, therefore, the osteons) run predominantly along the length of the bone. The canals branch and anastomose with each other. They also communicate with the marrow cavity, and with the external surface of the bone through channels that are called the ***canals of***

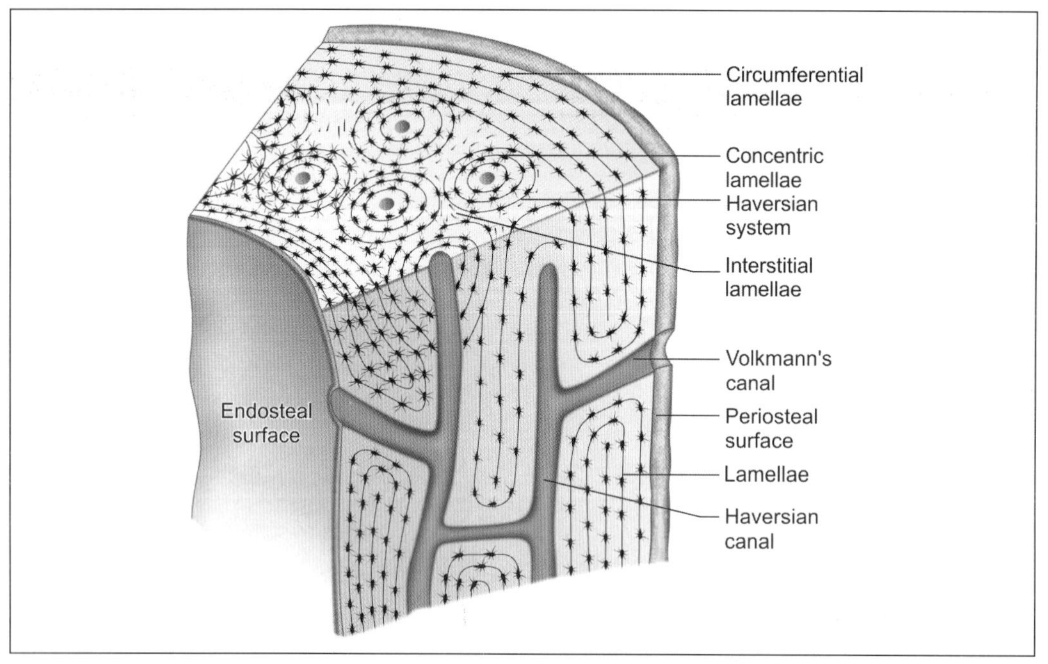

Fig. 7.9: Transverse section and longitudinal section through compact bone to show Haversian canals and the Volkmann's canal (Schematic representation)

Volkmann. Blood vessels and nerves pass through all these channels so that compact bone is permeated by a network of blood vessels that provide nutrition to it.

Comparision between cancellous and compact bone

There is an essential similarity in the structure of cancellous and compact bone. Both are made up of lamellae. The difference lies in the relative volume occupied by bony lamellae and by the spaces. In compact bone the spaces are small and the solid bone is abundant; whereas in cancellous bone the spaces are large and actual bone tissue is sparse (Plate 7.1 and 7.2).

Added Information

Osteons

During bone formation the first formed osteons do not have a clear lamellar structure, but consist of woven bone. Such osteons are described as primary osteons (or atypical Haversian systems). Subsequently, the primary osteons are replaced by secondary osteons (or typical Haversian systems) having the structure already described.

We have seen that osteons run in a predominantly longitudinal direction (i.e., along the long axis of the shaft). However, this does not imply that osteons lie parallel to each other. They may follow a spiral course, may branch, or may join other osteons. In transverse sections osteons may appear circular, oval or ellipsoid.

The number of lamellae in each osteon is highly variable. The average number is six.

In an osteon collagen fibres in one lamellus usually run either longitudinally or circumferentially (relative to the long axis of the osteon). Typically, the direction of fibres in adjoining lamellae is

Contd...

Plate 7.2: Compact Bone

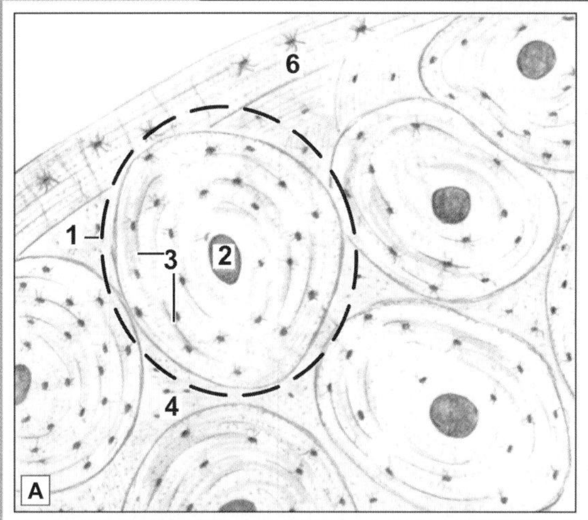

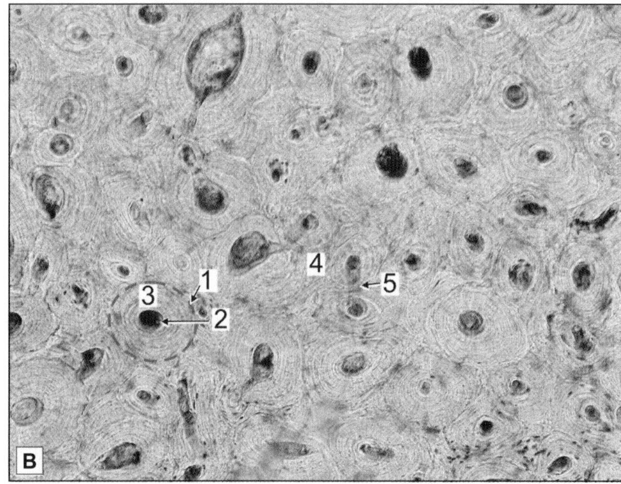

Structure of compact bone as seen in a transverse section. A. As seen in drawing; B. Photomicrograph.

- A transverse section through compact bone shows ring-like osteons (or Haversian systems)
- At the centre of each osteon there is a Haversian canal
- Around the canal there are concentric lamellae of bone amongst which there are small spaces called lacunae (containing osteocytes)
- Delicate canaliculi radiate from the lacunae; these contain cytoplasmic processes of osteocytes
- Interstitial lamellae fill intervals between Haversian systems
- Near the surface of compact bone, the lamellae are arranged in a parallel manner. These are circumferential lamellae
- Volkmann's canal interconnecting the adjacent Haversian canal may be seen.

Note: The appearance of compact bone is so characteristic that you are not likely to confuse it with any other tissue.

Key

1. Haversian system (osteon)
2. Haversian canal
3. Concentric lamellae
4. Interstitial lamellae
5. Volkmann's canal
6. Circumferential lamellae

alternately longitudinal and circumferential. It is because of this difference in direction of fibres that the lamellar arrangement is obvious in sections. It has been claimed that in fact fibres in all lamellae follow a spiral course, the 'longitudinal' fibres belonging to a spiral of a long 'pitch'; and the 'circumferential' fibres to spirals of a short 'pitch'.

The place where the periphery of one osteon comes in contact with another osteon (or with interstitial lamellae) is marked by the presence of a cement line. Along this line there are no collagen fibres, the line consisting mainly of inorganic matrix. The various lacunae within an osteon are connected with one another through canaliculi that also communicate with the Haversian canal.

Contd...

The peripheral canaliculi of the osteon do not (as a rule) communicate with those of neighbouring osteons: they form loops and turn back into their own osteon. A few canaliculi that pass through the cement line provide communications leading to interstitial lamellae.

The lacunae and canaliculi are only partially filled in by osteocytes and their processes. The remaining space is filled by a fluid that surrounds the osteocytes. This space is in communication with the Haversian canal and provides a pathway along which substances can pass from blood vessels in the Haversian canal to osteocytes.

When a transverse section of compact bone is examined with polarised light each osteon shows two bright bands that cross each other. This phenomenon is called birefringence. It is an indication of the very regular arrangement of collagen fibres (and the crystals related to them) within the lamellus.

FORMATION OF BONE

All bones are of mesodermal origin. The process of bone formation is called ***ossification***. We have seen that formation of most bones is preceded by the formation of a cartilaginous model, which is subsequently replaced by bone. This kind of ossification is called ***endochondral ossification***; and bones formed in this way are called ***cartilage bones***. In some situations (e.g., the vault of the skull) formation of bone is not preceded by formation of a cartilaginous model. Instead bone is laid down directly in a fibrous membrane. This process is called ***intramembranous ossification***; and bones formed in this way are called ***membrane bones***. The bones of the vault of the skull, the mandible, and the clavicle are membrane bones.

Intramembranous Ossification

The various stages in intramembranous ossification are as follows:

- At the site where a membrane bone is to be formed the mesenchymal cells become densely packed (i.e., a ***mesenchymal condensation*** is formed).
- The region becomes highly vascular.
- Some of the mesenchymal cells lay down bundles of collagen fibres in the mesenchymal condensation. In this way a membrane is formed.
- Some mesenchymal cells (possibly those that had earlier laid down the collagen fibres) enlarge and acquire a basophilic cytoplasm, and may now be called osteoblasts (Fig. 7.10A). They come to lie along the bundles of collagen fibres. These cells secrete a gelatinous matrix in which the fibres get embedded. The fibres also swell up. Hence the fibres can no longer be seen distinctly. This mass of swollen fibres and matrix is called ***osteoid*** (Fig.7.10B).
- Under the influence of osteoblasts calcium salts are deposited in osteoid. As soon as this happens the layer of osteoid can be said to have become one lamellus of bone (Fig. 7.10C).
- Over this lamellus, another layer of osteoid is laid down by osteoblasts. The osteoblasts move away from the lamellus to line the new layer of osteoid. However, some of them get caught between the lamellus and the osteoid (Fig. 7.10D). The osteoid is now ossified to form another lamellus. The cells trapped between the two lamellae become osteocytes (Fig. 7.10D).
- In this way a number of lamellae are laid down one over another, and these lamellae together form a trabeculus of bone (Fig. 7.10E).

The first formed bone may not be in the form of regularly arranged lamellae. The elements are irregularly arranged and form woven bone.

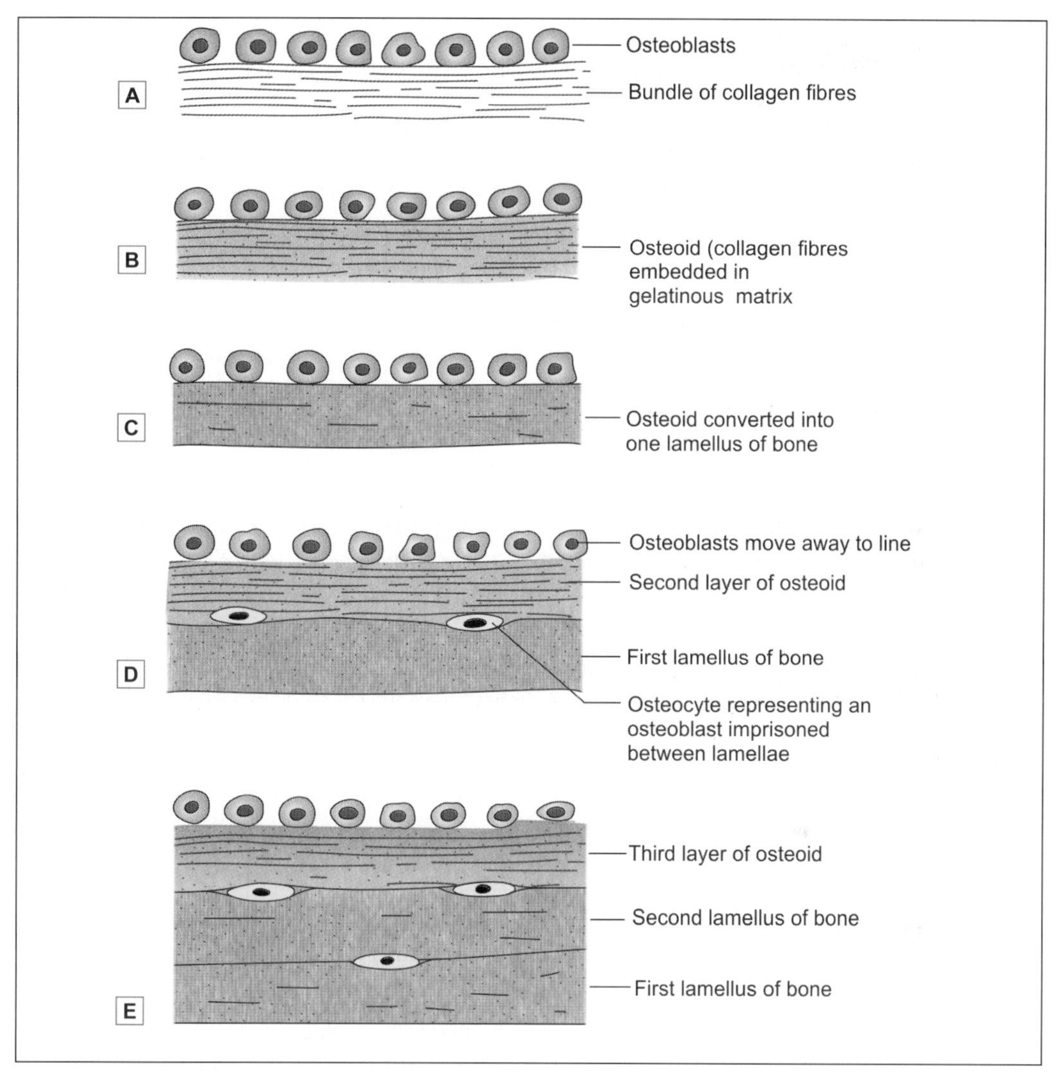

Fig. 7.10: Scheme to show how bony lamellae are laid down over one another (Schematic representation)

Endochondral Ossification

The essential steps in the formation of bone by endochondral ossification are as follows:

- At the site where the bone is to be formed, the mesenchymal cells become closely packed to form a mesenchymal condensation (Fig. 7.11A).
- Some mesenchymal cells become chondroblasts and lay down hyaline cartilage (Fig. 7.11B). Mesenchymal cells on the surface of the cartilage form a membrane called the perichondrium. This membrane is vascular and contains osteoprogenitor cells.
- The cells of the cartilage are at first small and irregularly arranged. However, in the area where bone formation is to begin, the cells enlarge considerably (Fig.7.11C).
- The intercellular substance between the enlarged cartilage cells becomes calcified, under the influence of alkaline phosphatase, which is secreted by the cartilage cells. The nutrition

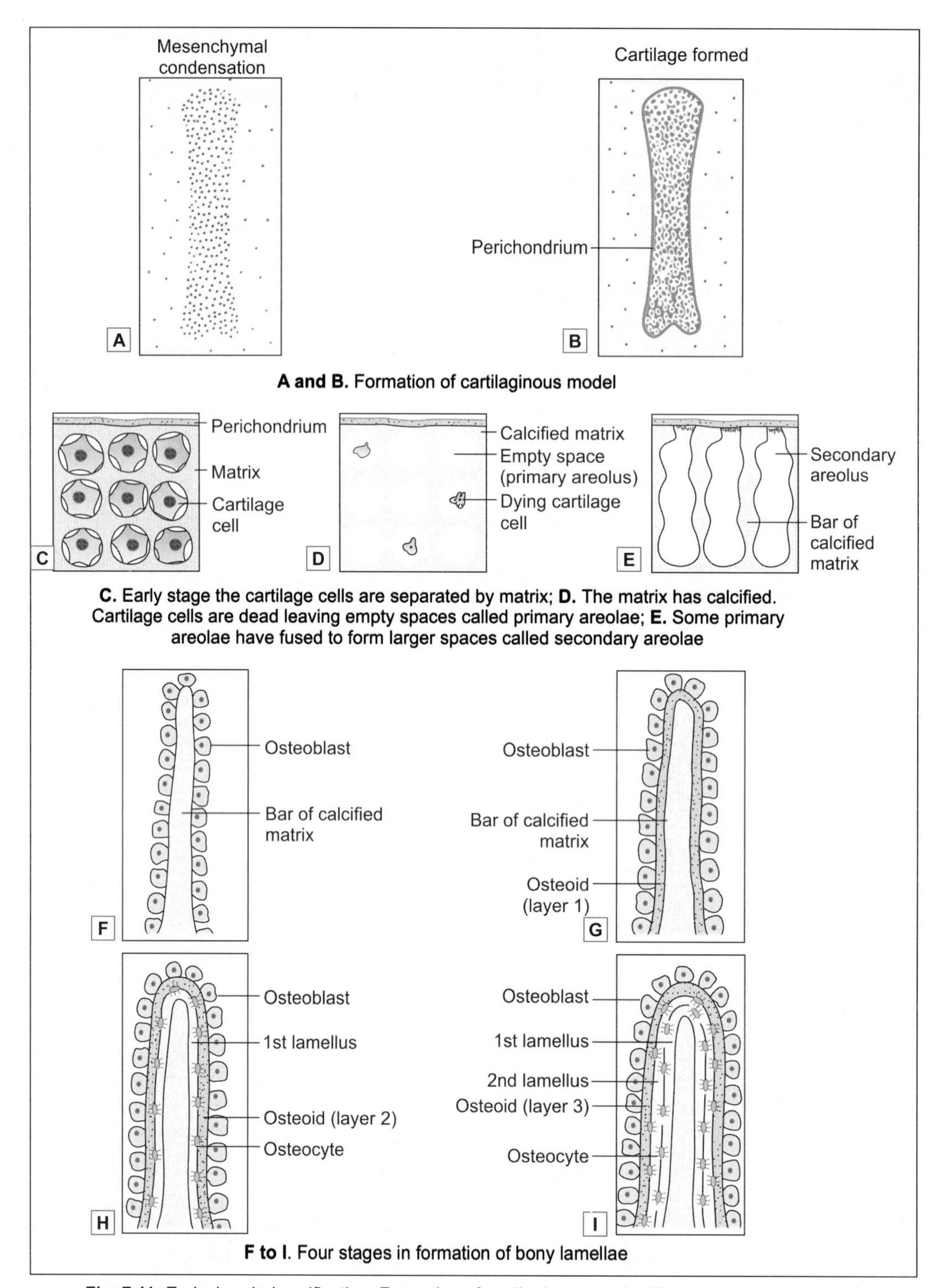

Fig. 7.11: Endochondral ossification. Formation of cartilaginous model (Schematic representation)

to the cells is thus cut off and they die, leaving behind empty spaces called ***primary areolae*** (Fig. 7.11C).

- Some blood vessels of the perichondrium (which may be called periosteum as soon as bone is formed) now invade the cartilaginous matrix. They are accompanied by osteoprogenitor cells. This mass of vessels and cells is called the ***periosteal bud***. It eats away much of the calcified matrix forming the walls of the primary areolae, and thus creates large cavities called ***secondary areolae*** (Fig. 7.11D).
- The walls of secondary areolae are formed by thin layers of calcified matrix that have not dissolved. The osteoprogenitor cells become osteoblasts and arrange themselves along the surfaces of these bars, or plates, of calcified matrix (Fig. 7.11E).
- These osteoblasts now lay down a layer of ossein fibrils embedded in a gelatinous ground substance (i.e., osteoid), exactly as in intramembranous ossification (Fig. 7.11F). This osteoid is calcified and a lamellus of bone is formed (Fig. 7.11G).
- Osteoblasts now lay down another layer of osteoid over the first lamellus. This is also calcified. Thus two lamellae of bone are formed. Some osteoblasts that get caught between the two lamellae become osteocytes. As more lamellae are laid down bony trabeculae are formed (Fig. 7.11H).
- It may be noted that the process of bone formation in endochondral ossification is exactly the same as in intramembranous ossification. The calcified matrix of cartilage only acts as a support for the developing trabeculae and is not itself converted into bone.
- At this stage the ossifying cartilage shows a central region where bone has been formed. As we move away from this area we see:
  - A region where cartilaginous matrix has been calcified and surrounds dead and dying cartilage cells
  - A zone of hypertrophied cartilage cells in an uncalcified matrix
  - Normal cartilage in which there is considerable mitotic activity.

  In this way formation of new cartilage keeps pace with the loss due to replacement by bone. The total effect is that the ossifying cartilage progressively increases in size.

Conversion of Cancellous Bone to Compact Bone

All newly formed bone is cancellous. It is converted into compact bone as follows.

Each space between the trabeculae of cancellous bone comes to be lined by a layer of osteoblasts. The osteoblasts lay down lamellae of bone as already described. The first lamellus is formed over the inner wall of the original space and is, therefore, shaped like a ring. Subsequently, concentric lamellae are laid down inside this ring thus forming an osteon. The original space becomes smaller and smaller and persists as a Haversian canal.

The first formed Haversian systems are called ***atypical Haversian systems*** or ***primary osteons***. These osteons do not have a typical lamellar structure, and their chemical composition may also be atypical.

Primary osteons are soon invaded by blood vessels and by osteoclasts that bore a new series of spaces through them. These new spaces are again filled in by bony lamellae, under the influence of osteoblasts, to form ***secondary osteons*** (or ***typical Haversian systems***). The process of formation and destruction of osteons takes place repeatedly as the bone enlarges in size; and continues even after birth. In this way the internal structure of the bone can be repeatedly remodelled to suit the stresses imposed on the bone.

Interposed in between osteons of the newest series there will be remnants of previous generations of osteons. The interstitial lamellae of compact bone represent such remnants.

When a newly created cavity begins to be filled in by lamellae of a new osteon, the first formed layer is atypical in that it has a very high density of mineral deposit. This layer can subsequently be identified as a ***cement line*** that separates the osteon from previously formed elements. As the cement line represents the line at which the process of bone erosion stops and at which the process of bone formation begins, it is also called a ***reversal line.*** From the above it will be clear why cement lines are never present around primary osteons, but are always present around subsequent generations of osteons.

HOW BONES GROW

A hard tissue like bone can grow only by deposition of new bone over existing bone i.e., by apposition. We will now consider some details of the method of bone growth in some situations.

Growth of Bones of Vault of Skull

In the bones of the vault of the skull (e.g., the parietal bone) ossification begins in one or more small areas called ***centres of ossification*** and forms following the usual process. At first it is in the form of narrow trabeculae or spicules. These spicules increase in length by deposition of bone at their ends. As the spicules lengthen they radiate from the centre of ossification to the periphery. Gradually the entire mesenchymal condensation is invaded by this spreading process of ossification and the bone assumes its normal shape. However, even at birth the radiating arrangement of trabeculae is obvious.

The mesenchymal cells lying over the developing bone differentiate to form the periosteum. The embryonic parietal bone, formed as described above, has to undergo considerable growth. After ossification has extended into the entire membrane representing the embryonic parietal bone, this bone is separated from neighbouring bones by intervening fibrous tissue (in the region of the sutures). Growth in size of the bone can occur by deposition of bone on the edges adjoining sutures (Figs. 7.12, 7.13). Growth in thickness and size of the bone also occurs when the overlying periosteum forms bone (by the process of intramembranous ossification described above) over the outer surface of the bone.

Simultaneously, there is removal of bone from the inner surface. In this way, as the bone grows in size, there is simultaneous increase in the size of the cranial cavity.

Development of a Typical Long Bone

In the region where a long bone is to be formed the mesenchyme first lays down a cartilaginous model of the bone (Figs. 7.14A to C). This cartilage is covered by perichondrium. Endochondral ossification starts in the central part of the cartilaginous model (i.e., at the centre of the future shaft).

This area is called the ***primary centre of ossification*** (Fig. 7.14D). Gradually, bone formation extends from the primary centre towards the ends of shaft. This is accompanied by progressive enlargement of the cartilaginous model.

Soon after the appearance of the primary centre, and the onset of endochondral ossification in it, the perichondrium (which may now be called periosteum) becomes active. The osteoprogenitor cells in its deeper layer lay down bone on the surface of the cartilaginous

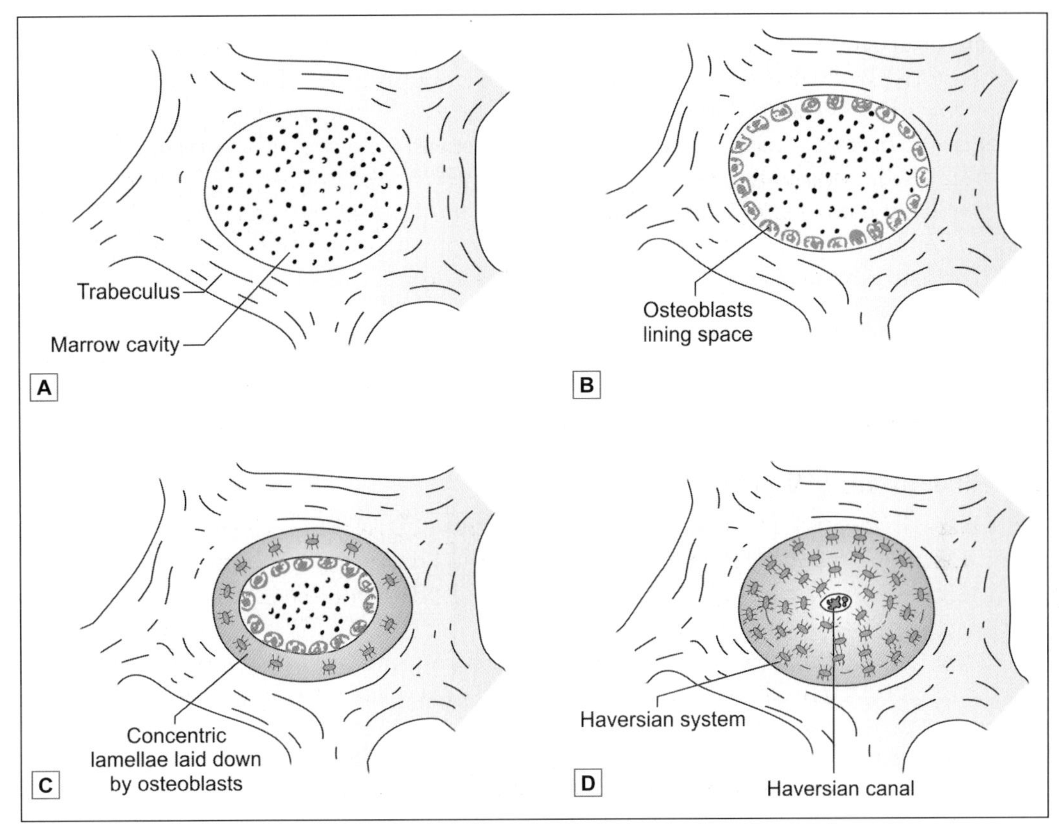

Fig. 7.12: Scheme to show how skull bones grow

model by ***intramembranous ossification***. This periosteal bone completely surrounds the cartilaginous shaft and is, therefore, called the ***periosteal collar*** (Figs. 7.14D and E).

The periosteal collar is first formed only around the region of the primary centre, but rapidly extends towards the ends of the cartilaginous model (Figs. 7.14F and G). It acts as a splint, and gives strength to the cartilaginous model at the site where it is weakened by the formation of secondary areolae. We shall see that most of the shaft of the bone is derived from this periosteal collar and is, therefore, membranous in origin.

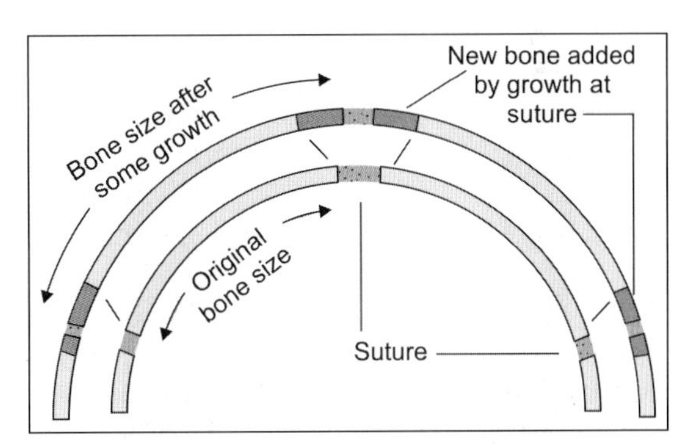

Fig. 7.13: Growth of skull bones at sutures

At about the time of birth the developing bone consists of a part called the ***diaphysis*** (or shaft), that is bony, and has been formed by extension of the primary centre of ossification, and ends that are cartilaginous (Fig. 7.14F). At varying times after birth ***secondary centres*** of endochondral ossification appear in the cartilages forming the ends of the bone (Fig. 7.14G).

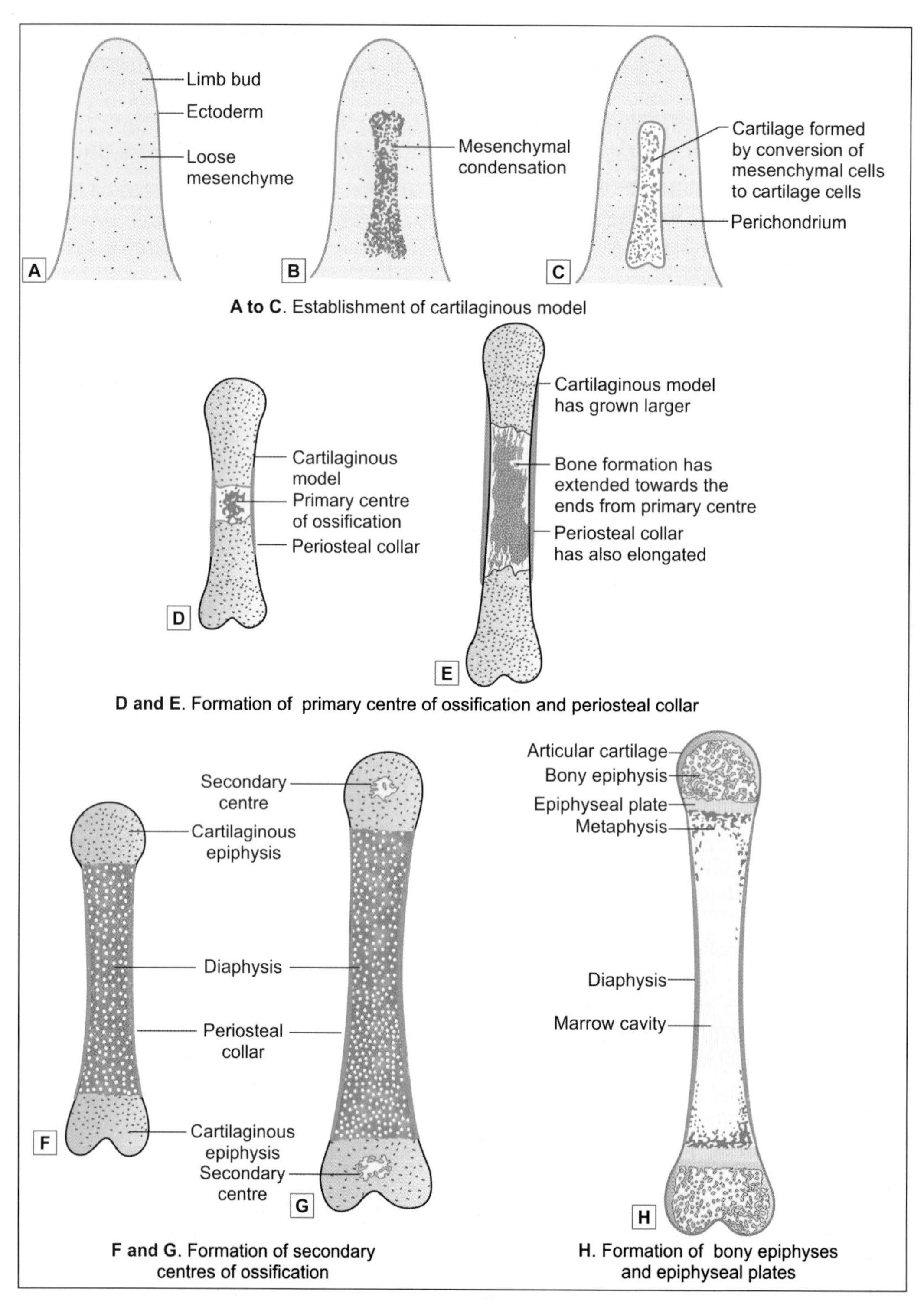

Fig. 7.14: Formation of a typical long bone (Schematic representation)

These centres enlarge until the ends become bony (Fig. 7.14H). More than one secondary centre of ossification may appear at either end. The portion of bone formed from one secondary centre is called an ***epiphysis***.

For a considerable time after birth the bone of the diaphysis and the bone of any epiphysis are separated by a plate of cartilage called the ***epiphyseal cartilage***, or ***epiphyseal plate***. This is formed by cartilage into which ossification has not extended either from the diaphysis or from the epiphysis. We shall see that this plate plays a vital role in growth of the bone.

Growth of a Long Bone

A growing bone increases both in length and in girth.

The periosteum lays down a layer of bone around the shaft of the cartilaginous model. This periosteal collar gradually extends to the whole length of the diaphysis. As more layers of bone are laid down over it, the periosteal bone becomes thicker and thicker. However, it is neither necessary nor desirable for it to become too thick. Hence, osteoclasts come to line the internal surface of the shaft and remove bone from this aspect. As bone is laid down outside the shaft it is removed from the inside. The shaft thus grows in diameter, and at the same time its wall does not become too thick. The osteoclasts also remove trabeculae lying in the centre of the bone that were formed by endochondral ossification. In this way, a ***marrow cavity*** is formed. As the shaft increases in diameter there is a corresponding increase in the size of the marrow cavity. This cavity also extends towards the ends of the diaphysis, but does not reach the epiphyseal plate. Gradually most of the bone formed from the primary centre (i.e., of endochondral origin) is removed, except near the bone ends, so that the wall of the shaft is ultimately made up entirely of periosteal bone formed by the process of intramembranous ossification.

To understand how a bone grows in length, we will now take a closer look at the ***epiphyseal plate***. Depending on the arrangement of cells, three zones can be recognised (Fig. 7.15).

- ***Zone of resting cartilage***. Here the cells are small and irregularly arranged.
- ***Zone of proliferating cartilage***. This is also called the ***zone of cartilage growth***. In this zone the cells are larger, and undergo repeated mitosis. As they multiply, they come to be arranged in parallel columns, separated by bars of intercellular matrix.

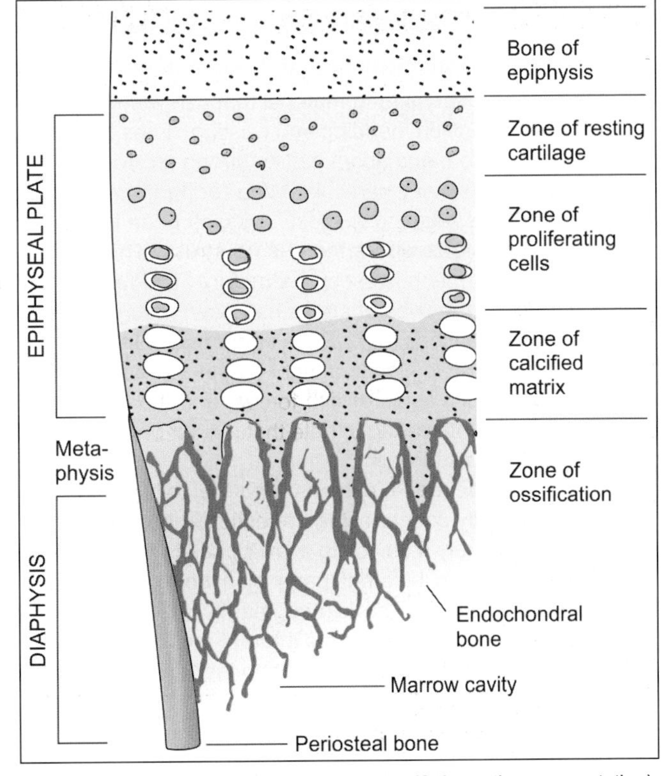

Fig. 7.15: Structure of an epiphyseal plate (Schematic representation)

- ***Zone of calcification***. This is also called the ***zone of cartilage transformation***. In this zone the cells become still larger and the matrix becomes calcified.

Next to the zone of calcification, there is a zone where cartilage cells are dead and the calcified matrix is being replaced by bone. Growth in length of the bone takes place by continuous transformation of the epiphyseal cartilage to bone in this zone (i.e., on the diaphyseal surface of the epiphyseal cartilage). At the same time, the thickness of the epiphyseal cartilage is maintained by the active multiplication of cells in the zone of proliferation. When the bone has attained its full length, cells in the cartilage stop proliferating. The process of calcification, however, continues to extend into it until the whole of the epiphyseal plate is converted into bone. The bone substance of the diaphysis and that of the epiphysis then become continuous. This is called ***fusion of the epiphysis***.

Metaphysis

The portion of the diaphysis adjoining the epiphyseal plate is called the metaphysis. It is a region of active bone formation and, for this reason, it is highly vascular. The metaphysis does not have a marrow cavity. Numerous muscles and ligaments are usually attached to the bone in this region. Even after bone growth has ceased, the calcium turnover function of bone is most active in the metaphysis, which acts as a store house of calcium. The metaphysis is frequently the site of infection (osteomyelitis) because blood vessels show hairpin bends and blood flow is sluggish.

Added Information

Correlation of Bone Structure and Some of its Mechanical Properties

Living bone is a very strong material that can withstand severe stresses. These stresses include compression, tension, bending and twisting. It has been estimated that bone is about three times as strong as wood, and about half as strong as steel.

The stresses on any particular bone (or part of it) can be very complex. The structure of each part of a bone is adapted to these stresses bone being thickest along lines of maximum stress, and absent in areas where there is no stress. This applies not only to the gross structure of a bone, but also to its microscopic structure. The fact that the trabeculae of cancellous bone are arranged along the lines of stress has been recognised since long: this fact is known as Wolff's law. In some situations (e.g., at the upper end of the femur) the trabeculae appear to be arranged predominantly in two planes at right angles to each other. It has been suggested that trabeculae in one plane resist compressive forces, and those in the other direction resist tensile forces. In this context it is interesting to note that a clear pattern of trabeculae is seen in the femur of a child only after it begins to walk.

Within compact bone, osteons (and the collagen fibres within them) are also arranged so as to most efficiently counteract the stresses imposed on them. The spiral arrangement of fibres in osteons is probably a device to allow bones to withstand severe twisting strains.

The stresses on different parts of a bone can be greatly altered in abnormal conditions. Following a fracture, if the two segments of a bone unite at an abnormal angle, it is seen that (over a period of time) the entire internal architecture of the bone becomes modified to adjust to the changed directions of stresses imposed.

Chapter 8

Muscular Tissue

Muscle tissue is composed predominantly of cells that are specialised to shorten in length by contraction. This contraction results in movement. It is in this way that virtually all movements within the body, or of the body in relation to the environment, are ultimately produced.

Muscle tissue is made up basically of cells that are called ***myocytes***. Myocytes are elongated in one direction and are, therefore, often referred to as ***muscle fibres***. Myocytes are mesodermal in origin.

Each muscle fibre is an elongated cell which contains contractile proteins actin and myosin. The various cell organelles of muscle fibres have been given special terms, plasma membrane-sarcolemma, cytoplasm-sarcoplasm, smooth endoplasmic reticulum (ER)-sarcoplasmic reticulum and mitochondria-sarcosome (Fig. 8.1).

Each muscle fibre is closely invested by connective tissue that is continuous with that around other muscle fibres. Because of this fact the force generated by different muscle fibres gets added together. In some cases a movement may be the result of simultaneous contraction of thousands of muscle fibres.

The connective tissue framework of muscle also provides pathways along which blood vessels and nerves reach muscle fibres.

TYPES OF MUSCULAR TISSUE

From the point of view of its histological structure, muscle is of three types:

1. Skeletal muscle
2. Smooth muscle
3. Cardiac muscle.

- ***Skeletal muscle*** is present mainly in the limbs and in relation to the body wall. Because of its close relationship to the bony skeleton, it is called ***skeletal muscle***. When examined under a microscope, fibres of skeletal muscle show prominent transverse striations. Skeletal muscle is, therefore, also called ***striated muscle***. Skeletal muscle can normally be made to contract under our will (to perform movements we desire). It is, therefore, also called ***voluntary muscle***. Skeletal muscle is supplied by somatic motor nerves.
- ***Smooth muscle*** is present mainly in relation to viscera. It is seen most typically in the walls of hollow viscera. As fibres of this variety do not show transverse striations, it is called ***smooth muscle***, or ***non-striated muscle***. As a rule, contraction of smooth muscle is not under our control; and smooth muscle is, therefore, also called ***involuntary muscle***. It is supplied by autonomic nerves.

- ***Cardiac muscle*** is present exclusively in the heart. It resembles smooth muscle in being involuntary; but it resembles striated muscle in that the fibres of cardiac muscle also show transverse striations. Cardiac muscle has an inherent rhythmic contractility the rate of which can be modified by autonomic nerves that supply it.

Note: It will be obvious that the various terms described above are not entirely satisfactory, there being numerous contradictions. Some 'skeletal' muscle has no relationship to the skeleton being present in situations such as the wall of the oesophagus, or of the anal canal. The term striated muscle is usually treated as being synonymous with skeletal muscle, but we have seen that cardiac muscle also has striations. In many instances the contraction of skeletal muscle may not be strictly voluntary (e.g., in sneezing or coughing; respiratory movements; maintenance of posture). Conversely, contraction of smooth muscle may be produced by voluntary effort as in passing urine.

SKELETAL MUSCLE

Microscopic Features

Skeletal muscle is made up essentially of long, cylindrical 'fibres.' The length of the fibres is highly variable, the longest being as much as 30 cm in length. The diameter of the fibres also varies considerably (10–60 μm; usually 50–60 μm).

Each 'fibre' is really a syncytium (Fig. 8.1A) with hundreds of nuclei along its length (The 'fibre' is formed, during development, by fusion of numerous myoblasts). The nuclei are elongated and lie along the periphery of the fibre, just under the cell membrane (which is called the ***sarcolemma***). The cytoplasm (or ***sarcoplasm***) is filled with numerous longitudinal fibrils that are called ***myofibrils*** (Fig. 8.1B).

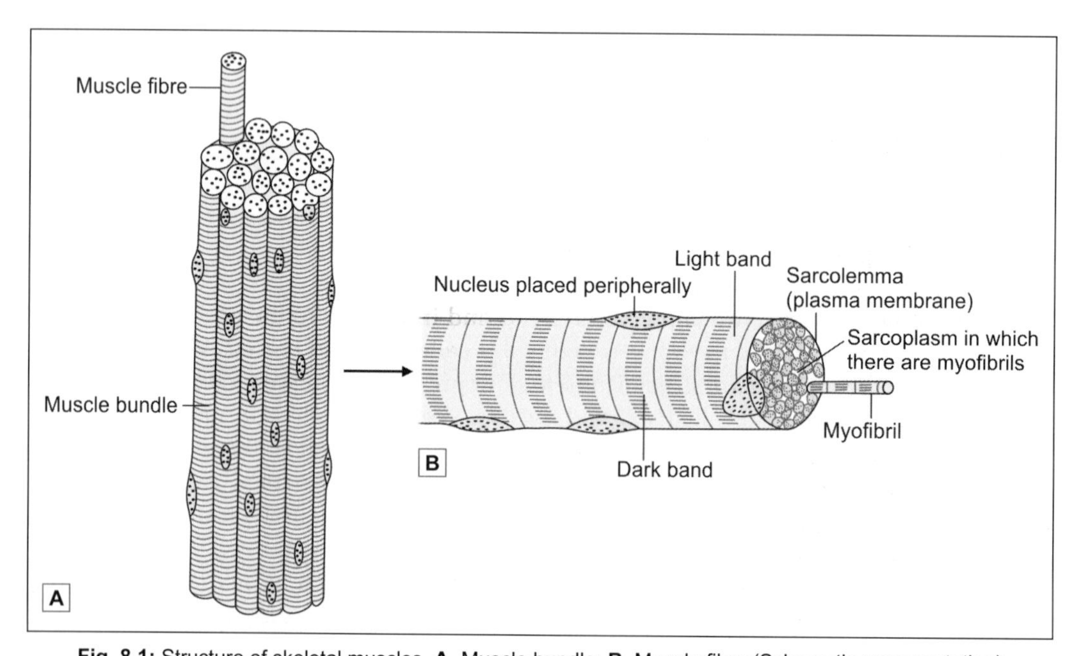

Fig. 8.1: Structure of skeletal muscles. **A.** Muscle bundle; **B.** Muscle fibre (Schematic representation)

In transverse sections through muscle fibres, prepared by routine methods, the myofibrils often appear to be arranged in groups that are called the ***fields of Cohneim's***. The appearance is now known to be an artefact. The myofibrils are in fact distributed uniformly throughout the fibre.

The most striking feature of skeletal muscle fibres is the presence of transverse striations in them (Plate 8.1). After staining with haematoxylin the striations are seen as alternate dark and light bands that stretch across the muscle fibre. The dark bands are called ***A-bands*** (anisotropic), while the light bands are called ***I-bands*** (isotropic) (Fig. 8.3). (As an aid to memory note that 'A' and 'I' correspond to the second letters in the words d**a**rk and l**i**ght).

In addition to myofibrils the sarcoplasm of a muscle fibre contains the usual cell organelles that tend to aggregate near the nuclei. Mitochondria are numerous. Substantial amounts of glycogen are also present. Glycogen provides energy for contraction of muscle.

Plate 8.1: Skeletal Muscle seen in Longitudinal Section

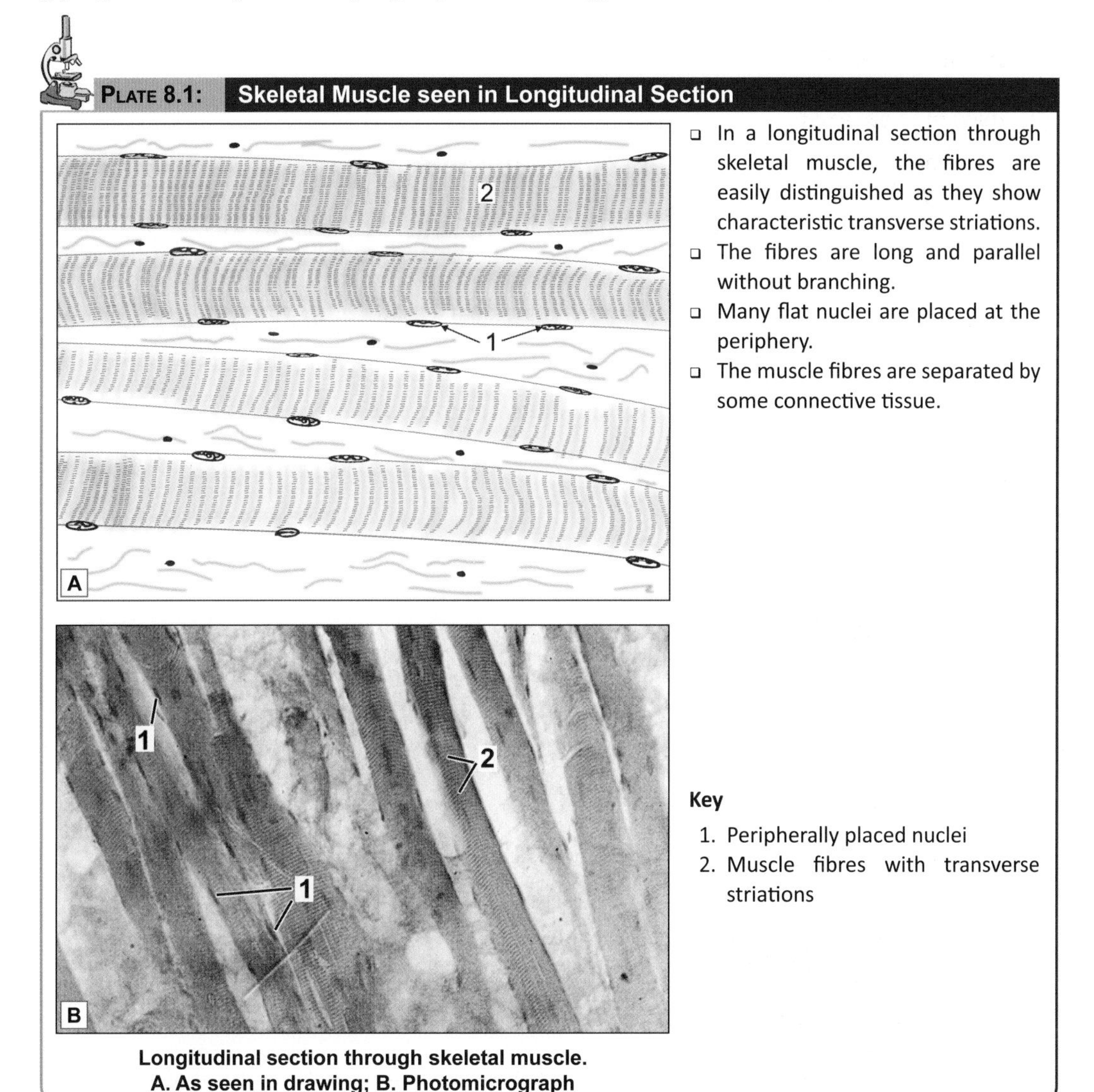

- In a longitudinal section through skeletal muscle, the fibres are easily distinguished as they show characteristic transverse striations.
- The fibres are long and parallel without branching.
- Many flat nuclei are placed at the periphery.
- The muscle fibres are separated by some connective tissue.

Key

1. Peripherally placed nuclei
2. Muscle fibres with transverse striations

Longitudinal section through skeletal muscle.
A. As seen in drawing; B. Photomicrograph

Added Information

Origin of terms I-Band and A-Band

We have seen that the light and dark bands of myofibrils (or of muscle fibres) are designated I-bands and A-bands, respectively. The letters 'I' and 'A' stand for the terms ***isotropic*** and ***anisotropic***, respectively. These terms refer to the way in which any material (e.g., a crystal) behaves with regard to the transmission of light through it. Some materials refract light equally in all directions: they are said to be isotropic. Other materials that do not refract light equally in different planes are anisotropic. These qualities depend on the arrangement of the elements making up the material. In the case of muscle fibres the precise reason for alternate bands being isotropic and anisotropic is not understood. The phenomenon is most probably due to peculiarities in arrangement of molecules within them.

Although striations can be made out in unstained material using ordinary light, they are much better seen through a microscope using polarised light.

Organisation

Within a muscle, the muscle fibres are arranged in the form of bundles or ***fasciculi***. The number of fasciculi in a muscle, and the number of fibres in each fasciculus, are both highly variable.

In small muscles concerned with fine movements (like those of the eyeball, or those of the vocal folds) the fasciculi are delicate and few in number. In large muscles (in which strength of contraction is the main consideration) fasciculi are coarse and numerous.

Each muscle fibre is closely invested by connective tissue. This connective tissue supports muscle fibres and unites them to each other.

- Each muscle fibres is surrounded by delicate connective tissue that is called the ***endomysium***.
- Individual fasciculi are surrounded by a stronger sheath of connective tissue called the ***perimysium***.
- The entire muscle is surrounded by connective tissue called the ***epimysium*** (Fig. 8.2 and Plate 8.2).

At the junction of a muscle with a tendon the fibres of the endomysium, the perimysium and the epimysium become continuous with the fibres of the tendon.

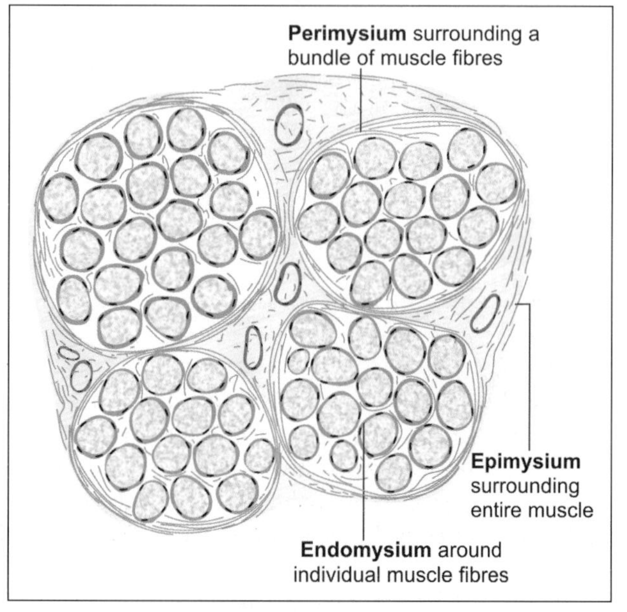

Fig. 8.2: Connective tissue present in relation to skeletal muscle (Schematic representation)

Ultrastructure of Striated Muscle

The ultrastructure of a muscle fibre can be seen under electron microscope (EM). Each muscle fibre is covered by a plasma membrane that is called the ***sarcolemma***. The sarcolemma is covered on

PLATE 8.2: Skeletal Muscle seen in Transverse Section

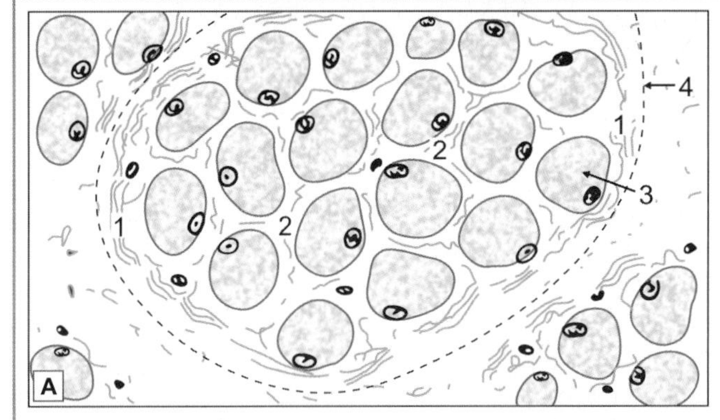

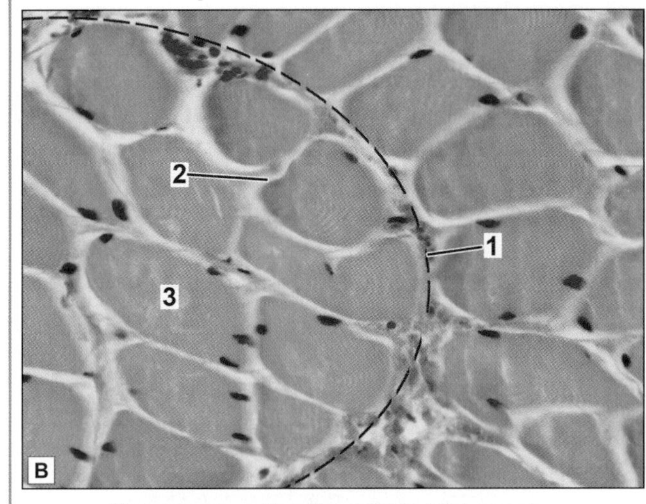

Transverse section through skeletal muscle; A. As seen in drawing; B. Photomicrograph

The transverse section of a skeletal muscle fibre is characterised by:

- Fibres seen as irregularly round structures with peripheral nuclei
- Muscle fibres grouped into numerous fasciculi
- Dots within the fibres are myofibrils which are seen at higher magnification.

The connective tissue of the muscle consists of:

- **Epimysium:** Connective tissue sheath of muscle (not seen in photomicrograph)
- **Perimysium:** Connective tissue covering of each fascicle
- **Endomysium:** Loose connective tissue surrounding each muscle fibre.

Key

1. Perimysium
2. Endomysium
3. Muscle fibre
4. Fasciculus

the outside by a ***basement membrane*** (also called the ***external lamina***) that establishes an intimate connection between the muscle fibre and the fibres (collagen, reticular) of the endomysium.

The cytoplasm (***sarcoplasm***) is permeated with myofibrils that push the elongated nuclei to a peripheral position. Between the myofibrils there is an elaborate system of membrane-lined tubes called the ***sarcoplasmic reticulum***. Elongated mitochondria (***sarcosomes***) and clusters of glycogen are also scattered amongst the myofibrils.

Perinuclear Golgi bodies, ribosomes, lysosomes, and lipid vacuoles are also present.

Structure of Myofibrils (Fig. 8.3)

When examined by EM each myofibril is seen to be made of fine myofilaments (discussed later).

Each myofibril has alternating I and A bands seen under light microscopy. In good preparations (specially if the fibres are stretched) some further details can be made out.

Running across the middle of each I-band there is a thin dark line called the ***Z-band***. The middle of the A-band is traversed by a lighter band, called the ***H-band*** (or ***H-zone***).

Running through the centre of the H-band a thin dark line can be made out. This is the ***M-band***. These bands appear to run transversely across the whole muscle fibre because corresponding bands in adjoining myofibrils lie exactly in alignment with one another.

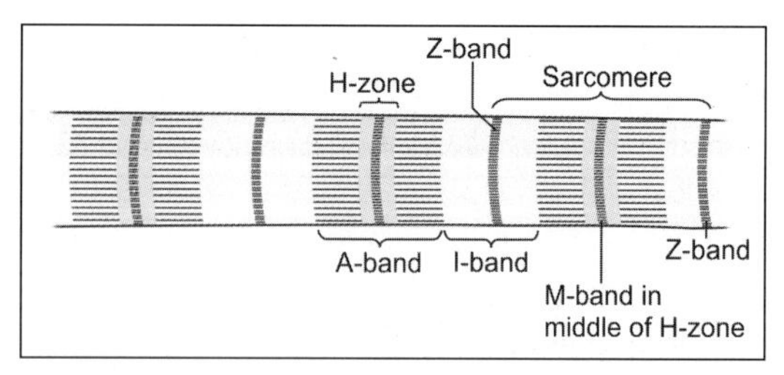

Fig. 8.3: Terminology of transverse bands in a myofibril. Note that the A-band is confined to one sarcomere, but the I-band is made up of parts of two sarcomeres that meet at the Z-band (Schematic representation)

The part of a myofibril situated between two consecutive Z-bands is called a ***sarcomere***. The cross striations seen in a myofibril under electron microscopy is due to the presence of orderly arrangement of ***myofilaments*** (contractile protein filaments) within it.

Added Information

Significance of letters Z, H, M

We have seen that the part of a myofibril between the two Z-bands is called a sarcomere. In other words a Z band is a plate lying between two sarcomeres. The letter 'Z' is from the German word ***zwischenschiebe*** (zwischen = between; schiebe = disc). The M-band is a plate lying in the middle of the sarcomere. The letter 'M' is from the German word ***mittleschiebe*** (mittle = middle). The H-band (or zone) is named after Hensen who first described it.

Structure of Myofilaments

There are two types of myofilaments in skeletal muscle, ***myosin*** and ***actin***, made up of molecules of corresponding proteins (Each myosin filament is about 12 nm in diameter, while an actin filament is about 8 nm in diameter. They are therefore referred to as thick and thin filaments respectively).

The arrangement of ***myosin*** and ***actin*** filaments within a sarcomere is shown in Figure 8.4.

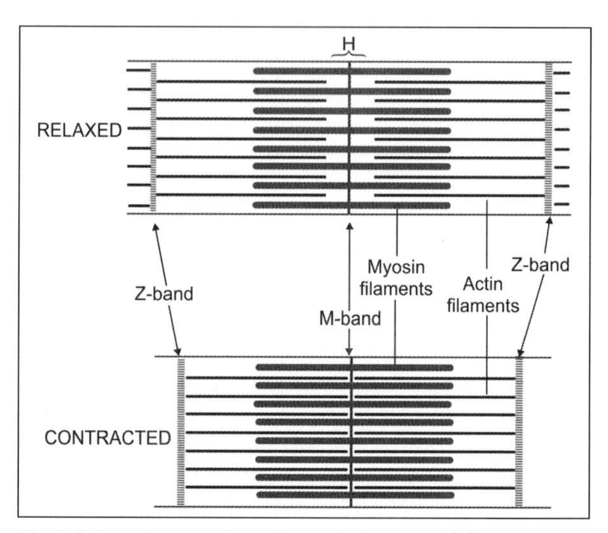

Fig. 8.4: Arrangement of myofilaments in sarcomere. Note that the width of the I-band becomes less, and that the H-zone disappears when the myofibril contracts (Schematic representation)

It will be seen that myosin filaments are confined to the A-band, the width of the band being equal to the length of the myosin filaments. The actin filaments are attached at one end to the Z-band.

From here they pass through the I-band and extend into the 'outer' parts of the A-band, where they interdigitate with the myosin filaments. Note that the I-band is made up of actin filaments alone. The H-band represents the part of the A-band into which actin filaments do not extend. The Z-band is really a complicated network at which the actin filaments of adjoining sarcomeres meet. The M-band is produced by fine interconnections between adjacent myosin filaments.

Pathological Correlation

Each muscle fibre contains a cytoskeleton. The fibres of the cytoskeleton are linked to actin fibres. The cytoskeleton is also linked to the external lamina through glycoproteins present in the cell membrane. Forces generated within the fibre are thus transmitted to the external lamina. The external lamina is in turn attached to connective tissue fibres around the muscle fibre. A number of proteins are responsible for these linkages. Genetic defects in these proteins can result in abnormalities in muscle (muscle dystrophy).

Added Information

Each ***myosin filament*** is made up of a large number of myosin molecules. Each molecule is made up of two units, each unit having a head and a tail (Fig. 8.5). The tails are coiled over each other. A myosin filament is a 'bundle' of the tails of such molecules (Fig. 8.6). The heads project outwards from the bundle as projections of the myosin filament. The projecting heads are arranged in a regular helical manner.

Because of the manner in which it is formed each myosin fibril can be said to have a head end and a tail end. The tail end is attached to the M-line.

Each ***actin*** filament is really composed of two subfilaments that are twisted round each other (Fig. 8.7). Each subfilament is a chain of globular (rounded) molecules. These globular molecules are G-actin, and the chain formed by them is designated as ***f-actin***. Each actin filament has a head end (that extends into the A-band) and a tail end that is anchored to the Z-line (through a protein called α-actinin). The filament also contains two other proteins called ***tropomyosin*** and ***troponin***. Tropomyosin is in the form of a long fibre that winds around actin and stabilises it. Troponin is a complex made up of several fractions. These complexes are arranged regularly over the actin fibre and represent sites at which myosin binds to actin.

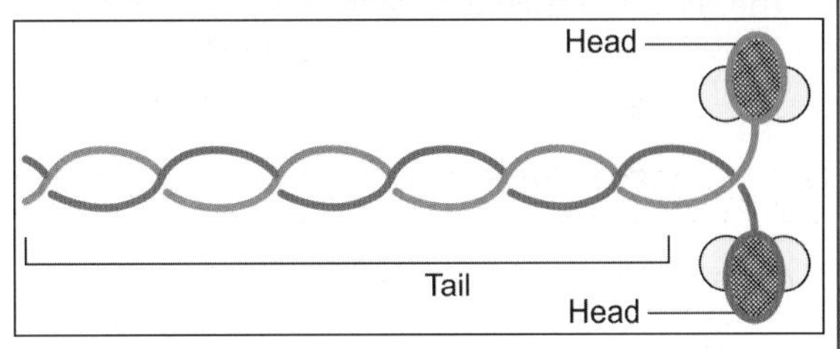

Fig. 8.5: Structure of a myosin molecule. Each molecule has two components (shown in red and blue) each consisting of a head and a tail. The tails are coiled over each other. The parts shown in red or blue are heavy myosin. Light myosin is shaded yellow (Schematic representation)

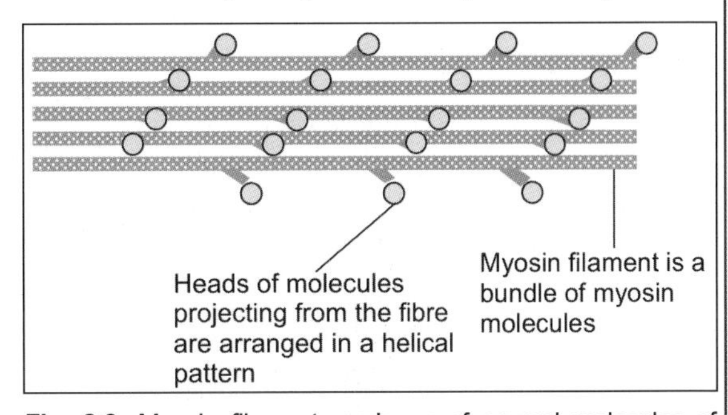

Fig. 8.6: Myosin filament made up of several molecules of myosin (Schematic representation)

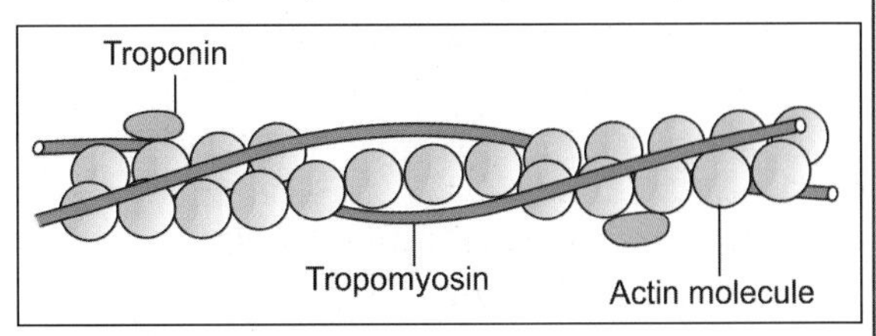

Fig. 8.7: Actin filament (F-actin) made up of globular molecules of G-actin (Schematic representation)

Contraction of Skeletal Muscle

During contraction, there is no shortening of individual thick and thin myofilaments; but there is an increase in the degree of overlap between the filaments.

In an uncontracted myofibril, overlap between actin and myosin filaments is minimal.

During contraction under the influence of energy released by adenosine triphosphate (ATP) and calcium released from sarcoplasmic reticulum, the fibril shortens by sliding in of actin filaments more and more into the intervals between the myosin filaments. As a result the width of the I-band decreases, but that of the A-band is unchanged. The H-bands are obliterated in a contracted fibril.

Other Proteins Present in Skeletal Muscle

Several proteins other than actin and myosin are present in muscle. Some of them are as follows.

- ***Actinin*** is present in the region of Z discs. It binds the tail ends of actin filaments to this disc.
- ***Myomesin*** is present in the region of the M disc. It binds the tail ends of myosin filaments to this disc.
- ***Titin*** links the head ends of myosin filaments to the Z disc. This is a long and elastic protein that can lengthen and shorten as required. It keeps the myosin filament in proper alignment.
- ***Desmin*** is present in intermediate filaments of the cytoskeleton. It links myofibrils to each other, and also to the cell membrane.

Sarcoplasmic Reticulum

In the intervals between myofibrils, the sarcoplasm contains an elaborate system of tubules called the ***sarcoplasmic reticulum*** (Fig. 8.8).

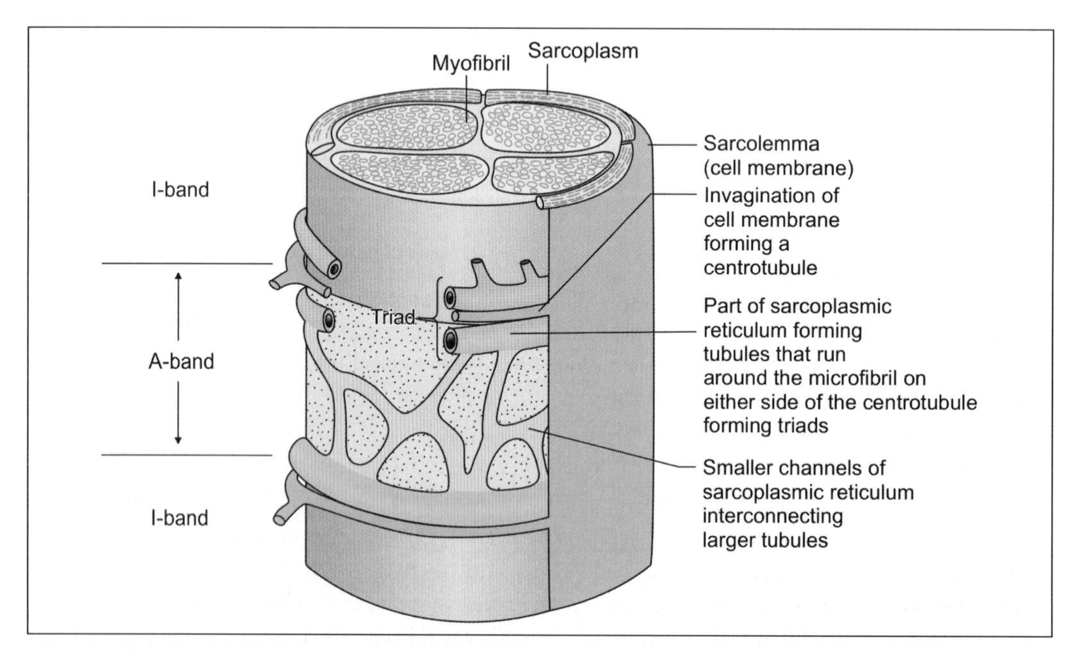

Fig. 8.8: Relationship of the sarcoplasmic reticulum and the T-tubes to a myofibril (Schematic representation)

The larger elements of this reticulum run in planes at right angles to the long axes of the myofibrils, and form rings around each myofibril. At the level of every junction between an A and I band the myofibril is encircled by a set of three closely connected tubules that constitute a ***muscle triad***.

Muscle Triad

For purposes of description each such triad can be said to be composed of an upper, a middle, and a lower tubule (Fig. 8.8). The upper and lower tubules of the triad are connected to the tubules of adjoining triads through a network of smaller tubules. There is one such network opposite each A-band, and another opposite each I-band. These networks, along with the upper and lower tubules of the triad, constitute the sarcoplasmic reticulum. This reticulum is a closed system of tubes.

The middle tube of the triad is an entity independent of the sarcoplasmic reticulum. It is called a ***centrotubule*** and belongs to what is called the ***T-system*** of membranes. The centrotubules are really formed by invagination of the sarcolemma into the sarcoplasm. Their lumina are, therefore, in communication with the exterior of the muscle fibre.

Contraction of muscle is dependent on release of calcium ions into myofibrils. In a relaxed muscle these ions are strongly bound to the membranes of the sarcoplasmic reticulum. When a nerve stimulus reaches a motor end plate the sarcolemma is depolarised. The wave of depolarisation is transmitted to the interior of the muscle fibre through the centrotubules. As a result of this wave calcium ions are released from the sarcoplasmic reticulum into the myofibrils causing their contraction.

Types of Skeletal Muscle Fibres

From morphological, histochemical and functional point of view, skeletal muscle fibres are of two types:

1. Red muscle fibre (because they are red in colour)
2. White muscle fibre.

The colour of red fibres is due to the presence (in the sarcoplasm) of a pigment called ***myoglobin***. This pigment is similar (but not identical with) haemoglobin. It is present also in white fibres, but in much lesser quantity.

As compared to white fibres the contraction of red fibres is relatively slow. Hence red fibres are also called ***slow twitch fibres***, or ***type I fibres***; while white fibres are also called ***fast twitch fibres*** or ***type II fibres***.

In addition to colour and speed of contraction there are several other differences between red and white fibres. In comparison to white fibres red fibres differ as follows.

Red fibres are narrower than white fibres. Relative to the volume of the myofibrils the sarcoplasm is more abundant. Probably because of this fact the myofibrils, and striations, are less well defined; and the nuclei are not always at the periphery, but may extend deeper into the fibre. Mitochondria are more numerous in red fibres, but the sarcoplasmic reticulum is less extensive. The sarcoplasm contains more glycogen. The capillary bed around red fibres is richer than around white fibres. Differences have also been described in enzyme systems and the respiratory mechanisms in the two types of fibres. Fibres intermediate between red and white fibres have also been described.

Table 8.1: Characteristics of red and white muscle fibres	
Red muscle fibre	**White muscle fibre**
Rich in myoglobin and cytochrome hence Red in colour	Less myoglobin and cytochrome hence White in colour
Narrow in diameter with less defined striations and nuclei not always placed at the periphery	Broader in diameter, with well defined striations and nuclei placed at the periphery
Volume of sarcoplasm more than myofibrils. Sarcoplasm contains more glycogen	Volume of myofibrils more. Sarcoplasm contains less glycogen
Numerous mitochondria but sarcoplasmic reticulum less extensive	Few mitochondria with extensive sarcoplasmic reticulum
Slow and continuous contraction (not easily fatigued)	Rapid contraction (easily fatigued)
Rich blood supply	Poor blood supply
e.g. Predominate in postural muscles which have to remain contracted over long periods	e.g. Predominate in muscles responsible for sharp active movements, e.g. extraocular muscles and flight muscles in birds

In some animals complete muscles may consist exclusively of red or white fibres, but in most mammals, including man, muscles contain an admixture of both types. Although red fibres contract slowly their contraction is more sustained, and they fatigue less easily. They predominate in the so called postural muscles (which have to remain contracted over long periods), while white fibres predominate in muscles responsible for sharp active movements.

Type II (white) fibres may be divided into type IIA and type IIB, the two types differing in their enzyme content, and in the chemical nature of their myosin molecules.

Blood Vessels and Lymphatics of Skeletal Muscle

Skeletal muscle is richly supplied with blood vessels. The main artery to the muscle enters it at the neurovascular hilus. Several other arteries may enter the muscle at its ends or at other places along its length. The arteries form a plexus in the epimysium and in the perimysium, and end in a network of capillaries that surrounds each muscle fibre. This network is richer in red muscle than in white muscle.

Veins leaving the muscle accompany the arteries.

A lymphatic plexus extends into the epimysium and the perimysium, but not into the endomysium.

Innervation of Skeletal Muscle

The nerve supplying a muscle enters it (along with the main blood vessels) at an area called the ***neurovascular hilus.*** This hilus is usually situated nearer the origin of the muscle than the insertion.

After entering the muscle the nerve breaks up into many branches that run through the connective tissue of the perimysium and endomysium to reach each muscle fibre. The nerve fibres supplying skeletal muscle are axons arising from large neurons in the anterior (or ventral) grey columns of the spinal cord (or of corresponding nuclei in the brainstem). These ***alpha-efferents*** have a large diameter and are myelinated. Because of repeated branching of its axon, one anterior grey column neuron may supply many muscle fibres all of which contract when this neuron 'fires'.

One anterior grey column neuron and the muscle fibres supplied by it constitute one ***motor unit*** (Fig. 8.9). The number of muscle fibres in one motor unit is variable. The units are smaller where precise control of muscular action is required (as in ocular muscles), and much larger in limb muscles where force of contraction is more important. The strength with which a muscle contracts at a particular moment depends on the number of motor units that are activated.

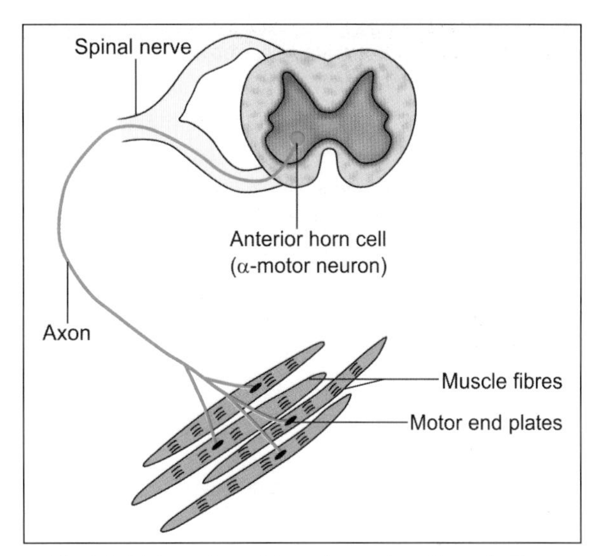

Fig. 8.9: Motor unit consisting of a number of muscle fibres innervated by a single motor neuron (Schematic representation)

The junction between a muscle fibre and the nerve terminal that supplies it is highly specialised and is called a ***motor end plate*** (discussed in detail later).

Apart from the alpha efferents described above every muscle receives smaller myelinated ***gamma-efferents*** that arise from gamma neurons in the ventral grey column of the spinal cord. These fibres supply special muscle fibres that are present within sensory receptors called ***muscle spindles***. These special muscle fibres are called ***intrafusal fibres***. Nerves to muscles also carry autonomic fibres that supply smooth muscle present in the walls of blood vessels.

Structure of Motor End Plate

It is the most common type of neuromuscular junctions. As the motor neuron enters a skeletal muscle, it branches many times. The number of branches depends on the size of the motor unit. On reaching the muscle fibre, the nerve loses its myelin sheath and breaks up into number of branches. Each branch is a naked axon terminal and forms the neural element of the motor end plate (Fig. 8.10).

The axon is slightly expanded here. At the site of the motor end plate, the surface of the muscle fibre is slightly elevated to form the muscular element of the plate, referred to as the ***soleplate***.

The naked axon terminal and the soleplate together form ***motor end plate*** (Fig. 8.11).

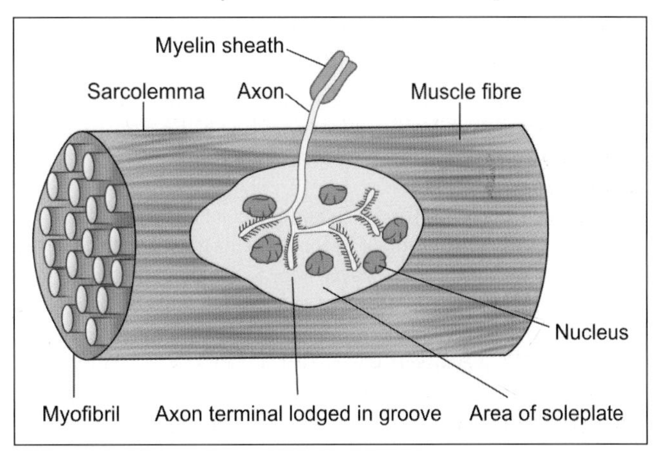

Fig. 8.10: Motor end plate seen in relation to a muscle fibre (surface view) (Schematic representation)

In the region of the motor end plate, axon terminals are lodged

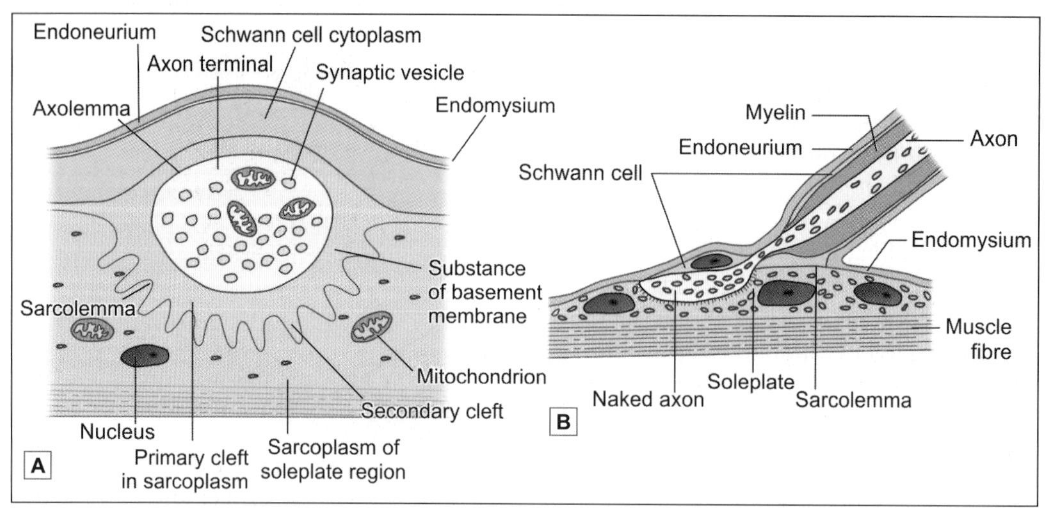

Fig. 8.11: A. A skeletal neuromuscular junction; **B.** Enlarged view of a muscle fibre showing the terminal naked axon lying in the surface groove of the muscle fibre (Schematic representation)

in grooves in the sarcolemma covering the soleplate. Between the axolemma (over the axon) and the sarcolemma (over the muscle fibre), there is a narrow gap (about 40 nm) occupied by various proteins that form a basal lamina. It follows that there is no continuity between axoplasm and sarcoplasm.

Axon terminals are lodged in grooves in the sarcolemma covering the soleplate. In Figure 8.11B, this groove is seen as a semicircular depression. This depression is the ***primary cleft***. The sarcolemma in the floor of the primary cleft is thrown into numerous small folds resulting in the formation of ***secondary (or subneural) clefts***.

In the region of the soleplate, the sarcoplasm of the muscle fibre is granular. It contains a number of nuclei and is rich in mitochondria, ER, and Golgi complexes.

Axon terminals are also rich in mitochondria. Each terminal contains vesicles similar to those seen in presynaptic boutons. The vesicles contain the neurotransmitter acetyl choline. Acetylcholine is released when nerve impulses reach the neuromuscular junction. It initiates a wave of depolarization in the sarcolemma resulting in contraction of the entire muscle fibre. Thereafter, the acetylcholine is quickly destroyed by the enzyme acetylcholine esterase. The presence of acetylcholine receptors has been demonstrated in the sarcolemma of the soleplate.

Clinical Correlation

Myasthenia gravis (MG) is a neuromuscular disorder of autoimmune origin in which the acetylcholine receptors (AChR) in the motor endplates of the muscles are damaged (Fig. 8.12). The term '***myasthenia***' means 'muscular weakness' and '***gravis***' implies 'serious'; thus both together denote the clinical characteristics of the disease. MG may be found at any age but adult women are affected more often than adult men in the ratio of 3:2. The condition presents clinically with muscular weakness and fatigability, initially in the ocular musculature but later spreads to involve the trunk and limbs.

Muscle Spindle

These are spindle-shaped sensory end organs located within striated muscle (Fig. 8.13). The spindle is bounded by a fusiform connective tissue capsule (forming an external capsule)

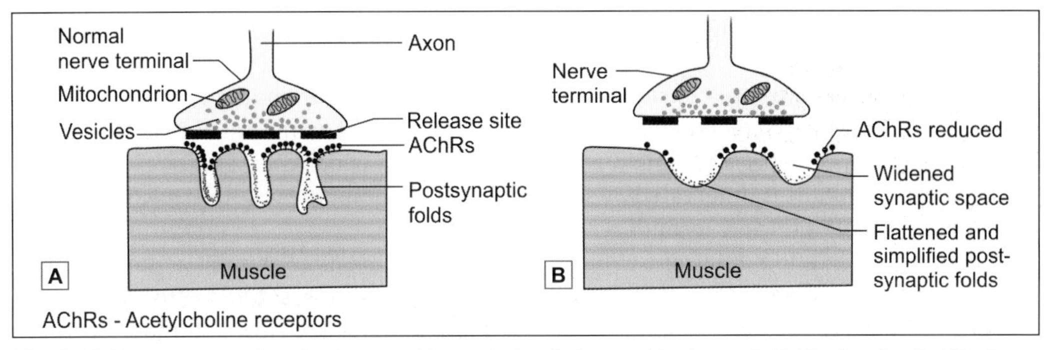

Fig. 8.12: Neuromuscular junction in normal transmission **A.** In myasthenia gravis **B.** The junction in MG shows reduced number of AChRs, flattened and simplified postsynaptic folds, a widened synaptic space but a normal nerve terminal (Schematic representation); AChRs-acetylcholine receptors

within which there are a few specialised muscle fibres. These are called ***intrafusal fibres*** in contrast to ***extrafusal fibres*** that constitute the main bulk of the muscle. Each spindle contains six to fourteen intrafusal fibres. Each intrafusal fibre is surrounded by an internal capsule of flattened fibroblasts and collagen.

Intrafusal fibres contain several nuclei that are located near the middle of the fibre. In some fibres, this region is dilated into a bag and are called ***nuclear bag fibres***. In other intrafusal

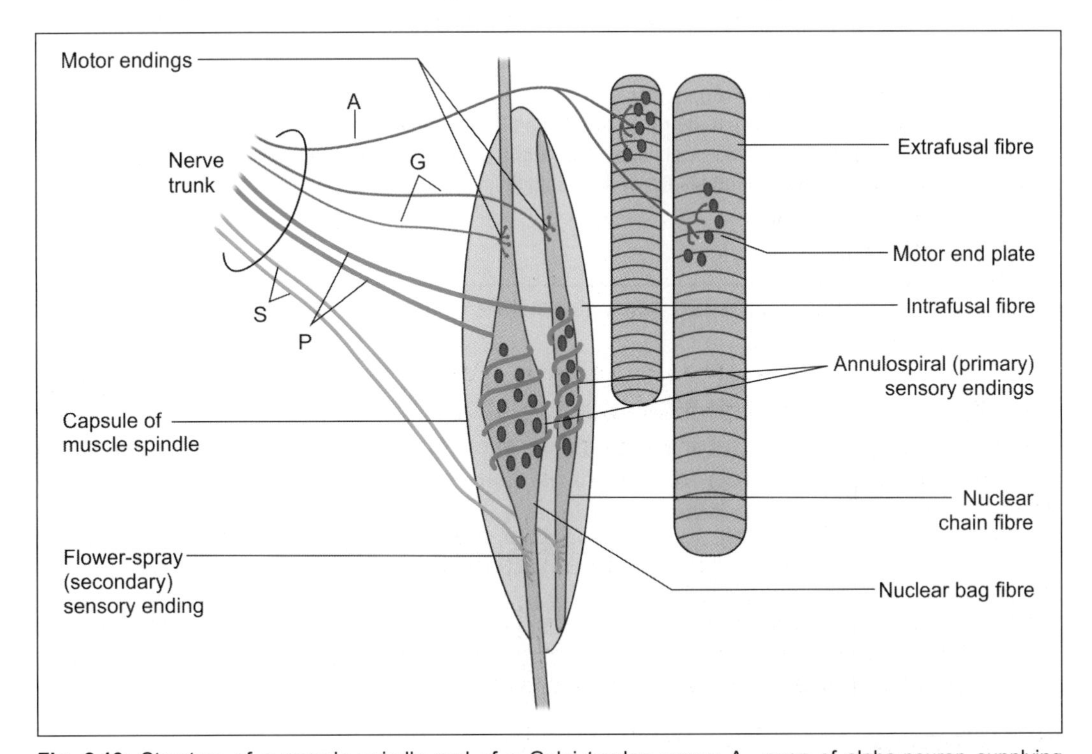

Fig. 8.13: Structure of a muscle spindle and of a Golgi tendon organ; A—axon of alpha-neuron supplying extrafusal fibre; G—axons of gamma neurons supplying intrafusal fibres; P and S—afferents from primary and secondary sensory endings, respectively (Schematic representation)

fibres, the nuclei lie in a single row, there being no dilatation. These are ***nuclear chain fibres***.

The nuclear bag fibres are considerably larger than the nuclear chain fibres. They extend beyond the capsule and gain attachment to the endomysium of extrafusal fibres. The nuclear chain fibres, on the other hand, remain within the capsule to which their ends are attached.

Each muscle spindle is innervated by sensory as well as motor nerves. The sensory endings are of two types, primary and secondary.

The motor innervation of intrafusal fibres is (mainly) by axons of gamma neurons located in the ventral grey column of the spinal cord. The sensory endings respond to stretch. Primary sensory endings are rapidly adapting while secondary endings are slow adapting. However, the precise role of these receptors is complex and varies in different types of fibres.

Spindles provide information to the CNS about the extent and rate of changes in length of muscle. Nuclear bag fibres are stimulated by rapid changes, while nuclear chain fibres react more slowly. Contraction of intrafusal fibres makes the spindle more sensitive to stretch.

Pathological Correlation

Skeletal muscle tumors

Rhabdomyoma and rhabdomyosarcoma are the benign and malignant tumours respectively of striated muscle.

- **Rhabdomyoma** is a rare benign soft tissue tumour. It should not be confused with glycogen-containing lesion of the heart designated as cardiac rhabdomyoma which is probably a hamartomatous lesion and not a true tumour. Soft tissue rhabdomyomas are predominantly located in the head and neck, most often in the upper neck, tongue, larynx and pharynx.
- **Rhabdomyosarcoma** is a much more common soft tissue tumour than rhabdomyoma, and is the commonest soft tissue sarcoma in children and young adults. It is a highly malignant tumour arising from rhabdomyoblasts in varying stages of differentiation with or without demonstrable cross-striations.

CARDIAC MUSCLE

The structure of cardiac muscle has many similarities to that of skeletal muscle; but there are important differences as well.

Similarities between Cardiac and Skeletal Muscle

Like skeletal muscle, cardiac muscle is made up of elongated 'fibres' within which there are numerous myofibrils (Plate 8.3). The myofibrils (and, therefore, the fibres) show transverse striations similar to those of skeletal muscle. A, I, Z and H bands can be made out in the striations. The connective tissue framework, and the capillary network around cardiac muscle fibres are similar to those in skeletal muscle.

With the EM it is seen that myofibrils of cardiac muscle have the same structure as those of skeletal muscle and are made up of actin and myosin filaments. A sarcoplasmic reticulum, T-system of centrotubules, numerous mitochondria and other organelles are present.

Differences between Cardiac and Skeletal Muscle

- The fibres of cardiac muscle do not run in strict parallel formation, but branch and anastomose with other fibres to form a network.

PLATE 8.3: **Cardiac Muscle**

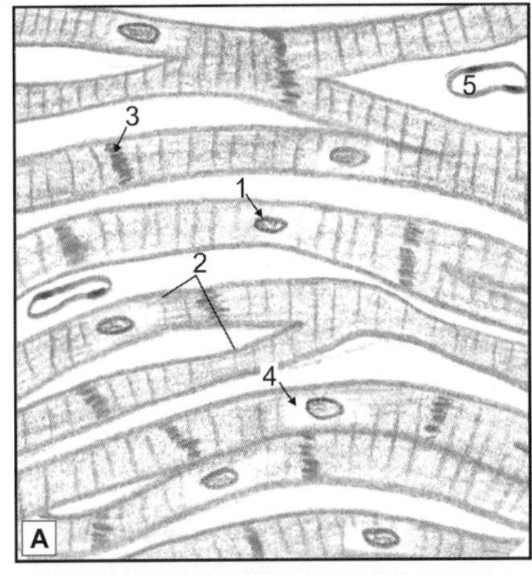

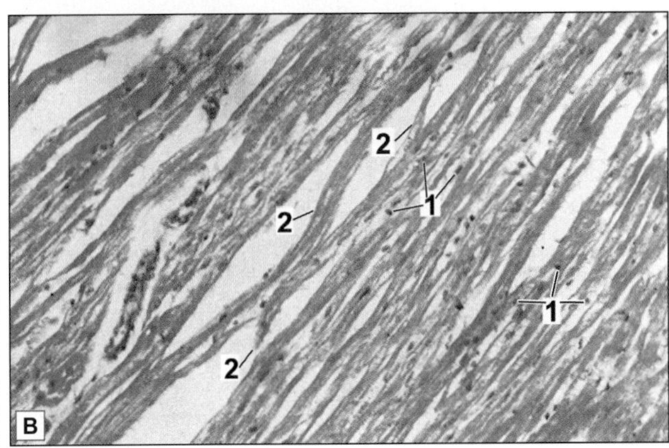

Cardiac muscle. A. As seen in drawing; B. Photomicrograph.

- The fibres are made up of 'cells' each of which has a centrally placed nucleus and transverse striations
- A clear space called perinuclear halo is seen around the nucleus
- Adjacent cells are separated from one another by transverse lines called intercalated discs
- Fibres show branching
- Blood vessels are also seen.

Key

1. Central nucleus
2. Branching fibres
3. Intercalated disc
4. Perinuclear halo
5. Capillary

- Each fibre of cardiac muscle is not a multinucleated syncytium as in skeletal muscle, but is a chain of cardiac muscle cells (or ***cardiac myocytes***) each having its own nucleus. Each myocyte is about 80 μm long and about 15 μm broad.
- The nucleus of each myocyte is located centrally (and not peripherally as in skeletal muscle).
- The sarcoplasm of cardiac myocytes is abundant and contains numerous large mitochondria. The myofibrils are relatively few. At places, the myofibrils merge with each other. As a result of these factors, the myofibrils and striations of cardiac muscle are not as distinct as those of skeletal muscle. In this respect cardiac muscle is closer to the red variety of skeletal muscle than to the white variety. Other similarities with red muscle are the presence of significant amounts of glycogen and of myoglobin, and the rich density of the capillary network around the fibres.
- With the EM it is seen that the sarcoplasmic reticulum is much less prominent than in skeletal muscle. The centrotubules of the T-system lie opposite the Z-bands (and not at

the junctions of A and I-bands as in skeletal muscle). The tubules are much wider than in skeletal muscle. Typical triads are not present. They are often replaced by ***dyads*** having one T-tube and one tube of the sarcoplasmic reticulum.

- With the light microscope the junctions between adjoining cardiac myocytes are seen as dark staining transverse lines running across the muscle fibre. These lines are called ***intercalated discs***. Sometimes these discs do not run straight across the fibres, but are broken into a number of 'steps' (Plate 8.3). The discs always lie opposite the I-bands.
- Cardiac muscle is involuntary and is innervated by autonomic fibres (in contrast to skeletal muscle that is innervated by cerebrospinal nerves). Nerve endings terminate near the cardiac myocytes, but motor end plates are not seen.
- Isolated cardiac myocytes contract spontaneously in a rhythmic manner. In the intact heart the rhythm of contraction is determined by a pacemaker located in the sinoatrial node. From here the impulse spreads to the entire heart through a conducting system made up of a special kind of cardiac muscle. From the above it will be appreciated that a nerve supply is not necessary for contraction of cardiac muscle. Nervous influences do, however, affect the strength and rate of contraction of the heart.

Added Information

With the EM it is seen that the intercalated discs are formed by cell membranes of adjacent myocytes, and by a layer of particularly dense cytoplasm present next to the cell membrane. The ends of actin filaments are embedded in this dense cytoplasm (Fig. 8.14). The cell membranes of adjoining myocytes are connected by numerous desmosomes, gap junctions, and tight junctions. Desmosomes link intermediate filaments present in the cytoskeleton of adjacent cells. Actin filaments of the cells end in relation to tight junctions. Gap junctions allow electrical continuity between adjacent myocytes, and thus convert the cardiac muscle into a ***physiological syncytium***.

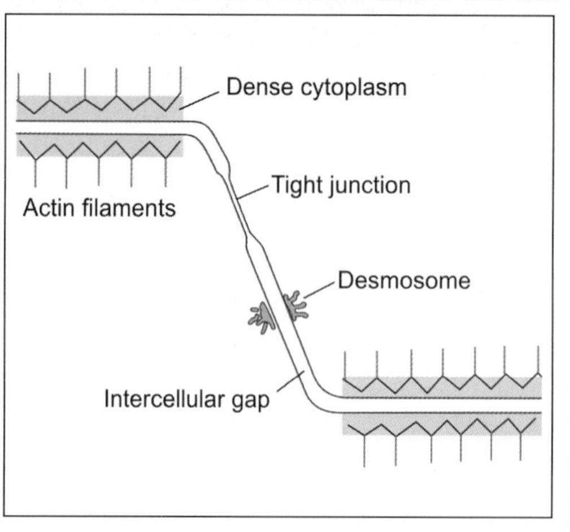

Fig. 8.14: Electrone microscopic structure of part of an intercalated disc (Schematic representation)

SMOOTH MUSCLE

Smooth muscle (also called ***non-striated, involuntary*** or ***plain muscle***) is made up of long spindle-shaped cells (myocytes) having a broad central part and tapering ends. The nucleus, which is oval or elongated, lies in the central part of the cell. The length of smooth muscle cells (often called fibres) is highly variable (15 μm to 500 μm).

With the light microscope the sarcoplasm appears to have indistinct longitudinal striations, but there are no transverse striations.

Smooth muscle cells are usually aggregated to form bundles, or fasciculi, that are further aggregated to form layers of variable thickness. In such a layer the cells are so arranged that the thick central part of one cell is opposite the thin tapering ends of adjoining cells (Fig. 8.15).

Aggregations of smooth muscle cells into fasciculi and layers is facilitated by the fact that each myocyte is surrounded by a network of delicate fibres (collagen, reticular, elastic) that holds the myocytes together (Plate 8.4).

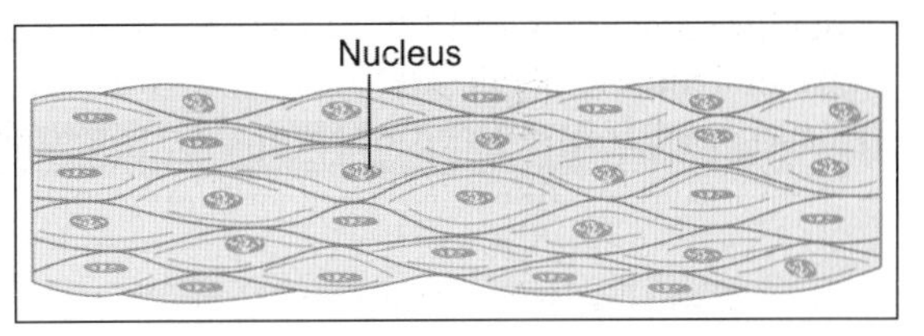

Fig. 8.15: Smooth muscle cells (Schematic representation)

The fibres between individual myocytes become continuous with the more abundant connective tissue that separates fasciculi or layers of smooth muscle.

The differences between skeletal, cardiac and smooth muscles are highlighted in Table 8.2.

Table 8.2: Differences between skeletal, cardiac and smooth muscles

Characteristics	Skeletal muscle	Cardiac muscle	Smooth muscle
Muscle fibre	Long, cylindrical and unbranched	Short, narrow and branched	Fusiform/spindle shaped and unbranched
Control	Voluntary	Involuntary	Involuntary
Location	Muscle of skeleton, tongue, oesophagus, diaphragm	Heart, pulmonary veins, superior and inferior vena cavae	Vessels, organs and viscera
Striations	Present (well defined)	Present (poorly defined)	Absent
Nuclei	Multiple, flat, located at periphery	Single, oval, present in centre	Single, oval, present in centre
Sarcoplasmic reticulum	Present (form triads)	Present (form dyad)	Absent
Intercalated disc	Absent	Present	Absent
Regeneration after injury	Seen (limited)	Not seen	Seen

Ultrastructure

Each smooth muscle cell is bounded by a plasma membrane. Outside the plasma membrane there is an external lamina to which the plasma membrane is adherent. Connective tissue fibres are attached to the lamina (through special proteins). Adjacent smooth muscle cells communicate through gap junctions. The longitudinal striations (seen with the light microscope) are due to the presence of delicate myofilaments. These myofilaments are composed mainly of the proteins actin and myosin, but these do not have the highly ordered arrangement seen in striated muscle. The actin filaments are also different from those in skeletal muscle. Troponin is not present. Apart from myofibrils the sarcoplasm also contains mitochondria, a Golgi complex, some granular endoplasmic reticulum, free ribosomes, and intermediate filaments. A sarcoplasmic reticulum, similar to that in skeletal muscle, is present, but is not as developed. Numerous invaginations (caveolae) resembling endocytic vesicles are seen near the surface of each myocyte, but no endocytosis occurs here.

PLATE 8.4: Smooth Muscle

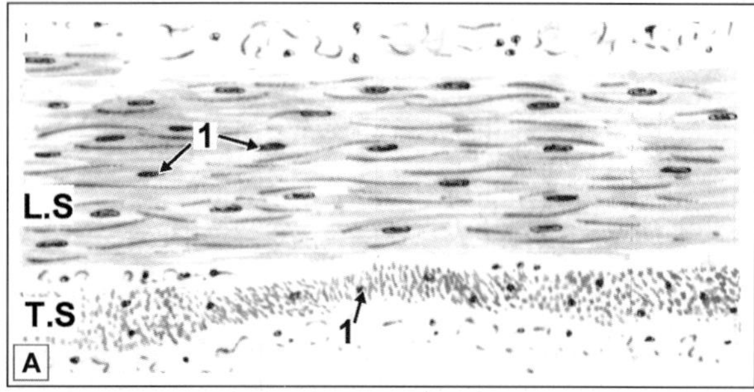

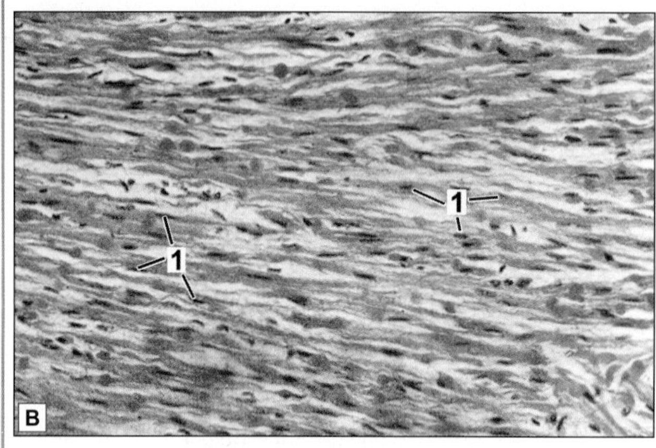

Smooth muscle. A. As seen in drawing; B. Photomicrograph of longitudinal section

In the drawing, muscle is seen cut longitudinally as well as transversely.

- Loose connective tissue is seen above and below the layers of muscle.

In longitudinal section:

- The smooth muscle fibres are ***spindle-shaped cells*** with tapering ends
- The nucleus is elongated and centrally placed
- ***No striations*** are seen.

In transverse section:

- The spindle-shaped cells are cut at different places along the length resulting in various shapes and sizes of the cells
- The nucleus is seen in those cells which are cut through the centre. Others do not show nuclei.

Key

1. Oval centrally placed nuclei

Contraction of Smooth Muscle

The mechanism of contraction of smooth muscle is different from that of skeletal muscle as follows:

- The myosin is chemically different from that in skeletal muscle. It binds to actin only if its light chain is phosphorylated. This phosphorylation of myosin is necessary for contraction of smooth muscle.
- As compared to skeletal muscle, smooth muscle needs very little ATP for contraction.
- The mechanisms regulating the flow of calcium ions into smooth muscle are different from those for skeletal muscle. Caveolae present on the surface of smooth muscle cells play a role in this process.
- Actin and myosin form bundles that are attached at both ends to the points on the cell membrane called anchoring points (or focal densities). When the muscle contracts these points are drawn closer to each other. This converts an elongated smooth muscle cell in one that is oval (Fig. 8.16).

Distribution

- Smooth muscle is seen most typically in the walls of hollow viscera including the stomach, the intestines, the urinary bladder and the uterus.
- It is present in the walls of several structures that are in the form of narrow tubes, e.g., arteries, veins, bronchi, ureters, deferent ducts, uterine tubes, and the ducts of several glands.
- The muscles that constrict and dilate the pupil are made up of smooth muscle.
- Some smooth muscle is present in the orbit (orbitalis); in the upper eyelid (Muller's muscle); in the prostate; in the skin of the scrotum (Dartos muscle). In the skin delicate bundles of smooth muscle are present in relation to hair follicles. These bundles are called the ***arrector pili*** muscles.

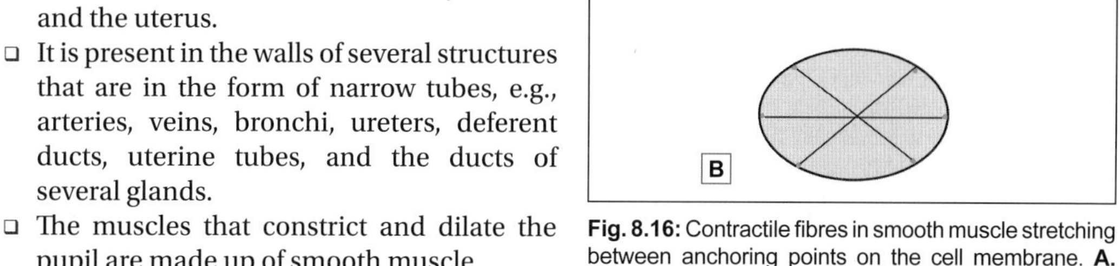

Fig. 8.16: Contractile fibres in smooth muscle stretching between anchoring points on the cell membrane. **A.** Cell in the relaxed state; **B.** Cell in the contracted state

Variations in Arrangement of Smooth Muscle

Smooth muscle fibres may be arranged in a variety of ways depending on functional requirements.

In some organs (e.g., the gut) smooth muscle is arranged in the form of two distinct layers: an inner circular and an outer longitudinal. Within each layer the fasciculi lie parallel to each other. Such an arrangement allows peristaltic movements to take place for propulsion of contents along the tube.

In some organs (e.g., the ureter) the arrangement of layers may be reversed, the longitudinal layer being internal to the circular one. In yet other situations there may be three layers: inner and outer longitudinal with a circular layer in between (e.g., the urinary bladder and vas deferens).

- In some regions (e.g., urinary bladder, uterus) the smooth muscle is arranged in layers, but the layers are not distinctly demarcated from each other. Even within layers the fasciculi tend to run in various directions and may form a network.
- In some tubes (e.g., the bile duct) a thick layer of circular muscle may surround a segment of the tube forming a ***sphincter***. Contraction of the sphincter occludes the tube.
- In the skin, and in some other places, smooth muscle occurs in the form of narrow bands.

Innervation

Smooth muscle is innervated by autonomic nerves, both sympathetic and parasympathetic. The two have opposite effects. For example, in the iris, parasympathetic stimulation causes constriction of the pupil, and sympathetic stimulation causes dilatation. It may be noted that sympathetic or parasympathetic nerves may cause contraction of muscle at some sites, and relaxation at other sites.

Blood Vessels and Lymphatics

Blood vessels and lymphatics are present in smooth muscle, but the density of blood vessels is much less than in skeletal muscle (in keeping with much less activity).

Pathological Correlation

- All varieties of muscle can hypertrophy when exposed to greater stress. Hypertrophy takes place by enlargement of existing fibres, and not by formation of new fibres. Skeletal muscle hypertrophies with exercise. Cardiac muscle hypertrophies if the load on a chamber of the heart is increased for any reason. An example is the hypertrophy of muscle in the wall of the left ventricle in hypertension. Hypertrophy of smooth muscle is seen most typically in the uterus where myocytes may increase from a length of about 15 to 20 µm at the beginning of pregnancy to as much as 500 µm towards the end of pregnancy.
- Smooth muscle and cardiac muscle have very little capacity for regeneration. Any defects produced by injury or disease are usually repaired by formation of fibrous tissue.
- Skeletal muscle fibres can undergo some degree of regeneration. They cannot divide to form new fibres. However, satellite cells present in relation to them (just deep to the external lamina) can give rise to new muscle fibres. Satellite cells are regarded as persisting myoblasts. When large segments of a muscle are destroyed the gap is filled in by fibrous tissue.

MYOEPITHELIAL CELLS

Apart from muscle myoepithelial cells show the presence of contractile proteins (actin and myosin). ***Myoepitheliocytes*** (or ***myoepithelial cells***) are present in close relation to secretory elements of some glands. They help to squeeze secretions out of secreting elements. Myoepithelial cells may be ***stellate***, forming baskets around acini, or may be ***fusiform***.

Myoepitheliocytes are seen in salivary glands, the mammary glands, and sweat glands. These cells are of ectodermal origin. With the EM they are seen to contain actin and myosin filaments. They can be localised histochemically, because they contain the protein ***desmin*** that is specific to muscle. Myoepithelial cells are innervated by autonomic nerves.

Clinical Correlation

- Excessive activity of smooth muscle is responsible for many symptoms. Constriction of bronchi leads to asthma. Spasm of smooth muscle can give rise to severe pain (colic) that may originate in the intestines (intestinal colic), ureter (renal colic), or bile duct (biliary colic). These symptoms can be relieved by drugs that cause relaxation of smooth muscle.
- Some diseases of muscle (referred to as muscular dystrophy) are caused by genetic defects in proteins that link fibres of the cytoskeleton to the external lamina. One such protein is dystrophin, and its absence is associated with a disease called **Duchenne muscular dystrophy**.

Chapter 9

Lymphatics and Lymphoid Tissue

The lymphoid system includes lymphatic vessels and lymphoid tissue.

LYMPHATIC VESSEL

When circulating blood reaches the capillaries, part of its fluid content passes into the surrounding tissues as tissue fluid. Most of this fluid re-enters the capillaries at their venous ends. Some of it is, however, returned to the circulation through a separate system of ***lymphatic vessels*** (usually called ***lymphatics***).

The fluid passing through the lymphatic vessels is called ***lymph***.

The smallest lymphatic (or lymph) vessels are lymphatic capillaries that join together to form larger lymphatic vessels.

The largest lymphatic vessel in the body is the ***thoracic duct***. It drains lymph from the greater part of the body. The thoracic duct ends by joining the left subclavian vein at its junction with the internal jugular vein. On the right side there is the ***right lymphatic duct*** that has a similar termination.

LYMPHOID TISSUE

Lymphoid tissue may be broadly classified as:
- Diffuse lymphoid tissue
- Dense lymphoid tissue

Diffuse Lymphoid Tissue

Diffuse lymphoid tissue consists of diffusely arranged lymphocytes and plasma cells in the mucosa of large intestine, trachea, bronchi and urinary tract.

Dense Lymphoid Tissue

It consists of an aggregation of lymphocytes arranged in the form of nodules. These nodules are found either as discrete encapsulated organs or in close association to the lining epithelium of the gut. Dense lymphoid tissue can therefore be divided as:
- ***Discrete lymphoid organs:*** These include thymus, lymph nodes, spleen and tonsils.
- ***Mucosa-associated lymphoid tissue (MALT):*** Small numbers of lymphocytes may be present almost anywhere in the body, but significant aggregations are seen in relation to the

mucosa of the respiratory, alimentary and urogenital tracts. These aggregations are referred to as ***MALT***.

- ***Mucosa-associated lymphoid tissue in the respiratory system:*** In the respiratory system the aggregations are relatively small and are present in the walls of the trachea and large bronchi. The term ***bronchial-associated lymphoid tissue*** (***BALT***) is applied to these aggregations.
- ***Mucosa-associated lymphoid tissue in the alimentary system:*** This is also called ***gut-associated lymphoid tissue*** (GALT) and includes Peyer's patches of ilium, adenoids (located in the roof of pharynx), lingual tonsils in posterior 1/3rd of tongue, palatine tonsils and lymphoid nodules in vermiform appendix.

Note: Thymus and bone marrow are primary lymphoid organs while others are secondary lymphoid organs.

LYMPH

Lymph is a transudate from blood and contains the same proteins as in plasma, but in smaller amounts, and in somewhat different proportions. Suspended in lymph there are cells that are chiefly lymphocytes. Most of these lymphocytes are added to lymph as it passes through lymph nodes, but some are derived from tissues drained by the nodes.

Large molecules of fat (chylomicrons) that are absorbed from the intestines enter lymph vessels. After a fatty meal these fat globules may be so numerous that lymph becomes milky (and is then called ***chyle***). Under these conditions the lymph vessels can be seen easily as they pass through the mesentery.

LYMPHOCYTES

The cell population of a lymph node is made up (overwhelmingly) of lymphocytes.

In the embryo lymphocytes are derived from mesenchymal cells present in the wall of the yolk sac, in the liver and in the spleen. These stem cells later migrate to bone marrow. Lymphocytes formed from these stem cells (in bone marrow) enter the blood. Depending on their subsequent behaviour they are classified into two types.

- Some of them travel in the bloodstream to reach the thymus. Here they divide repeatedly and undergo certain changes. They are now called ***T-lymphocytes*** ('T' from thymus). These T-lymphocytes, that have been 'processed' in the thymus re-enter the circulation to reach lymphoid tissue in lymph nodes, spleen, tonsils and intestines.

  In lymph nodes T-lymphocytes are found in the diffuse tissue around lymphatic nodules. In the spleen they are found in white pulp. From these masses of lymphoid tissue many lymphocytes pass into lymph vessels, and through them they go back into the circulation. In this way lymphocytes keep passing out of blood into lymphoid tissue (and bone marrow), and back from these into the blood. About 85% of lymphocytes seen in blood are T-lymphocytes (Fig. 9.1).
- Lymphocytes of a second group arising from stem cells in bone marrow enter the bloodstream, but do not go to the thymus. They go directly to lymphoid tissues (other than the thymus). Such lymphocytes are called ***B-lymphocytes*** ('B' from ***bursa of Fabricus***, a diverticulum of the cloaca in birds: in birds B-lymphocytes are formed here). In contrast

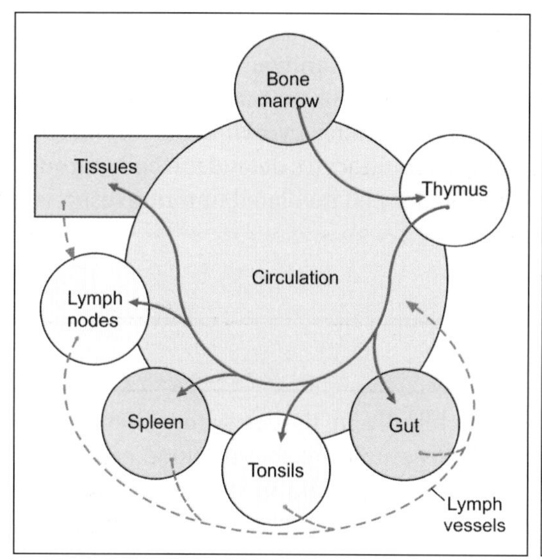

Fig. 9.1: Circulation of T-lymphocytes (Schematic representation)

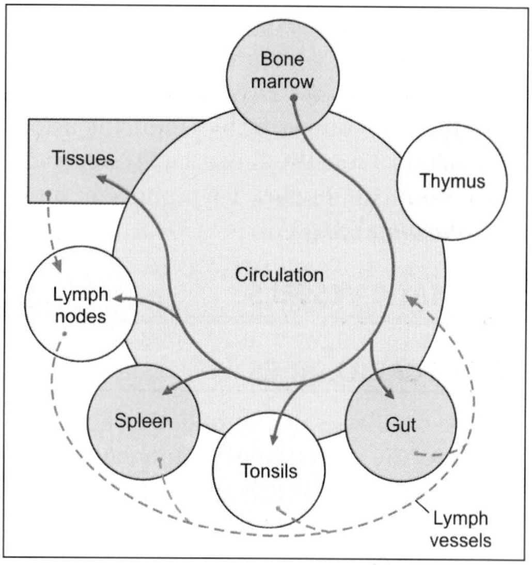

Fig. 9.2: Circulation of B-lymphocytes (Schematic representation)

to T-lymphocytes that lie in the diffuse lymphoid tissue of the lymph nodes and spleen, B-lymphocytes are seen in lymphatic nodules. The germinal centres are formed by actively dividing B-lymphocytes, while the dark rims of lymphatic nodules are formed by dense aggregations of B-lymphocytes. Like T-lymphocytes, B-lymphocytes also circulate between lymphoid tissues and the bloodstream (Fig. 9.2).

Lymphocytes and the Immune System

Lymphocytes are an essential part of the ***immune system*** of the body that is responsible for defence against invasion by bacteria and other organisms. In contrast to granulocytes and monocytes that directly attack invading organisms, lymphocytes help to destroy them by producing substances called ***antibodies***. These are protein molecules that have the ability to recognise a 'foreign' protein (i.e., a protein not normally present in the individual). The foreign protein is usually referred to as an ***antigen***.

Every antigen can be neutralised only by a specific antibody. It follows that lymphocytes must be capable of producing a very wide range of antibodies; or rather that there must be a very wide variety of lymphocytes, each variety programmed to recognise a specific antigen and to produce antibodies against it.

This function of antibody production is done by ***B-lymphocytes***. When stimulated by the presence of antigen the cells enlarge and get converted to plasma cells. The plasma cells produce antibodies.

Antibodies are also called ***immunoglobulins***. Immunoglobulins are of five main types viz., IgG, IgM, IgA, IgE and IgD.

T-lymphocytes are also concerned with immune responses, but their role is somewhat different from that of B-lymphocytes. T-lymphocytes specialise in recognising cells that are foreign to the host body. These may be fungi, virus infected cells, tumour cells, or cells of another individual. T-lymphocytes have surface receptors that recognise specific antigens (there being many varieties of T-lymphocytes each type recognising a specific antigen). When exposed to a suitable stimulus the T-lymphocytes multiply and form large cells that can destroy abnormal cells by direct contact, or by producing cytotoxic substances called ***cytokines*** or ***lymphokines*** (See below). From the above it will be seen that while B-lymphocytes defend the body through blood borne antibodies, T-lymphocytes are responsible for cell mediated immune responses (***cellular immunity***).

LYMPHATIC VESSELS

LYMPH CAPILLARIES

Lymph capillaries (or lymphatic capillaries) begin blindly in tissues where they form a network. The structure of lymph capillaries is basically similar to that of blood capillaries, but is adapted for much greater permeability. There is an inner lining of endothelium. The basal lamina is absent or poorly developed. Pericytes or connective tissue are not present around the capillary.

As compared to blood capillaries, much larger molecules can pass through the walls of lymph capillaries. These include colloidal material, fat droplets, and particulate matter such as bacteria. It is believed that these substances pass into lymph capillaries through gaps between endothelial cells lining the capillary; or by pinocytosis.

Lymph capillaries are present in most tissues of the body. They are absent in avascular tissues (e.g., the cornea, hair, nails); in the splenic pulp; and in the bone marrow. It has been held that lymphatics are not present in nervous tissue, but we now know that some vessels are present.

LARGER LYMPH VESSELS

The structure of the thoracic duct and of other larger lymph vessels is similar to that of veins (Fig. 9.3).

A tunica intima, media and adventitia can be distinguished. Elastic fibres are prominent and can be seen in all three layers.

The media, and also the adventitia contain some smooth muscle.

In most vessels, the smooth muscle is arranged circularly, but in the thoracic duct the muscle is predominantly longitudinal.

Numerous valves, similar to those in veins, are present in small as well as large lymphatic vessels. They are more numerous than in veins. The valves often give lymph vessels a beaded appearance.

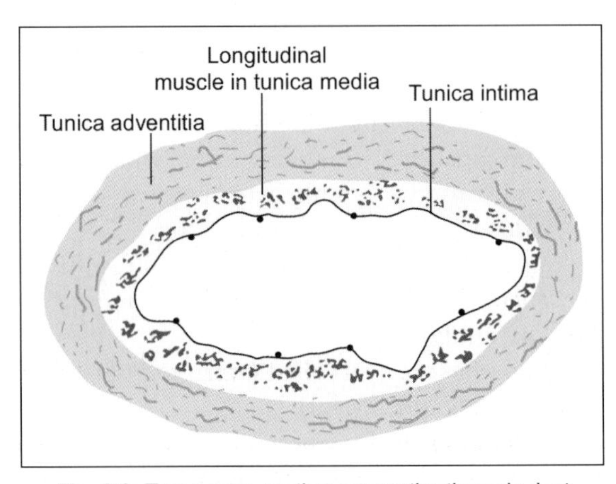

Fig. 9.3: Transverse section across the thoracic duct (Schematic representation)

Clinical Correlation

- **Lymphangitis** is inflammation of the lymphatics. Lymphangitis may be acute or chronic.
  - **Acute lymphangitis** occurs in the course of many bacterial infections. The most common organisms are β-haemolytic streptococci and staphylococci. When this happens in vessels of the skin, the vessels are seen as red lines that are painful.
  - **Chronic lymphangitis** occurs due to persistent and recurrent acute lymphangitis or from chronic infections like tuberculosis, syphilis and actinomycosis.
- **Lymphoedema** is swelling of soft tissues due to localised increase in the quantity of lymph. It may be primary (idiopathic) or secondary (obstructive). Secondary lymphoedema is more common and may be due to lymphatic invasion by malignant tumour. Post-irradiation fibrosis. Parasitic infestations, e.g., in filariasis of lymphatics producing elephantiasis. Obstructive lymphoedema occurs only when the obstruction is widespread as otherwise collaterals develop. The affected area consists of dilatation of lymphatics distal to obstruction with increased interstitial fluid. Rupture of dilated large lymphatics may result in escape of milky chyle into the peritoneum **(chyloperitoneum)**, into the pleural cavity **(chylothorax)**, into pericardial cavity **(chylopericardium)** and into the urinary tract **(chyluria)**.

LYMPH NODES

General Features

Scattered along the course of lymphatic vessels there are numerous small masses of lymphoid tissue called ***lymph nodes*** that are usually present in groups.

As a rule lymph from any part of the body passes through one or more lymph nodes before entering the blood stream. (There are some exceptions to this rule. For example, some lymph from the thyroid gland drains directly into the thoracic duct). Lymph nodes act as filters removing bacteria and other particulate matter from lymph. Lymphocytes are added to lymph in these nodes.

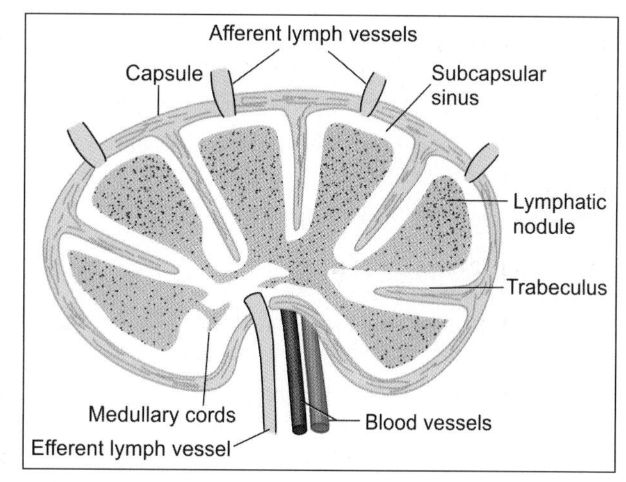

Fig. 9.4: Some features of the structure of a lymph node (Schematic representation)

Each group of lymph nodes has a specific area of drainage.

Each lymph node consists of a connective tissue framework; and of numerous lymphocytes and other cells, that fill the interstices of the network. The entire node is bean-shaped, the concavity constituting a hilum through which blood vessels enter and leave the node. Several lymph vessels enter the node on its convex aspect. Usually, a single lymph vessel leaves the node through its hilum (Fig. 9.4).

Microscopic Features

When a section through a lymph node is examined (at low magnification) it is seen that the

PLATE 9.1: Lymph Node

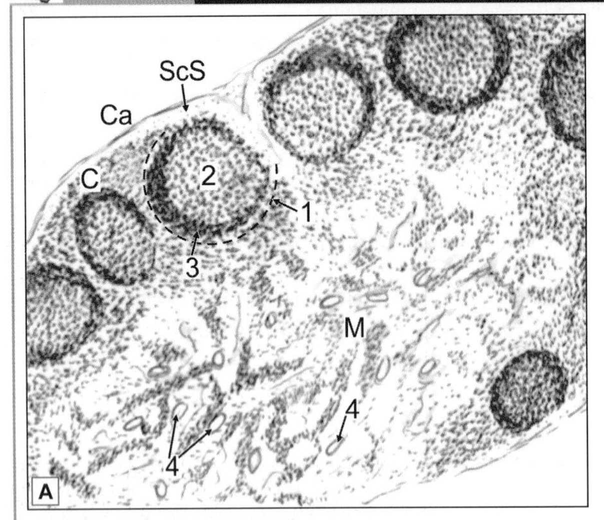

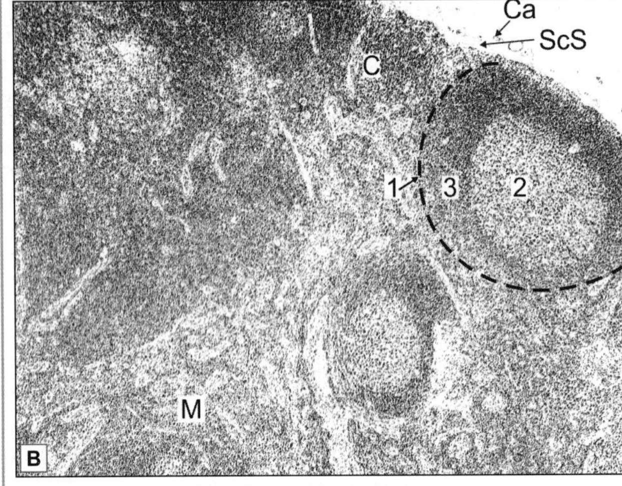

Lymph node. A. As seen in drawing; B. Photomicrograph

- A thin capsule surrounds the lymph node and sends in trabeculae
- Just beneath the capsule a clear space is seen. This is the subcapsular sinus
- A lymph node has an outer cortex and an inner medulla
- The cortex is packed with lymphocytes. A number of rounded lymphatic follicles (or nodules) are present. Each nodule has a pale staining germinal centre surrounded by a zone of densely packed lymphocytes
- Within the medulla the lymphocytes are arranged in the form of anastomosing cords. Several blood vessels can be seen in the medulla.

Note: All lymphoid tissue are easily recognised due to presence of aggregation of dark staining nuclei. The nuclei belong to lymphocytes.

Key

1. Lymphatic nodule
2. Germinal centre
3. Zone of dense lymphocytes
4. Blood vessels

C. Cortex
M. Medulla
Ca. Capsule
Scs. Subcapsular sinus

node has an outer zone that contains densely packed lymphocytes, and therefore stains darkly: this part is the ***cortex***. The cortex does not extend into the hilum.

Surrounded by the cortex, there is a lighter staining zone in which lymphocytes are fewer: this area is the ***medulla*** (Plate 9.1).

Cortex

Within the cortex there are several rounded areas that are called ***lymphatic follicles*** or ***lymphatic nodules***. Each nodule has a paler staining ***germinal centre*** surrounded by a zone of densely packed lymphocytes.

Medulla

Within the medulla, the lymphocytes are arranged in the form of branching and anastomosing cords.

Connective Tissue Framework

A lymph node is surrounded by a ***capsule***. The capsule consists mainly of collagen fibres. Some elastic fibres and some smooth muscle may be present.

Just below the capsule is the ***subcapsular sinus*** (Fig. 9.4).

A number of ***septa*** (or ***trabeculae***) extend into the node from the capsule and divide the node into lobules.

The hilum is occupied by a mass of dense fibrous tissue.

A delicate network of reticular fibres occupies the remaining spaces forming a fibrous/reticular framework within the node. Associated with the network there are reticular cells that have traditionally been regarded as macrophages. However, it is now believed that they are fibroblasts and do not have phagocytic properties.

Circulation of Lymph through Lymph Nodes

The entire lymph node is pervaded by a network of reticular fibres. Most of the spaces of this network are packed with lymphocytes. At some places, however, these spaces contain relatively few cells, and form channels through which lymph circulates. These channels known as sinuses are lined by endothelium, but their walls allow free movement of lymphocytes and macrophages into and out of the channels.

Afferent lymphatics reaching the convex outer surface of the node enter an extensive ***subcapsular sinus*** (Fig. 9.4). From this sinus a number of radial cortical sinuses run through the cortex towards the medulla.

Reaching the medulla the sinuses join to form larger medullary sinuses. In turn the medullary sinuses join to form (usually) one, or more than one, efferent lymph vessel through which lymph leaves the node.

Note: The afferent vessels to a lymph node enter the cortex, while the efferent vessel emerges from the medulla. The sinuses are lined by endothelium.

Lymph passing through the system of sinuses comes into intimate contact with macrophages present in the node. Bacteria and other particulate matter are removed from lymph by these cells. Lymphocytes freely enter or leave the node through these channels.

Blood Supply of Lymph Nodes

Arteries enter the lymph node at the hilum. They pass through the medulla to reach the cortex where they end in arterioles and capillaries. These arterioles and capillaries are arranged as loops that drain into venules.

Postcapillary venules in lymph nodes are unusual in that they are lined by cuboidal endothelium. (They are, therefore, called ***high endothelial venules***). This 'high' endothelium readily allows the passage of lymphocytes between the bloodstream and the surrounding tissue.

These endothelial cells bear receptors that are recognised by circulating lymphocytes. Contact with these receptors facilitates passage of lymphocytes through the vessel wall.

Functions

- Lymph nodes are centres of lymphocyte production. Both B-lymphocytes and T-lymphocytes are produced here by multiplication of preexisting lymphocytes. These lymphocytes pass into lymph and thus reach the bloodstream.
- Bacteria and other particulate matter are removed from lymph through phagocytosis by macrophages. Antigens thus carried into these cells are 'presented' to lymphocytes stimulating their proliferation. In this way lymph nodes play an important role in the immune response to antigens.
- Plasma cells (representing fully mature B-lymphocytes) produce antibodies against invading antigens, while T-lymphocytes attack cells that are 'foreign' to the host body.

Clinical Correlation

- Infection in any part of the body can lead to enlargement and inflammation of lymph nodes draining the area. Inflammation of lymph nodes is called **lymphadenitis**.
- Carcinoma (cancer) usually spreads from its primary site either by growth of malignant cells along lymph vessels, or by 'loose' cancer cells passing through lymph to nodes into which the area drains. This leads to enlargement of the lymph nodes of the region. Examination of lymph nodes gives valuable information about the spread of cancer. In surgical excision of cancer, lymph nodes draining the region are usually removed.

THE SPLEEN

General Features

The spleen is the largest lymphoid organ of the body. Normally it is a blood-forming organ in foetal life and blood-destroying organ in postnatal life (graveyard of RBCs). Since it is in the bloodstream, it filters the blood from blood-borne antigens and microorganisms.

Connective Tissue Framework

Except at the hilum, the surface of the spleen is covered by a layer of peritoneum (referred to as the ***serous coat***). Deep to the serous layer the organ is covered completely by a ***capsule*** (Plate 9.2).

Trabeculae arising from the capsule extend into the substance of the spleen. As they do so the trabeculae divide into smaller divisions that form a network.

The capsule and trabeculae are made up of fibrous tissue in which elastic fibres are abundant. In some animals they contain much smooth muscle, but this is not a prominent feature of the human spleen.

The spaces between the trabeculae are pervaded by a network of reticular fibres, embedded in an amorphous matrix.

Fibroblasts (reticular cells) and macrophages are also present in relation to the reticulum. The interstices of the reticulum are pervaded by lymphocytes, blood vessels and blood cells, and by macrophages.

Parenchyma

On examination with naked eye, the interior of the spleen shows round white areas (white pulp) surrounded by red matrix (red pulp).

PLATE 9.2: Spleen

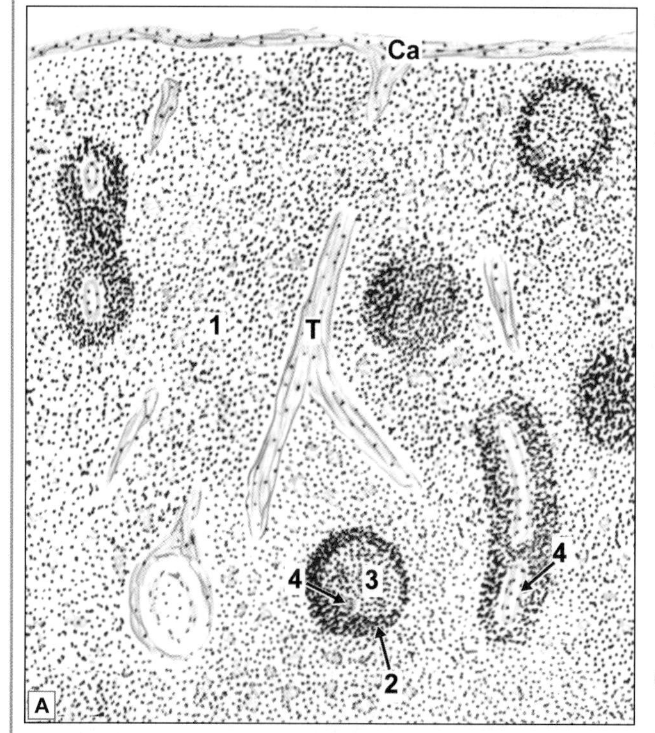

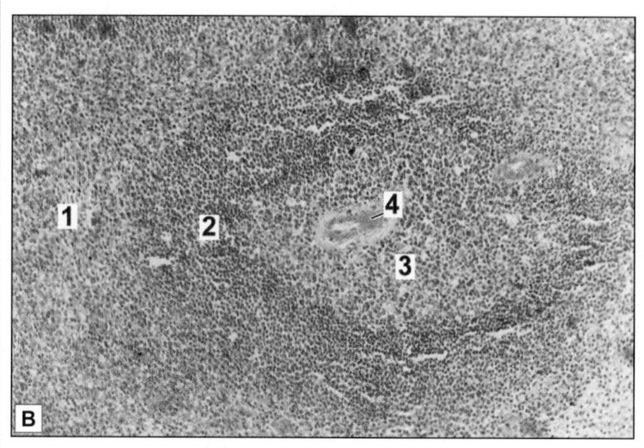

Spleen. A. As seen in drawing; B. Photomicrograph

- The spleen is characterised by a thick capsule with trabeculae extending from it into the organ (not shown in photomicrograph)
- The substance of the organ is divisible into the red pulp in which there are diffusely distributed lymphocytes and numerous sinusoids; and the white pulp in which dense aggregations of lymphocytes are present. The latter are in the form of cords surrounding arterioles
- When cut transversely the cords resemble the lymphatic nodules of lymph nodes, and like them they have germinal centres surrounded by rings of densely packed lymphocytes. However, the nodules of the spleen are easily distinguished from those of lymph nodes because of the presence of an arteriole in each nodule
- This arteriole occupies an eccentric position in the nodule.

Note: Observe that this organ is full of lymphocytes. In the drawing also see the cords of densely packed lymphocytes around arteriole.

Key

1. Red pulp
2. White pulp
3. Germinal centre
4. Arteriole

Ca. Capsule
T. Trabeculus

The White Pulp

The white pulp is made up of aggregations of lymphocytes that surround a small artery or arterioles. As a result it is in the form of cord-like aggregations of lymphocytes that follow the branching pattern of the arterioles. The cords appear to be circular in transverse section.

At places the cords are thicker than elsewhere and contain lymphatic nodules similar to those seen in lymph nodes. These nodules are called ***Malpighian bodies***.

Each nodule has a germinal centre and a surrounding cuff of densely packed lymphocytes.

The nodules are easily distinguished from those of lymph nodes because of the presence of an arteriole in each of them. The arteriole is placed eccentrically at the margin of the germinal centre (between it and the surrounding cuff of densely packed cells). More than one arteriole may be present in relation to one germinal centre (Plate 9.2).

The functional significance of the white pulp is similar to that of cortical tissue of lymph nodes. Lymphatic nodules of the white pulp are aggregations of B-lymphocytes. The germinal centres are areas where B-lymphocytes are dividing. Lymphocytes surrounding the arteriole are referred to as peri-arterial lymphatic sheath (PALS). Lymphocytes of the PALS are chiefly T-lymphocytes.

The Red Pulp

Red pulp is a modified lymphoid tissue infiltrated with all the cells of circulating blood. It is like a sponge permeated by spaces (splenic sinusoids) lined by reticular cells. The intervals between the spaces are filled by B-lymphocytes as well as T-lymphocytes, macrophages, and blood cells. These cells appear to be arranged as cords (***splenic cords***, of Billroth). The cords form a network.

The zone of red pulp immediately surrounding white pulp is the ***marginal zone***. This zone has a rich network of sinusoids. Numerous antigen-presenting cells are found close to the sinusoids. This region seems to be specialised for bringing antigens confined to circulating blood (e.g., some bacteria) into contact with lymphocytes in the spleen so that an appropriate immune response can be started against the antigens (Such contact does not take place in lymph nodes. Antigens reach lymph nodes from tissues, through lymph). Surgical removal of the spleen (splenectomy) reduces the ability of the body to deal with blood borne infections.

Circulation through the Spleen

On reaching the hilum of the spleen, the splenic artery divides into about five branches that enter the organ independently.

Each branch divides and subdivides as it travels through the trabecular network. Arterioles arising from this network leave the trabeculae to pass into the inter-trabecular spaces. For some distance each arteriole is surrounded by a dense sheath of lymphocytes. These lymphocytes constitute the ***white pulp*** of the spleen.

The arteriole then divides into a number of straight vessels that are called ***penicilli***. Penicillar arterioles either open into the red pulp (***Open circulation***) or they open into the splenic sinusoids (***Closed circulation***) (Fig. 9.5). Veins from these sinusoids and the red pulp end in the trabecular veins.

The sinusoids of the spleen are lined by a somewhat modified endothelium. The endothelial cells here are elongated and are shaped like bananas. They are referred to as ***stave cells***. With

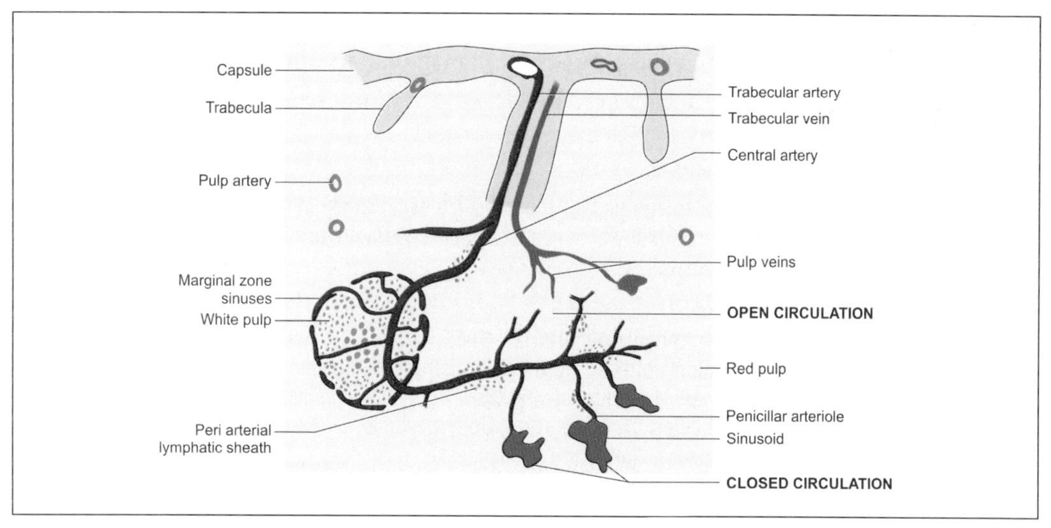

Fig. 9.5: The splenic circulation (Schematic representation)

the EM a system of ultramicroscopic fibrils is seen to be present in their cytoplasm. The fibrils may help to alter the shape of the endothelial cells thus opening or closing gaps between adjoining cells.

The spleen acts as a filter for worn out red blood cells. Normal erythrocytes can change shape and pass easily through narrow passages in penicilli and ellipsoids. However, cells that are aged are unable to change shape and are trapped in the spleen where they are destroyed by macrophages.

Lymph Vessels of the Spleen

Traditionally, it has been held that in the spleen lymph vessels are confined to the capsule and trabeculae. Recent studies have shown, however, that they are present in all parts of the spleen. Lymphocytes produced in the spleen reach the bloodstream mainly through the lymph vessels.

Functions

- Like other lymphoid tissues the spleen is a centre where both B-lymphocytes and T-lymphocytes multiply, and play an important role in immune responses. As stated above, the spleen is the only site where an immune response can be started against antigens present in circulating blood (but not present in tissues).
- The spleen contains the largest aggregations of macrophages of the mononuclear phagocyte system. In the spleen the main function of these cells is the destruction of red blood corpuscles that have completed their useful life. This is facilitated by the intimate contact of blood with the macrophages because of the presence of an open circulation. Macrophages also destroy worn out leukocytes, and bacteria.
- In fetal life the spleen is a centre for production of ***all*** blood cells. In later life only lymphocytes are produced here.

- The spleen is often regarded as a store of blood that can be thrown into the circulation when required. This function is much less important in man than in some other species.

Clinical Correlation

Splenomegaly

Enlargement of spleen is termed as splenomegaly. It occurs in a wide variety of disorders which increase the cellularity and vascularity of the organ. Many of the causes are exaggerated forms of normal splenic function.

Causes of splenomegaly

- Infections: Malaria, Leishmaniasis.
- Disorders of immunoregulation: Rheumatoid arthritis, SLE.
- Altered splenic blood flow: Cirrhosis of liver, Portal vein obstruction, Splenic vein obstruction.
- Lymphohaematogenous malignancies: Hodgkin's disease, Non-Hodgkin's lymphomas.
- Diseases with abnormal erythrocytes: Thalassaemias, Spherocytosis, Sickle cell disease.
- Storage diseases: Gaucher's disease, Niemann-Pick's disease.
- Miscellaneous: Amyloidosis, Primary and metastatic splenic tumours.

THE THYMUS

The thymus is an organ that is a hazy entity for most students. This is because of the fact that the organ is not usually seen in dissection hall cadavers (because of atrophy in old people, and because of rapid autolysis after death). The organ is also not accessible for clinical examination (as it lies deep to the manubrium sterni). At birth the thymus weighs 10-15 grams. The weight increases to 30–40 grams at puberty. Subsequently, much of the organ is replaced by fat. However, the thymus is believed to produce T-lymphocytes throughout life.

The thymus consists of right and left lobes that are joined together by fibrous tissue. Each lobe has a connective tissue capsule. Connective tissue septa passing inwards from the capsule incompletely subdivide the lobe into a large number of lobules.

Each lobule is about 2 mm in diameter. It has an outer cortex and an inner medulla. Both the cortex and medulla contain cells of two distinct lineages as described below. The medulla of adjoining lobules is continuous.

The thymus has a rich blood supply. It does not receive any lymph vessels, but gives off efferent vessels.

Epithelial Cells (Epitheliocytes)

Embryologically these cells are derived from endoderm lining the third pharyngeal pouch (It is possible that some of them may be of ectodermal origin). The cells lose all contact with the pharyngeal wall. In the fetus their epithelial origin is obvious. Later they become flattened and may branch. The cells join to form sheets that cover the internal surface of the capsule, the surfaces of the septa, and the surfaces of blood vessels. The epithelial cells lying deeper in the lobule develop processes that join similar processes of other cells to form a reticulum. It may be noted that this reticulum is cellular, and has no similarity to the reticulum formed by reticular fibres (and associated fibroblasts) in lymph nodes and spleen. Epithelial cells of the thymus are not phagocytic.

It has been suggested that the sheets of epithelial cells present deep to the capsule, around septa, and around blood vessels form an effective ***blood-thymus barrier*** that prevents antigens (present in blood) from reaching lymphocytes present in the thymus. Epitheliocytes also promote T-cell differentiation and proliferation.

On the basis of structural differences several types of epitheliocytes are recognised. Type 1 epitheliocytes line the inner aspect of the capsule, the septa and blood vessels. These are the cells forming the partial haemothymic barrier mentioned above. Type 2 and type 3 cells are present in the outer and inner parts of the cortex respectively. Type 4 cells lie in the deepest parts of the cortex, and also in the medulla. They form a network containing spaces that are occupied by lymphocytes. Type 5 cells are present around corpuscles of Hassall (see below).

Cortical epitheliocytes are also described as ***thymic nurse cells***. They destroy lymphocytes that react against self antigens.

Lymphocytes of the Thymus (Thymocytes)

In the cortex of each lobule of the thymus the reticulum formed by epithelial cells is densely packed with lymphocytes. Stem cells formed in bone marrow travel to the thymus. Here they come to lie in the superficial part of the cortex, and divide repeatedly to form small lymphocytes. Lymphatic nodules are not present in the normal thymus.

The medulla of each lobule also contains lymphocytes, but these are less densely packed than in the cortex. As a result the epithelial reticulum is more obvious in the medulla than in the cortex. As thymocytes divide they pass deeper into the cortex, and into the medulla. Ultimately, they leave the thymus by passing into blood vessels and lymphatics. For further details of thymic lymphocytes see below.

Macrophages

Apart from epithelial cells and lymphocytes the thymus contains a fair number of macrophages (belonging to the mononuclear phagocyte system). They are placed subjacent to the capsule, at the corticomedullary junction, and in the medulla. The subcapsular macrophages are highly phagocytic. Deeper lying macrophages are dendritic cells. Their significance is considered below.

Corpuscles of Hassall

These are small rounded structures present in the medulla of the thymus. Each corpuscle has a central core formed by epithelial cells that have undergone degeneration. These cells ultimately form a pink staining hyaline mass. Around this mass there is a wall formed by concentrically arranged epithelial cells. These cells also stain bright pink with haematoxylin and eosin. The central mass of the corpuscle may also contain degenerating macrophages. The functional significance of the corpuscles of Hassall is not understood.

Functions of The Thymus

- The role of the thymus in lymphopoiesis has been discussed earlier. Stem cells (from bone marrow) that reach the superficial part of the cortex divide repeatedly to form smaller lymphocytes. It has been postulated that during these mitoses the DNA of the lymphocytes undergoes numerous random mutations, as a result of which different lymphocytes acquire

Plate 9.3: Thymus

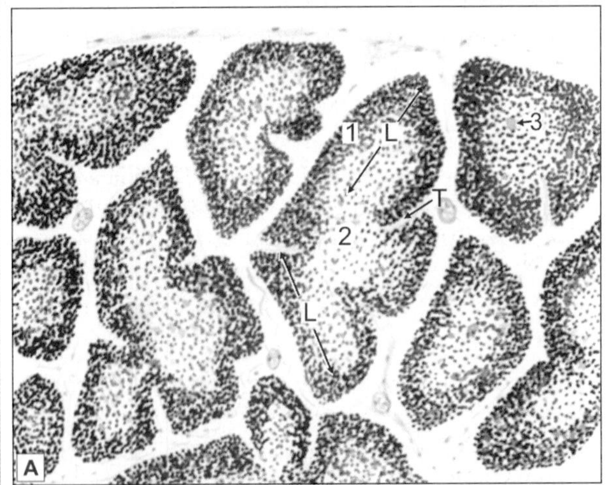

In the slide it can be seen that

- The thymus is made up of lymphoid tissue arranged in the form of distinct lobules. The presence of this lobulation enables easy distinction of the thymus from all other lymphoid organs
- The lobules are partially separated from each other by connective tissue septae
- In each lobule an outer darkly stained cortex (in which lymphocytes are densely packed); and an inner lightly stained medulla (in which the cells are diffuse) are present
- Whereas the cortex is confined to one lobule, the medulla is continuous from one lobule to another
- The medulla contains pink staining rounded masses called the corpuscles of Hassall.

Key

1. Cortex
2. Medulla
3. Hassall's corpuscle
4. Epithelial cell

L. Lobule
T. Trabeculae

Thymus. A. As seen in drawing (low magnification); B. As seen in drawing (high magnification); C. Photomicrograph

the ability to recognise a very large number of different proteins, and to react to them. As it is not desirable for lymphocytes to react against the body's own proteins, all lymphocytes that would react against them are destroyed. It is for this reason that 90% of lymphocytes formed in the thymus are destroyed within three to four days. The remaining lymphocytes, that react only against proteins foreign to the body, are thrown into the circulation as circulating, immunologically competent T-lymphocytes. They lodge themselves in secondary lymph organs like lymph nodes, spleen, etc., where they multiply to form further T-lymphocytes of their own type when exposed to the appropriate antigen.

From the above it will be understood why the thymus is regarded as a ***primary lymphoid organ*** (along with bone marrow). It has been held that, within the thymus, lymphocytes are not allowed to come into contact with foreign antigens, because of the presence of the blood-thymic barrier. It has also been said that because of this thymocytes do not develop into large lymphocytes or into plasma cells, and do not form lymphatic nodules. While these views may hold as far as the thymic cortex is concerned, they do not appear to be correct in respect of the medulla. Recently it has been postulated that the medulla of the thymus (or part of it) is a separate 'compartment'. After thymocytes move into this compartment they probably come into contact with antigens presented to them through dendritic macrophages. Such contact may be a necessary step in making T-lymphocytes competent to distinguish between foreign antigens and proteins of the body itself.

- The proliferation of T-lymphocytes and their conversion into cells capable of reacting to antigens, probably takes place under the influence of hormones produced by epithelial cells of the thymus. T-lymphocytes are also influenced by direct cell contact with epitheliocytes. Hormones produced by the thymus may also influence lymphopoiesis in peripheral lymphoid organs. This influence appears to be specially important in early life, as lymphoid tissues do not develop normally if the thymus is removed. Thymectomy has much less influence after puberty as the lymphoid tissues have fully developed by then.

  A number of hormones produced by the thymus have now been identified as follows. They are produced by epitheliocytes.

  - ***Thymulin*** enhances the function of various types of T-cells, specially that of suppressor cells.
  - ***Thymopoietin*** stimulates the production of cytotoxic T-cells. The combined action of thymulin and thymopoietin allows precise balance of the activity of cytotoxic and suppressor cells.
  - ***Thymosin alpha 1*** stimulates lymphocyte production, and also the production of antibodies.
  - ***Thymosin beta 4*** is produced by mononuclear phagocytes.
  - ***Thymic humoral factor*** controls the multiplication of helper and suppressor T-cells.

Apart from their actions on lymphocytes, hormones (or other substances) produced in the thymus probably influence the adenohypophysis and the ovaries. In turn, the activity of the thymus is influenced by hormones produced by the adenohypophysis, by the adrenal cortex, and by sex hormones.

Clinical Correlation

Enlargement of the thymus is often associated with a disease called **myasthenia gravis**. In this condition there is great weakness of skeletal muscle. In many such cases the thymus is enlarged and there may be a tumour in it. Removal of the thymus may result in considerable improvement in some cases.

Myasthenia gravis is now considered to be a disturbance of the immune system. There are some proteins to which acetyl choline released at motor end plates gets attached. In myasthenia gravis antibodies are produced against these proteins rendering them ineffective. Myasthenia gravis is, thus, an example of a condition in which the immune system begins to react against one of the body's own proteins. Such conditions are referred to as **autoimmune diseases**.

MUCOSA-ASSOCIATED LYMPHOID TISSUE

As discussed significant aggregations of lymphocytes seen in relation to the mucosa of the respiratory, alimentary and urogenital tracts are referred to as ***mucosa-associated lymphoid tissue (MALT).*** The total volume of MALT is more or less equal to that of the lymphoid tissue present in lymph nodes and spleen. Mucosa associated aggregations of lymphoid tissue have some features in common as follows:

- These aggregations are in the form of one or more lymphatic follicles (nodules) having a structure similar to nodules of lymph nodes. Germinal centres may be present. Diffuse lymphoid tissue (termed the ***parafollicular zone***) is present in the intervals between the nodules. The significance of the nodules and of the diffuse aggregations of lymphocytes are the same as already described in the case of lymph nodes. The nodules consist predominantly of B-lymphocytes, while the diffuse areas contain T-lymphocytes.
- These masses of lymphoid tissue are present in very close relationship to the lining epithelium of the mucosa in the region concerned, and lie in the substantia propria. Larger aggregations extend into the submucosa. Individual lymphocytes may infiltrate the epithelium and may pass through it into the lumen.
- The aggregations are not surrounded by a capsule, nor do they have connective tissue septa. A supporting network of reticular fibres is present.
- As a rule these masses of lymphoid tissue do not receive afferent lymph vessels, and have no lymph sinuses. They do not, therefore, serve as filters of lymph. However, they are centres of lymphocyte production. Lymphocytes produced here pass into lymph nodes of the region through efferent lymphatic vessels. Some lymphocytes pass through the overlying epithelium into the lumen.

Mucosa-associated Lymphoid Tissue in the Respiratory System

In the respiratory system the aggregations are relatively small and are present in the walls of the trachea and large bronchi. The term ***bronchial-associated lymphoid tissue*** (***BALT***) is applied to these aggregations.

Mucosa-associated Lymphoid Tissue in the Alimentary System

This is also called ***gut-associated lymphoid tissue*** (GALT). In the alimentary system examples of aggregations of lymphoid tissue are tonsils, Peyer's patches and lymphoid nodules in vermiform appendix.

PLATE 9.4: Palatine Tonsil

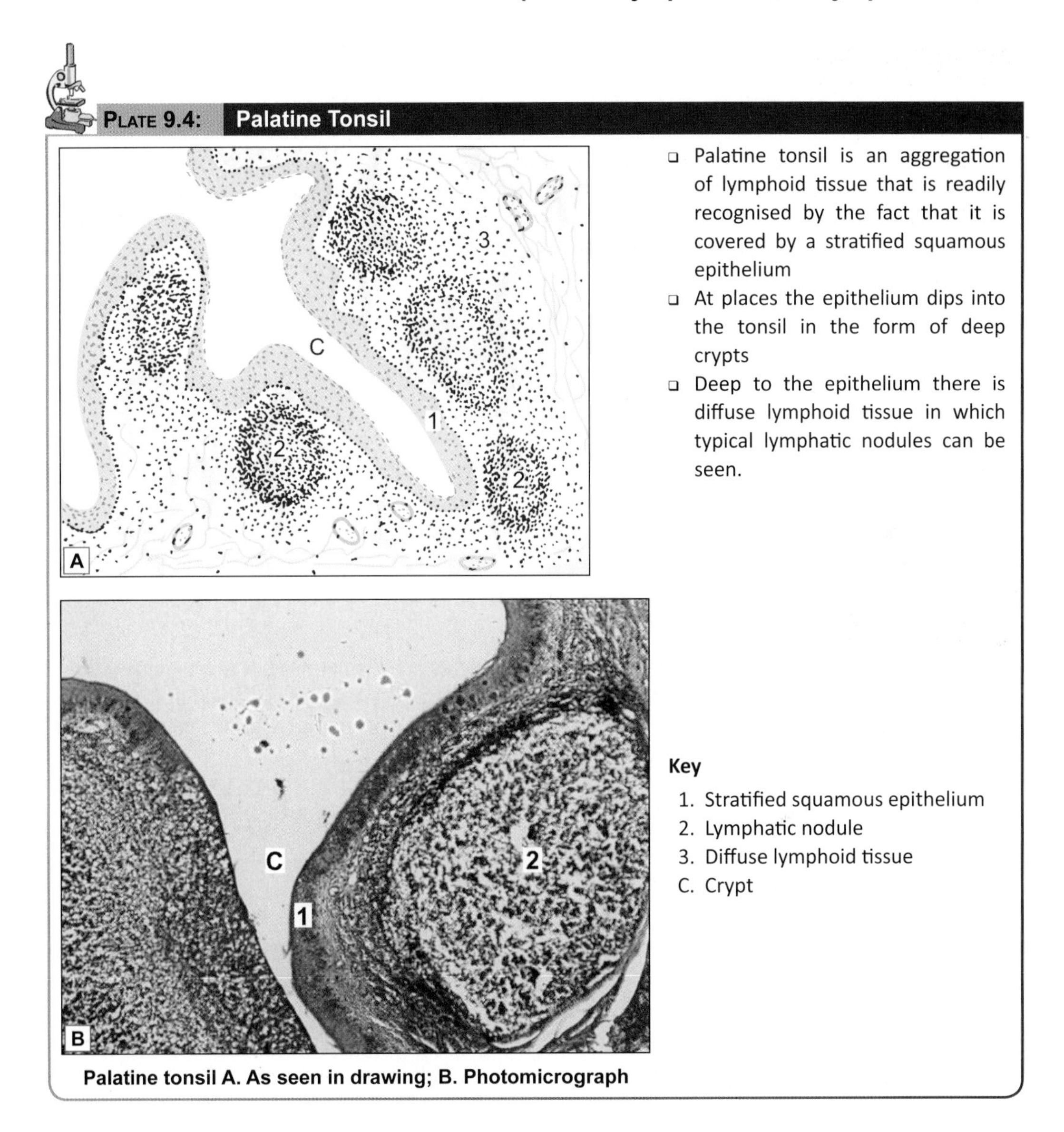

- Palatine tonsil is an aggregation of lymphoid tissue that is readily recognised by the fact that it is covered by a stratified squamous epithelium
- At places the epithelium dips into the tonsil in the form of deep crypts
- Deep to the epithelium there is diffuse lymphoid tissue in which typical lymphatic nodules can be seen.

Key

1. Stratified squamous epithelium
2. Lymphatic nodule
3. Diffuse lymphoid tissue

C. Crypt

Palatine tonsil A. As seen in drawing; B. Photomicrograph

TONSILS

Near the junction of the oral cavity with the pharynx there are a number of collections of lymphoid tissue that are referred to as ***tonsils***. The largest of these are the right and left ***palatine tonsils***, present on either side of the oropharyngeal isthmus (In common usage the word tonsils refers to the palatine tonsils). Another midline collection of lymphoid tissue, the ***pharyngeal tonsil,*** is present on the posterior wall of the pharynx. Smaller collections are present on the dorsum of the posterior part of the tongue (***lingual tonsils***), and around the pharyngeal openings of the auditory tubes (***tubal tonsils***).

The Palatine Tonsils

Each palatine tonsil (right or left) consists of diffuse lymphoid tissue in which lymphatic nodules are present. The lymphoid tissue is covered by stratified squamous epithelium continuous with that of the mouth and pharynx (Fig. 9.6 and Plate 9.4). This epithelium extends into the substance of the tonsil in the form of several ***tonsillar crypts***. The lumen of a crypt usually contains some lymphocytes that have travelled into it through the epithelium. Desquamated epithelial cells and bacteria are also frequently present in the lumen of the crypt.

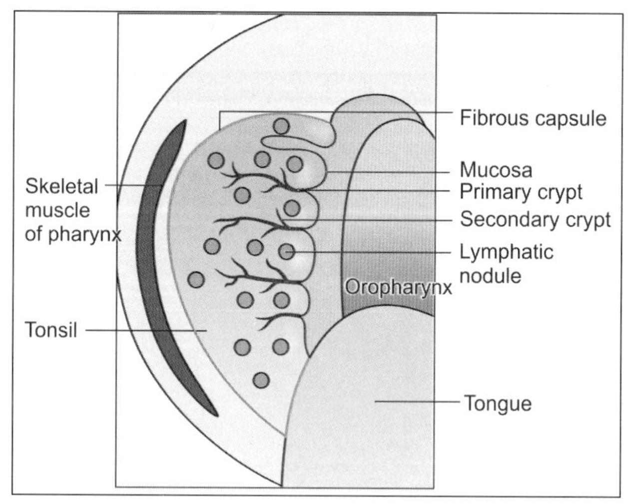

Fig. 9.6: Coronal section of palatine tonsil (Schematic representation)

Clinical Correlation

Tonsillitis is the infection of palatine tonsils. The palatine tonsils are often infected. It is a common cause of sore throat. Frequent infections can lead to considerable enlargement of the tonsils, specially in children. Such enlarged tonsils may become a focus of infection and their surgical removal (**tonsillectomy**) may then become necessary.

Chapter 10

The Blood and the Mononuclear Phagocyte System

Blood is regarded as a modified connective tissue because the cellular elements in it are separated by a considerable amount of 'intercellular substance' (see below); and because some of the cells in it have close affinities to cells in general connective tissue.

THE PLASMA

In contrast to all other connective tissues, the 'intercellular substance' of blood is a liquid called ***plasma***. The cellular elements float freely in the plasma. Plasma consists of water with dissolved ***colloids*** and ***crystalloids***. The colloids are proteins including prothrombin (associated with the clotting of blood), immunoglobulins (involved in immunological defence mechanisms), hormones, etc. The crystalloids are ions of sodium, chloride, potassium, calcium, magnesium, phosphate, bicarbonate, etc. Several other substances like glucose and amino acids are also present.

About 55% of the total volume of blood is plasma, the rest being constituted by the cellular elements described below.

CELLULAR ELEMENTS OF BLOOD

The cellular or formed elements of blood are of three main types. These are ***red blood corpuscles*** or ***erythrocytes***, ***white blood corpuscles*** or ***leucocytes***, and ***blood platelets***. We refer to them as 'cellular' or 'formed' elements rather than as cells because of the fact that red blood corpuscles are not strictly cells (see below). However, in practice, the terms red blood cells and white blood cells are commonly used.

We have seen that about 55% of the total volume of blood is accounted for by plasma. Most of the remaining 45% is made up of red blood corpuscles, the leucocytes and platelets constituting less than 1% of the volume. If we take one cubic millimetre (mm^3 = microlitre or μL) of blood we find that it contains about five million erythrocytes. In comparison there are only about 7000 leucocytes in the same volume of blood.

ERYTHROCYTES (RED BLOOD CORPUSCLES)

When seen in surface view each erythrocyte is a circular disc having a diameter of about 7 μm (6.5–8.5 μm). When viewed from the side it is seen to be biconcave, the maximum thickness being about 2 μm (Fig. 10.1). Erythrocytes are cells that have lost their nuclei (and other organelles). They are bounded by a plasma membrane. They contain a red coloured protein called ***haemoglobin***. It is because of the presence of haemoglobin that erythrocytes (and blood as a whole) are red in colour. Haemoglobin plays an important role in carrying oxygen from the lungs to all tissues of the body. In a healthy person there are about 15 g of haemoglobin in every 100 ml of blood.

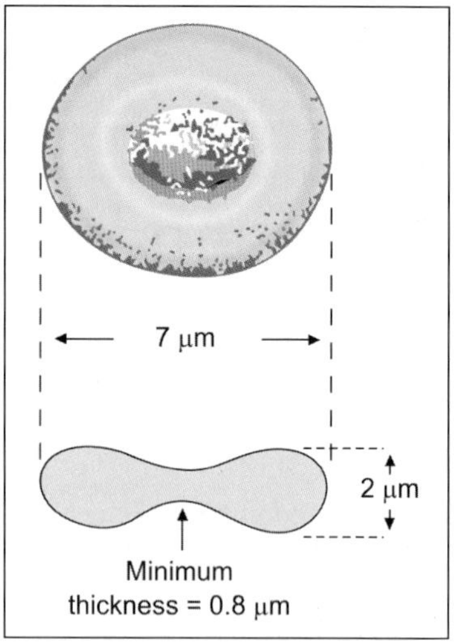

7 μm

2 μm

Minimum
thickness = 0.8 μm

Fig. 10.1: Average dimensions of an erythrocyte. **A.** Surface view; **B.** (Schematic representation)

When erythrocytes are seen in a film of blood spread out on a slide, they appear yellow (or pale red) in colour. Their rims (being thicker) appear darker than the central parts. When suspended in a suitable medium erythrocytes often appear to be piled over one another, this is described as ***rouleaux formation*** (Fig. 10.2). If a piece of transparent tissue (e.g., omentum) of a living animal is placed under a microscope, and a capillary focused, erythrocytes can be seen moving through the capillary. When thus examined it is seen that erythrocytes can alter their shape to pass through capillaries that are much narrower than the diameter of the erythrocytes.

Erythrocytes maintain their normal shape only if suspended in an isotonic solution. If the surrounding medium becomes hypotonic the cells absorb water, swell up, and ultimately burst, this is called ***haemolysis***. Alternatively, if erythrocytes are placed in a hypertonic solution, they shrink and their surfaces develop irregularities (***crenation***). Such cells are sometimes called ***echinocytes***.

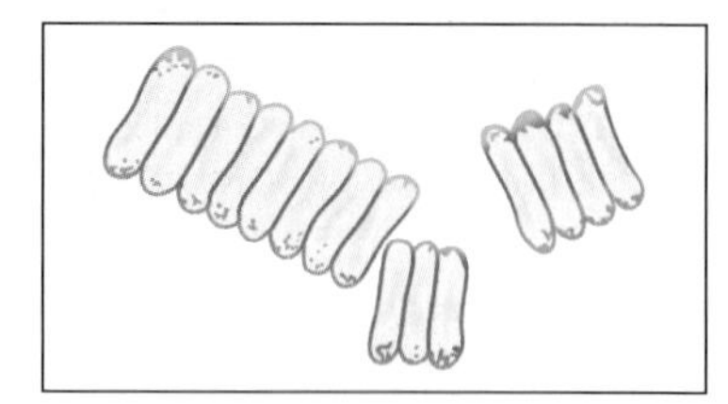

Fig. 10.2: Erythrocytes in rouleaux formation (Schematic representation)

Life span of Erythrocytes

Erythrocytes are formed in bone marrow from where they enter the blood stream. Each erythrocyte has a life of about 100 to 120 days at the end of which it is removed from blood by cells of the mononuclear phagocyte system (specially in the spleen and bone marrow). The constituents of erythrocytes are broken down and reused to form new erythrocytes.

Structure of Erythrocytes

Like cell membranes of other cells, the plasma membranes of erythrocytes are composed of lipids and proteins. Several types of proteins are present, including ***ABO antigens*** responsible for a person's blood group.

The shape of erythrocytes is maintained by a cytoskeleton made up of the protein ***spectrin***. Spectrin filaments are anchored to the cell membrane by another protein ***ankyrin***. Actin filaments and some other proteins are also present.

Haemoglobin

Haemoglobin consists of molecules of ***globulin*** bound to an iron containing porphyrin called ***haem***. Each globulin molecule is made up of a group of four polypeptide chains. The composition of the polypeptide chains is variable, and as a result several types of haemoglobin can exist. Most of normal adult haemoglobin is classified as haemoglobin A (HbA). Haemoglobin A_2 (HbA_2) is also present. Abnormal forms of haemoglobin include haemoglobin S (in sickle cell disease).

Fetal erythrocytes are nucleated and contain a different form of haemoglobin (HbF). However, in the later part of fetal life these erythrocytes are gradually replaced by those of the adult type.

Apart from haemoglobin, erythrocytes contain enzyme systems that control pH by adjusting sodium levels within the erythrocytes. They derive energy by anaerobic metabolism of glucose and by adenosine triphosphate (ATP) generation (via a hexose monophosphate shunt).

Pathological Correlation

Anaemia

Deficiency of haemoglobin in blood is called anaemia. Anaemia is commonly produced by deficiency of iron in diet. In this kind of anaemia red blood cells are small (microcytic) and pale staining (hypochromic). Another cause of anaemia is recurrent bleeding from any cause (e.g., excessive menstrual bleeding, infestation with hook worms etc). Anaemia can also result from excessive destruction of red blood cells (haemolytic anaemia). This is more likely to occur when red blood cells are abnormal. One such abnormality is caused by absence of ankyrin (see above) so that red blood cells become spherical (spherocytosis) rather than biconcave. Excessive cell destruction also occurs in sickle cell disease (see above), and in malaria.

LEUCOCYTES (WHITE BLOOD CORPUSCLES)

Differences between Erythrocytes and Leucocytes

Leucocytes are different from erythrocytes in several ways:

- They are true cells, each leucocyte having a nucleus, mitochondria, Golgi complex, and other organelles.
- They do not contain haemoglobin and, therefore, appear colourless in unstained preparations.
- Unlike erythrocytes that do not have any mobility of their own, leucocytes can move actively.
- Erythrocytes do not normally leave the vascular system, but leucocytes can move out of it to enter surrounding tissues. In fact blood is merely a route by which leucocytes travel from bone marrow to other destinations.
- Most leucocytes have a relatively short life span.

Features of Different types of Leucocytes

Leucocytes are of various types. Some of them have granules in their cytoplasm and are, therefore, called ***granulocytes***. Depending on the staining characters of their granules granulocytes are further divided into ***neutrophil leucocytes*** (or ***neutrophils***), ***eosinophil leucocytes*** (or ***eosinophils***), and ***basophil leucocytes*** (or ***basophils***).

Apart from these granulocytes there are two types of agranular leucocytes. These are ***lymphocytes*** and ***monocytes***.

Apart from the presence or absence of granules, and their nature, the different types of leucocytes show various other differences. In describing the differences it is usual for text-books to consider all features of one type of leucocyte together. However, in practice, it is more useful to take the features one by one and to compare each feature in the different types of leucocytes as given below.

Relative Number

There are about 7,000 leucocytes (range 5,000–10,000) in every cubic millimetre (=mm^3=μL) of blood. Of these about two thirds (60–70%) are neutrophils, and about one-fourth (20–30%) are lymphocytes. The remaining types are present in very small numbers. The eosinophils are about 3%, the basophils about 1%, and the monocytes about 5%. The relative and absolute numbers of the different types of leucocytes vary considerably in health; and to a more marked degree in disease. Estimations of their numbers provide valuable information for diagnosis of many diseases. In this connection it is to be stressed that absolute numbers are more significant than percentages. In a healthy individual neutrophils are 3,000–6,000/μL; lymphocytes 1500–2700/μL; monocytes 100–700/μL; eosinophils 100–400/μL; and basophils 25–200/μL.

Relative size

Leucocytes are generally examined in thin films of blood that are spread out on glass slides. In the process of making such films the cells are flattened and, therefore, appear somewhat larger than they are when suspended in a fluid medium. In a dry film all types of granulocytes, and monocytes are about 10 μm in diameter. Most lymphocytes are distinctly smaller (6–8 μm) and are called small lymphocytes, but some (called large lymphocytes) measure 12–15 μm.

Nuclei

In lymphocytes the nucleus is spherical, but may show an indentation on one side (Fig. 10.3). It stains densely in small lymphocytes, but tends to be partly euchromatic in large lymphocytes. In monocytes the nucleus is ovoid and may be indented it is placed eccentrically. In basophils the nucleus is ***S-shaped***. The nucleus of the eosinophil leucocyte is made up of two or three lobes that are joined by delicate strands. In neutrophil leucocytes the nucleus is very variable in shape and consists of several lobes (up to 6), that is why these cells are also called ***polymorphonuclear leucocytes***, or simply ***polymorphs***. The number of lobes increases with the life of the cell (Fig. 10.3).

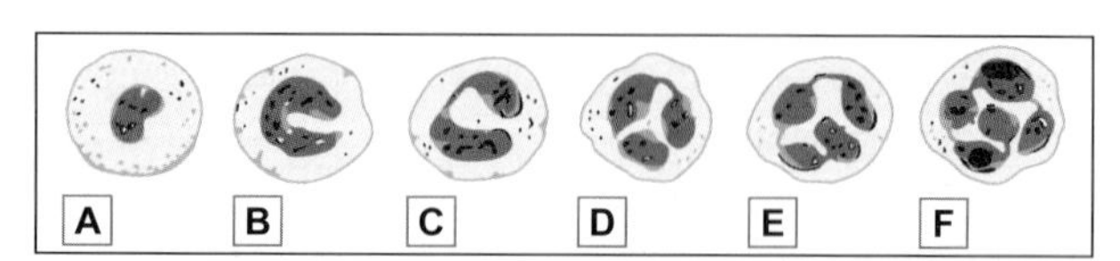

Fig. 10.3: Neutrophil leucocytes (B to F) showing varying numbers of lobes in their nuclei. A metamyelocyte is shown at 'A' for comparison (Schematic representation)

Cytoplasm

The cytoplasm of a lymphocyte is scanty and forms a thin rim around the nucleus. It is clear blue in stained preparations. In monocytes the cytoplasm is abundant. It stains blue, but in contrast to the 'transparent' appearance in lymphocytes the cytoplasm of monocytes is like frosted glass. Granules are not present in the cytoplasm of lymphocytes or of monocytes. The cytoplasm of granulocytes is marked by the presence of numerous granules. In neutrophils the granules are very fine and stain lightly with both acidic and basic dyes. The granules of neutrophils are really lysosomes, they are of various types depending upon the particular enzymes present in them. The granules of eosinophil leucocytes are large and stain brightly with acid dyes (like eosin). These are also lysosomes. With the electron microscope (EM) the granules are seen to contain a ***crystalloid*** (Fig. 10.4). The outer part or ***matrix*** of each granule contains lysosomal enzymes. The crystalline core has proteins that are responsible for red staining. In basophil leucocytes the cytoplasm contains large spherical granules that stain with basic dyes; are periodic acid schiff (PAS) positive; and stain metachromatically with some dyes.

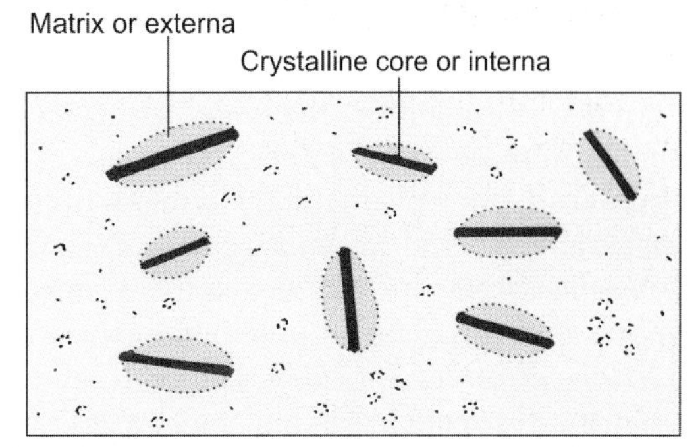

Fig. 10.4: Granules of eosinophil leucocyte as seen by electron microscope (Schematic representation)

Other cell organelles including mitochondria, Golgi complex and endoplasmic reticulum are present in the cytoplasm of leucocytes. Microtubules are present and probably play a role in movements of leucocytes. Mitochondria are particularly abundant in monocytes. Organelles are sparse in small lymphocytes, but are much more conspicuous in large lymphocytes.

Motility and Phagocytosis

All leucocytes are capable of amoeboid movement. Neutrophils and monocytes are the most active. The eosinophil and basophil leucocytes move rather slowly. Lymphocytes in blood show the least power of movement. However, when they settle on solid surfaces they become freely motile and can pass through various tissues.

Because of their motility leucocytes easily pass through capillaries into surrounding tissues, and can migrate through the latter. Neutrophils collect in large numbers at sites of infection. Here they phagocytose bacteria and use the enzymes in their lysosomes to destroy the bacteria. Eosinophils are phagocytic, but their ability to destroy bacteria is less than that of neutrophils. Monocytes are also actively phagocytic. Most monocytes in blood are, in fact, macrophages on their way to other tissues from bone marrow.

Leucocytes in circulating blood are in an inactive state. To leave the circulation a leucocyte first adheres to endothelium, and then traverses the vessel wall. Cytokines that are present in diseased areas greatly stimulate adhesion of leucocytes to endothelium and their migration through the vessel wall.

Life span

In contrast to erythrocytes which have a life span of about 100–120 days, the life of a neutrophil leucocyte is only about 15 hours. Eosinophils live for a few days, while basophils can live for 9–18 months. The life span of lymphocytes is variable. Some live only a few days (***short-lived lymphocytes***) while others may live several years (***long-lived lymphocytes***).

Neutrophils (Fig. 10.5)

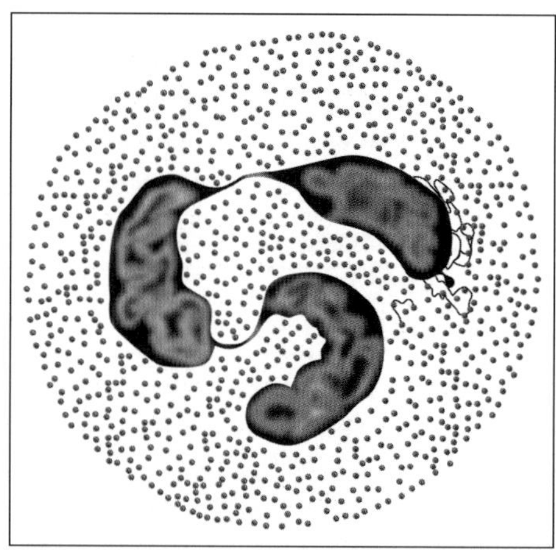

Fig. 10.5: Neutrophil (Schematic representation)

The granules of neutrophils are of three types. The first formed, or ***primary, granules*** are similar to those in lysosomes. They contain acid hydrolases. They also contain ***myeloperoxidase*** that is antibacterial. ***Secondary granules*** contain substances that are secreted into extracellular spaces and stimulate an inflammatory reaction. ***Tertiary granules*** are concerned with adhesion of leucocytes to other cells and their phagocytosis.

Neutrophils contain abundant glycogen that provides energy to the cells after they leave the blood stream.

Neutrophils die soon after they have phagocytosed materials. Lysosomal enzymes are then released into surrounding tissue causing liquefaction. Pus thus formed consists of dead neutrophils and fluid.

Neutrophils bear receptors that can recognise and adhere to foreign particles and bacteria. After adhesion, the neutrophil sends out pseudopodia that surround the foreign matter and phagocytose it. They are attracted by ***chemotaxins*** that are produced by dead cells present in areas where there is infection. The motility of neutrophils depends on changing patterns of actin filaments in their cytoplasm.

Basophils (Fig. 10.6)

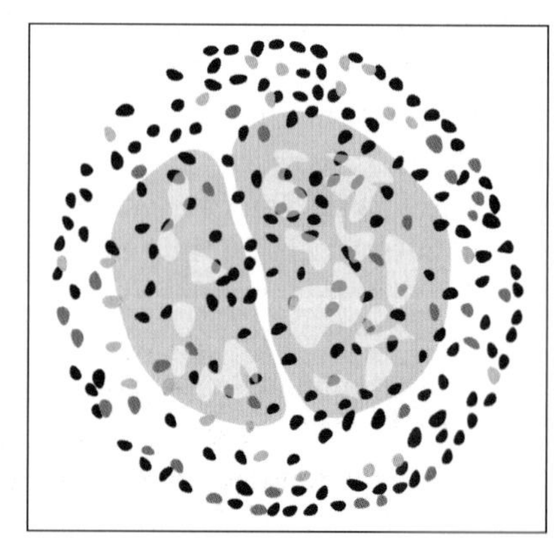

Fig. 10.6: Basophil (Schematic representation)

The staining characters of the granules of basophils are very similar to those of mast cells. Like the granules of mast cells basophils contain histamine. Many authorities, therefore, regard basophils to be precursors of mast cells. (However, differences in reactions to monoclonal antibodies suggest that even though basophils and mast cells are closely related they may be distinct cell types).

In addition to histamine, basophils contain heparin, chondroitin sulphate and leukotriene 3 (see below) which are also present in mast cells. Antibodies (produced by lymphocytes in the presence of antigens) get attached to the cell membranes of basophils and mast cells, and this leads to release of histamine into surrounding tissue. This phenomenon is described as the ***immediate hypersensitivity reaction***.

Eosinophils (Fig. 10.7)

The number of eosinophils (in blood and tissues) is greatly increased in some allergic conditions, and in parasitic infestations. Eosinophils have a functional correlation with basophils (and mast cells) as follows.

When stimulated by antigens mast cells release histamine (and other substances) into tissues, and set up an allergic reaction. Eosinophils are attracted (chemotactically) to the matter released by mast cells. They try to reduce and localise the allergic reaction by neutralising histamine. They also release factors that prevent further degranulation of mast cells. Leukotriene 3 released by mast cells is also inhibited by eosinophils.

In addition to the main granules (which are large), EM studies reveal small granules in mature eosinophils. These granules have been shown to contain the enzymes acid phosphatase and aryl sulphatase. These enzymes are probably secreted into surrounding tissues.

The number of eosinophils (in blood and in tissues) shows a circadian rhythm, being greatest in the morning and least in the afternoon.

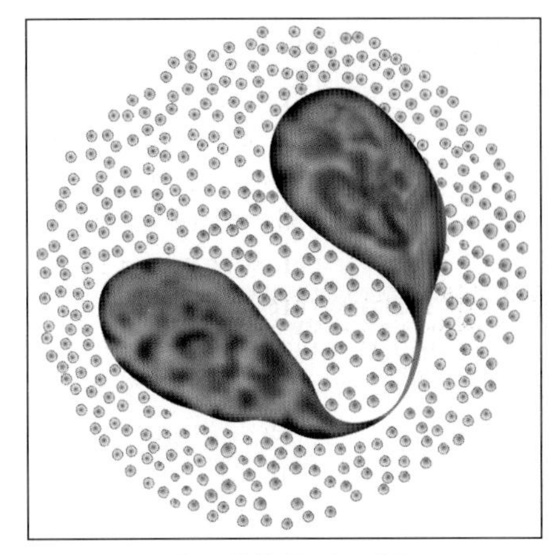

Fig. 10.7: Eosinophil (Schematic representation)

Lymphocytes (Fig. 10.8)

Lymphocytes are numerous and constitute about 20–30% of all leucocytes in blood. Large numbers of lymphocytes are also present in bonc marrow, and as aggregations in various lymphatic tissues.

Small lymphocytes with dense nuclei, and sparse cytoplasm and organelles, are regarded as resting cells. Large lymphocytes include two types of cells. Some of them are ***lymphoblasts*** that are capable of dividing to form small lymphocytes. Other large lymphocytes are mature cells that have been stimulated because of the presence of antigens.

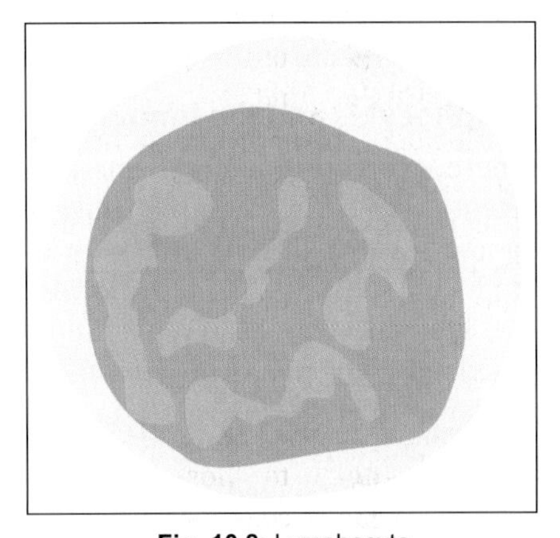

Fig. 10.8: Lymphocyte (Schematic representation)

Monocytes (Fig. 10.9)

The monocyte is the largest mature leucocyte in the peripheral blood measuring 12–20 μm in diameter. It possesses a large, central, oval, notched or indented or horseshoe-shaped nucleus which has characteristically fine reticulated chromatin network. The cytoplasm is abundant, pale blue and contains many fine dust-like granules and vacuoles.

The main functions of monocytes are as under:

- ***Phagocytosis*** of antigenic material or micro-organisms.

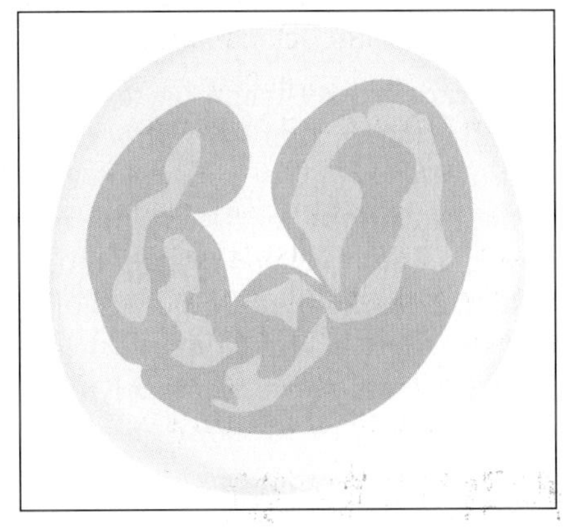

Fig. 10.9: Monocyte (Schematic representation)

- Immunologic function as ***antigen-presenting*** *cells* and present the antigen to lymphocytes to deal with further.
- As ***mediator of inflammation***, they are involved in release of prostaglandins, stimulation of the liver to secrete acute phase reactants.

Tissue macrophages of different types included in RE system are derived from blood monocytes

Pathological Correlation

Leukaemia

It is a condition in which there is uncontrolled production of leucocytes by bone marrow. It is a malignant, life threatening condition. Leucocyte precursors, normally confined to bone marrow are seen in large numbers in peripheral blood.

Leukaemias are of different types depending on the type of leucocytes, or leucocyte precursors, that are proliferating. When progress of the disease is slow (chronic Leukaemia) the proliferating cells get time to differentiate, and they can be recognised. However, in acute Leukaemias the peripheral blood is flooded with undifferentiated precursors of leucocytes.

Lymphocytes and the Immune System

Lymphocytes are an essential part of the ***immune system*** of the body that is responsible for defence against invasion by bacteria and other organisms. In contrast to granulocytes and monocytes that directly attack invading organisms, lymphocytes help to destroy them by producing substances called ***antibodies***.

B-lymphocytes, plasma cells, and the antibodies produced by them are the basis of what is described as the ***humoral immune response*** to antigens. Initially, all B-lymphocytes produce similar antibodies, but subsequently specialisation takes place to produce antibodies for specific antigens only. Those B-lymphocytes that produce antibodies against normal body proteins are eliminated.

T-lymphocytes are also concerned with immune responses, but their role is somewhat different from that of B-lymphocytes. T-lymphocytes specialise in recognising cells that are foreign to the host body. These may be fungi, virus infected cells, tumour cells, or cells of another individual. T-lymphocytes have surface receptors that recognise specific antigens (there being many varieties of T-lymphocytes each type recognising a specific antigen). When exposed to a suitable stimulus the T-lymphocytes multiply and form large cells that can destroy abnormal cells by direct contact, or by producing cytotoxic substances called ***cytokines*** or ***lymphokines***. While B-lymphocytes defend the body through blood borne antibodies, T-lymphocytes are responsible for cell mediated immune responses (***cellular immunity***). T-lymphocytes can also influence the immune responses of B-lymphocytes as well as those of other T-lymphocytes; and also those of non-lymphocytic cells.

Like B-lymphocytes some T-lymphocytes also retain a memory of antigens encountered by them, and they can respond more strongly when the same antigens are encountered again.

Pathological Correlation

The destruction of foreign cells by T-lymphocytes is responsible for the 'rejection' of tissues or organs grafted from one person to another. Such rejection is one of the major problems in organ transplantation.

Added Information

Investigations using sophisticated techniques have established that T-lymphocytes can be divided into several categories as follows:

- Cytotoxic T-cells (TC cells) attack and destroy other cells by release of lysosomal proteins. They have the ability to recognise proteins that are foreign to the host body.
- Delayed type hypersensitivity related T-cells synthesise and release lymphokines when they come in contact with antigens. Lymphokines attract macrophages into the area. They also stimulate macrophages to destroy the antigen. One type of cytokine called interleukin-2 stimulates the proliferation of both B-lymphocytes and T-lymphocytes.
- Helper T-cells (or TH cells) play a rather indirect role in stimulating production of antibodies by B-lymphocytes. When a macrophage ingests an antigen some products of antigen break-down pass to the cell surface. Here they combine with special molecules (Class II MHC molecules) present in the cell membrane. This complex of antigen remnant and MHC protein, present on the macrophage, can be recognised by helper T-cells. When helper T-cells come in contact with this complex they look for B-lymphocytes capable of producing antibody against the particular antigen. They then stimulate these B-lymphocytes to multiply so that large numbers of B-lymphocytes capable of producing antibody against the particular antigen are produced. This is how immunity against the particular antigen is acquired. Helper T-cells are specifically destroyed by the virus responsible for acquired immune deficiency syndrome (AIDS) resulting in a loss of immunity.
- Suppressor T-cells (or TS cells) have a role opposite to that of helper T-cells. They suppress the activities of B-lymphocytes and of other T-lymphocytes. The possibility of such suppression allows fine control of the activities of lymphocytes.
- Natural killer cells are similar to cytotoxic T-cells, but their actions are less specific than those of the latter. These cells can destroy virus infected cells and some tumour cells. Their structure is somewhat different from that of typical lymphocytes.

  Natural killer cells are regarded by some as a third variety of lymphocytes (in addition to B-lymphocytes and T-lymphocytes) as markers in them are different from those in typical T-lymphocytes.

Markers for Cells of Immune System

Many proteins that are specific to cells of the immune system are now recognised. These may be cytosolic or cell membrane proteins. Using antibodies specific to these proteins it is possible to identify many varieties of lymphocytes and macrophages. The proteins are called ***cluster designation (CD) molecules***, and are designated by number (CD-1, CD-2 etc).

Pathological Correlation

The virus (HIV) that causes the disease called AIDS attaches itself to the protein CD-4 present on the cell membrane, and destroys cells bearing this protein. Reduction in the number of CD-4 bearing lymphocytes is an important indicator of the progress of AIDS.

Cytokines (Lymphokines)

We have seen that T-lymphocytes produce cytokines that affect other cells. The main function of these cytokines is to stimulate production of blood cells and their precursors. Apart from T-cells, cytokines are also produced by monocytes, macrophages, some fibroblasts, and some endothelial cells. Some cytokines that have been identified are as follows:

Table 10.1: Cytokines produced by different cells

Cell Type	Cytokines produced										
	IL-1	IL-2	IL-3	IL-5	IL-6	IL-8	IL-9	G-CSF	M-CSF	GM-CSF	Stem cell factor
T-Lymphocytes		+	+	+	+					+	
Monocytes	+				+	+	+	+	+		
Endothelium								+	+		+
Fibroblasts	+				+	+		+	+		+

Table 10.2: Cytokines that stimulates production of blood cells or activate them

Cell Type	Cytokines stimulating them										
	IL-2	IL-3	IL-5	IL-6	IL-8	IL-9	IL-11	G-CSF	M-CSF	GM-CSF	Erythro-poietin
Granulocyte precursors								+		+	
Eosinophil & basophil precursors			+								
Neutrophil activation					+						
Monocyte activation	+										
Monocyte precursors							+		+	+	
Erythrocyte precursors						+					+
Megakaryocyte precursors		+		+		+					+
T-cell production	+										

- Interleukins are of eleven known types (IL-1 to IL-11)
- Granulocyte macrophage colony stimulating factor (GM-CSF)
- Granulocyte colony stimulating factor (G-CSF)
- Macrophage colony stimulating factor (M-CSF)
- Stem cell factor
- Erythropoietin

The cytokines produced by different types of cells are summarised in Table 10.1. The cytokines stimulating production of different blood cells are given in Table 10.2.

BLOOD PLATELETS

Blood platelets are round, oval, or irregular discs about 3 μm in diameter. They are also known as ***thrombocytes***. The discs are biconvex. Each disc is bounded by a plasma membrane within which there are mitochondria and membrane bound vesicles. There is no nucleus. In ordinary blood films the platelets appear to have a clear outer zone (***hyalomere***) and a granular central part (***granulomere***).

Platelets are concerned with the clotting of blood. As soon as blood is shed from a vessel, platelets stick to each other and to any available surfaces (specially to collagen fibres). Platelets break down into small granules and threads of fibrin appear around them.

There are about 250,000 to 500,000 platelets per μl of blood. The life of a platelet is about 10 days.

Electronmicroscopy shows that the cell membrane of a platelet is covered by a thick coat of glycoprotein. The presence of this coat facilitates adhesion of the platelet to other surfaces. Apart from the usual organelles the cytoplasm contains microtubules, actin filaments, and myosin filaments that are of importance in clot retraction (see below).

The cytoplasm also contains three main types of membrane bound vesicles (or granules) as follows. ***Alpha granules*** contain ***fibrin*** and various other proteins. They also contain a ***platelet derived growth factor. Delta granules*** (or dense granules) contain 5-hydroxytryptamine, calcium ions, adenosine diphosphate (ADP) and ATP. ***Lambda granules*** contain lysosomal enzymes (acid hydrolases). A few peroxisomes are also present.

Clotting of Blood

During clotting of blood, platelets adhere to one another forming a ***platelet plug*** that assists in stopping the bleeding. Fibrin is released from alpha granules. Under the influence of various factors fibrin is converted into filaments that form a clot. The actin and myosin filaments (present in platelets) cause the fibrin clot to contract (***clot retraction***). At later stages clot removal is aided by lysosomal enzymes present in platelets.

Clinical Correlation

Thrombocytopenia is defined as a reduction in the peripheral blood platelet count below the lower limit of normal, i.e. below 150,000/μl. Thrombocytopenia is associated with abnormal bleeding that includes spontaneous skin purpura and mucosal haemorrhages as well as prolonged bleeding after trauma. However, spontaneous haemorrhagic tendency becomes clinically evident only after severe depletion of the platelet count to level below 20,000/μl.

FORMATION OF BLOOD (HAEMOPOIESIS) (FIG. 10.10)

In embryonic life, blood cells are first formed in relation to mesenchymal cells surrounding the yolk sac. After the second month of intrauterine life blood formation starts in the liver; later in the spleen; and still later in the bone marrow. At first lymphocytes are formed along with other cells of blood in bone marrow, but later they are formed mainly in lymphoid tissues. In postnatal life blood formation is confined to bone marrow and lymphoid tissue. However, under conditions in which the bone marrow is unable to meet normal requirements, blood cell formation may start in the liver and spleen. This is referred to as ***extramedullary haemopoiesis***. There has been considerable controversy regarding the origin of various types of blood cells. The ***monophyletic theory*** holds that all types of blood cells are derived from a common ***stem cell***; while according to the ***polyphyletic theory*** there are several independent types of stem cells. There is no doubt that in the embryo all blood forming cells are derived from mesenchyme,

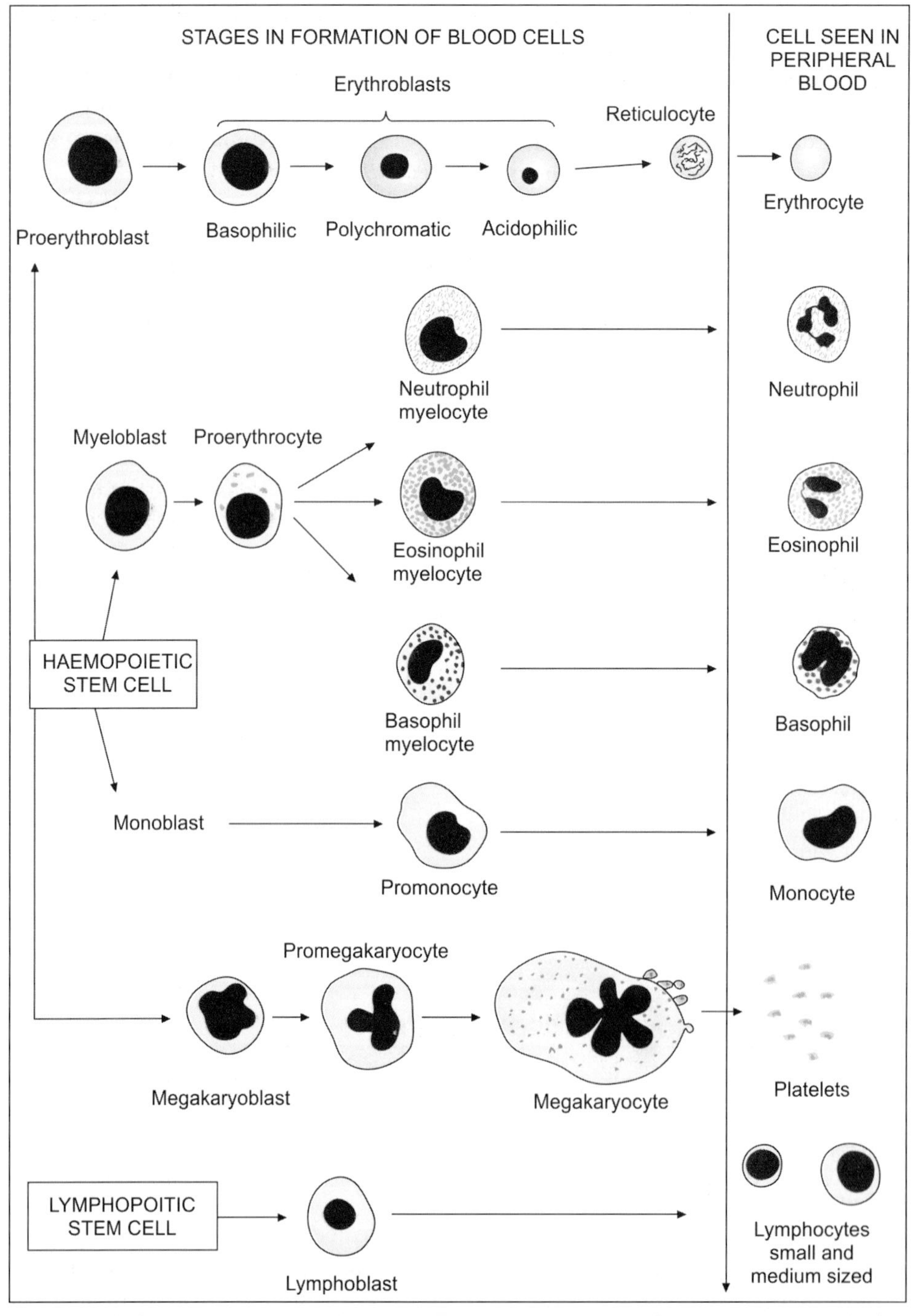

Fig. 10.10: Stages in formation of blood cells (Traditional terms used) (Schematic representation)

and that the earliest stem cells are capable of forming all types of blood cells. Subsequently the potency of stem cells becomes restricted.

Details of the structure of developing blood cells are beyond the scope of this book. The purpose of the account that follows is mainly to make the student familiar with numerous terms that are used in this connection. We will first study the terms traditionally used for developing blood cells. Some terms based on recent researches will be introduced later.

We have already seen that there are embryonic stem cells that are pleuripotent. Arising from them they are the following:

- ***Haemopoietic stem cells*** or ***haemocytoblasts***, that are present (in postnatal life) only in bone marrow and give rise to all blood cells other than lymphocytes.
- ***Lymphopoietic stem cells*** that are present in bone marrow and in lymphoid tissues and give rise to lymphocytes. (Stages in the development of blood cells are illustrated in Fig. 10.10).

Formation of Erythrocytes

Precursor cells of the erythrocyte series are called ***erythroblasts*** or ***normoblasts***. The earliest identifiable precursors of erythrocytes are large cells called ***proerythrocytes***. These are succeeded by slightly smaller cells called ***early erythroblasts,*** these cells do not contain haemoglobin and their cytoplasm is basophil. As haemoglobin begins to be formed some areas of cytoplasm become eosinophil while others remain basophil, such cells are called ***intermediate erythroblasts***. As more and more haemoglobin is formed the entire cytoplasm becomes acidophil, these cells are the ***final erythroblasts***. Note that during these changes the erythroblasts become progressively smaller. At this stage their nuclei shrink and are thrown out of the cells. The cytoplasm now has a reticular appearance [produced by ribonucliec acid (RNA) remaining within it], these cells are, therefore, called ***reticulocytes***. Reticulocytes leave the bone marrow and enter the blood stream. Here they lose their reticulum within a day or two and become mature erythrocytes.

Formation of Granulocytes

It is believed that neutrophil, eosinophil and basophil leucocytes arise from a common early derivative of the haemopoietic stem cell that is called a ***myeloblast***. (It is, however, possible that there may be separate types of myeloblasts for each type of granulocyte). The myeloblast matures into a larger cell called a ***promyelocyte*** that is marked by the presence of large granules (lysosomes) in its cytoplasm. The promyelocyte now gives rise to ***myelocytes*** in which the granules are smaller and ***specific*** so that at this stage neutrophil, eosinophil and basophil myelocytes can be recognised. The nucleus now undergoes transformation dividing into two or three lobes in eosinophil cells; and forming up to six lobes in neutrophils. With the nuclei assuming their distinctive appearance the myelocytes become mature granulocytes.

Formation of Monocytes

Monocytes are also formed in bone marrow from haemopoietic stem cells. Recent evidence (see below) suggests a common origin of monocytes and granulocytes. Early stages in the formation of monocytes are referred to as ***monoblasts***. These change into ***promonocytes*** and finally into mature ***monocytes***. We have already noted that the stem cells that give rise to monocytes also give rise to other cells of the mononuclear phagocyte system.

Formation of Platelets

The precursor cells of blood platelets are called ***megakaryoblasts***. The megakaryoblast enlarges to form a ***promegakaryocyte***. Still further enlargement converts it into a ***megakaryocyte,*** this cell may be 50 to 100 µm in diameter, and has a multi-lobed nucleus. The nucleus is polyploid because of repeated divisions that are not accompanied by divisions of cytoplasm. Platelets are formed by separation of small masses of cytoplasm from this large cell. Each mass of cytoplasm is covered by cell membrane and contains some endoplasmic reticulum. In this way the cytoplasm of a megakaryocyte becomes subdivided into many small portions each forming one platelet.

Added Information

Haemopoiesis

Blood cell formation is currently a subject of great interest to research workers. The methods being used to investigate the subject include cell or organ culture, and experiments based on replacement of bone marrow. Some facts that have emerged as a result of these studies are as follows.

- Adult bone marrow contains a small number of cells that are totipotent i.e., they can give origin to all types of blood cells. Totipotent stem cells can also divide to form new cells of their own type. Totipotent stem cells give rise to lymphocytic stem cells and to pleuripotent haemal stem cells.
- Lymphocytic stem cells given origin to all lymphocytes (T and B).
- Pleuripotent haemal stem cells give origin to three types of cells as follows.
  - Cells that are precursors of erythrocytes.
  - Cells that are precursors of granulocytes as well as those of monocytes.
  - Cells that are precursors of megakaryoblasts, and hence of platelets.

A complex terminology has developed for various precursors of blood cells. Unfortunately, there is lack of uniformity in its use, and much of it is based on experiments on rodents. Applicability to humans requires confirmation. In spite of these reservations, it is necessary for medical students to be aware of these developments. Some of the terms used are shown in Fig. 10.11. We will not go into details of these terms here. However, two terms need explanation.

An important question engaging the attention of reseach workers is the stage up to which blood cell precursors are self renewing (i.e., they can divide to forms more cells of their own type). Some authorities hold that only totipotent stem cells have this power, while others believe that some cell types lower down the line may also be self renewing. In leukaemias, one line of treatment is to destroy all cells in the bone marrow of the patient by radiation, and to then transfuse marrow cells from a normal donor. The procedure has so far had only limited success because most of the transplanted cells survive only for a limited period (as they are not self renewing).

MONONUCLEAR PHAGOCYTE SYSTEM

Distributed widely through the body there are a series of cells that share the property of being able to phagocytose unwanted matter including bacteria and dead cells. These cells also play an important role in defence mechanisms, and in carrying out this function they act in close collaboration with lymphocytes. In the past some of the cells of this system have been included under the term ***reticulo-endothelial system***, but this term has now been discarded as it is established that most endothelial cells do not act as macrophages. The term ***macrophage***

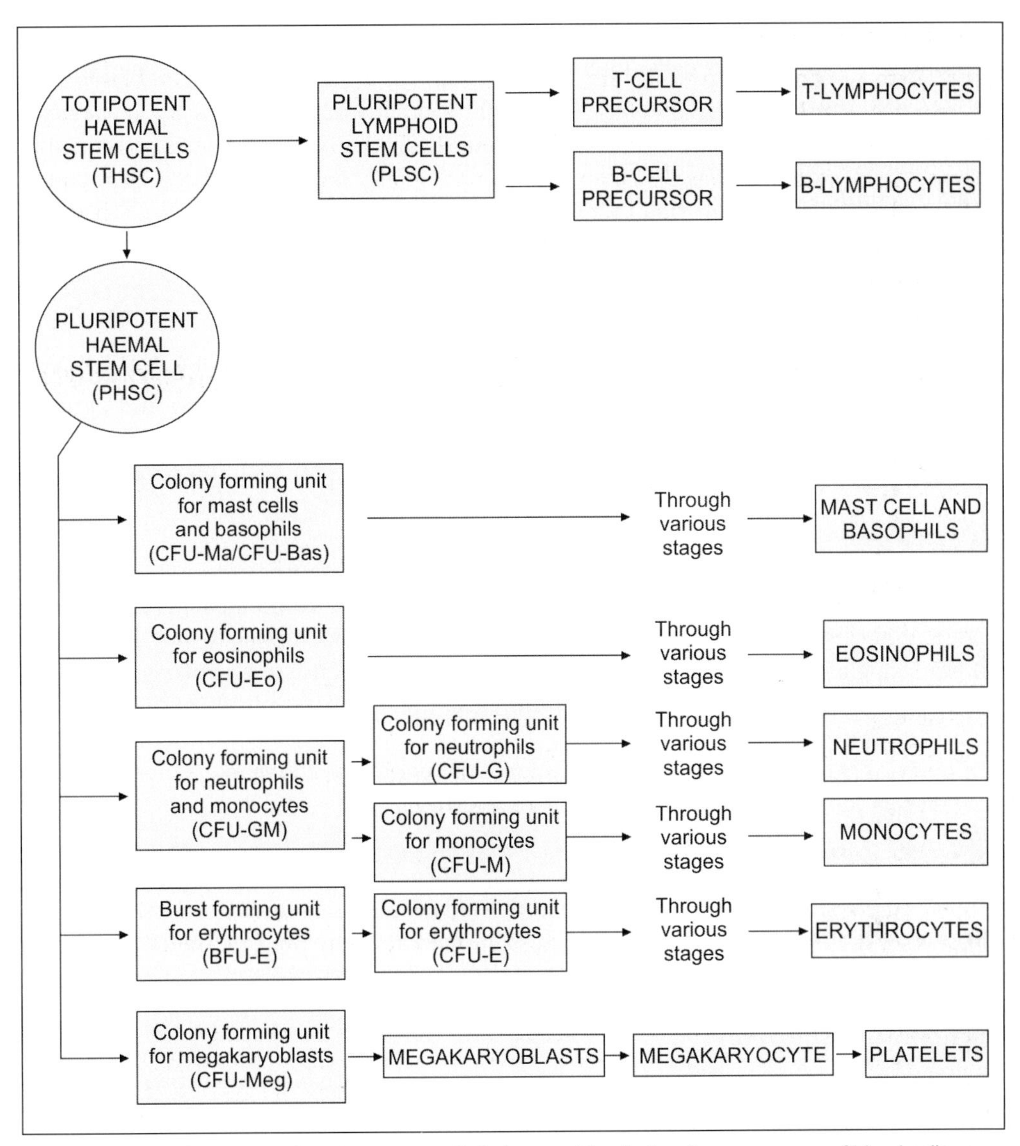

Fig. 10.11: Scheme to show terms currently being used for designating precursors of blood cells

system has also been used for cells of the system, but with the discovery of a close relationship between these cells and mononuclear leucocytes of blood the term ***mononuclear phagocyte system*** (or ***monocyte phagocyte system***) has come into common usage. It is now known that all macrophages are derived from stem cells in bone marrow that also give origin to mononuclear cells of blood.

Cells of Mononuclear Phagocyte System

The various cells that are usually included in the mononuclear phagocyte system are as follows.

- Monocytes of blood, and their precursors in bone marrow (monoblasts, promonocytes).

- Macrophage cells (histiocytes) of connective tissue.
- Littoral cells (von Kupffer cells) interspersed among cells lining the sinusoids of the liver; and cells in walls of sinusoids in the spleen and lymph nodes.
- Microglial cells of the central nervous system.
- Macrophages in pleura, peritoneum, alveoli of lungs, spleen, and in synovial joints.
- Free macrophages present in pleural, peritoneal and synovial fluids.
- Dendritic cells of the epidermis, and similar highly branched cells in lymph nodes, spleen and thymus. These are now grouped as ***antigen presenting cells*** (see below).

Structure of Cells

All cells of the mononuclear phagocyte system have some features in common. They are large cells (15–25 μm) in diameter. The nucleus is euchromatic. Granular and agranular endoplasmic reticulum, Golgi complex and mitochondria are present, as are endocytic vesicles and lysosomes. The cells have irregular surfaces that bear filopodia (irregular microvilli). Most of the cells are more or less oval in shape, but the dendritic cells are highly branched.

Macrophages often form aggregations. In relation to the peritoneum and pleura such aggregations are seen as ***milky spots***; and in the spleen they form ellipsoids around small arteries. When they come in contact with large particles, macrophages may fuse to form multinuclear giant cells (***foreign body giant cells***). In the presence of organisms like tubercle bacilli the cells may transform to ***epithelioid cells.*** (These are involved in T-cell mediated immune responses).

Monocytes can be classified into various types depending on the markers present in them, and their relationship to cytokines (Tables 10.1 and 10.2).

Origin of Cells

All cells of the system are believed to arise from stem cells in bone marrow (classified as CFU-G/M stem cells). From bone marrow they pass into blood where they are seen as monocytes. From blood they pass into the tissues concerned.

Functional Classification

From a functional point of view mononuclear phagocytes are divided into two main types.

- With the exception of dendritic cells all the cell types are classified as ***highly phagocytic cells***.
- The dendritic mononuclear phagocytes (now called dendritic antigen presenting cells) are capable of phagocytosis, but their main role is to initiate immune reactions in lymphocytes present in lymph nodes, spleen and thymus (in the manner discussed above). It has been postulated that ***all*** dendritic cells are primarily located in the skin. From here they pick up antigens and migrate to lymphoid tissues where they stimulate lymphocytes against these antigens. They are therefore referred to as antigen presenting cells (APCs). Most APCs are derived from monocytes, but some are derived from other sources.

Functions of the Mononuclear Phagocyte System

Participation in Defence Mechanisms

As already stated the cells have the ability to phagocytose particulate matter, dead cells, and organisms. In the lungs, alveolar macrophages engulf inhaled particles and are seen as ***dust cells***. In the spleen and liver, macrophages destroy aged and damaged erythrocytes.

Role in Immune Responses

- All mononuclear phagocytes bear antigens on their surface (class II MHC antigens). Antigens phagocytosed by macrophages are partially digested by lysosomes. Some remnants of these pass to the cell surface where they form complexes with the MHC antigens. This complex has the ability to stimulate T-lymphocytes.
- Certain T-lymphocytes produce macrophage activating factors (including interleukin-2) that influence the activity of macrophages. Macrophages when thus stimulated synthesise and secrete cytokines that stimulate the proliferation and maturation of further lymphocytes.
- When foreign substances (including organisms) enter the body, antibodies are produced against them (by lymphocytes). These antibodies adhere to the organisms. Macrophages bear receptors (on their surface) that are able to recognise these antibodies. In this way macrophages are able to selectively destroy such matter by phagocytosis, or by release of lysosomal enzymes.

  From the above it will be seen that lymphocytes and macrophages constitute an integrated immune system for defence of the body.
- When suitably stimulated mononuclear phagocytes secrete a tumour necrosing factor (TNF) which is able to kill some neoplastic cells.
- Macrophages influence the growth and differentiation of tissues by producing several growth factors and differentiation factors.

Chapter 11

Nervous System

The nervous system may be divided into the ***central nervous system (CNS)***, made up of the brain and spinal cord, the ***peripheral nervous system (PNS)***, consisting of the peripheral nerves and the ganglia associated with them, and the ***autonomic nervous system***, consisting of the ***sympathetic*** and the ***parasympathetic nervous systems***. The brain consists of the ***cerebrum*** (made up of two large cerebral hemispheres), ***cerebellum***, ***midbrain***, ***pons***, and ***medulla oblongata***. The midbrain, pons, and medulla together form the ***brainstem***. The medulla is continuous below with the spinal cord (Flow chart 11.1).

TISSUES CONSTITUTING THE NERVOUS SYSTEM

The nervous system is made up of, predominantly, by a tissue that has the special property of being able to conduct impulses rapidly from one part of the body to another. The specialised cells that constitute the functional units of the nervous system are called ***neurons***. Within the brain and spinal cord neurons are supported by special kind of connective tissue cells that are called ***neuroglia.*** Nervous tissue, composed of neurons and neuroglia, is richly supplied with blood. It has been taught that lymph vessels are not present, but the view has recently been challenged.

Flow chart 11.1: Anatomical classification of nervous system

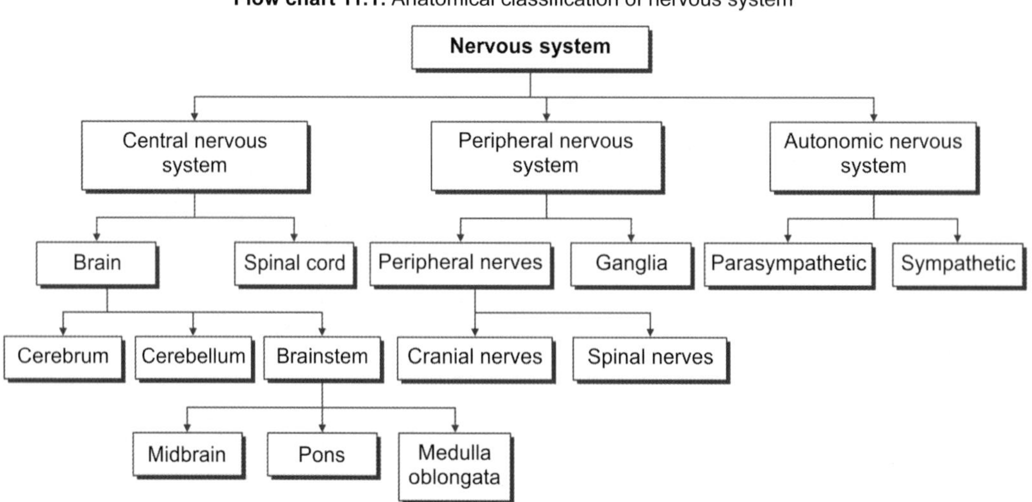

The nervous system of man is made up of innumerable neurons. The total number of neurons in the human brain is estimated at more than 1 trillion. Neurons are regarded not merely as simple conductors, but as cells that are specialised for the reception, integration, interpretation and transmission of information.

Nerve cells can convert information obtained from the environment into codes that can be transmitted along their axons. By such coding the same neuron can transmit different kinds of information.

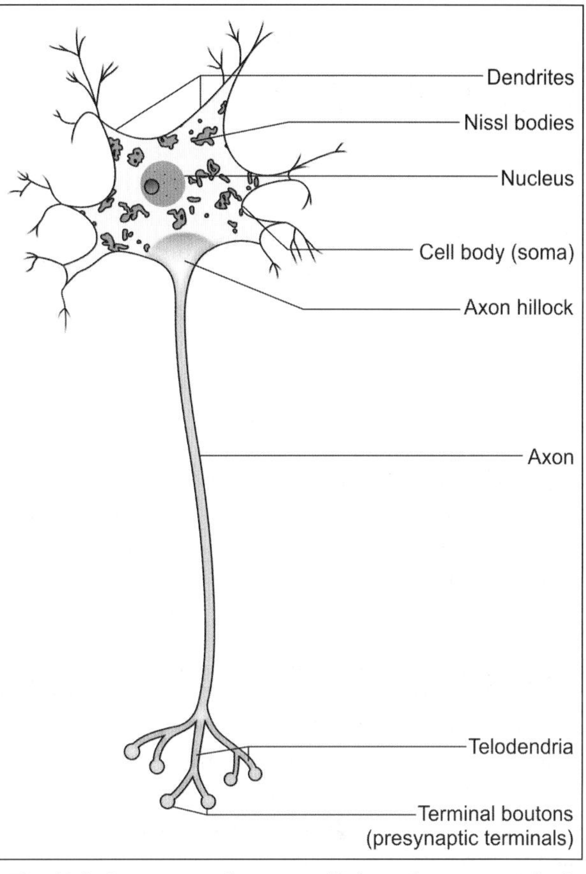

Fig. 11.1: Some parts of a neuron (Schematic representation)

STRUCTURE OF A NEURON

Neurons vary considerably in size, shape and other features. However, most of them have some major features in common and these are described below (Figs 11.1 to 11.5).

Cell Body/Perikaryon

A neuron consists of a ***cell body*** that gives off a variable number of ***processes*** (Fig. 11.1). The cell body is also called the ***soma*** or ***perikaryon***. Like a typical cell it consists of a mass of cytoplasm surrounded by a cell membrane.

Cytoplasm

The cytoplasm contains a large central nucleus (usually with a prominent nucleolus), numerous mitochondria, lysosomes and a Golgi complex (Fig. 11.2). In the past it has stated that centrioles are not present in neurons, but studies with the electron microscope have shown that centrioles are present. In addition to these features, the cytoplasm of a neuron has some distinctive characteristics not seen in other cells.

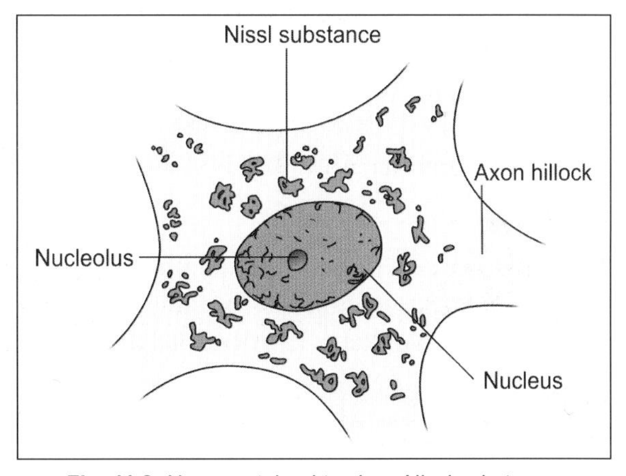

Fig. 11.2: Neuron stained to show Nissl substance. Note that the Nissl substance extends into the dendrites but not into the axon

Nissl Substance

The cytoplasm shows the presence of a granular material that stains intensely with basic dyes; this material is the ***Nissl substance*** (also called Nissl bodies or granules) (Fig. 11.2). When examined by EM, these bodies are seen to be composed of rough surfaced endoplasmic reticulum (Fig. 11.3). The presence of abundant granular endoplasmic reticulum is an indication of the high level of protein synthesis in neurons. The proteins are needed for maintenance and repair, and for production of neurotransmitters and enzymes.

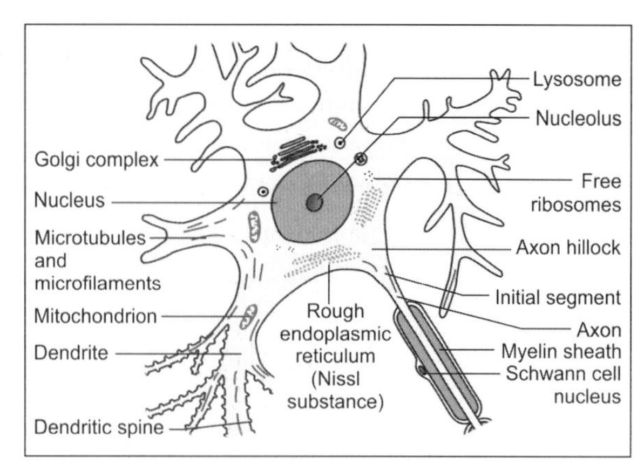

Fig. 11.3: Some features of the structure of a neuron as seen by electron microscope (Schematic representation)

Neurofibrils

Another distinctive feature of neurons is the presence of a network of fibrils permeating the cytoplasm (Fig. 11.4). These ***neurofibrils*** are seen, with the EM, to consist of microfilaments and microtubules. (The centrioles present in neurons may be concerned with the production and maintenance of microtubules).

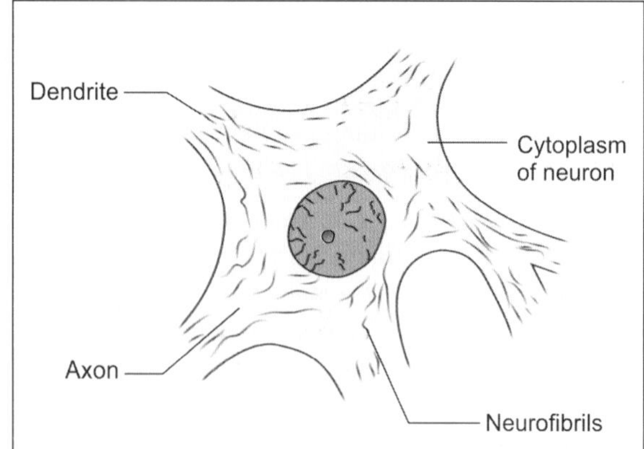

Fig. 11.4: Neuron stained to show neurofibrils. Note that the fibrils extend into both axons and dendrites

Pigment Granules

Some neurons contain pigment granules (e.g., neuromelanin in neurons of the substantia nigra). Ageing neurons contain a pigment lipofuscin (made up of residual bodies derived from lysosomes).

Neurites

The processes arising from the cell body of a neuron are called ***neurites.*** These are of two kinds. Most neurons give off a number of short branching processes called ***dendrites*** and one longer process called an ***axon.***

Dendrites

The dendrites are characterised by the fact that they terminate near the cell body. They are irregular in thickness, and Nissl granules extend into them. They bear numerous small spines that are of variable shape. In a dendrite, the nerve impulse travels ***towards the cell body***.

Axon

The axon may extend for a considerable distance away from the cell body.

Table 11.1: Difference between Axons and Dendrites

Axon	Dendrites
Axon is a single, long, thin process of a nerve cell, which terminates away from the nerve cell body	Dendrites are multiple, short, thick and tapering processes of the nerve cell which terminate near the nerve cell body
Axon rarely branches at the right angle (axon collaterals) but ends by dividing into many fine processes called axon terminals.	Dendrites are highly branched. Their branching pattern forms a dendritic tree.
It has uniform diameter and smooth surface	The thickness of dendrite reduces as it divides repeatedly. Its surface is not smooth, but it bears many small spine-like projections for making synaptic contacts with the axons of other nerve cells
It is free of Nissl granules	Nissl granules are present in dendrites
The nerve impulses travel away from the cell body	The nerve impulses travel towards the cell body

- The longest axons may be as much as a metre long.
- Each axon has a uniform diameter, and is devoid of Nissl substance.
- The cytoplasm within the axon is called axoplasm and its cell membrane is called axolemma.
- The axoplasm contains all the cell organelles of neurons cell body except ribosomes. Hence, proteins synthesized in the cell body are continuously transported towards the axon terminals by a process called axoplasmic transport.
- In an axon the impulse travels ***away from the cell body.***
- Axons constitute what are commonly called ***nerve fibres.*** The bundles of nerve fibres found in CNS are called as ***nerve tracts,*** while the bundles of nerve fibres found in PNS are called ***peripheral nerves.***
- An axon (or its branches) can terminate in two ways. Within the central nervous system, it always terminates by coming in intimate relationship with another neuron, the junction between the two neurons being called a ***synapse***. Outside the central nervous system, the axon may end in relation to an effector organ (e.g., muscle or gland), or may end by synapsing with neurons in a peripheral ganglion.

Added Information

- The proteins present in dendrites and axons are not identical. This fact is used for immunocytochemical identification of dendrites in tissue sections. A protein MAP-2 is present exclusively in dendrites and helps in their identification.

Axon Hillock and Initial Segment

The axon is free of Nissl granules. The Nissl-free zone extends for a short distance into the cell body: this part of the cell body is called the ***axon hillock.*** The part of the axon just beyond the axon hillock is called the ***initial segment*** (Fig. 11.2).

Added Information

Axon hillock and initial segment:

The axon hillock and the initial segment of the axon are of special functional significance. This is the region where action potentials are generated (spike generation) resulting in conduction along the axon. The initial segment is unmyelinated. It often receives axo-axonal synapses that are inhibitory. The plasma membrane here is rich in voltage sensitive channels.

Axoplasmic flow:

The cytoplasm of neurons is in constant motion. Movements of various materials occurs through axons. This ***axoplasmic flow*** takes place both away from and towards the cell body. The flow away from the cell body is greater. Some materials travel slowly (0.1 to 2 mm a day) constituting a ***slow transport.*** In contrast other materials (mainly in the form of vesicles) travel 100 to 400 mm a day constituting a ***rapid transport.***

Myelinated and Unmyelinated Axons

Axon may be myelinated or unmyelinated. Axons having a myelin sheath are called ***myelinated axons.***

Myelin Sheath

- Myelin sheath when present is seen outside the axolemma.
- The cells providing this sheath for axons lying outside in the peripheral nervous system are called ***Schwann cells.*** Axons lying within the central nervous system are provided a similar covering by a kind of neuroglial cell called an ***oligodendrocyte.***
- The nature of this sheath is best understood by considering the mode of its formation. An axon lying near a Schwann cell invaginates into the cytoplasm of the Schwann cell. In this process the axon comes to be suspended by a fold of the cell membrane of the Schwann cell: this fold is called the ***mesaxon*** (Fig. 11.5). In some situations the mesaxon becomes greatly elongated and comes to be spirally wound around the axon, which is thus surrounded by several layers of cell membrane. Lipids are deposited between adjacent layers of the membrane. These layers of the mesaxon, along with the lipids, form the ***myelin sheath.*** Outside the myelin sheath a thin layer of Schwann cell cytoplasm persists to form an additional sheath that is called the ***neurilemma*** (also called the neurilemmal sheath or Schwann cell sheath).

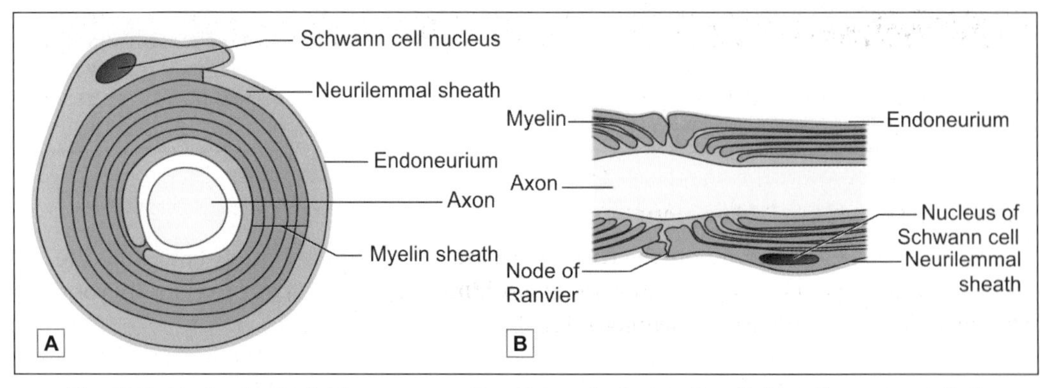

Fig. 11.5: Myelin sheath. **A.** Transverse section; **B.** Longitudinal section (Schematic representation)

- The presence of a myelin sheath increases the velocity of conduction (for a nerve fibre of the same diameter). It also reduces the energy expended in the process of conduction.
- An axon is related to a large number of Schwann cells over its length (Fig. 11.6). Each Schwann cell provides the myelin sheath for a short segment of the axon. At the junction of any two such segments there is a short gap in the myelin sheath. These gaps are called the ***nodes of Ranvier.***
- The nodes of Ranvier have great physiological importance. When an impulse travels down a nerve fibre it does not proceed uniformly along the length of the axis cylinder, but jumps from one node to the next. This is called ***saltatory conduction*** (In unmyelinated neurons the impulse travels along the axolemma. Such conduction is much slower than saltatory conduction and consumes more energy).
- The segment of myelin sheath between two nodes of Ranvier is called ***internode***.

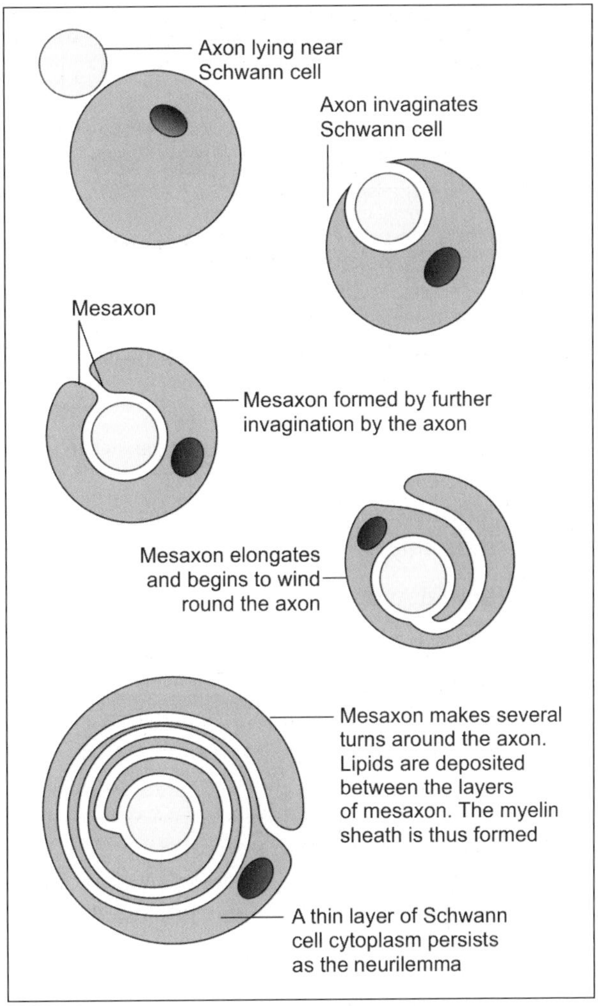

Fig. 11.6: Stages in the formation of the myelin sheath by a Schwann cell (Schematic representation)

Composition of Myelin Sheath

Myelin contains protein, lipids, and water. The main lipids present include cholesterol, phospholipids, and glycosphingolipids. Other lipids are present in smaller amounts.

Pathological Correlation

Myelination can be seriously impaired, and there can be abnormal collections of lipids, in disorders of lipid metabolism. Various proteins have been identified in myelin sheaths and abnormality in them can be the basis of some neuropathies.

The composition and structure of myelin sheaths formed by oligodendrocytes show differences from those formed by Schwann cells. The two are different in protein content and can be distinguished by immunocytochemical methods. As damage to neurons within the CNS is not followed by regeneration, oligodendrocytes have no role to play in this respect. Also note that, in multiple sclerosis, myelin formed by oligodendrocytes undergoes degeneration, but that derived from Schwann cells is spared.

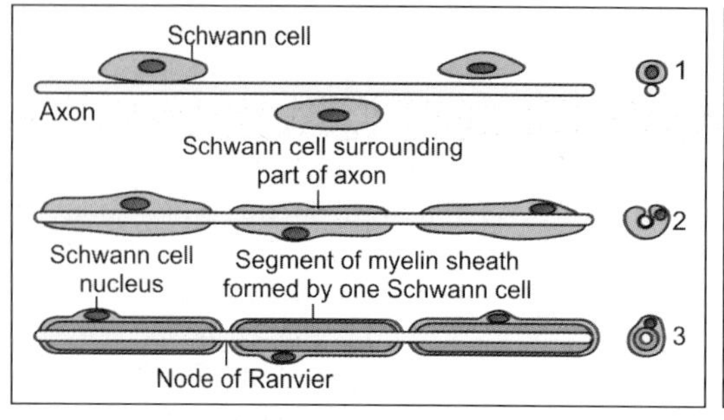

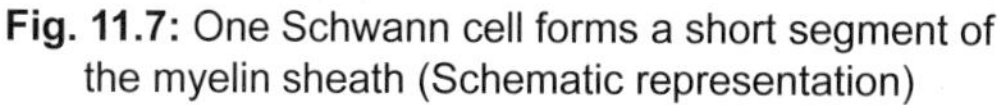

Fig. 11.7: One Schwann cell forms a short segment of the myelin sheath (Schematic representation)

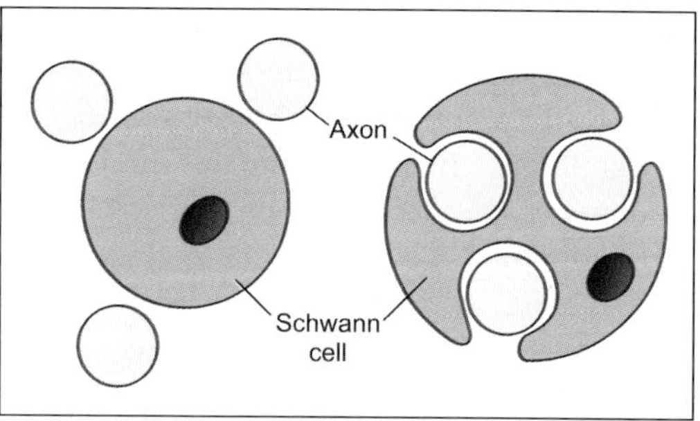

Fig. 11.8: Relationship of unmyelinated axons to Schwann cells (Schematic representation)

Functions of the Myelin Sheath

- The presence of a myelin sheath increases the velocity of conduction (for a nerve fibre of the same diameter)
- It reduces the energy expended in the process of conduction
- It is responsible for the colour of the white matter of the brain and spinal cord.

Unmyelinated Axons

There are some axons that are devoid of myelin sheaths. These ***unmyelinated axons*** invaginate into the cytoplasm of Schwann cells, but the mesaxon does not spiral around them (Figs. 11.7 and 11.8). Another difference is that several such axons may invaginate into the cytoplasm of a single Schwann cell.

TYPES OF NEURONS

- ***On the basis of number of processes***
  - ***Unipolar neurons:*** These neurons have single process (which is highly convoluted). After a very short course, this process divides into two. One of the divisions represents the axon; the other is functionally a dendrite, but its structure is indistinguishable from that of an axon.
    E.g. Neurons in dorsal root ganglion (Fig. 11.9).
  - ***Bipolar neurons:*** These neurons have only one axon and one dendrite.
    E.g. Neurons in vestibular and spiral ganglia.

Table 11.2: Difference between myelinated and unmyelinated nerve fibres

Myelinated nerves	Unmyelinated nerves
Myelinated nerves have axons of large diameter	Unmyelinated nerves have axons of small diameter
Axons surrounded by concentric layers of Schwann cell plasma membrane	Axons surrounded by cytoplasm of Schwann cells
Nerve impulse jumps from one node to other node called as saltatory conduction	Nerve impulse travels uniformly along the axolemma
Saltatory conduction seen in myelinated nerves is fast and consumes less energy	Conduction seen in unmyelinated nerves is slow and consumes more energy

PLATE 11.1: **Single Nerve Fibre**

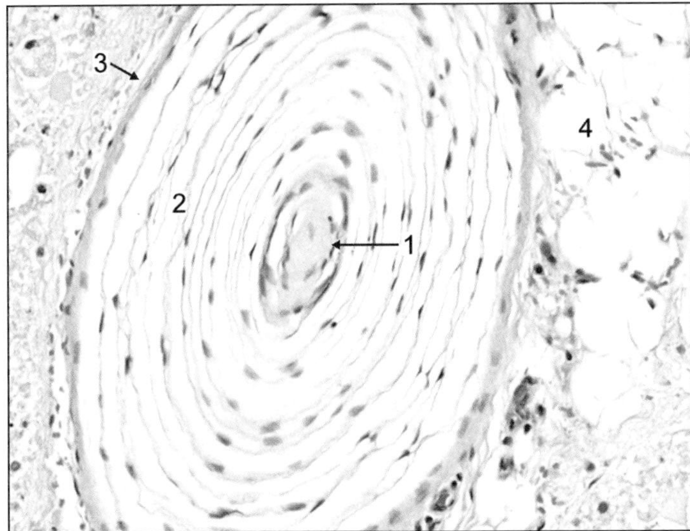

The axis cylinder is at the centre (arrow). In the region of the myelin sheath (surrounding the axis cylinder) there are a series of concentric membranes. They represent the greatly elongated mesaxon. Actual myelin is not seen as it has been dissolved away during the preparation of the section. The neurilemma can be seen. Some fat cells are also present.

Photomicrograph of single nerve fibre seen at high magnification

Key

1. Axis cylinder
2. Myelin sheath
3. Neurilemma
4. Fat cells

   - ***Multipolar neurons:*** It is most common type of neurons; the neuron gives off several processes, i.e. these neurons have one axon and many dendrites.
     E.g. Motor neurons (Fig. 11.9)
- ***On the basis of function***
   - ***Sensory neuron:*** They carry impulses from receptor organ to the central nervous system (CNS).
   - ***Motor neuron:*** They transmit impulses from the CNS to the muscles and glands
- ***On the basis of length of axons***
   - ***Golgi type I:*** These neurons have long axons, and connect remote regions.
     E.g. pyramidal cells of motor cortex in cerebrum
   - ***Golgi type II:*** These neurons have short axons which end near the cell body.
     E.g. Cerebral and cerebellar cortex.

PERIPHERAL NERVES

Peripheral nerves are collections of nerve fibres (axons). These may be myelinated or unmyelinated.

- Some nerve fibres carry impulses from the spinal cord or brain to peripheral structures like muscle or gland, these are called ***efferent*** or ***motor*** fibres. Efferent fibres are axons of neurons (the cell bodies of which are) located in the grey matter of the spinal cord or of the brainstem.

- Other nerve fibres carry impulses from peripheral organs to the brain or spinal cord: these are called ***afferent*** fibres. Afferent nerve fibres are processes of neurons that are located (as a rule) in sensory ganglia. In the case of spinal nerves these ganglia are located on the dorsal nerve roots. In the case of cranial nerves they are located on ganglia situated on the nerve concerned. The neurons in these ganglia are usually of the unipolar type.

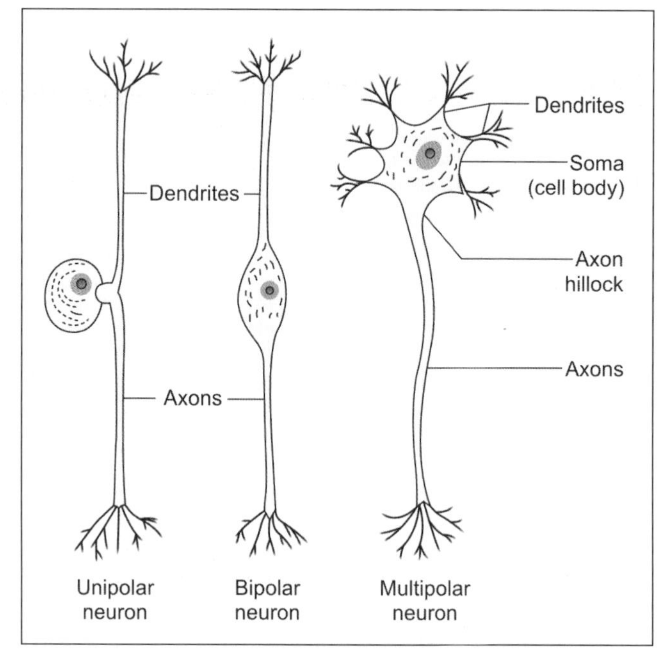

Fig. 11.9: Unipolar, bipolar and multipolar neurons (Schematic representation)

Basic Structure of Peripheral Nerves

- In the peripheral nerves each nerve fibre with its Schwann cell and basal lamina is surrounded by a layer of connective tissue called the ***endoneurium*** (Fig. 11.10 and Plate 11.2). The endoneurium contains collagen, fibroblasts, Schwann cells, endothelial cells and macrophages.

  Many nerve fibres together form bundles or ***fasciculi.*** Endoneurium holds adjoining nerve fibres together and facilitates their aggregation to form ***fasciculi***.
- Each fasciculus is surrounded by a thicker layer of connective tissue called the ***perineurium***. The perineurium is made up of layers of flattened cells separated by layers of collagen fibres. The perineurium probably controls diffusion of substances in and out of axons. A very thin nerve may consist of a single fasciculus, but usually a nerve is made up of several fasciculi.

Table 11.3: Morphological classification of neurons

Morphology	Location and example
According to polarity	
• Unipolar/pseudounipolar	Posterior root ganglia of spinal nerves, sensory ganglia of cranial nerves
• Bipolar	Bipolar cells of retina, sensory ganglia of cochlear and vestibular nerves
• Multipolar	Motor cells forming fibre tracts
According to size of nerve fibre	
• Golgi type I	Purkinje cells of cerebellum, Anterior horn cells of spinal cord, Pyramidal cells of cerebral cortex
• Golgi type II	Cerebral and cerebellar cortex

PLATE 11.2: Single Nerve Fibre

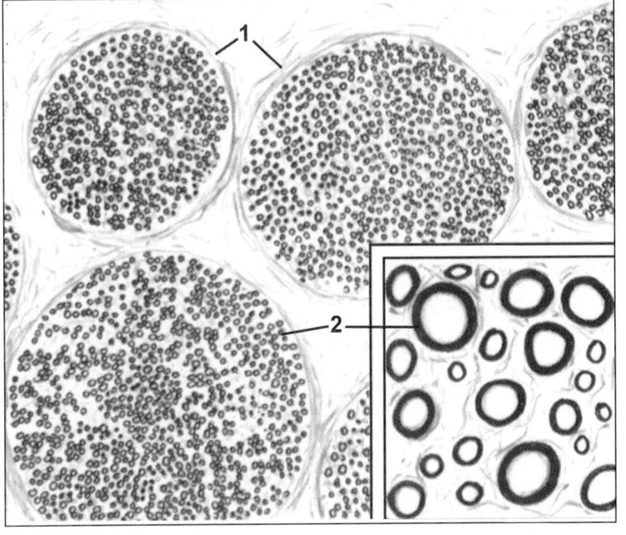

Peripheral nerve, special stain (as seen in drawing)

- The photomicrograph shows transverse section of a peripheral nerve.
- The nerve has been fixed in osmic acid that stains the myelin sheaths (around nerve fibres) black. The myelin sheaths, therefore, appear as black rings. Note that the nerve fibres are arranged in bundles that are held by connective tissue (called perineurium). In the inset we see nerve fibres at higher magnification. Note the varying sizes of the fibres. The connective tissue around individual nerve fibres is called endoneurium.

Key

1. Perineurium
2. Endoneurium

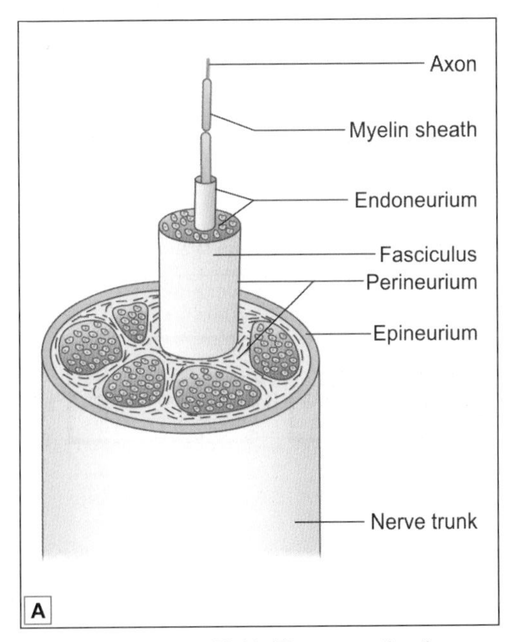

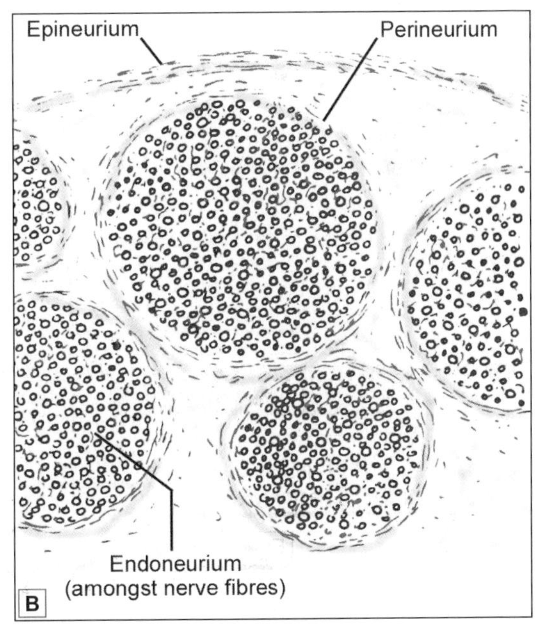

Fig. 11.10: The connective tissue supporting nerve fibres of a peripheral nerve. **A.** Longitudinal section; **B.** Transverse section (Schematic representation)

- The fasciculi are held together by a fairly dense layer of connective tissue that surrounds the entire nerve and is called the ***epineurium.***

Clinical Correlation

- The epineurium contains fat that cushions nerve fibres. Loss of this fat in bedridden patients can lead to pressure on nerve fibres and paralysis.
- Blood vessels to a nerve travel through the connective tissue that surrounds it. Severe reduction in blood supply can lead to ***ischaemic neuritis*** and pain.

NEUROGLIA

In addition to neurons, the nervous system contains several types of supporting cells called neuroglia (Flow chart 11.2).

Types of Neuroglia

- Astrocytes, oligodendrocytes, and microglia found in the parenchyma of the brain and spinal cord.
- Ependymal cells lining the ventricular system.
- Schwann cells lemmocytes or peripheral glia forming myelin sheaths around axons of peripheral nerves. It is important to note that both neurilemma and myelin sheaths are components of Schwann cells.
- Capsular cells (also called satellite cells or capsular gliocytes) that surround neurons in peripheral ganglia.
- Various types of supporting cells found in relation to motor and sensory terminals of nerve fibres.

Some workers use the term neuroglia for all these categories, while others restrict the term only to supporting cells present within the brain and spinal cord. The latter convention is used in the description that follows.

Flow chart 11.2: Types of neuroglia found in central and peripheral nervous systems

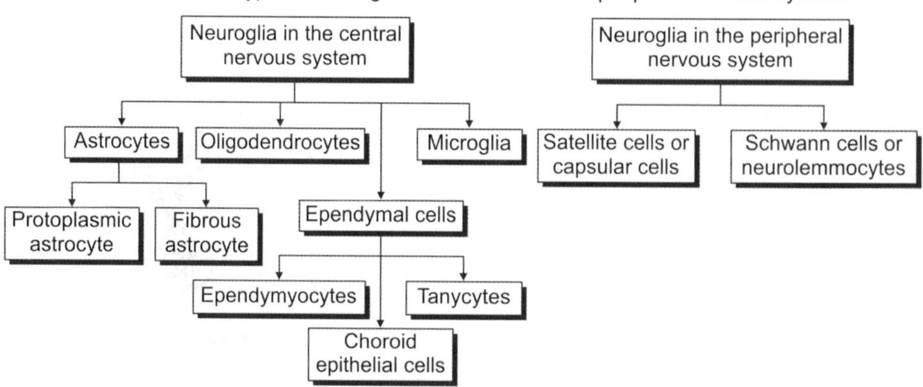

Neuroglia of Brain and Spinal Cord

Neuroglial cells present in the parenchyma of brain and spinal cord are mainly of four types:

- Astrocytes, that may be subdivided into fibrous and protoplasmic astrocytes.
- Oligodendrocytes
- Ependymal cells
- Microglia.

All neuroglial cells are much smaller in size than neurons. However, they are far more numerous. It is interesting to note that the number of glial cells in the brain and spinal cord is 10–50 times as much as that of neurons. Neurons and neuroglia are separated by a very narrow extracellular space.

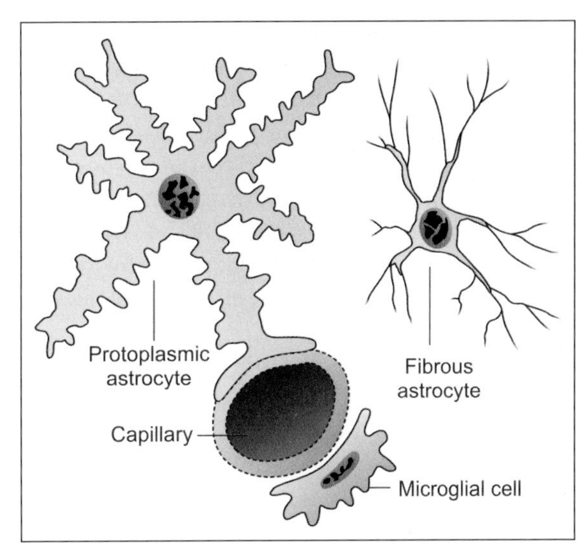

Fig. 11.11: Astrocytes and microglial cells (Schematic representation)

In ordinary histological preparations, only the nuclei of neuroglial cells are seen. Their processes can be demonstrated by special techniques.

Astrocytes

These are small star-shaped cells that give off a number of processes (Fig. 11.11). The processes are often flattened into leaf-like laminae that may partly surround neurons and separate them from other neurons. The processes frequently end in expansions in relation to blood vessels or in relation to the surface of the brain. Small swellings called **gliosomes** are present on the processes of astrocytes. These swellings are rich in mitochondria.

Astrocytes are of two types: fibrous and protoplasmic.

- ***Fibrous astrocytes*** are seen mainly in white matter. Their processes are thin and are asymmetrical.
- ***Protoplasmic astrocytes*** are seen mainly in grey matter. Their processes are thicker than those of fibrous astrocytes and are symmetrical.
- ***Intermediate forms*** between fibrous and protoplasmic astrocytes are also present. Protoplasmic extensions of astrocytes surround nodes of Ranvier.

The processes of astrocytes are united to those of other astrocytes through gap junctions. Astrocytes communicate with one another through calcium channels. Such communication plays a role in regulation of synaptic activity and in the metabolisms of neurotransmitters and neuromodulators.

Function

- They provide mechanical support to neurons.
- In view of their nonconducting nature, they serve as insulators and prevent neuronal impulses from spreading in unwanted directions.

- They are believed to help in neuronal function by playing an important role in maintaining a suitable metabolic environment for the neurons. They can absorb neurotransmitters from synapses, thus, terminating their action
- They help in the formation of blood-brain barrier.
- Substances secreted by end feet of astrocytes probably assist in maintaining a membrane, the ***glia limitans externa***, which covers the exposed surfaces of the brain. They also help to maintain the basal laminae of blood vessels that they come in contact with.
- Astroglial cells are also responsible for repair of damaged areas of nervous tissue. They proliferate in such regions (**gliosis**). The microglia act as macrophages to engulf and destroy unwanted material.

Oligodendrocytes

These cells are rounded or pear-shaped bodies with relatively few processes (oligo—scanty) (Fig. 11.12).

These cells provide myelin sheaths to nerve fibres that lie within the brain and spinal cord. Their relationship to nerve fibres is basically similar to that of Schwann cells to peripheral nerve fibres. However, in contrast to a Schwann cell that ensheaths only one axon, an oligodendrocyte may enclose several axons.

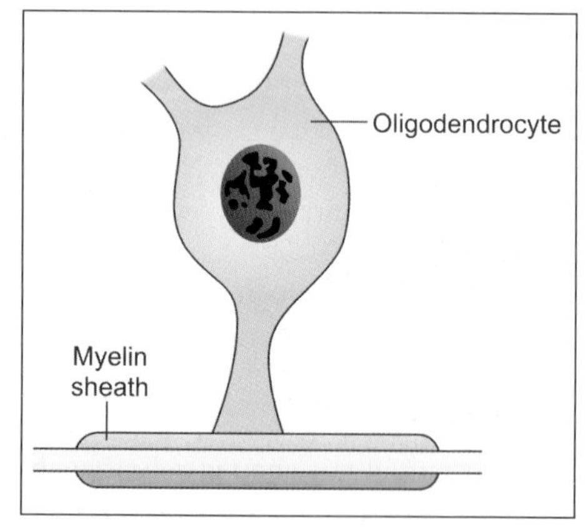

Fig. 11.12: Oligodendrocyte giving off a process that forms a segment of the myelin sheath of an axon (Schematic representation)

Oligodendrocytes are classified into several types depending on the number of neurons they provide sheaths to. As a rule, oligodendrocytes present in relation to large diameter axons provide sheaths to fewer axons than those related to axons of small diameter. The plasma membrane of oligodendrocytes comes into contact with axolemma at nodes of Ranvier.

Function

Oligodendrocytes provide myelin sheaths to nerve fibres within the CNS for fast conduction of nerve impulses.

Ependymal Cells

Ependymal cells line the ventricles of the brain and central canal of the spinal cord. Ependymal cells are mainly of three types:

- Ependymocytes
- Choroid epithelial cells
- Tanycytes.

The ependymocytes constitute the majority of the ependymal cells. The specialised ependymal cells in choroid plexuses (choroidal epithelial cells) secrete cerebrospinal fluid. The ependymal cells lining the floor of the fourth ventricle have long basal processes and are termed '***tanycytes***'.

Function

Ependymal cells are concerned in exchanges of material between the brain and the cerebrospinal fluid at the brain-cerebrospinal fluid barrier. The blood in the capillaries of the choroid plexus is filtered through choroid epithelial cells at the blood–cerebrospinal fluid barrier to secrete cerebrospinal fluid.

Microglia

These are the smallest neuroglial cells (Fig. 11.11). The cell body is flattened. The processes are short. These cells are frequently seen in relation to capillaries. As already stated, they differ from other neuroglial elements in being mesodermal in origin. They are probably derived from monocytes that invade the brain during fetal life. They are more numerous in grey matter than in white matter.

Function

They act as phagocytes and become active after damage to nervous tissue by trauma or disease. The microglia act as macrophages to engulf and destroy unwanted material.

Pathological Correlation

Tumours of nervous tissue

- Precursors of neural cells can give rise to medulloblastomas. Once mature neurons are formed they lose the power of mitosis and do not give origin to tumours
- Certain tumors called germinomas appear near the midline, mostly near the third ventricle. They arise from germ cells that also give rise to teratomas
- Most tumours of the brain arise from neuroglial cells. Astrocytomas are most common. Oligodendromas are also frequent
- Tumours can also arise from ependyma and from Schwann cells.

THE SYNAPSE

Synapses are sites of junction between neurons.

Synapses can be broadly classified into:

- Chemical synapses
- Electrical synapses.

Synapses involving the release of neurotransmitters are referred to as ***chemical synapses***.

At some sites one cell may excite another without the release of a transmitter. At such sites adjacent cells have direct channels of communication through which ions can pass from one cell to another altering their electrical status. Such synapses are called ***electrical synapses***.

At the site of an electrical synapse plasma membranes (of the two elements taking part) are closely applied, the gap between them being about 4 nm. Proteins called ***connexins*** project into this gap from the membrane on either side of the synapse. The proteins are so arranged that small open channels are created between the two synaptic elements.

Electrical synapses are common in lower vertebrates and invertebrates. They have been demonstrated at some sites in the brains of mammals (for example, in the inferior olive and cerebellum).

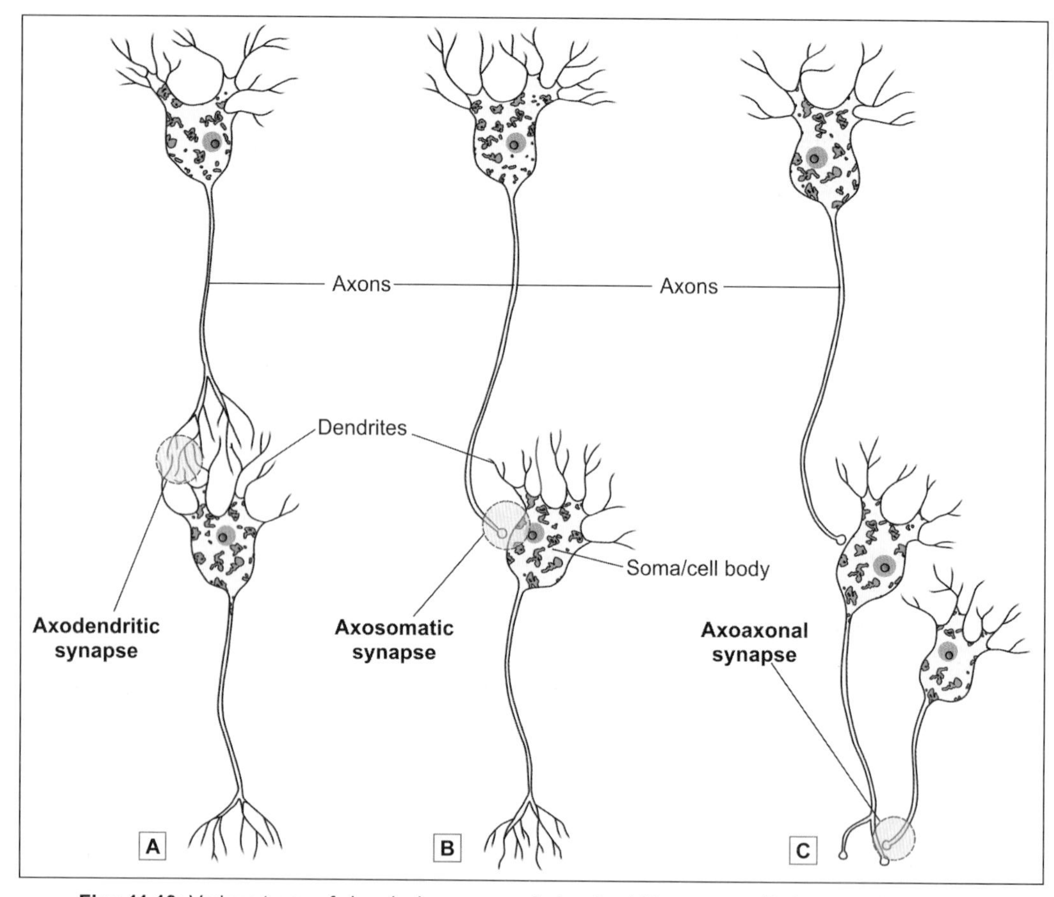

Figs 11.13: Various types of chemical synapses. **A.** Axodendritic synapse; **B.** Axosomatic synapse; **C.** Axoaxonal synapse (Schematic representation)

Junctions between receptors and neurons, or between neurons and effectors, share some of the features of typical synapses and may also be regarded as synapses. Junctions between cardiac myocytes and between smooth muscle cells, are regarded as electrical synapses.

Classification of a Chemical Synapse Based on Neuronal Elements Taking Part

Synapses may be of various types depending upon the parts of the neurons that come in contact.

- ***Axodendritic synapse:*** It is the most common type of synapse. In this type, an axon terminal establishes contact with the dendrite of a receiving neuron to form a synapse (Fig. 11.13A).
- ***Axosomatic synapse:*** The axon terminal synapses with the cell body (Fig. 11.13B).
- ***Axoaxonal synapse:*** The axon terminal synapses with the axon of the receiving neuron. An axoaxonal synapse may be located either on the initial segment (of the receiving axon) or just proximal to an axon terminal (Fig. 11.13C).

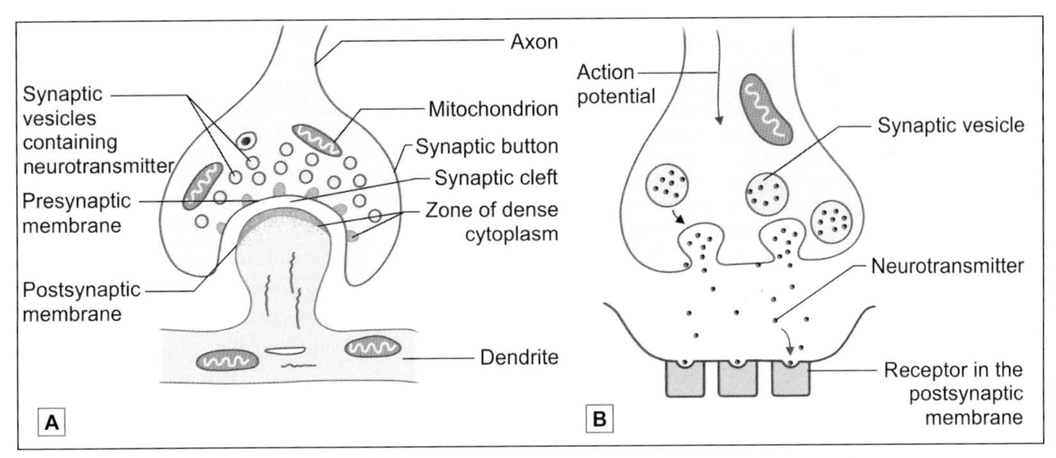

Fig. 11.14: A. Structure of a typical chemical synapse as seen under electron microscope; **B.** Mechanism of synaptic transmission (Schematic representation)

- ***Dendroaxonic synapse:*** In some parts of the brain (for example, the thalamus), we see some synapses in which the presynaptic element is a dendrite, instead of an axon, which synapses with the axon of the receiving neuron.
- ***Dendrodendritic synapse:*** Synapse between two dendrites.
- ***Somatosomatic synapse:*** The soma of a neuron may synapse with the soma of another neuron.
- ***Somatodendritic synapse:*** Synapse between a soma and a dendrite.

Structure of a Chemical Synapse

A synapse transmits an impulse only in one direction. The two elements taking part in a synapse can, therefore, be spoken of as ***presynaptic*** and ***postsynaptic*** (Fig. 11.14).

In an axodendritic synapse, the terminal enlargement of the axon may be referred to as the ***presynaptic bouton*** or ***synaptic bag***. The region of the dendrite receiving the axon terminal is the ***postsynaptic process***. The two are separated by a space called the ***synaptic cleft***. Delicate fibres or granular material may be seen within the cleft. On either side of the cleft, there is a region of dense cytoplasm. On the presynaptic side, this dense cytoplasm is broken up into several bits. On the postsynaptic side, the dense cytoplasm is continuous and is associated with a meshwork of filaments called the ***synaptic web***.

The thickened areas of membrane on the presynaptic and postsynaptic sides constitute the ***active zone*** of a synapse. Neurotransmission takes place through this region.

GANGLIA

Aggregations of cell bodies of neurons, present outside the brain and spinal cord are known as ganglia. Ganglia are of two main types: ***sensory*** and ***autonomic***.

Sensory ganglia (Plate 11.3) are present on the dorsal nerve roots of spinal nerves, where they are called dorsal nerve root ganglia or spinal ganglia. They are also present on the 5th, 7th, 8th, 9th and 10th cranial nerves. We have seen that the neurons in these ganglia are of the unipolar type (except in the case of ganglia associated with the vestibulocochlear nerve

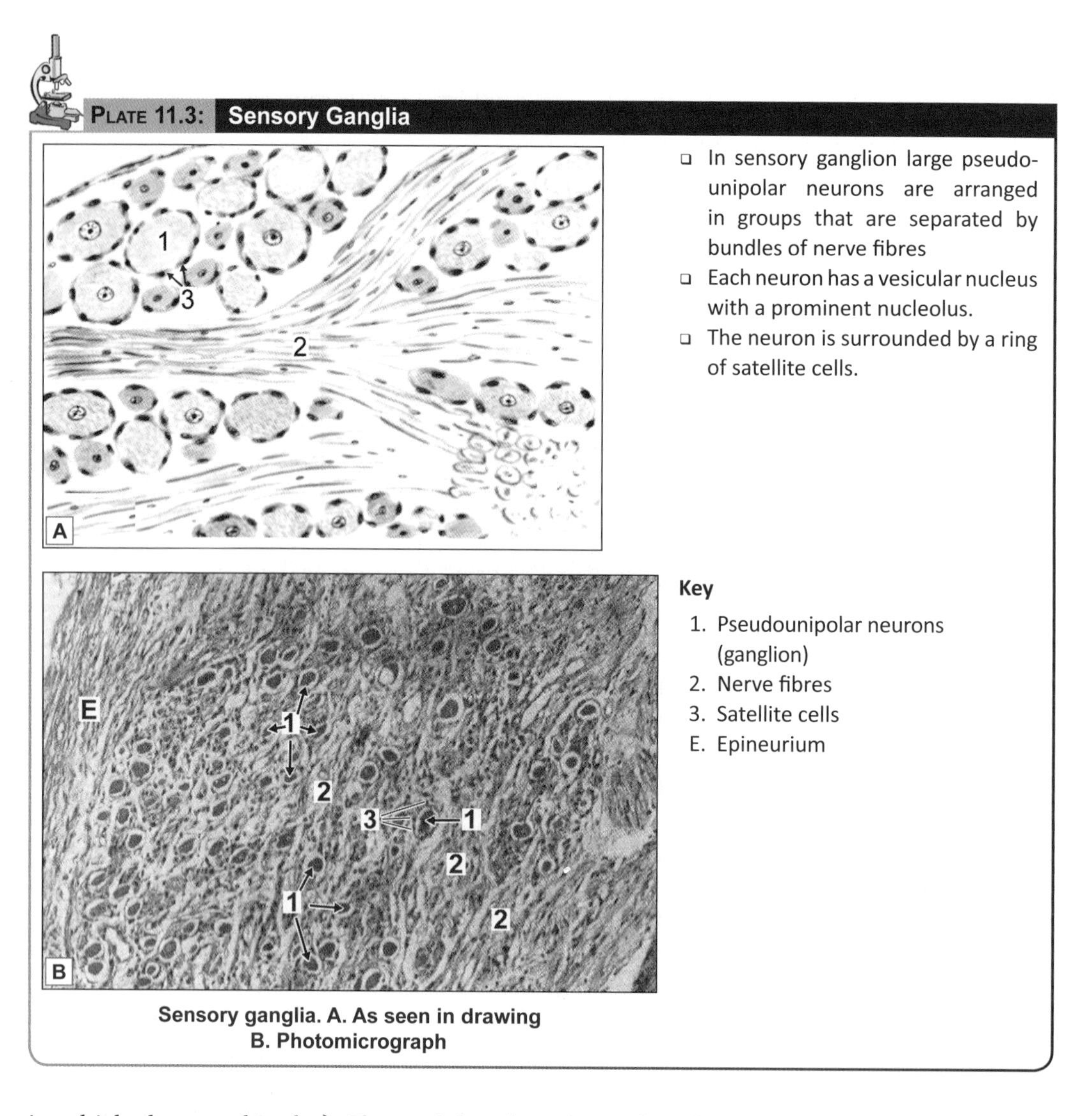

Sensory ganglia. A. As seen in drawing B. Photomicrograph

in which they are bipolar). The peripheral process of each neuron forms an afferent (or sensory) fibre of a peripheral nerve. The central process enters the spinal cord or brainstem. (For further details of the connections of these neurons see the author's Textbook of Human Neuroanatomy).

Autonomic ganglia (Plate 11.4) are concerned with the nerve supply of smooth muscle or of glands. The pathway for this supply always consists of two neurons: preganglionic and postganglionic. The cell bodies of preganglionic neurons are always located within the spinal cord or brainstem. Their axons leave the spinal cord or brainstem and terminate by synapsing with postganglionic neurons, the cell bodies of which are located in autonomic ganglia. Autonomic ganglia are, therefore, aggregations of the cell bodies of postganglionic neurons. These neurons are multipolar. Their axons leave the ganglia as postganglionic fibres to reach and supply smooth muscle or gland. Autonomic ganglia are subdivisible into two major

PLATE 11.4: Autonomic Ganglia

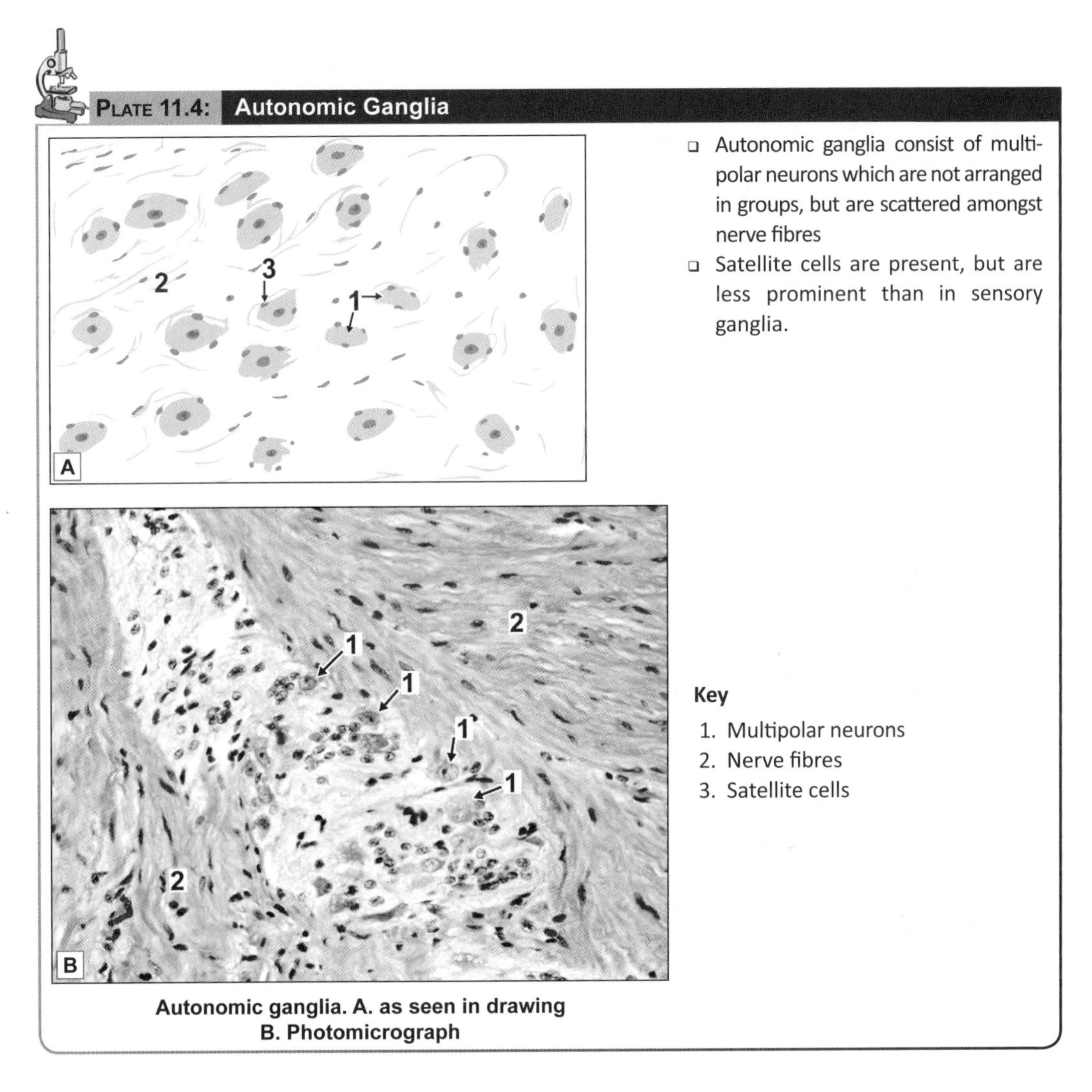

Autonomic ganglia. A. as seen in drawing
B. Photomicrograph

- Autonomic ganglia consist of multipolar neurons which are not arranged in groups, but are scattered amongst nerve fibres
- Satellite cells are present, but are less prominent than in sensory ganglia.

Key

1. Multipolar neurons
2. Nerve fibres
3. Satellite cells

types: sympathetic and parasympathetic. Sympathetic ganglia are located on the right and left sympathetic trunks. Parasympathetic ganglia usually lie close to the viscera supplied through them (For further details of the connections of sympathetic and parasympathetic ganglia see the author's Textbook of Human Neuroanatomy).

Structure of Sensory Ganglia

In haematoxylin and eosin stained sections the neurons of sensory ganglia are seen to be large and arranged in groups chiefly at the periphery of the ganglion (Plate 11.3). In sections stained by silver impregnation the neurons can be seen to be unipolar. The groups of cells are separated by groups of myelinated nerve fibres.

The cell body of each neuron is surrounded by a layer of flattened capsular cells or satellite cells. Outside the satellite cells there is a layer of delicate connective tissue (The satellite cells are continuous with the Schwann cells covering the processes arising from the neuron. The connective tissue covering each neuron is continuous with the endoneurium).

The entire ganglion is pervaded by fine connective tissue. The ganglion is covered on the outside by a connective tissue capsule.

Structure of Autonomic Ganglia

The neurons of autonomic ganglia are smaller than those in sensory ganglia (Plate 11.4). With silver impregnation they are seen to be multipolar. The neurons are not arranged in definite groups as in sensory ganglia, but are scattered throughout the ganglion. The nerve fibres are non-myelinated and thinner. They are, therefore, much less conspicuous than in sensory ganglia.

Satellite cells are present around neurons of autonomic ganglia, but they are not so well defined. The ganglion is permeated by connective tissue that also provides a capsule for it (just as in sensory ganglia).

The Nissl substance of the neurons is much better defined in autonomic ganglia than in sensory ganglia. In sympathetic ganglia the neuronal cytoplasm synthesises catecholamines; and in parasympathetic ganglia it synthesises acetylcholine. These neurotransmitters travel down the axons to be released at nerve terminals.

SPINAL CORD; CEREBELLAR CORTEX; CEREBRAL CORTEX

GREY AND WHITE MATTER

Sections through the spinal cord or through any part of the brain show certain regions that appear whitish, and others that have a darker greyish colour. These constitute the ***white*** and ***grey matter*** respectively. Microscopic examination shows that the cell bodies of neurons are located only in grey matter that also contains dendrites and axons starting from or ending on the cell bodies. Most of the fibres within the grey matter are unmyelinated. On the other hand the white matter consists predominantly of myelinated fibres. It is the reflection of light by myelin that gives this region its whitish appearance. Neuroglia and blood vessels are present in both grey and white matter.

The arrangement of the grey and white matter differs at different situations in the brain and spinal cord. In the spinal cord and brainstem the white matter is on the outside whereas the grey matter forms one or more masses embedded within the white matter. In the cerebrum and cerebellum there is an extensive, but thin, layer of grey matter on the surface. This layer is called the ***cortex.*** Deep to the cortex there is white matter, but within the latter several isolated masses of grey matter are present. Such isolated masses of grey matter present anywhere in the central nervous system are referred to as ***nuclei.*** As grey matter is made of cell bodies of neurons (and the processes arising from or terminating on them) nuclei can be defined as groups of cell bodies of neurons. The axons arising in one mass of grey matter very frequently terminate by synapsing with neurons in other masses of grey matter. The axons connecting two (or more) masses of grey matter are frequently numerous enough to form recognisable bundles. Such aggregations of fibres are called ***tracts.*** Larger collections of fibres are also referred to as ***funiculi, fasciculi*** or ***lemnisci.*** Large bundles of fibres connecting the cerebral or cerebellar hemispheres to the brainstem are called ***peduncles.***

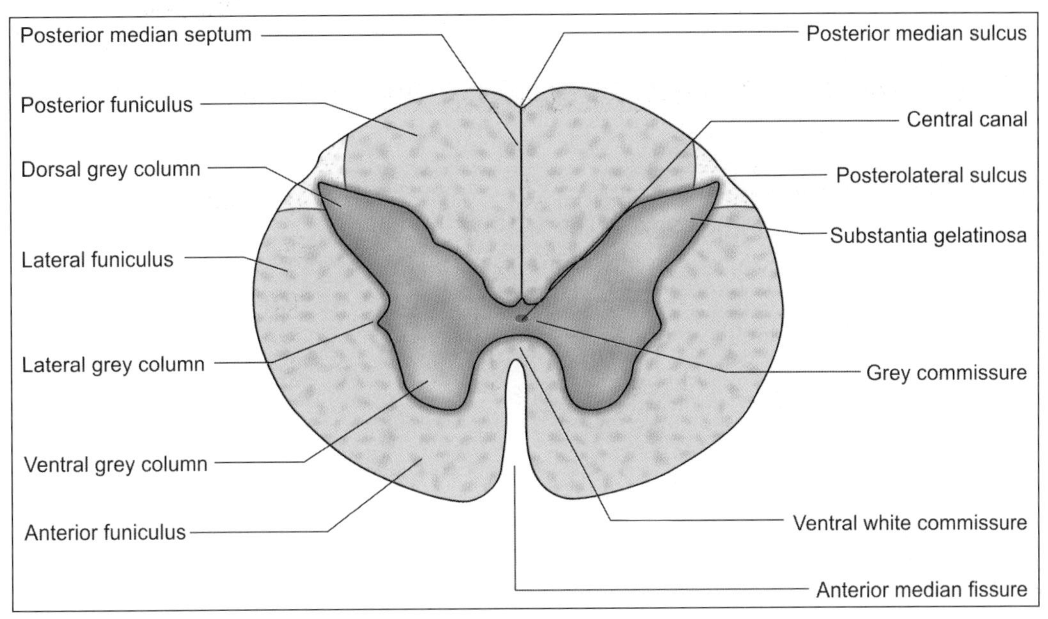

Fig. 11.15: Main features to be seen in a transverse section through the spinal cord (Schematic representation)

Aggregations of processes of neurons outside the central nervous system constitute ***peripheral nerves.***

THE SPINAL CORD

The spinal cord is the most important content of the vertebral canal. The upper end of the spinal cord becomes continuous with the medulla oblongata. The lowest part of the spinal cord is conical and is called the ***conus medullaris***. The conus is continuous, below, with a fibrous cord called the ***filum terminale (modification of pia mater)***.

When seen in transverse section the grey matter of the spinal cord forms an H-shaped mass (Fig. 11.15 and Plate 11.5). In each half of the cord the grey matter is divisible into a larger ventral mass, the ***anterior (or ventral) grey column***, and a narrow elongated ***posterior (or dorsal) grey column***. In some parts of the spinal cord a small lateral projection of grey matter is seen between the ventral and dorsal grey columns. This is the ***lateral grey column***. The grey matter of the right and left halves of the spinal cord is connected across the middle line by the ***grey commissure*** that is traversed by the ***central canal.*** The central canal of the spinal cord contains cerebrospinal fluid. The canal is lined by ependyma.

The cell bodies of neurons differ in size and in prominence of Nissl substance in different regions of spinal grey matter. They are most prominent in the anterior grey column.

The white matter of the spinal cord is divided into right and left halves, in front by a deep ***anterior median fissure***, and behind by the ***posterior median septum***. In each half of the cord the white matter medial to the dorsal grey column forms the ***posterior funiculus*** (or posterior white column). The white matter medial and ventral to the anterior grey column forms the ***anterior funiculus*** (or anterior white column), while the white matter lateral to the anterior

PLATE 11.5: Spinal Cord

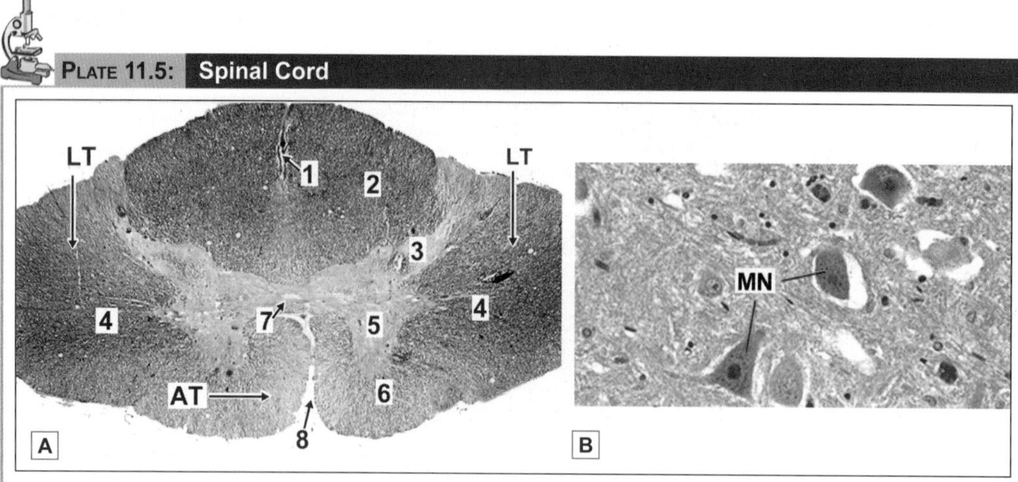

Spinal cord. A. Panoramic view (Photomicrograph); B. Grey matter (Photomicrograph)

The spinal cord has a characteristic oval shape. It is made up of white matter (containing mainly of myelinated fibres), and grey matter (containing neurons and unmyelinated fibres). The grey matter lies towards the centre and is surrounded all round by white matter. The grey matter consists of a centrally placed mass and projections (horns) that pass forwards and backwards.

Note: The stain used for the slide is Luxol Fast Blue.

Key

1. Posterior median septum
2. Posterior white column
3. Posterior grey column
4. Lateral white column
5. Anterior grey column
6. Anterior white column
7. Central canal lying in grey commissure. The fibres in front of the grey commissure form the anterior white commissure
8. Anterior median sulcus

AT. Anterior motor tracts
LT. Lateral motor tracts
MN. Multipolar neurons in grey matter

and posterior grey columns forms the ***lateral funiculus*** (The anterior and lateral funiculi are collectively referred to as the ***anterolateral funiculus***).

The white matter of the right and left halves of the spinal cord is continuous across the middle line through the ***ventral white commissure*** which lies anterior to the ***grey commissure***. The white matter contains tracts (ascending or descending) that connect grey matter at different levels of the spinal cord. Some tracts ascend into (or descend from) the brainstem, the cerebellum or the cerebral cortex. Details of such tracts are given in books of neuroanatomy.

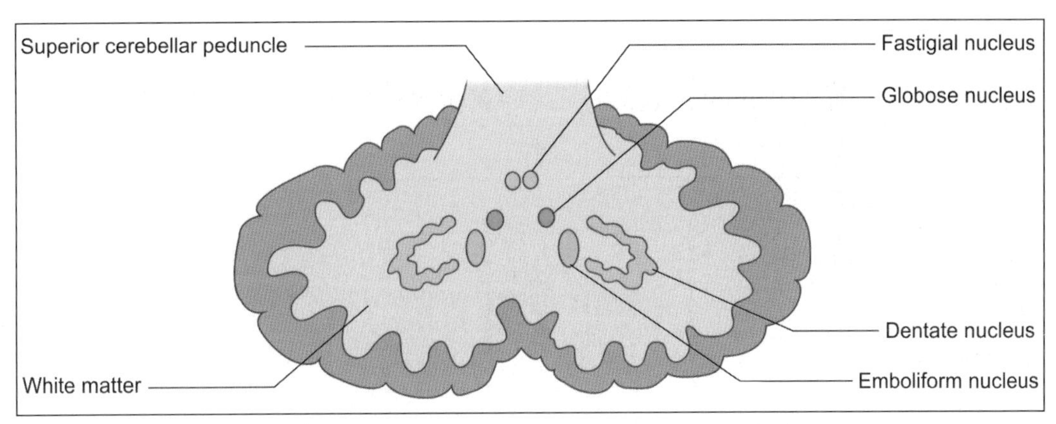

Fig. 11.16: Cerebellar nuclei (Schematic representation)

THE CEREBELLAR CORTEX

General Features

The cerebellum (or small brain) lies in the posterior cranial fossa. Like the cerebrum, the cerebellum has a superficial layer of grey matter, the cerebellar cortex. Because of the presence of numerous fissures, the cerebellar cortex is extensive.

Grey Matter of the Cerebellum

Most of the grey matter of the cerebellum is arranged as a thin layer covering the central core of white matter. This layer is the ***cerebellar cortex***. The subdivisions of the cerebellar cortex correspond to the subdivisions of the cerebellum.

Embedded within the central core of white matter there are masses of grey matter that constitute the ***cerebellar nuclei.*** These are (Fig. 11.16).

- The ***dentate nucleus*** lies in the centre of each cerebellar hemisphere. It is made up of a thin lamina of grey matter that is folded upon itself so that it resembles a crumpled purse.
- The ***emboliform nucleus*** lies on the medial side of the dentate nucleus.
- The ***globose nucleus*** lies medial to the emboliform nucleus.
- The ***fastigial nucleus*** lies close to the middle line in the anterior part of the superior vermis.

White Matter of the Cerebellum

The central core of each cerebellar hemisphere is formed by white matter. The peduncles are continued into this white matter. The white matter of the two sides is connected by a thin lamina of fibres that are closely related to the fourth ventricle. The upper part of this lamina forms the superior medullary velum, and its inferior part forms the inferior medullary velum. Both these take part in forming the roof of the fourth ventricle.

Structure of the Cerebellar Cortex

In striking contrast to the cortex of the cerebral hemispheres, the cerebellar cortex has a uniform structure in all parts of the cerebellum. It may be divided into three layers (Fig. 11.17 and Plate 11.6) as follows.

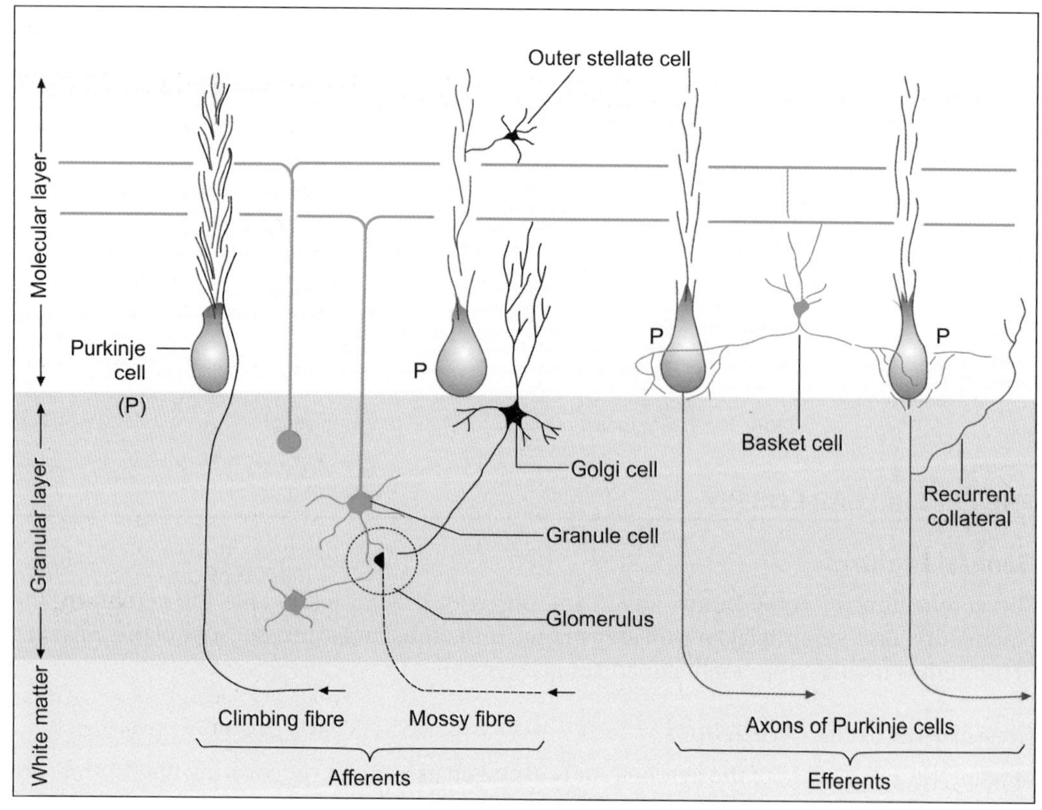

Fig. 11.17: Arrangement of neurons in the cerebellar cortex (Schematic representation)

- ***Molecular layer*** (most superficial)
- ***Purkinje cell layer***
- ***Granular layer***, which rests on white matter.

The neurons of the cerebellar cortex are of five main types:

- ***Purkinje cells***, forming the layer named after them
- ***Granule cells***, forming the granular layer
- ***Outer (external) stellate cells***, lying in molecular layer
- ***Basket cells***, lying in the molecular layer
- ***Golgi cells***, present in the granular layer.

Molecular Layer

The molecular layer is the superficial layer of the cortex and situated just below the pia mater. It stains lightly with haematoxylin eosin and is featureless as it consists of few cells and more of myelinated and unmyelinated fibres.

Two types of cells are found in this layer:

- Stellate cells—situated in the superficial part of the molecular layer
- Basket cells—situated in the deeper layer.

Plate 11.6: Cerebellar Cortex

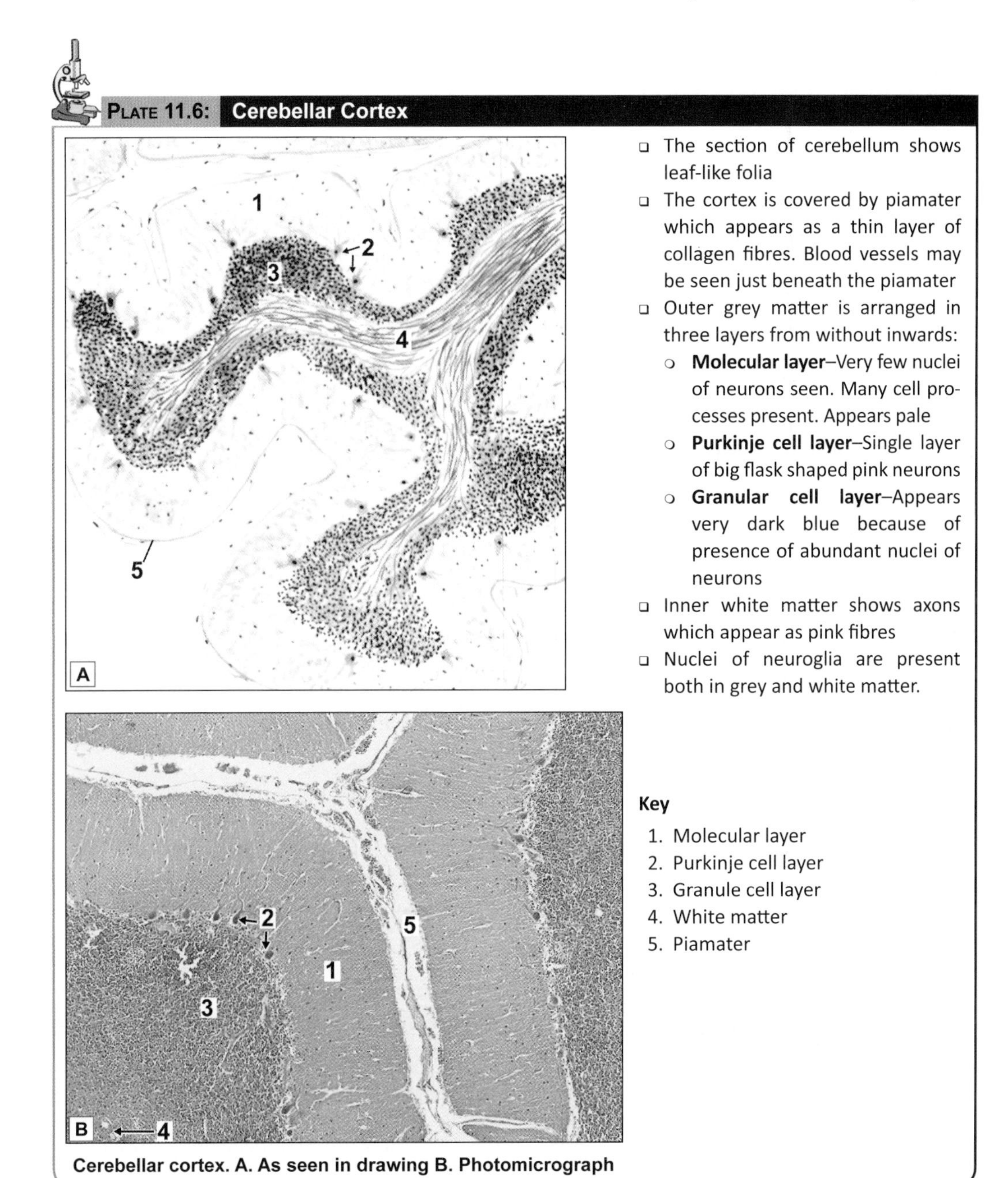

- The section of cerebellum shows leaf-like folia
- The cortex is covered by piamater which appears as a thin layer of collagen fibres. Blood vessels may be seen just beneath the piamater
- Outer grey matter is arranged in three layers from without inwards:
  - **Molecular layer**–Very few nuclei of neurons seen. Many cell processes present. Appears pale
  - **Purkinje cell layer**–Single layer of big flask shaped pink neurons
  - **Granular cell layer**–Appears very dark blue because of presence of abundant nuclei of neurons
- Inner white matter shows axons which appear as pink fibres
- Nuclei of neuroglia are present both in grey and white matter.

Key

1. Molecular layer
2. Purkinje cell layer
3. Granule cell layer
4. White matter
5. Piamater

Cerebellar cortex. A. As seen in drawing B. Photomicrograph

Outer Stellate Cells

These cells and their processes are confined to the molecular layer of the cerebellar cortex. Their dendrites (which are few) synapse with parallel fibres (of granule cells) while their axons synapse with dendrites of Purkinje cells (near their origin).

Basket Cells

These cells lie in the deeper part of the molecular layer of the cerebellar cortex. Their dendrites (which are few) ramify in the molecular layer and are intersected by parallel fibres with which they synapse. They also receive recurrent collaterals from Purkinje cells, climbing fibres, and mossy fibres. The axons of these cells branch and form networks (or baskets) around the cell bodies of Purkinje cells. Their terminations synapse with Purkinje cells at the junction of the cell body and axon (preaxon).

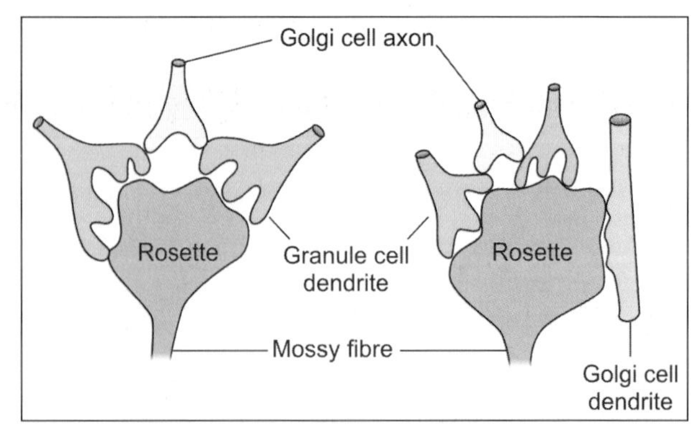

Fig. 11.18: Structure of cerebellar glomeruli. The outer capsule is not shown (Schematic representation)

Purkinje Cell Layer

The Purkinje cell layer contains flask-shaped cell bodies of Purkinje cells. This layer is unusual in that it contains only one layer of neurons. The Purkinje cells are evenly spaced. A dendrite arises from the 'neck' of the 'flask' and passes 'upwards' into the molecular layer. Here, it divides and subdivides to form an elaborate dendritic tree. The branches of this 'tree' all lie in one plane.

The axon of each Purkinje cell passes 'downwards' through the granular layer to enter the white matter. As described later, these axons constitute the only efferents of the cerebellar cortex. They end predominantly by synapsing with neurons in cerebellar nuclei. They are inhibitory to these neurons.

Granular Layer

It is the innermost layer and consists of numerous granule cells and a few golgi cells and brush cells. The granular layer stains deeply with haematoxylin because it is densely packed with granular cell.

Granule Cells

These are very small, numerous, spherical neurons that occupy the greater part of the granular layer. The spaces not occupied by them are called ***cerebellar islands***. These islands are occupied by special synaptic structures called ***glomeruli*** (Fig. 11.18).

Each granule cell gives off three to five short dendrites. These end in claw-like endings, which enter the glomeruli where they synapse with the terminals of mossy fibres (see below). The granule cells receive impulses from afferent mossy fibres. The dendrites of granule cells and axons of golgi cells synapse with terminals of mossy fibres to form lightly stained areas called glomeruli.

Golgi Neurons

These are large, stellate cells lying in the granular layer, just deep to the Purkinje cells. They are GABAergic inhibitory neurons. Their dendrites enter the molecular layer, where they branch profusely, and synapse with the parallel fibres.

Afferent Fibres Entering the Cerebellar Cortex

The afferent fibres to the cerebellar cortex are of two different types:
- Mossy fibres
- Climbing fibres.

Mossy Fibres

All fibres entering the cerebellum, other than olivocerebellar, end as mossy fibres. Mossy fibres originate from the vestibular nuclei (vestibulocerebellar), pontine nuclei (pontocerebellar), and spinal cord (spinocerebellar) and terminate in the granular layer of the cortex within glomeruli. Before terminating, they branch profusely within the granular layer, each branch ends in an expanded terminal called a ***rosette*** (Fig. 11.18).

Afferent inputs through mossy fibres pass through granule cells to reach the Purkinje cells.

Climbing Fibres

These fibres represent terminations of axons reaching the cerebellum from the inferior olivary complex (Fig. 11.17). They pass through the granular layer and the Purkinje cell layer to reach the molecular layer. Each climbing fibre exerts specific influence on one Purkinje cell.

Note: Both mossy and climbing fibres are excitatory.

Efferent Fibres

The efferent fibres from the cerebellar cortex are axons of Purkinje cells, which terminate in the cerebellar (central) nuclei (Fig. 11.17). Axons of the Purkinje cells are inhibitory to cerebellar nuclei.

The fibres from dentate, emboliform, and globose nuclei leave cerebellum through the superior cerebellar peduncle. The fibres from the fastigial nucleus leave the cerebellum through inferior cerebellar peduncle.

Table 11.4: Nerve cells in the cerebral cortex

Pyramidal cells	Cells of Martinoti
Stellate cells	Basket cells
Fusiform cells	Chandelier cells
Horizontal cells of Cajal	Double bouquet cells

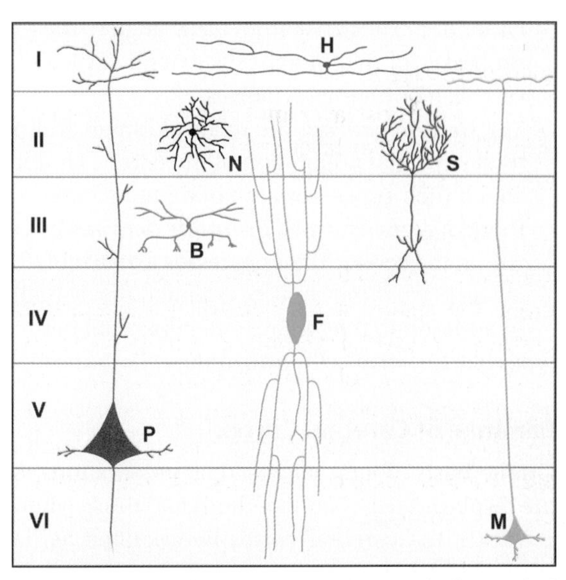

Fig. 11.19: Some of the cell types to be seen in the cerebral cortex (Schematic representation). B-basket cells; F-fusiform cells; H-horizontal cell of Cajal; N-neuroglia form cell; S-stellate cell; P-pyramidal cell; M-cells of Martinoti

THE CEREBRAL CORTEX

General Features

The surface of the cerebral hemisphere is covered by a thin layer of grey matter called the ***cerebral cortex***.

Like other masses of grey matter the cerebral cortex contains the cell bodies of an innumerable number of neurons along with their processes, neuroglia and blood vessels. The neurons are of various sizes and shapes. They establish extremely intricate connections with each other and with axons reaching the cortex from other masses of grey matter.

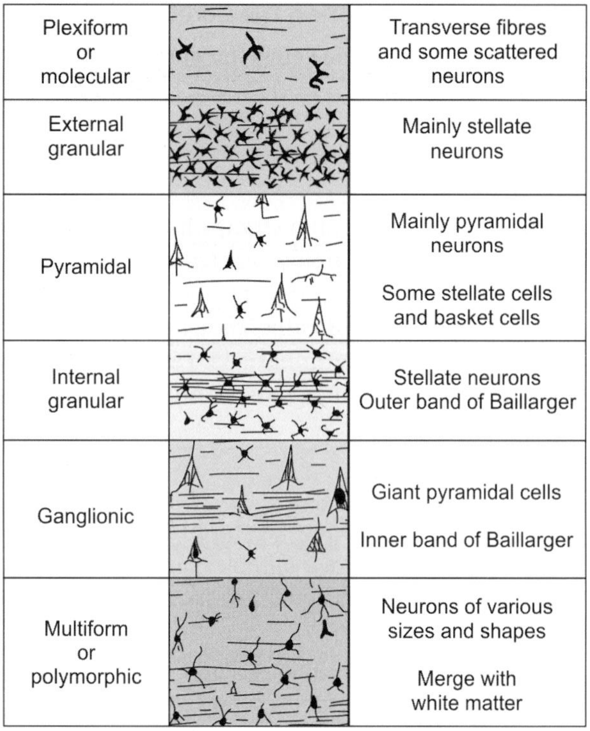

Layer		Features
Plexiform or molecular		Transverse fibres and some scattered neurons
External granular		Mainly stellate neurons
Pyramidal		Mainly pyramidal neurons Some stellate cells and basket cells
Internal granular		Stellate neurons Outer band of Baillarger
Ganglionic		Giant pyramidal cells Inner band of Baillarger
Multiform or polymorphic		Neurons of various sizes and shapes Merge with white matter

Fig. 11.20: Laminae of cerebral cortex (Schematic representation)

Neurons in the Cerebral Cortex

Cortical neurons vary in size, in the shape of their cell bodies, and in the lengths, branching patterns and orientation of their processes. Some of these are described below (Fig. 11.19).

- The most abundant type of cortical neurons are the ***pyramidal cells***. About two thirds of all cortical neurons are pyramidal. Their cell bodies are triangular, with the apex generally directed towards the surface of the cortex. A large dendrite arises from the apex. Other dendrites arise from basal angles. The axon arises from the base of the pyramid. The processes of pyramidal cells extend vertically through the entire thickness of cortex and establish numerous synapses.
- The ***stellate neurons*** are relatively small and multipolar. They form about one-third of the total neuronal population of the cortex. Under low magnifications (and in preparations in which their processes are not demonstrated) these neurons look like granules. They have, therefore, been termed ***granular neurons*** by earlier workers. Their axons are short and end within the cortex. Their processes extend chiefly in a vertical direction within the cortex, but in some cases they may be oriented horizontally.

In addition to the stellate and pyramidal neurons, the cortex contains numerous other cell types like fusiform cells, horizontal cells and cells of Martinoti.

Laminae of Cerebral Cortex

On the basis of light microscopic preparations stained by methods in which the cell bodies are displayed (e.g., Nissl method) and those where myelinated fibres are stained (e.g., Weigert method), the cerebral cortex is described as having six layers or laminae (Fig. 11.20 and Plate 11.7).

From the superficial surface downwards these laminae are as follows.

- Plexiform or molecular layer
- External granular layer

PLATE 11.7: Cerebral Cortex

Cerebral cortex. A. As seen in drawing; B. Photomicrograph

A slide of cerebral cortex shows outer grey and inner white matter. Multipolar neurons of various shapes are arranged in six layers in the grey matter. Axons of these neurons are present in the white matter. Neurologlia and blood vessels are present in grey and white matter.

Key

1. Molecular layer consisting of a few neurons and many cell processes
2. External granular layer with densely packed nuclei
3. Pyramidal cell layer with large triangular cells (these cells are characteristic of cerebral cortex)
4. Internal granular layer
5. Ganglionic layer with large pale cells
6. Multiform layer with cells of varied shapes

W. White matter containing axons
P. Piamater.

- Pyramidal cell layer
- Internal granular layer
- Ganglionic layer
- Multiform layer

The plexiform layer is made up predominantly of fibres although a few cells are present. All the remaining layers contain both stellate and pyramidal neurons as well as other types of neurons. The external and internal granular layers are made up predominantly of stellate (granular) cells. The predominant neurons in the pyramidal layer and in the ganglionic layer are pyramidal. The largest pyramidal cells (giant pyramidal cells of Betz) are found in the ganglionic layer. The multiform layer contains cells of various sizes and shapes.

In addition to the cell bodies of neurons the cortex contains abundant nerve fibres. Many of these are vertically oriented. In addition to the vertical fibres the cortex contains transversely running fibres that form prominent aggregations in certain situations. One such aggregation, present in the internal granular layer is called the ***external band of Baillarger***. Another, present in the ganglionic layer is called the ***internal band of Baillarger***.

Added Information

Variations in Cortical Structure

The structure of the cerebral cortex shows considerable variation from region to region, both in terms of thickness and in the prominence of the various laminae described above. Finer variations form the basis of the subdivisions into Brodmann's areas. Other workers divide the cortex into five broad varieties. These are as follows.

- In the ***agranular cortex*** the external and internal granular laminae are inconspicuous. This type of cortex is seen most typically in the precentral gyrus (area 4) and is, therefore, believed to be typical of 'motor' areas. It is also seen in some other areas.
- In the ***granular cortex*** the granular layers are highly developed while the pyramidal and ganglionic layers are poorly developed or absent. In the visual area the external band of Baillarger is prominent and forms a white line that can be seen with the naked eye when the region is freshly cut across. This ***stria of Gennari*** gives the name ***striate cortex*** to the visual cortex.
  Between the two extremes represented by the agranular and granular varieties of cortex, three intermediate types are described as follows.
- ***Frontal cortex***
- ***Parietal cortex*** and
- ***Polar cortex.***

The frontal type is nearest to the agranular cortex, the pyramidal cells being prominent, while the polar type is nearest to the granular cortex.

Chapter 12

Skin and its Appendages

SKIN

The skin forms the external covering of the body. It is the largest organ constituting 15–20% of total body mass.

TYPES OF SKIN

There are two types of skin.

- ***Thin or hairy skin:*** In this type of skin, epidermis is very thin. It contains hair and is found in all others parts of body except palms and soles (Plate 12.1).
- ***Thick or glabrous skin:*** In this type of skin, epidermis is very thick with a thick layer of stratum corneum. It is found in palms of hands and soles of feet and has no hair (Plate 12.2).

STRUCTURE OF SKIN

The skin consists of two layers

- A superficial layer the ***epidermis***, made up of stratified squamous epithelium
- A deeper layer, the ***dermis***, made up of connective tissue (Fig. 12.1).

The dermis rests on subcutaneous tissue (***subcutis***). This is sometimes described as a third layer of skin.

In sections through the skin the line of junction of the two layers is not straight, but is markedly wavy because of the presence of numerous finger-like projections of dermis upwards into the epidermis. These projections are called ***dermal papillae***. The downward projections of the epidermis (in the intervals between the dermal papillae) are sometimes called ***epidermal papillae*** (Fig. 12.2).

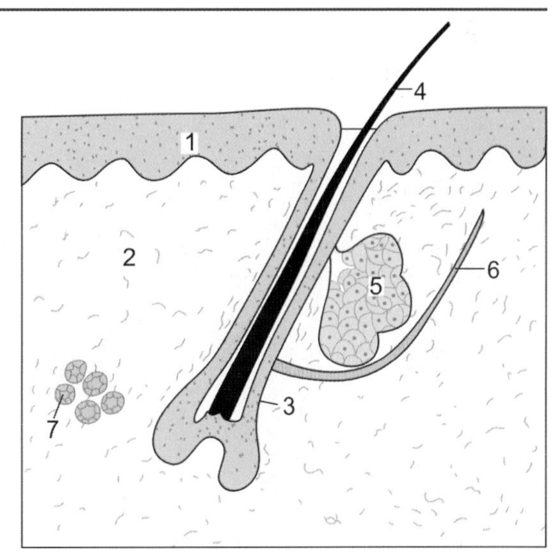

Fig. 12.1: Thin skin (Schematic representation) 1—epidermis, 2—dermis, 3—hair follicle, 4—hair, 5—sebaceous gland, 6—arrector pili muscle, 7—sweat glands

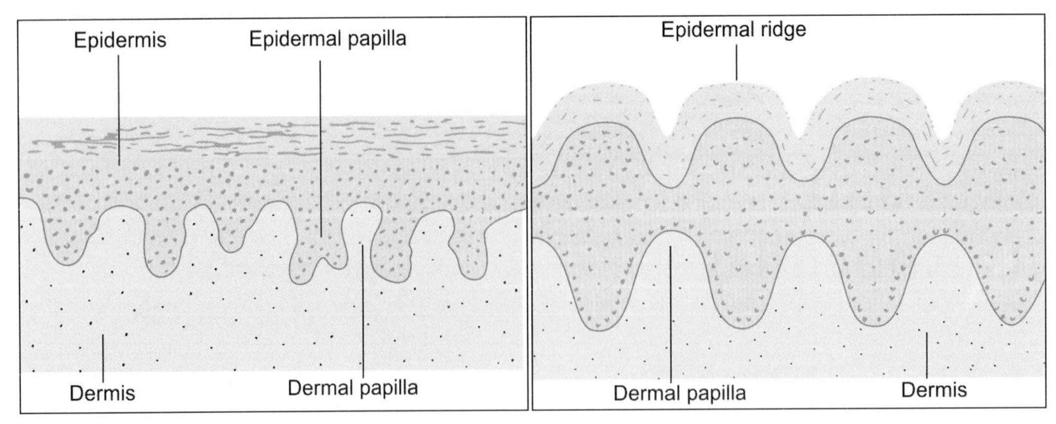

Fig. 12.2: Dermal and epidermal papillae (Schematic representation)

Fig. 12.3: Epidermal ridges (Schematic representation)

Note: The surface of the epidermis is also often marked by elevations and depressions. These are most prominent on the palms and ventral surfaces of the fingers, and on the corresponding surfaces of the feet. Here the elevations form characteristic ***epidermal ridges*** or ***rete ridges*** (Fig. 12.3) that are responsible for the highly specific fingerprints of each individual.

The Epidermis

The epidermis consists of stratified squamous keratinised epithelium (Fig. 12.4).

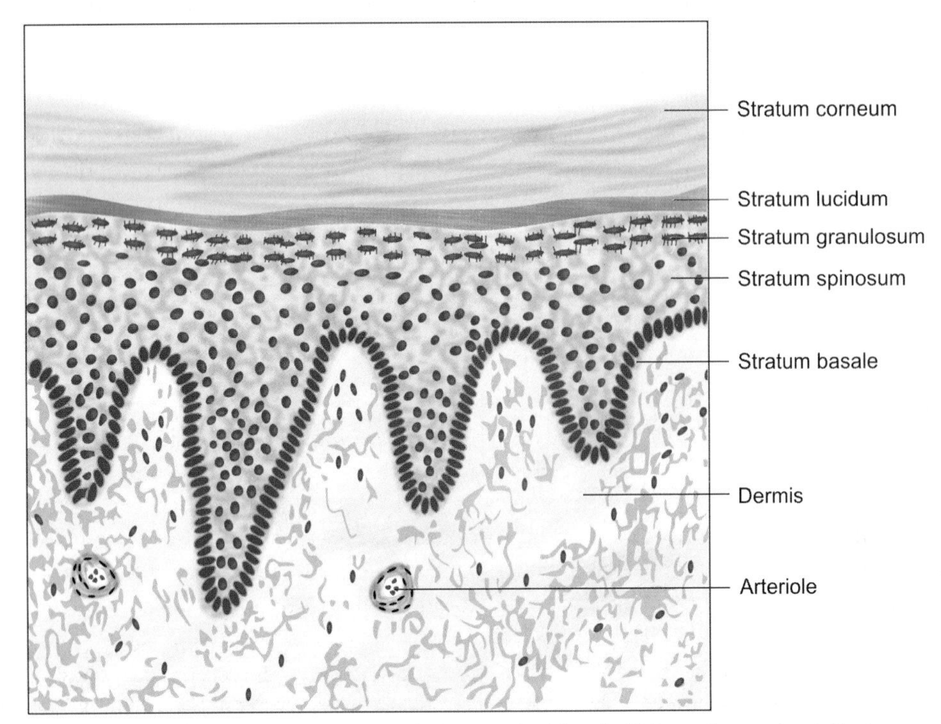

Fig. 12.4: Section through showing the layers of epidermis (Schematic representation)

Layers of Epidermis (Fig. 12.4)

- ***Stratum basale:*** It is the deepest or ***basal layer*** of epidermis. It is made up of a single layer of columnar cells that rest on a basal lamina. The basal layer contains stem cells that undergo mitosis to give off cells called ***keratinocytes***. Keratinocytes form the more superficial layers of the epidermis. The basal layer is, therefore, also called the ***germinal layer*** (***stratum germinativum***).
- ***Stratum spinosum:*** Above the basal layer there are several layers of polygonal keratinocytes that constitute the ***stratum spinosum*** (or ***Malpighian layer***). The cells of this layer are attached to one another by numerous desmosomes. During routine preparation of tissue for sectioning the cells often retract from each other except at the desmosomes. As a result the cells appear to have a number of 'spines': this is the reason for calling this layer the stratum spinosum (Fig. 12.5). For the same reason the keratinocytes of this layer are also called ***prickle cells***.

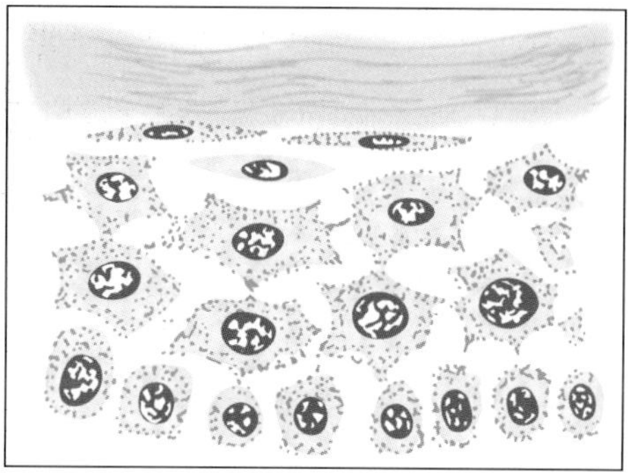

  The cytoplasm of cells in the stratum spinosum is permeated with fibrils (made up of bundles of keratin filaments). The fibrils are attached to the cell wall at desmosomes. Some mitoses may be seen in the deeper cells of the stratum spinosum. Because of this fact the stratum spinosum is included, along with the basal cell layer, in the ***germinative zone*** of the epidermis.

Fig. 12.5: Cells of the stratum spinosum showing typical spines (Schematic representation)

- ***Stratum granulosum:*** Overlying the stratum spinosum there are a few (1 to 5) layers of flattened cells that are characterised by the presence of deeply staining granules in their cytoplasm. These cells constitute the ***stratum granulosum***. The granules in them consist of a protein called ***keratohyalin*** (precursor of keratin). The nuclei of cells in this layer are condensed and dark staining (pyknotic).

  With the EM it is seen that, in the cells of this layer, keratin filaments are more numerous, and are arranged in the form of a thick layer.
- ***Stratum lucidum:*** Superficial to the stratum granulosum there is the ***stratum lucidum*** (lucid = clear). This layer is so called because it appears homogeneous, the cell boundaries being extremely indistinct. Traces of flattened nuclei are seen in some cells.
- ***Stratum corneum:*** It is most superficial layer of the epidermis. This layer is acellular. It is made up of flattened scale-like elements (squames) containing keratin filaments embedded in protein. The squames are held together by a glue-like material which contains lipids and carbohydrates. The presence of lipid makes this layer highly resistant to permeation by water.

  The thickness of the stratum corneum is greatest where the skin is exposed to maximal friction, e.g., on the palms and soles. The superficial layers of the epidermis are being constantly shed off, and are replaced by proliferation of cells in deeper layers.

Note: The stratum corneum, the stratum lucidum, and the stratum granulosum are collectively referred to as the ***zone of keratinisation***, or as the ***cornified zone*** (in distinction to the germinative zone described above). The stratum granulosum and the stratum lucidum are well formed only in thick non-hairy skin (e.g., on the palms). They are usually absent in thin hairy skin.

Plate 12.1: Thin Skin

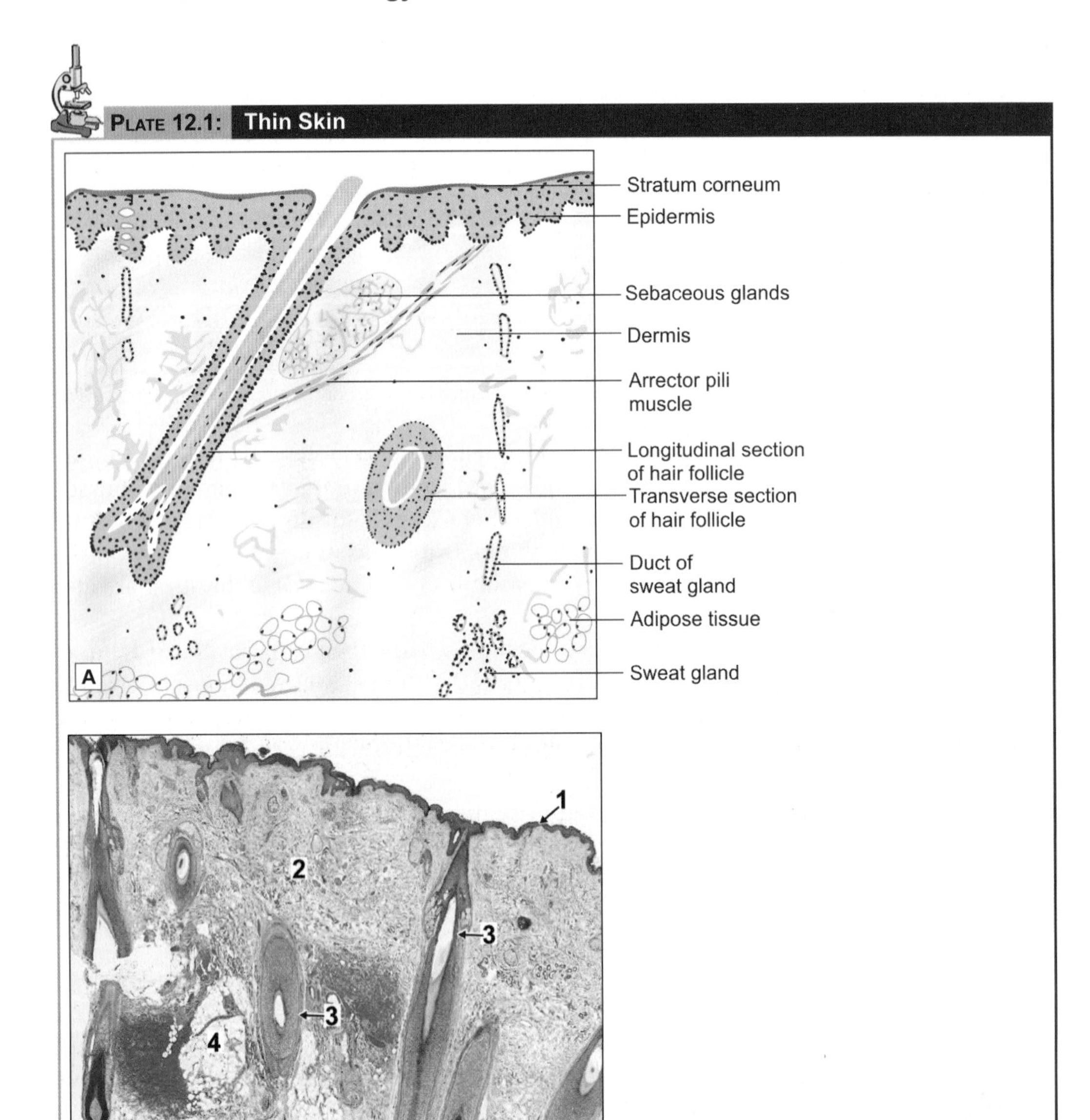

Thin Skin A. As seen in drawing; B. Photomicrograph.

Thin skin or hairy skin is characterised by:

- Presence of thin epidermis made up of keratinised stratified squamous epithelium (stratum corneum is thin)
- Hair follicles, sebaceous glands and sweat glands are present in the dermis
- It is found in all others parts of body except palms and soles.

Key

1. Epidermis
2. Dermis
3. Hair follicle
4. Sebaceous gland

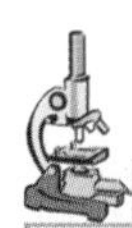

PLATE 12.2: Thick or Glabrous Skin

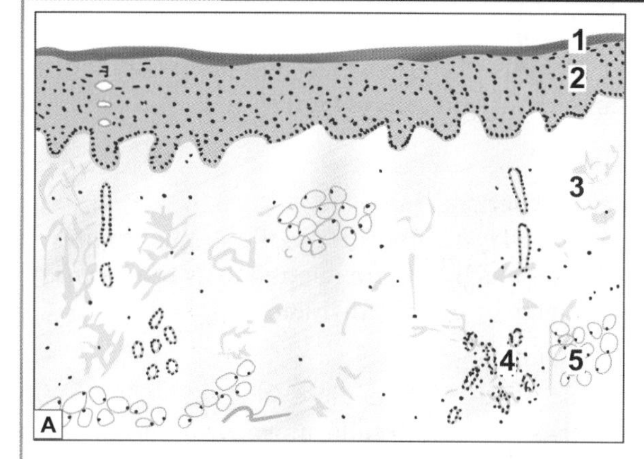

Thick or glabrous skin is characterised by:

- Presence of thick epidermis made up of keratinised stratified squamous epithelium (stratum corneum is very thick)
- Hair follicles and sebaceous glands are absent in dermis
- Sweat glands are present in the dermis
- It is found in palms of hands and soles of feet.

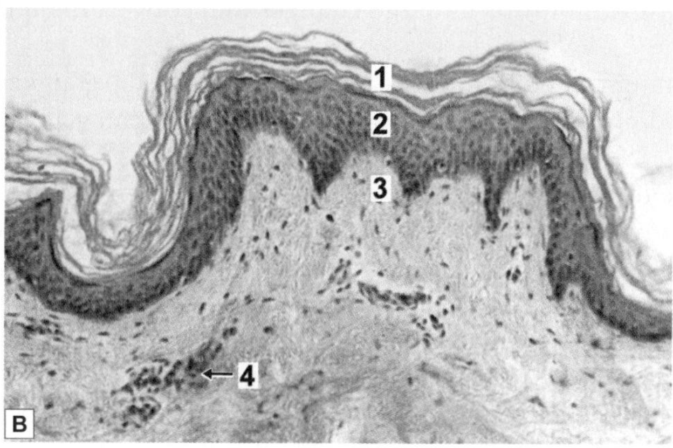

Key

1. Keratin
2. Epidermis (stratified squamous epithelium)
3. Dermis
4. Sweat glands
5. Adipocytes

Thick skin A. As seen in drawing; B. Photomicrograph

Pathological Correlation

- **Basal cell carcinoma:** It affects the basal cells of stratum basale. Typically, the basal cell carcinoma is a locally invasive, slow-growing tumour of middle-aged that rarely metastasises. It occurs exclusively on hairy skin, the most common location (90%) being the face, usually above a line from the lobe of the ear to the corner of the mouth.
- **Squamous cell carcinoma:** It affects the squamous cells of stratum spinosum. Squamous cell carcinoma may arise on any part of the skin and mucous membranes lined by squamous epithelium but is more likely to occur on sun-exposed parts in older people. Although squamous carcinomas can occur anywhere on the skin, most common locations are the face, pinna of the ears, back of hands and mucocutaneous junctions such as on the lips, anal canal and glans penis. Cutaneous squamous carcinoma arising in a pre-existing inflammatory and degenerative lesion has a higher incidence of developing metastases.

Cells of Epidermis

Although the epidermis is, by tradition, described as a stratified squamous epithelium, it has been pointed out that the majority of cells in it are not squamous (flattened). Rather the stratum corneum is not cellular at all.

The epidermis consists of two types of cells—keratinocytes and nonkeratinocytes including melanocytes, dendritic cell of Langerhans and cells of Merkel.

Keratinocytes

Keratinocytes are the predominant cell type of epidermis.

They are formed from stem cells present in basal layer. After entering the stratum spinosum some keratinocytes may undergo further mitoses. Such cells are referred to as ***intermediate stem cells***. Thereafter, keratinocytes do not undergo further cell division.

Essential steps in the formation of keratin are as follows:

- Basal cells of the epidermis contain numerous intermediate filaments. These are called cytokeratin filaments or tonofibrils. As basal cells move into the stratum spinosum the proteins forming the tonofibrils undergo changes that convert them to keratin filaments.
- When epidermal cells reach the stratum granulosum, they synthesise keratohyalin granules. These granules contain specialised proteins (which are rich in sulphur containing amino acids e.g., histidine, cysteine).
- Keratin consists of keratin filaments embedded in keratohyalin. Cells of the superficial layers of the stratum granulosum are packed with keratin. These cells die leaving behind the keratin mass in the form of an acellular layer of thin flakes.
- Cells in the granular layer also show membrane bound, circular, granules that contain glycophospholipids. These granules are referred to as lamellated bodies, or keratosomes. When these cells die the material in these granules is released and acts as a glue that holds together flakes of keratin. The lipid content of this material makes the skin resistant to water. However, prolonged exposure to water causes the material to swell. This is responsible for the altered appearance of the skin after prolonged exposure to water (more so if the water is hot, or contains detergents).

Added Information

The time elapsing between the formation of a keratinocyte in the basal layer of the epidermis, and its shedding off from the surface of the epidermis is highly variable. It is influenced by many factors including skin thickness, and the degree of friction on the surface. On the average it is 40-50 days.

In some situations it is seen that flakes of keratin in the stratum corneum are arranged in regular columns (one stacked above the other). It is believed that localised areas in the basal layer of the epidermis contain groups of keratinocytes all derived from a single stem cell. It is also believed that all the cells in the epidermis overlying this region are derived from the same stem cell. Such groups of cells, all derived from a single stem cell, and stacked in layers passing from the basal layer to the surface of the epidermis, constitute ***epidermal proliferation units***. One dendritic cell (see below) is present in close association with each such unit.

Melanocytes

Melanocytes are derived from melanoblasts that arise from the neural crest.

These cells are responsible for synthesis of melanin.

They may be present amongst the cells of the germinative zone, or at the junction of the epidermis and the dermis. Each melanocyte gives off many processes each of which is applied to a cell of the germinative zone.

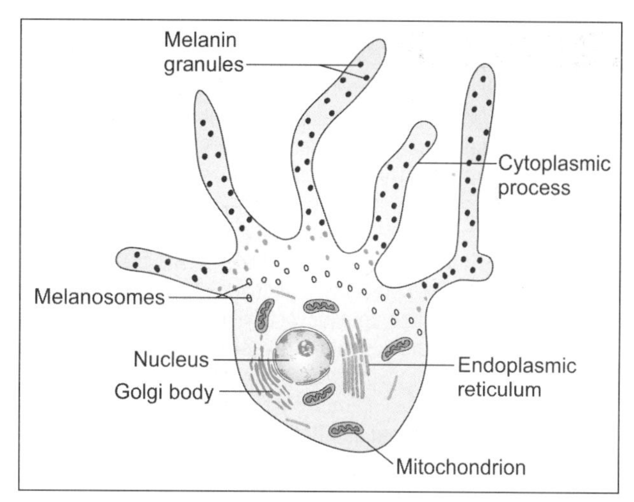

Fig. 12.6: Melanocyte showing dendritic processes (Schematic representation)

Melanin granules formed in the melanocyte are transferred to surrounding non-melanin-producing cells through these processes (Fig. 12.6). Because of the presence of processes melanocytes are also called ***dendritic cells*** (to be carefully distinguished from the dendritic macrophages described below).

Melanin

The cells of the basal layer of the epidermis, and the adjoining cells of the stratum spinosum contain a brown pigment called ***melanin***. The pigment is much more prominent in dark skinned individuals.

Melanin (***eumelanin***) is derived from the amino acid tyrosine. Tyrosine is converted into dihydroxy-phenylalanine (DOPA) that is in turn converted into melanin. Enzymes responsible for transformation of DOPA into melanin can be localised histochemically by incubating sections with DOPA that is converted into melanin. This is called the ***DOPA reaction***. It can be used to distinguish between true melanocytes and other cells that only store melanin. (In the past the term melanocyte has sometimes been applied to epithelial cells that have taken up melanin produced by other cells. However, the term is now used only for cells capable of synthesising melanin).

With the EM melanin granules are seen to be membrane bound organelles that contain pigment. These organelles are called ***melanosomes***. Melanosomes bud off from the Golgi complex. They enter the dendrites of the melanocytes. At the ends of the dendrites melanosomes are shed off from the cell and are engulfed by neighbouring keratinocytes. This is the manner in which most cells of the germinative zone acquire their pigment.

Added Information

The colour of skin is influenced by the amount of melanin present. It is also influenced by some other pigments present in the epidermis; and by pigments haemoglobin and oxyhaemoglobin present in blood circulating through the skin. The epidermis is sufficiently translucent for the colour of blood to show through, specially in light skinned individuals. That is why the skin becomes pale in anaemia; blue when oxygenation of blood is insufficient; and pink while blushing.

Clinical Correlation

- **Vitiligo:** It is a common skin disease in which the melanocytes are destroyed due to an autoimmune reaction. This results in bilateral depigmentation of skin.
- **Naevocellular naevi:** Pigmented naevi or moles are extremely common lesions on the skin of most individuals. They are often flat or slightly elevated lesions; rarely they may be papillomatous or pedunculated. Most naevi appear in adolescence and in early adulthood due to hormonal influence but rarely may be present at birth.
- **Malignant melanoma:** Malignant melanoma or melanocarcinoma arising from melanocytes is one of the most rapidly spreading malignant tumour of the skin that can occur at all ages but is rare before puberty. The tumour spreads locally as well as to distant sites by lymphatics and by blood. The aetiology is unknown but there is role of excessive exposure of white skin to sunlight. Besides the skin, melanomas may occur at various other sites such as oral and anogenital mucosa, oesophagus, conjunctiva, orbit and leptomeninges. The common sites on the skin are the trunk (*in men*), legs (*in women*); other locations are face, soles, palms and nail-beds.

Dendritic Cells of Langerhans

Apart from keratinocytes and dendritic melanocytes the stratum spinosum also contains other dendritic cells that are quite different in function from the melanocytes. These are the dendritic cells of Langerhans.

These cells are also found in oral mucosa, vagina and thymus. These cells belong to the mononuclear phagocyte system.

The dendritic cells of Langerhans originate in bone marrow.

They are believed to play an important role in protecting the skin against viral and other infections. It is believed that the cells take up antigens in the skin and transport them to lymphoid tissues where the antigens stimulate T-lymphocytes. Under the EM dendritic cells are seen to contain characteristic elongated vacuoles that have been given the name ***Langerhans bodies***, or ***Birbeck bodies***. The contents of these vacuoles are discharged to the outside of the cell through the cell membrane.

The dendritic cells of Langerhans also appear to play a role in controlling the rate of cell division in the epidermis. They increase in number in chronic skin disorders, particularly those resulting from allergy.

Cells of Merkel

The basal layer of the epidermis also contains specialised sensory cells called the cells of Merkel. Sensory nerve endings are present in relation to these cells.

The Dermis

The dermis is made up of connective tissue (Plate 12.1). It is divided into two layers.

- ***Papillary layer:*** The papillary layer forms the superficial layers of dermis and includes the dense connective tissue of the dermal papillae. These papillae are best developed in the thick skin of the palms and soles. Each papilla contains a capillary loop. Some papillae contain tactile corpuscles.
- ***Reticular layer:*** The reticular layer of the dermis is the deep layer of dermis and consists mainly of thick bundles of collagen fibres. It also contains considerable numbers of elastic fibres. Intervals between the fibre bundles are usually occupied by adipose tissue. The dermis rests on the superficial fascia through which it is attached to deeper structures.

Clinical Correlation

- The fibre bundles in the reticular layer of the dermis mostly lie parallel to one another. In the limbs the predominant direction of the bundles is along the long axis of the limb; while on the trunk and neck the direction is transverse. The lines along which the bundles run are often called ***cleavage lines*** as they represent the natural lines along which the skin tends to split when penetrated. The cleavage lines are of importance to the surgeon as incisions in the direction of these lines gape much less than those at right angles to them.
- The dermis contains considerable amounts of elastic fibres. Atrophy of elastic fibres occurs with age and is responsible for loss of elasticity and wrinkling of the skin.
- If for any reason the skin in any region of the body is rapidly stretched, fibre bundles in the dermis may rupture. Scar tissue is formed in the region and can be seen in the form of prominent white lines. Such lines may be formed on the anterior abdominal wall in pregnancy: they are known as ***linea gravidarum***.

BLOOD SUPPLY OF THE SKIN

Blood vessels to the skin are derived from a number of arterial plexuses. The deepest plexus is present over the deep fascia. There is another plexus just below the dermis (***rete cutaneum*** or ***reticular plexus***); and a third plexus just below the level of the dermal papillae (***rete subpapillare***, or ***papillary plexus***). Capillary loops arising from this plexus pass into each dermal papilla.

Blood vessels do not penetrate into the epidermis. The epidermis derives nutrition entirely by diffusion from capillaries in the dermal papillae. Veins from the dermal papillae drain (through plexuses present in the dermis) into a venous plexus lying on deep fascia.

A special feature of the blood supply of the skin is the presence of numerous arteriovenous anastomoses that regulate blood flow through the capillary bed and thus help in maintaining body temperature.

NERVE SUPPLY OF THE SKIN

The skin is richly supplied with sensory nerves. Dense networks of nerve fibres are seen in the superficial parts of the dermis. Sensory nerves end in relation to various types of specialised terminals like free nerve endings, Meissner's corpuscles, Pacinian corpuscles and Ruffini's corpuscles.

In contrast to blood vessels some nerve fibres do penetrate into the deeper parts of the epidermis.

Apart from sensory nerves the skin receives autonomic nerves that supply smooth muscle in the walls of blood vessels; the arrectores pilorum muscles; and myoepithelial cells present in relation to sweat glands. They also provide a secretomotor supply to sweat glands. In some regions (nipple, scrotum) nerve fibres innervate smooth muscle present in the dermis.

FUNCTIONS OF THE SKIN

- The skin provides mechanical protection to underlying tissues. In this connection we have noted that the skin is thickest over areas exposed to greatest friction.

  The skin also acts as a physical barrier against entry of microorganisms and other substances. However, the skin is not a perfect barrier and some substances, both useful (e.g., ointments) or harmful (e.g., poisons), may enter the body through the skin.

- The skin prevents loss of water from the body. The importance of this function is seen in persons who have lost extensive areas of skin through burns. One important cause of death in such cases is water loss.
- The pigment present in the epidermis protects tissues against harmful effects of light (specially ultraviolet light). This is to be correlated with the heavier pigmentation of skin in races living in the tropics; and with increase in pigmentation after exposure to sunlight. However, some degree of exposure to sunlight is essential for synthesis of vitamin D. Ultraviolet light converts 7-dehydrocholesterol (present in skin) to vitamin D.
- The skin offers protection against damage of tissues by chemicals, by heat, and by osmotic influences.
- The skin is a very important sensory organ, containing receptors for touch and related sensations. The presence of relatively sparse and short hair over most of the skin increases its sensitivity.
- The skin plays an important role in regulating body temperature. Blood flow through capillaries of the skin can be controlled by numerous arteriovenous anastomoses present in it. In cold weather blood flow through capillaries is kept to a minimum to prevent heat loss. In warm weather the flow is increased to promote cooling. In extreme cold, when some peripheral parts of the body (like the digits, the nose and the ears) are in danger of being frozen the blood flow through these parts increases to keep them warm.

In warm climates cooling of the body is facilitated by secretion of sweat and its evaporation. Sweat glands also act as excretory organs.

APPENDAGES OF THE SKIN

The appendages of the skin are the hair, nails, sebaceous glands and sweat glands. The mammary glands may be regarded as highly specialised appendages of the skin.

HAIR

Hair are present on the skin covering almost the whole body. The sites where they are not present include the palms, the soles, the ventral surface and sides of the digits, and some parts of the male and female external genitalia.

Differences in the length and texture of hair over different parts of the body, and the differences in distribution of hair in the male and female, are well known. It has to be emphasised, however, that many areas that appear to be hairless (e.g., the eyelids) have very fine hair, some of which may not even appear above the surface of the skin.

In animals with a thick coat of hair (fur) the hair help to keep the animal warm. In man this function is performed by subcutaneous fat. The relative hairlessness of the human skin is an adaptation to make the skin a more effective sensory surface. The presence of short, sparsely distributed hair, with a rich nerve supply of their roots, increases the sensitivity of the skin.

Parts of Hair

Each hair consists of a part (of variable length) that is seen on the surface of the body; and a part anchored in the thickness of the skin. The visible part is called the ***shaft***, and the embedded part is called the ***root***. The root has an expanded lower end called the ***bulb***. The bulb is invaginated from below by part of the dermis that constitutes the ***hair papilla***. The root of each hair is surrounded by a tubular sheath called the ***hair follicle*** (Fig. 12.7). The follicle is made up of several layers of cells that are derived from the layers of the skin.

Hair roots are always attached to skin obliquely. As a result the emerging hair is also oblique and easily lies flat on the skin surface.

Structure of Hair Shaft

A hair may be regarded as a modified part of the stratum corneum of the skin. It consists of three layers (Fig. 12.7).

- ***Cuticle:*** The surface of the hair is covered by a thin membrane called the ***cuticle***, that is formed by flattened cornified cells. Each of these cells has a free edge (directed distally) that overlaps part of the next cell.
- ***Cortex:*** It lies deep to the cuticle. The cortex is acellular and is made up of keratin.
- ***Medulla:*** An outer cortex and an inner medulla can be made out in large hair, but there is no medulla in thin hair. In thick hair the medulla consists of cornified cells of irregular shape.

The cornified elements making up the hair contain melanin that is responsible for their colour. Both in the medulla and in the cortex of a hair minute air bubbles are present: they influence its colour. The amount of air present in a hair increases with age and, along with loss of pigment, is responsible for greying of hair.

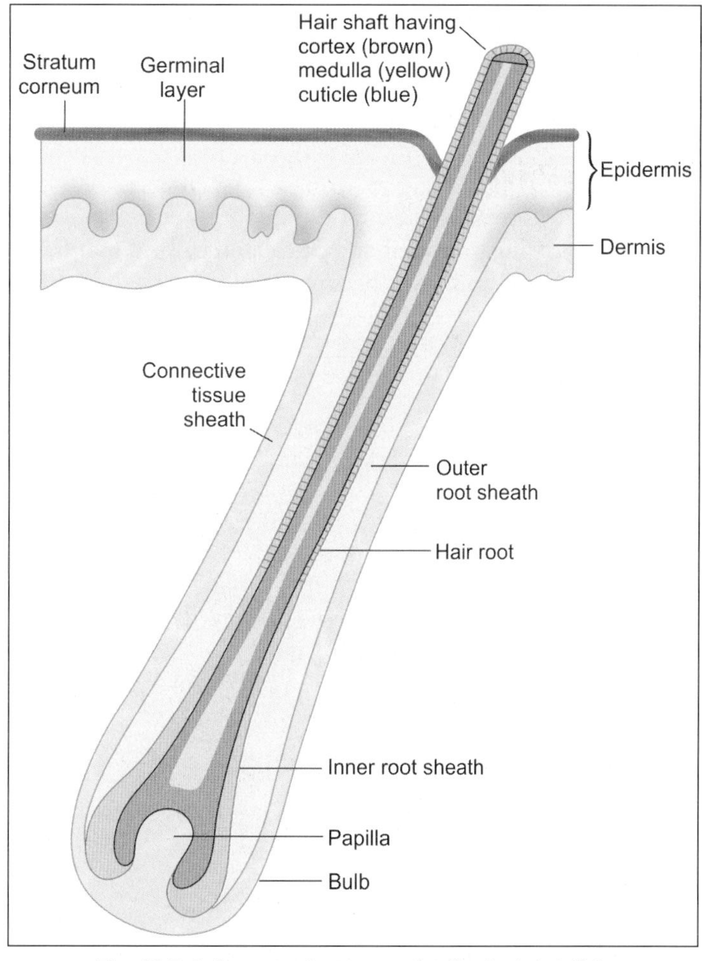

Fig. 12.7: Scheme to show some details of a hair follicle (Schematic representation)

Structure of Hair Follicle

The hair follicle may be regarded as a part of the epidermis that has been invaginated into the dermis around the hair root. Its innermost layer, that immediately surrounds the hair root is, therefore, continuous with the surface of the skin; while the outermost layer of the follicle is continuous with the dermis.

The wall of the follicle consists of three main layers. Beginning with the innermost layer they are as follows.

- The ***inner root sheath*** present only in the lower part of the follicle.
- The ***outer root sheath*** that is continuous with the stratum spinosum.
- A connective tissue sheath derived from the dermis.

Note: The inner and outer root sheath are derived from epidermis.

Inner Root Sheath

The inner root sheath is further divisible into the following (Fig.12.8).

- The innermost layer is called the ***cuticle***. It lies against the cuticle of the hair, and consists of flattened cornified cells.
- Next there are one to three layers of flattened nucleated cells that constitute ***Huxley's layer***, or the ***stratum epitheliale granuloferum***. Cells of this layer contain large eosinophilic granules (***trichohyaline granules***).
- The outer layer (of the inner root sheath) is made up of a single layer of cubical cells with flattened nuclei. This is called ***Henle's layer***, or the ***stratum epitheliale pallidum***.

Outer Root Sheath

The outer root sheath is continuous with the stratum spinosum of the skin, and like the latter it consists of living, rounded and nucleated cells. When traced towards the lower end of the follicle the cells of this layer become continuous with the hair bulb (at the lower end of the

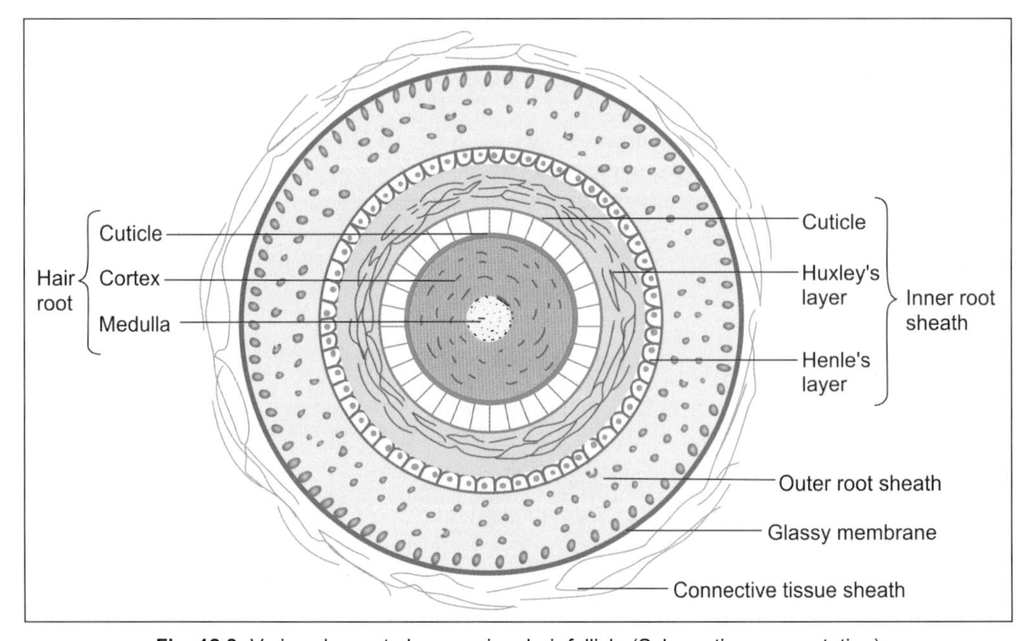

Fig. 12.8: Various layers to be seen in a hair follicle (Schematic representation)

hair root). The cells of the hair bulb also correspond to those of the stratum spinosum, and constitute the ***germinative matrix***. These cells show great mitotic activity. Cells produced here pass superficially and undergo keratinisation to form the various layers of the hair shaft already described. They also give rise to cells of the inner root sheath. The cells of the papilla are necessary for proper growth in the germinative matrix. The outermost layer of cells of the outer root sheath, and the lowest layer of cells of the hair bulb (that overlie the papilla) correspond to the basal cell layer of the skin.

The outer root sheath is separated from the connective tissue sheath by a basal lamina that appears structureless and is, therefore, called the ***glassy membrane*** (This membrane is strongly eosinophilic and PAS positive).

Connective Tissue Sheath

The ***connective tissue sheath*** is made up of tissue continuous with that of the dermis. The tissue is highly vascular, and contains numerous nerve fibres that form a basket-like network round the lower end of the follicle.

Note: Present in close association with hair follicles there are sebaceous glands (described below). One such gland normally opens into each follicle near its upper end. The arrector pili muscles (described below), pass obliquely from the lower part of the hair follicle towards the junction of the epidermis and dermis.

Added Information

Some other terms used in relation to the hair follicle may be mentioned here. Its lower expanded end is the fundus. The region above the opening of the sebaceous duct is the infundibulum. Below the infundibulum the isthmus extends up to the attachment of the arrector pili. The part of the follicle below this point is the inferior segment.

Clinical Correlation

Alopecia Areata

It is characterized by patchy or generalized hair loss on scalp, face, or body occurring gradually over a period of weeks to months. New patches of alopecia may appear while other resolve. The patient does not experience any pain, itching or burning. Physical examination reveals well-circumscribed round to oval patches of hair loss. The scalp appears normal without erythema, scale, scarring, or atrophy. At periphery of alopecia–"exclamation point" hair, short, broken hair with distal ends broader than proximal ends, are noted.

Arrector Pili Muscles

These are bands of smooth muscle attached at one end to the dermis, just below the dermal papillae; and at the other end to the connective tissue sheath of a hair follicle. The arrector pili muscles, pass obliquely from the lower part of the hair follicle towards the junction of the epidermis and dermis. It lies on that side of the hair follicle that forms an obtuse angle with the skin surface (Fig. 12.1, Plate 12.1). A sebaceous gland (see below) lies in the angle between the hair follicle and the arrector pili.

Contraction of the muscle has two effects. Firstly, the hair follicle becomes almost vertical (from its original oblique position) relative to the skin surface. Simultaneously the skin surface overlying the attachment of the muscle becomes depressed while surrounding areas become raised. These reactions are seen during exposure to cold, or during emotional excitement, when the 'hair stand on end' and the skin takes on the appearance of 'goose flesh'. The second effect of contraction of the arrector pili muscle is that the sebaceous gland is pressed upon and its secretions are squeezed out into the hair follicle. The arrector pili muscles receive a sympathetic innervation.

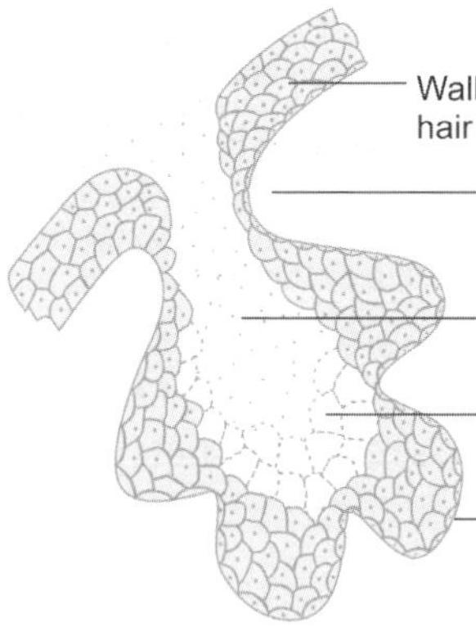

SEBACEOUS GLANDS

Sebaceous glands are present in dermis in close association with hair follicles. One such gland normally opens into each follicle near its upper end. Each gland consists of a number of alveoli that are connected to a broad duct that opens into a hair follicle (Fig. 12.1, Plate 12.3). Each alveolus is pear shaped. It consists of a solid mass of polyhedral cells and has hardly any lumen (Fig. 12.9).

Fig. 12.9: Sebaceous gland (Schematic representation)

The outermost cells are small and rest on a basement membrane. The inner cells are larger, more rounded, and filled with lipid. This lipid is discharged by disintegration of the innermost cells that are replaced by proliferation of outer cells. The sebaceous glands are, therefore, examples of holocrine glands.

The secretion of sebaceous glands is called ***sebum***. Its oily nature helps to keep the skin and hair soft. It helps to prevent dryness of the skin and also makes it resistant to moisture. Sebum contains various lipids including triglycerides, cholesterol, cholesterol esters and fatty acids.

In some situations sebaceous glands occur independently of hair follicles. Such glands open directly on the skin surface. They are found around the lips, and in relation to some parts of the male and female external genitalia.

The tarsal (Meibomian) glands of the eyelid are modified sebaceous glands. Montgomery's tubercles present in the skin around the nipple (areola) are also sebaceous glands. Secretion by sebaceous glands is not under nervous control.

Clinical Correlation

Acne vulgaris: Acne vulgaris is a very common chronic inflammatory dermatosis found predominantly in adolescents in both sexes. The lesions are seen more commonly on face, upper chest and upper back. The appearance of lesions around puberty is related to physiologic hormonal variations. The condition affects the pilosebaceous unit (consisting of hair follicle and its associated sebaceous gland), the opening of which is blocked by keratin material resulting in formation of *comedones*. Comedones may be open having central black appearance due to oxidation of melanin called ***black heads***, or they may be in closed follicles referred to as ***white heads***. A closed comedone may get infected and result in pustular acne.

PLATE 12.3: Hair Follicle and Sebaceous Gland

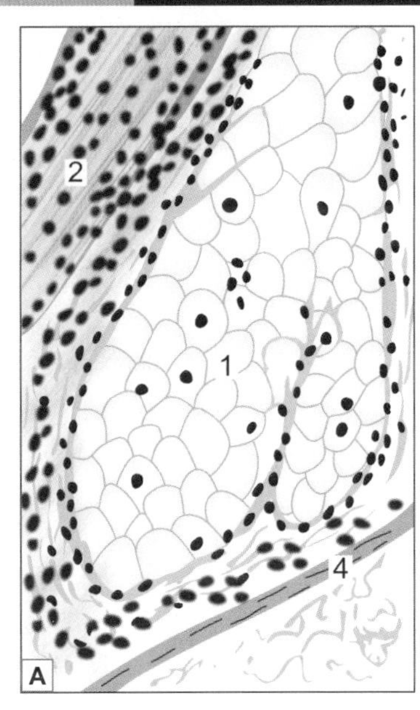

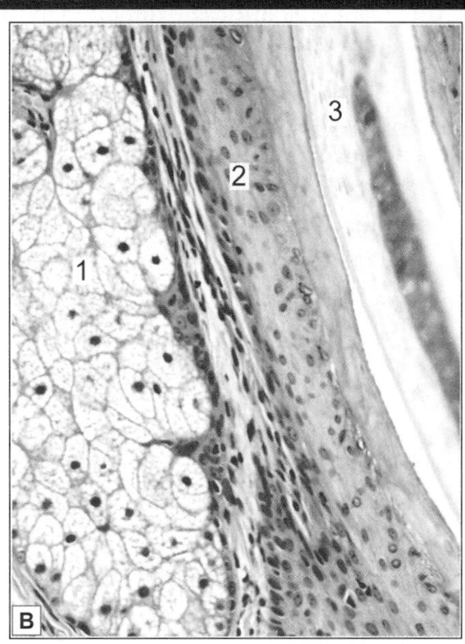

Hair follicle and sebaceous gland. A. As seen in drawing; B. Photomicrograph.

In figures small areas of skin at higher magnification are shown. The parts of a sebaceous gland and hair follicle containing a hair root can be seen. Each sebaceous gland consists of a number of alveoli that open into a hair follicle. Each alveolus is pear shaped. It consists mainly of a solid mass of polyhedral cells.

Key

1. Sebaceous gland
2. Wall of hair follicle
3. Hair shaft.
4. Arrector pili

SWEAT GLANDS

Sweat glands produce sweat or perspiration. They are present in the skin over most of the body. They are of two types:

- Typical or merocrine sweat glands
- Atypical or apocrine sweat glands.

Typical Sweat Glands

Typical sweat glands are of the merocrine variety. Their number and size varies in the skin over different parts of the body. They are most numerous in the palms and soles, the forehead and scalp, and the axillae.

The entire sweat gland consists of a single long tube (Fig. 12.10). The lower end of the tube is highly coiled on itself and forms the ***body*** (or ***fundus***) or the gland. The body is made up of the secretory part of the gland. It lies in the reticular layer of the dermis, or sometimes in subcutaneous tissue. The part of the tube connecting the secretory element to the skin surface is the ***duct***. It runs upwards through the dermis to reach the epidermis. Within the epidermis the duct follows a spiral course to reach the skin surface. The orifice is funnel shaped. On the palms, soles and digits the openings of sweat glands lie in rows on epidermal ridges.

The wall of the tube making up the gland consists of an inner epithelial lining, its basal lamina, and a supporting layer of connective tissue.

In the secretory part the epithelium is made up of a single layer of cubical or polygonal cells. Sometimes the epithelium may appear to be pseudostratified.

In larger sweat glands flattened contractile, ***myoepithelial cells*** (Fig. 12.11) are present between the epithelial cells and their basal lamina. They probably help in expressing secretion out of the gland.

In the duct the lining epithelium consists of two or more layers of cuboidal cells (constituting a stratified cuboidal epithelium). As the duct passes through the epidermis its wall is formed by the elements that make up the epidermis.

As is well known the secretion of sweat glands has a high water content. Evaporation of this water plays an important role in cooling the body. Sweat glands (including the myoepithelial cells) are innervated by cholinergic nerves.

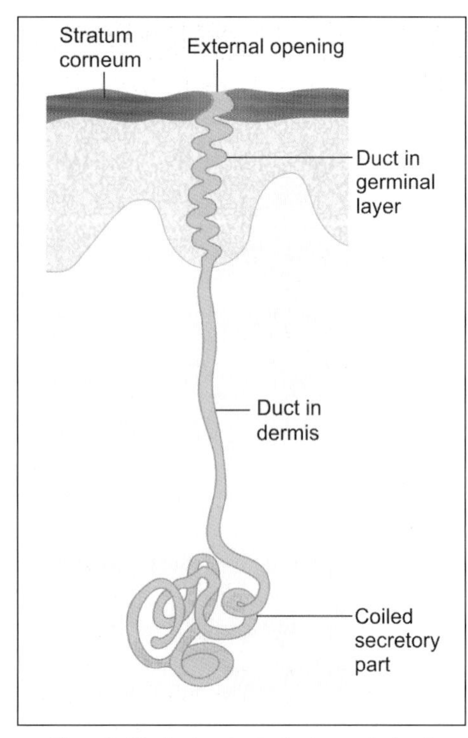

Fig. 12.10: Parts of a typical sweat gland (Schematic representation)

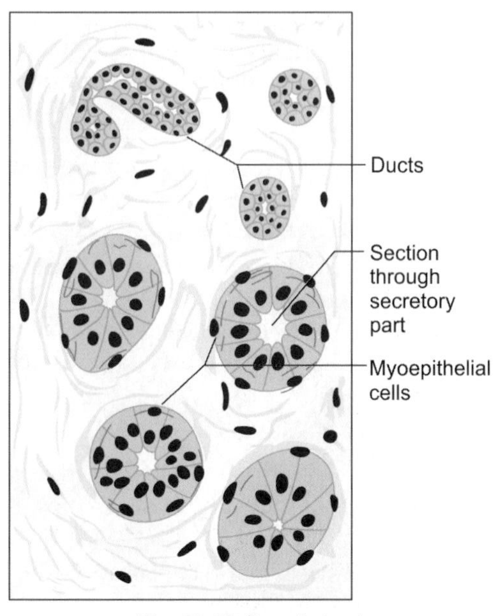

Fig. 12.11: Sweat gland (Schematic representation high power view)

Atypical Sweat Glands

Atypical sweat glands are of the apocrine variety. In other words the apical parts of the secretory cells are shed off as part of their secretion. Apocrine sweat glands are confined to some parts of the body including the axilla, the areola and nipple, the perianal region, the glans penis, and some parts of the female external genitalia.

Apart from differences in mode of secretion apocrine sweat glands have the following differences from typical (merocrine) sweat glands.

- Apocrine sweat glands are much larger in size. However, they become fully developed only after puberty.

Added Information

EM studies have shown that the lining cells are of two types, dark and clear. The bodies of ***dark cells*** are broad next to the lumen and narrow near the basement membrane. In contrast the ***clear cells*** are broadest next to the basement membrane and narrow towards the lumen. The dark cells are rich in RNA and in mucopolysaccharides (which are PAS positive). Their secretion is mucoid. The clear cells contain much glycogen. Their cytoplasm is permeated by canaliculi that contain microvilli. The secretion of clear cells is watery.

- The tubes forming the secretory parts of the glands branch and may form a network.
- Their ducts open not on the skin surface, but into hair follicles.
- The lumen of secretory tubules is large. The lining epithelium is of varying height: it may be squamous, cuboidal or columnar. When the cells are full of stored secretion they are columnar. With partial shedding of contents the cells appear to be cuboidal, and with complete emptying they become flattened. (Some workers describe a layer of flattened cells around the inner cuboidal cells). Associated with the apocrine mode of secretion (involving shedding of the apical cytoplasm) the epithelial surface is irregular, there being numerous projections of protoplasm on the luminal surface of the cells. Cell discharging their secretions in a merocrine or holocrine manner may also be present.
- The secretions of apocrine sweat glands are viscous and contain proteins. They are odourless, but after bacterial decomposition they give off body odours that vary from person to person.
- Conflicting views have been expressed regarding the innervation of apocrine sweat glands. According to some authorities the glands are not under nervous control. Others describe an adrenergic innervation (in contrast to cholinergic innervation of typical sweat glands); while still others describe both adrenergic and cholinergic innervation.

  Wax producing ***ceruminous glands*** of the external acoustic meatus, and ***ciliary glands*** of the eyelids are modified sweat glands.

NAILS

Nails are present on fingers and toes. Nails have evolved from the claws of animals. Their main function in man is to provide a rigid support for the finger tips. This support increases the sensitivity of the finger tips and increases their efficiency in carrying out delicate movements.

The nail represents a modified part of the zone of keratinisation of the epidermis. It is usually regarded as a much thickened continuation of the stratum lucidum, but it is more like the stratum corneum in structure. The nail substance consists of several layers of dead, cornified, 'cells' filled with keratin.

Structure of Nails

The main part of a nail is called its ***body***. The body has a free distal edge. The proximal part of the nail is implanted into a groove on the skin and is called the ***root*** (or ***radix***). The tissue on which the nail rests is called the ***nail bed***. The nail bed is highly vascular, and that is why the nails look pink in colour.

When we view a nail in longitudinal section (Fig. 12.12) it is seen that the nail rests on the cells of the germinative zone (stratum spinosum and stratum basale). The germinative

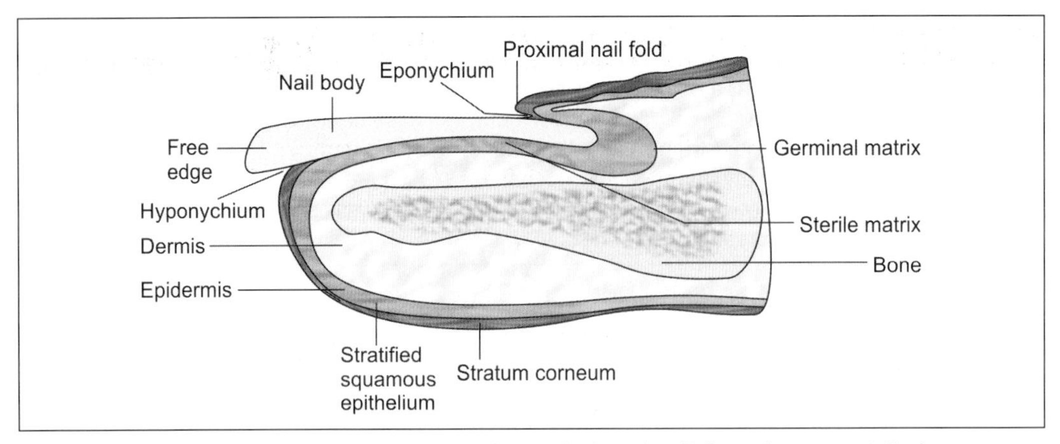

Fig. 12.12: Parts of a nail as seen in a longitudinal section (Schematic representation)

zone is particularly thick near the root of the nail where it forms the ***germinal matrix***. The nail substance is formed mainly by proliferation of cells in the germinal matrix. However, the superficial layers of the nail are derived from the proximal nail fold.

When viewed from the surface (i.e., through the nail substance) the area of the germinal matrix appears white (in comparison to the pink colour of the rest of the nail). Most of this white area is overlapped by the fold of skin (***proximal nail fold***) covering the root of the nail, but just distal to the nail fold a small semilunar white area called the ***lunule*** is seen (Fig. 12.13). The lunule is most conspicuous in the thumb nail. The germinal matrix is connected to the underlying bone (distal phalanx) by fibrous tissue.

The germinative zone underlying the body of the nail (i.e., the nail bed) is much thinner than the germinal matrix. It does not contribute to the growth of the nail; and is, therefore, called the ***sterile matrix***. As the nail grows it slides distally over the sterile matrix. The dermis that lies deep to the sterile matrix does not show the usual dermal papillae. Instead it shows a number of parallel, longitudinal ridges. These ridges look like very regularly arranged papillae in transverse sections through a nail.

The root of the nail is overlapped by a fold of skin called the proximal nail fold. The greater part of each lateral margin of the nail is also overlapped by a skin fold called the ***lateral nail fold***. The groove between the lateral nail fold and the nail bed (in which the lateral margin of the nail lies) is called the ***lateral nail groove***.

The stratum corneum lining the deep surface of the proximal nail fold extends for a short distance on to the surface of the nail. This extension of the stratum corneum is called the ***eponychium***. The stratum corneum lining the skin of the finger tip is also reflected onto the undersurface of the free distal edge of the nail: this reflection is called the ***hyponychium***.

The dermis underlying the nail bed is firmly attached to the distal phalanx. It is highly vascular and contains arteriovenous anastomoses. It also contains numerous sensory nerve endings.

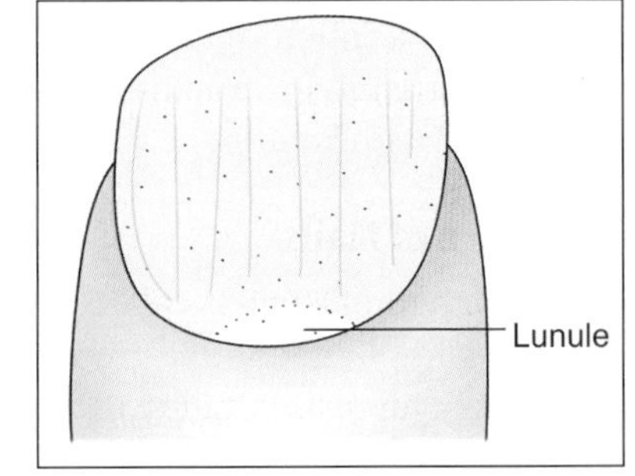

Fig. 12.13: Lunule of a nail (Schematic representation)

Growth of Nails

Nails undergo constant growth by proliferation of cells in the germinal matrix. Growth is faster in hot weather than in cold. Finger nails grow faster than toe nails. Nail growth can be disturbed by serious illness or by injury over the nail root, resulting in transverse grooves or white patches in the nails. These grooves or patches slowly grow towards the free edge of the nail. If a nail is lost by injury a new one grows out of the germinal matrix if the latter is intact.

Pathological Correlation

- **Onychia:** It is the inflammation of nail folds and shedding of nail resulting due to the introduction of microscopic pathogens through small wounds.
- **Onycholysis:** It is characterised by the loosening of exposed portion of nail from nail bed. It usually begins at the free edge and continues to lunula.
- **Paronychia:** It is caused due to bacterial or fungal infection producing change in the shape of nail plate.
- **Koilonychia:** It is caused due to iron deficiency or Vit B_{12} deficiency and is characterised by abnormal thinness and concavity (spoon-shape) of the nails.

Chapter 13

The Cardiovascular System

The cardiovascular system consists of the heart and blood vessels. The blood vessels that take blood from the heart to various tissues are called ***arteries***. The smallest arteries are called ***arterioles***. Arterioles open into a network of ***capillaries*** that pervade the tissues. Exchanges of various substances between the blood and the tissues take place through the walls of capillaries. In some situations, capillaries are replaced by slightly different vessels called ***sinusoids***. Blood from capillaries (or from sinusoids) is collected by small ***venules*** that join to form ***veins***. The veins return blood to the heart.

Blood vessels deliver nutrients, oxygen and hormones to the cells of the body and remove metabolic base products and carbon dioxide from them.

ENDOTHELIUM

The inner surfaces of the heart, and of all blood vessels are lined by flattened ***endothelial cells*** (also called ***endotheliocytes***). On surface view the cells are polygonal, and elongated along the length of the vessel. Cytoplasm is sparse.

The cytoplasm contains endoplasmic reticulum and mitochondria. Microfilaments and intermediate filaments are also present, and these provide mechanical support to the cell. Many endothelial cells show invaginations of cell membrane (on both internal and external surfaces). Sometimes the inner and outer invaginations meet to form channels passing right across the cell (seen typically in small arterioles). These features are seen in situations where vessels are highly permeable.

Adjoining endothelial cells are linked by tight junctions, and also by gap junctions. Externally, they are supported by a basal lamina.

Functions of Endothelium

Apart from providing a smooth internal lining to blood vessels and to the heart, endothelial cells perform a number of other functions as follows:

- Endothelial cells are sensitive to alterations in blood pressure, blood flow, and in oxygen tension in blood.
- They secrete various substances that can produce vasodilation by influencing the tone of muscle in the vessel wall.
- They produce factors that control coagulation of blood. Under normal conditions clotting is inhibited. When required, coagulation can be facilitated.

- Under the influence of adverse stimuli (e.g., by cytokines) endothelial cells undergo changes that facilitate passage of lymphocytes through the vessel wall. In acute inflammation, endothelium allows neutrophils to pass from blood into surrounding tissues.
- Under the influence of histamine (produced in allergic states) endothelium becomes highly permeable, allowing proteins and fluid to diffuse from blood into tissues. The resultant accumulation of fluid in tissues is called ***oedema***.

Note: Changes in properties of endothelium described above take place rapidly (within minutes).

ARTERIES

Basic Structure of Arteries

The histological structure of an artery varies considerably with its diameter. However, all arteries have some features in common which are as follows (Fig. 13.1):

- The wall of an artery is made up of three layers
    - The innermost layer is called the ***tunica intima*** (tunica = coat). It consists of:
        - An endothelial lining
        - A thin layer of glycoprotein which lines the external aspect of the endothelium and is called the ***basal lamina***

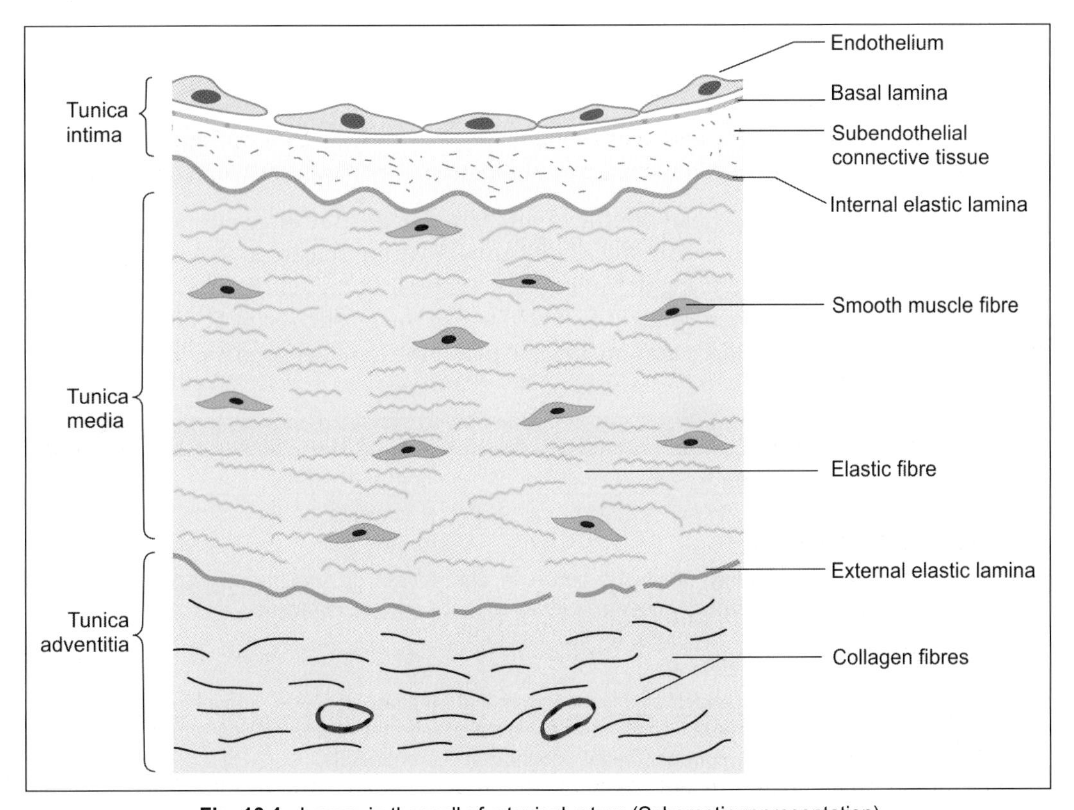

Fig. 13.1: Layers in the wall of a typical artery (Schematic representation)

- A delicate layer of subendothelial connective tissue
- A membrane formed by elastic fibres called the ***internal elastic lamina***.

- Outside the tunica intima there is the ***tunica media*** or middle layer. The media may consist predominantly of elastic tissue or of smooth muscle. Some connective tissue is usually present. On the outside the media is limited by a membrane formed by elastic fibres, this is the external elastic lamina.
- The outermost layer is called the ***tunica adventitia***. This coat consists of connective tissue in which collagen fibres are prominent. This layer prevents undue stretching or distension of the artery.

The fibrous elements in the intima and the adventitia (mainly collagen) run longitudinally (i.e., along the length of the vessel), whereas those in the media (elastic tissue or muscle) run circularly. Elastic fibres, including those of the internal and external elastic laminae are often in the form of fenestrated sheets (fenestrated = having holes in it).

Elastic and Muscular Arteries

On the basis of the kind of tissue that predominates in the tunica media, arteries are often divided into:

- Elastic arteries (large or conducting arteries)
- Muscular arteries (medium arteries)

Elastic arteries include the aorta and the large arteries supplying the head and neck (carotids) and limbs (subclavian, axillary, iliac). The remaining arteries are muscular (Table 13.1).

Although all arteries carry blood to peripheral tissues, elastic and muscular arteries play differing additional roles.

Elastic Arteries

When the left ventricle of the heart contracts, and blood enters the large elastic arteries with considerable force, these arteries distend significantly. They are able to do so because of much elastic tissue in their walls. During diastole (i.e., relaxation of the left ventricle) the walls of the arteries come back to their original size because of the elastic recoil of their walls. This recoil acts as an additional force that pushes the blood into smaller arteries. It is because of

Table 13.1: Comparison between elastic artery and muscular artery

Layers	Elastic artery	Muscular artery
Adventitia	It is relatively thin with greater proportion of elastic fibres.	It consists of thin layer of fibroelastic tissue.
Media	Made up mainly of elastic tissue in the form of fenestrated concentric membranes. There may be as many as fifty layers of elastic membranes.	Made up mainly of smooth muscles arranged circularly
Intima	It is made up of endothelium, subendothelial connective tissue and internal elastic lamina. The subendothelial connective tissue contains more elastic fibres. The internal elastic lamina is not distinct.	Intima is well developed, specially internal elastic lamina which stands out prominently.

this fact that blood flows continuously through arteries (but with fluctuation of pressure during systole and diastole).

The elastic arteries are also called as ***conducting vessels*** as their main function is to conduct the blood from heart to muscular arteries.

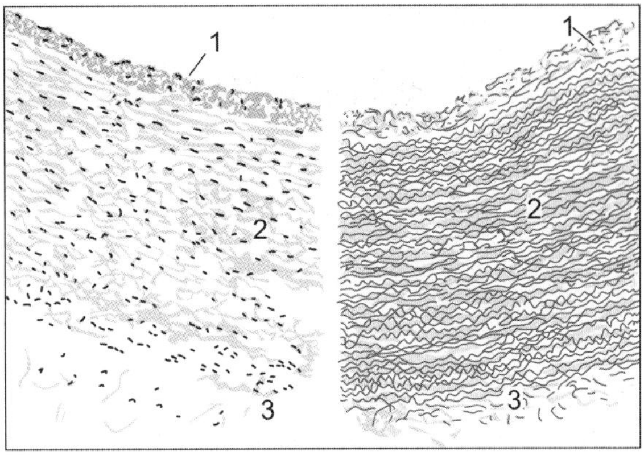

Fig. 13.2: Elastic artery (Schematic representation). The left half of the figure shows the appearance in a section stained with haematoxylin and eosin. The right half shows the appearance in a section stained by a special method that makes elastic fibres evident. (With this method the elastic fibres are stained black, muscle fibres are yellow, and collagen is pink). 1–tunica intima; 2–tunica media containing abundant elastic tissue arranged in the form of a number of membranes; 3– tunica adventitia

Structure of Elastic Arteries (Fig. 13.2 and Plate 13.1)

- ***Tunica intima:*** It is made up of endothelium, subendothelial connective tissue and internal elastic lamina. The subendothelial connective tissue contains more elastic fibres in the elastic arteries. The internal elastic lamina is not distinct from the media as it has the same structure as the elastic membranes of the media.
- ***Tunica media:*** The media is made up mainly of elastic tissue. The elastic tissue is in the form of a series of concentric membranes that are frequently fenestrated (Plate 13.1). In the aorta (which is the largest elastic artery) there may be as many as fifty layers of elastic membranes. Between the elastic membranes there is some loose connective tissue. Some smooth muscle cells may be present.
- ***Tunica adventitia:*** It is relatively thin in large arteries, in which a greater proportion of elastic fibres are present. These fibres merge with the external elastic lamina.

Muscular Arteries

A muscular artery has the ability to alter the size of its lumen by contraction or relaxation of smooth muscle in its wall. Muscular arteries can, therefore, regulate the amount of blood flowing into the regions supplied by them, hence they are also called as ***distributing arteries***.

Structure of Muscular Arteries

The muscular arteries differ from elastic arteries in having more smooth muscle fibres than elastic fibres. The transition from elastic to muscular arteries is not abrupt. In proceeding distally along the artery there is a gradual reduction in elastic fibres and increase in smooth muscle content in the media.

- ***Tunica intima:*** The internal elastic lamina in the muscular arteries stands out distinctly from the muscular media of smaller arteries.
- ***Tunica media:*** It is made up mainly of smooth muscles (Plate 13.2). This muscle is arranged circularly. Between groups of muscle fibres some connective tissue is present, which may contain some elastic fibres. Longitudinally arranged muscle is present in the media of arteries that undergo repeated stretching or bending. Examples of such arteries are the coronary, carotid, axillary and palmar arteries.
- ***Tunica adventitia.***

PLATE 13.1: Elastic Artery

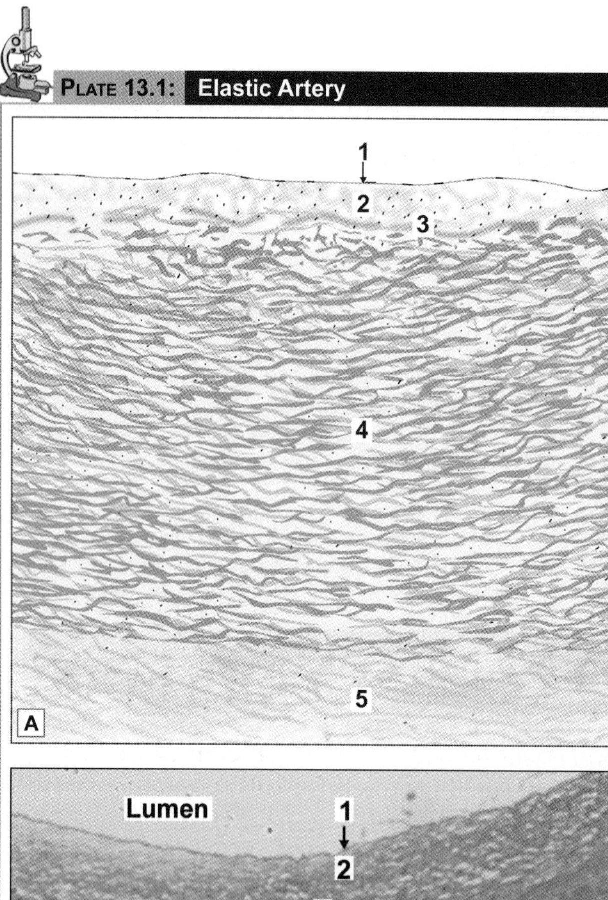

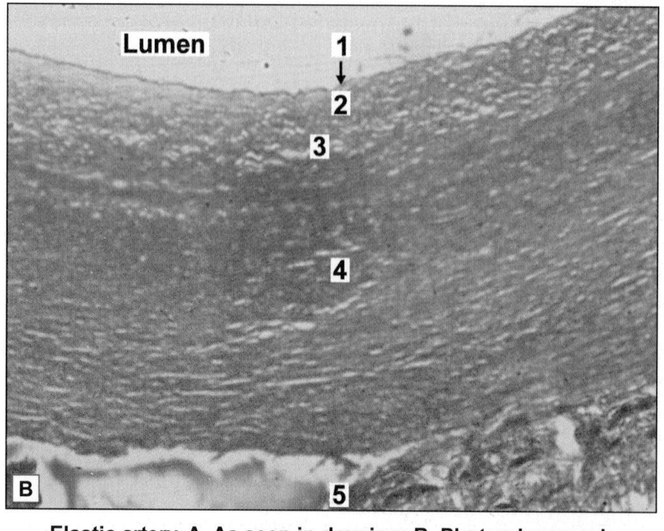

Elastic artery. A. As seen in drawing; B. Photomicrograph

Elastic artery is characterised by presence of:

- Tunica intima consisting of endothelium, subendothelial connective tissue and internal elastic lamina
- The first layer of elastic fibres is called the internal elastic lamina. The internal elastic lamina is not distinct from the elastic fibres of media
- Well developed subendothelial layer in tunica intima
- Thick tunica media with many elastic fibres and some smooth muscle fibres
- Tunica adventitia containing collagen fibres with several elastic fibres
- Vasa vasorum in the tunica adventitia (Not seen in this slide).

Key

1. Endothelium
2. Subendothelial connective tissue
3. Internal elastic lamina

(1–3: Tunica intima)

4. Tunica media
5. Tunica adventitia

Plate 13.2: Muscular (Medium Size) Artery

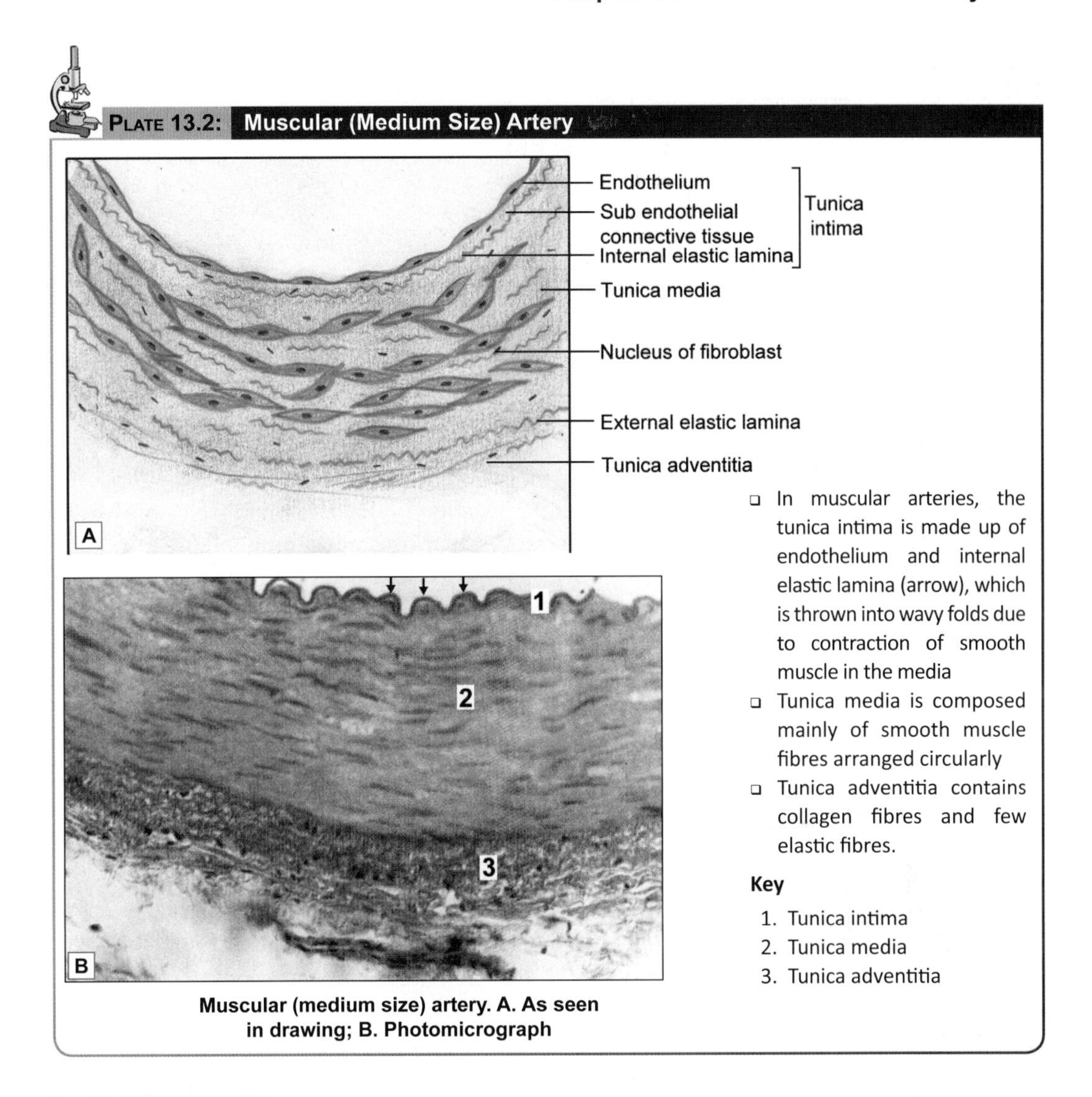

Muscular (medium size) artery. A. As seen in drawing; B. Photomicrograph

- In muscular arteries, the tunica intima is made up of endothelium and internal elastic lamina (arrow), which is thrown into wavy folds due to contraction of smooth muscle in the media
- Tunica media is composed mainly of smooth muscle fibres arranged circularly
- Tunica adventitia contains collagen fibres and few elastic fibres.

Key

1. Tunica intima
2. Tunica media
3. Tunica adventitia

Clinical Correlation

Atheroma

The most common disease of arteries is ***atheroma***, in which the intima becomes infiltrated with fat and collagen. The thickenings formed are ***atheromatous plaques***. Atheroma leads to narrowing of the arterial lumen, and consequently to reduced blood flow. Damage to endothelium can induce coagulation of blood forming a ***thrombus*** which can completely obstruct the artery. This leads to death of the tissue supplied. When this happens in an artery supplying the myocardium (***coronary thrombosis***) it leads to ***myocardial infarction*** (***manifesting as a heart attack***). In the brain (***cerebral thrombosis***) it leads to a ***stroke*** and ***paralysis***. An artery weakened by atheroma may undergo dilation (***aneurysm***), or may even rupture.

ARTERIOLES

When traced distally, muscular arteries progressively decrease in calibre till they have a diameter of about 100 μm. They then become continuous with arterioles. The larger or ***muscular arterioles*** are 100 to 50 μm in diameter (Fig. 13.3). Arterioles less than 50 μm in diameter are called ***terminal arterioles***. All the three layers, i.e. tunica adventitia, tunica media and tunica intima are thin as compared to arteries. In arterioles, the adventitia is made up of a thin network of collagen fibres.

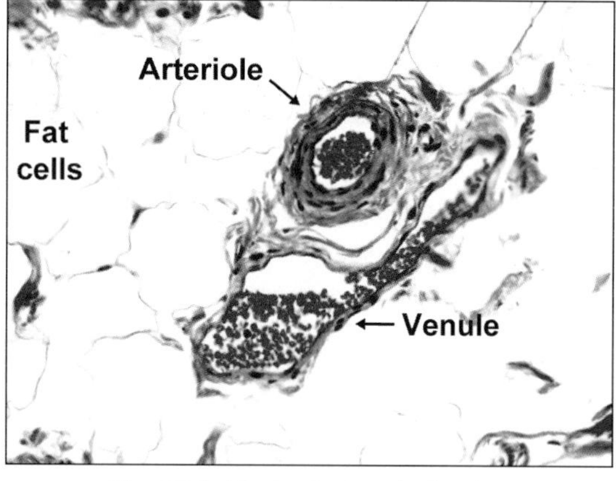

Fig. 13.3: Photomicrograph showing an arteriole and a venule

Arterioles are the main regulators of peripheral vascular resistance. Contraction and relaxation of the smooth muscles present in the walls of the arterioles can alter the peripheral vascular resistance (or blood pressure) and the blood flow.

Muscular arterioles can be distinguished from true arteries:
- By their small diameter
- They do not have an internal elastic lamina. They have a few layers of smooth muscle in their media.

Terminal arterioles can be distinguished from muscular arterioles as follows:
- They have a diameter less than 50 μm, the smallest terminal arterioles having a diameter as small as 12 μm.
- They have only a thin layer of muscle in their walls.
- They give off lateral branches (called meta-arterioles) to the capillary bed.

The initial segment of each lateral branch is surrounded by a few smooth muscle cells. These muscle cells constitute the ***precapillary sphincter***. This sphincter regulate the flow of blood to the capillaries.

CAPILLARIES

Terminal arterioles are continued into a capillary plexus that pervades the tissue supplied. Capillaries are the smallest blood vessels. The average diameter of a capillary is 8 μm. Exchanges (of oxygen, carbon dioxide, fluids and various molecules) between blood and tissue take place through the walls of the capillary plexus (and through postcapillary venules). The arrangement of the capillary plexus and its density varies from tissue to tissue, the density being greatest in tissues having high metabolic activity.

Structure of Capillaries

The wall of a capillary is formed essentially by endothelial cells that are lined on the outside by a basal lamina (glycoprotein). Overlying the basal lamina there may be isolated branching perivascular cells (pericytes), and a delicate network of reticular fibres and cells. Pericyte or adventitial cells contain contractile filaments in the cytoplasm and can transform into other cells.

Types of Capillaries

There are two types of capillaries:

1. Continuous
2. Fenestrated

Continuous Capillaries

Typically, the edges of endothelial cells fuse completely with those of adjoining cells to form a continuous wall. Such capillaries are called ***continuous capillaries*** (Fig. 13.4).

In continuous capillaries exchanges of material between blood and tissue take place through the cytoplasm of endothelial cells. This is suggested by the presence of numerous pinocytotic vesicles in the cytoplasm; and by the presence of numerous depressions (***caveolae***) on the cell surfaces, which may represent pinocytotic vesicles in the process of formation. Apart from transport through the cytoplasm, substances may also pass through the intercellular material separating adjoining endothelial cells.

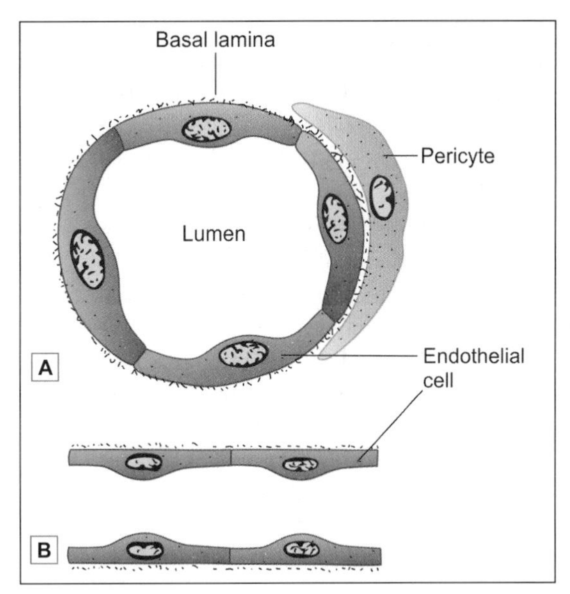

Fig. 13.4: Structure of continuous capillary. **A.** Circular section; **B.** Longitudinal section (Schematic representation)

Continuous capillaries are seen in the skin, connective tissue, muscle, lungs and brain.

Fenestrated Capillaries

In some organs the walls of capillaries appear to have apertures in their endothelial lining, these are, therefore, called ***fenestrated capillaries*** (Fig. 13.5). The 'apertures' are, however, always closed by a thin diaphragm (which may represent greatly thinned out cytoplasm of an endothelial cell, or only the basal lamina).

Some fenestrations represent areas where endothelial cell cytoplasm has pores passing through the entire thickness of the cell.

In the case of fenestrated capillaries diffusion of substances takes place through the numerous fenestrae in the capillary wall.

Fenestrated capillaries are seen in renal glomeruli, intestinal villi, endocrine glands and pancreas.

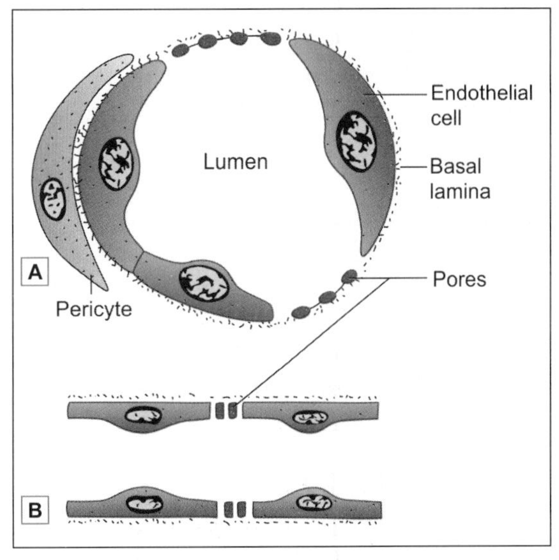

Fig. 13.5: Structure of fenestrated capillary. **A.** Circular section; **B.** Longitudinal section (Schematic representation)

SINUSOIDS

In some tissues the 'exchange' network is made up of vessels that are somewhat different from capillaries, and are called ***sinusoids*** (Fig. 13.6).

Sinusoids are found typically in organs that are made up of cords or plates of cells. These include the liver, the adrenal cortex, the hypophysis cerebri, and the parathyroid glands. Sinusoids are also present in the spleen, in the bone marrow, and in the carotid body.

The wall of a sinusoid consists only of endothelium supported by a thin layer of connective tissue. The wall may be incomplete at places, so that blood may come into direct contact with tissue cells. Deficiency in the wall may be in the form of fenestrations (***fenestrated sinusoids***) or in the form of long slits (***discontinuous sinusoids***, as in the spleen).

At some places the wall of the sinusoid consists of phagocytic cells instead of endothelial cells.

Sinusoids have a broader lumen (about 20 μm) than capillaries. The lumen may be irregular. Because of this fact blood flow through them is relatively sluggish.

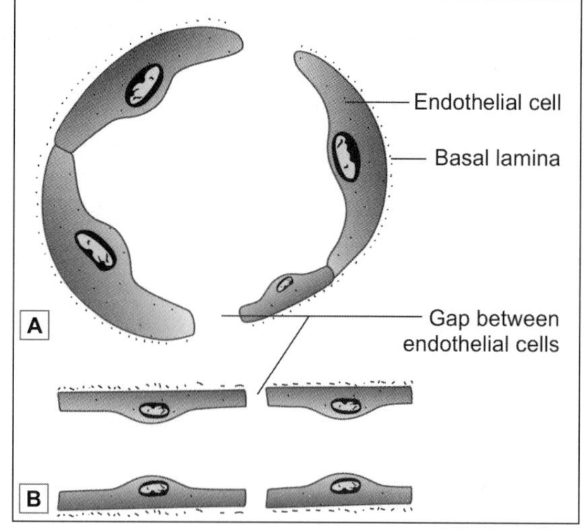

Fig. 13.6: Structure of sinusoid. **A.** Circular section; **B.** Longitudinal section (Schematic representation)

VEINS

The basic structure of veins is similar to that of arteries. The tunica intima, media and adventitia can be distinguished, specially in large veins. The structure of veins differs from that of arteries in the following respects (Fig. 13.7 and Plate 13.3):

- The wall of a vein is distinctly thinner than that of an artery having the same sized lumen.
- The tunica media contains a much larger quantity of collagen

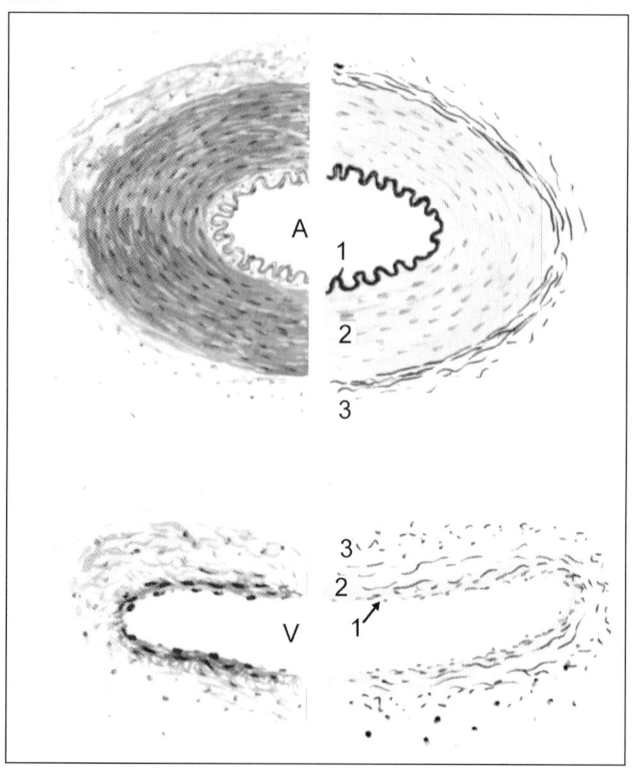

Fig. 13.7: Medium sized artery (above) and vein (below). The left half of the figure shows the appearance as seen with haematoxylin and eosin staining. The right half shows appearance when elastic fibres are stained black. 1–internal elastic lamina; 2–tunica media; 3–tunica adventitia, A–artery; V–vein; (Schematic representation)

than in arteries. The amount of elastic tissue or of muscle is much less.

- Because of the differences mentioned above, the wall of a vein is easily compressed. After death veins are usually collapsed. In contrast arteries retain their patency.
- In arteries the tunica media is usually thicker than the adventitia. In contrast the adventitia of veins is thicker than the media (specially in large veins). In some large veins (e.g., the inferior vena cava) the adventitia contains a considerable amount of elastic and muscle fibres that run in a predominantly longitudinal direction. These fibres facilitate elongation and shortening of the vena cava with respiration. This is also facilitated by the fact that collagen fibres in the adventitia form a meshwork that spirals around the vessel.
- A clear distinction between the tunica intima, media and adventitia cannot be made out in small veins as all these layers consist predominantly of fibrous tissue. Muscle is conspicuous by its complete absence in venous spaces of erectile tissue, in veins of cancellous bone, dural venous sinuses, retinal veins, and placental veins.

Valves of Veins

Most veins contain valves that allow the flow of blood towards the heart, but prevent its regurgitation in the opposite direction. Typically each valve is made up of two semilunar cusps. Each cusp is a fold of endothelium within which there is some connective tissue that is rich in elastic fibres. Valves are absent in very small veins; in veins within the cranial cavity, or within the vertebral canal; in the venae cavae; and in some other veins.

Flow of blood through veins is assisted by contractions of muscle in their walls. It is also assisted by contraction of surrounding muscles specially when the latter are enclosed in deep fascia.

Clinical Correlation

Varicose Veins

Varicose veins are permanently dilated and tortuous superficial veins of the lower extremities, especially the long saphenous vein and its tributaries. About 10–12% of the general population develops varicose veins of lower legs, with the peak incidence in 4th and 5th decades of life. Adult females are affected more commonly than the males, especially during pregnancy. This is attributed to venous stasis in the lower legs because of compression on the iliac veins by pregnant uterus.

VENULES

The smallest veins, into which capillaries drain, are called ***venules*** (Fig. 13.3). They are 20–30 μm in diameter. Their walls consist of endothelium, basal lamina, and a thin adventitia consisting of longitudinally running collagen fibres. Flattened or branching cells called ***pericytes*** may be present outside the basal laminae of small venules (called ***postcapillary venules***), while some muscle may be present in larger vessels (***muscular venules***).

Functionally, venules have to be distinguished from true veins. The walls of venules (specially those of postcapillary venules) have considerable permeability and exchanges between blood and surrounding tissues can take place through them. In particular venules are the sites at which lymphocytes and other cells may pass out of (or into) the blood stream.

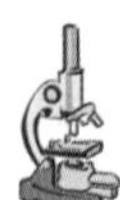

PLATE 13.3: Vein

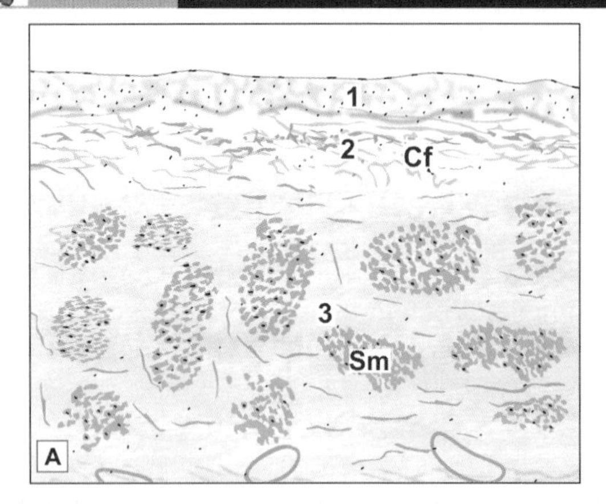

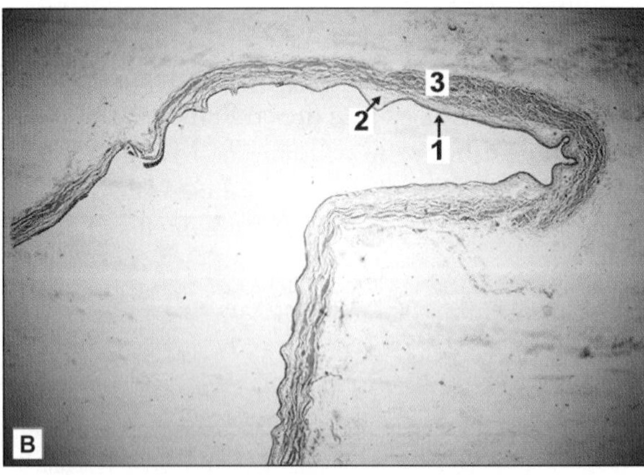

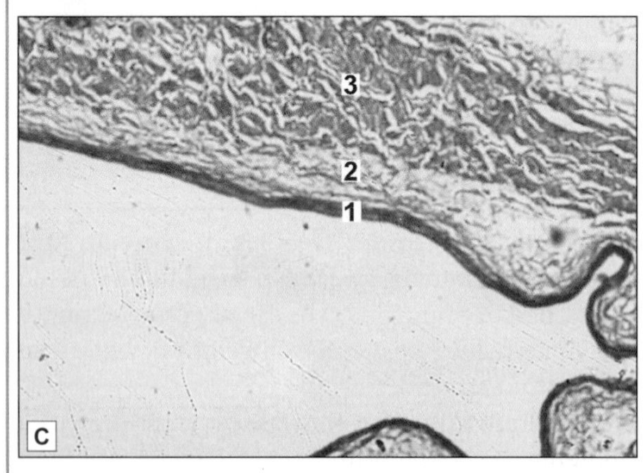

Vein. A. As seen in drawing; B. Photomicrograph (low magnification); C. Photomicrograph (high magnification).

- The vein has a thinner wall and a larger lumen than the artery
- The tunica intima, media and adventitia can be made out, but they are not sharply demarcated
- The media is thin and contains a much larger quantity of collagen fibres than arteries. The amount of elastic tissue or of muscle is much less
- The adventitia is relatively thick and contains considerable amount of elastic and muscle fibres.

Note: The luminal surface appears as a dark line, with an occasional nucleus along it.

Key

1. Tunica intima
2. Tunica media
3. Tunica adventitia

Cf. Collagen fibres
Sm. Smooth muscles

BLOOD VESSELS, LYMPHATICS AND NERVES SUPPLYING BLOOD VESSELS

The walls of small blood vessels receive adequate nutrition by diffusion from blood in their lumina. However, the walls of large and medium sized vessels are supplied by small arteries called ***vasa vasorum*** (literally 'vessels of vessels'; singular = ***vas vasis***). These vessels supply the adventitia and the outer part of the media. These layers of the vessel wall also contain many lymphatic vessels.

Blood vessels have a fairly rich supply by autonomic nerves (sympathetic). The nerves are unmyelinated. Most of the nerves are vasomotor and supply smooth muscle. Their stimulation causes vasoconstriction in some arteries, and vasodilatation in others. Some myelinated sensory nerves are also present in the adventitia.

MECHANISMS CONTROLLING BLOOD FLOW THROUGH THE CAPILLARY BED

The requirements of blood flow through a tissue may vary considerably at different times. For example, a muscle needs much more blood when engaged in active contraction, than when relaxed. Blood flow through intestinal villi needs to be greatest when there is food to be absorbed. The mechanisms that adjust blood flow through capillaries are considered below.

Blood supply to relatively large areas of tissue is controlled by contraction or relaxation of smooth muscle in the walls of muscular arteries and arterioles. Control of supply to smaller areas is effected through arteriovenous anastomoses, precapillary sphincters, and thoroughfare channels.

Arteriovenous Anastomoses

In many parts of the body, small arteries and veins are connected by direct channels that constitute arteriovenous anastomoses. These channels may be straight or coiled. Their walls have a thick muscular coat that is richly supplied with sympathetic nerves. When the anastomoses are patent blood is short circuited from the artery to the vein so that very little blood passes through the capillary bed. However, when the muscle in the wall of the anastomosing channel contracts its lumen is occluded so that all blood now passes through the capillaries. Arteriovenous anastomoses are found in the skin specially in that of the nose, lips and external ear; and in the mucous membrane of the alimentary canal and nose. They are also seen in the tongue, in the thyroid, in sympathetic ganglia, and in the erectile tissues of sex organs.

Arteriovenous anastomoses in the skin help in regulating body temperature, by increasing blood flow through capillaries in warm weather; and decreasing it in cold weather to prevent heat loss.

In some regions we see arteriovenous anastomoses of a special kind. The vessels taking part in these anastomoses are in the form of a rounded bunch covered by connective tissue. This structure is called a ***glomus*** (Fig. 13.8). Each glomus consists

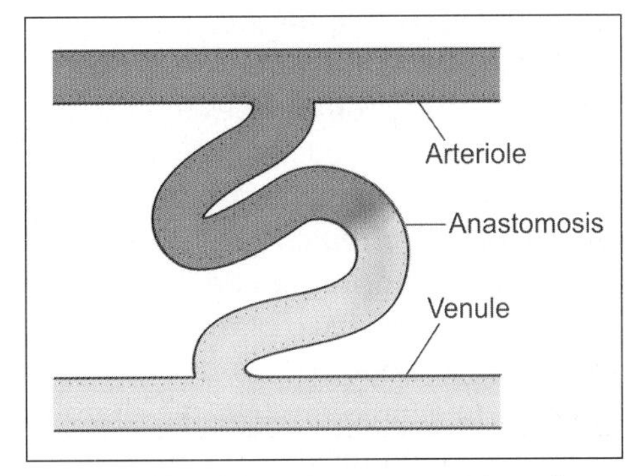

Fig. 13.8: An arteriovenous anastomosis (glomus) (Schematic representation)

of an afferent artery; one or more coiled (S-shaped) connecting vessels; and an efferent vein.

Blood flow through the glomus is controlled in two different ways:

- Firstly, the wall of the afferent artery has a number of elevations that project into the lumen; and probably have a valvular function. These projections are produced partly by endothelium, and partly by muscle.

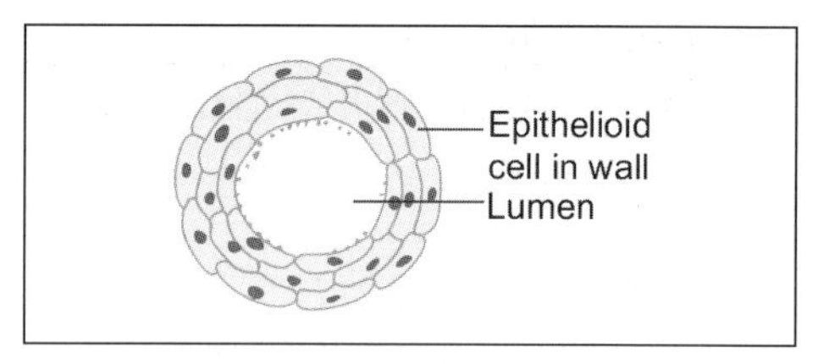

Fig. 13.9: Section across the connecting channel of an arteriovenous anastomosis showing epithelioid cells (Schematic representation)

- Secondly, the connecting vessels have thick muscular walls in which the muscle fibres are short and thick with central nuclei. These cells have some resemblance to epithelial cells and are, therefore, termed ***epithelioid cells*** (Fig. 13.9). They have similarities to pericytes present around capillaries. The lumen of the connecting channel can be occluded by contraction (or swelling) of epithelioid cells.

Glomera are found in the skin at the tips of the fingers and toes (specially in the digital pads and nailbeds); in the lips; the tip of the tongue; and in the nose. They are concerned with the regulation of the circulation in these areas in response to changes in temperature.

Added Information

Arteriovenous anastomoses are few and inefficient in the newborn. In old age, again, arteriovenous anastomoses of the skin decrease considerably in number. These observations are to be correlated with the fact that temperature regulation is not efficient in the newborn as well as in old persons.

Precapillary Sphincters and Thoroughfare Channels

Arteriovenous anastomoses control blood flow through relatively large segments of the capillary bed. Much smaller segments can be individually controlled as follows.

Capillaries arise as side branches of terminal arterioles. The initial segment of each such branch is surrounded by a few smooth muscle cells that constitute a ***precapillary sphincter*** (Fig. 13.10). Blood flow, through any part of the capillary bed, can be controlled by the precapillary sphincter.

In many situations, arterioles and venules are connected (apart from capillaries) by some channels that resemble capillaries, but have a larger calibre. These channels run a relatively direct course between the arteriole and venule. Isolated smooth muscle fibres may be present on their walls. These are called ***thoroughfare channels*** (Fig. 13.10). At times when most of the precapillary sphincters in the region are contracted (restricting flow through capillaries), blood is short circuited from arteriole to venule through the thoroughfare channels. A thoroughfare channel and the capillaries associated with it are sometimes referred to as a ***microcirculatory unit.***

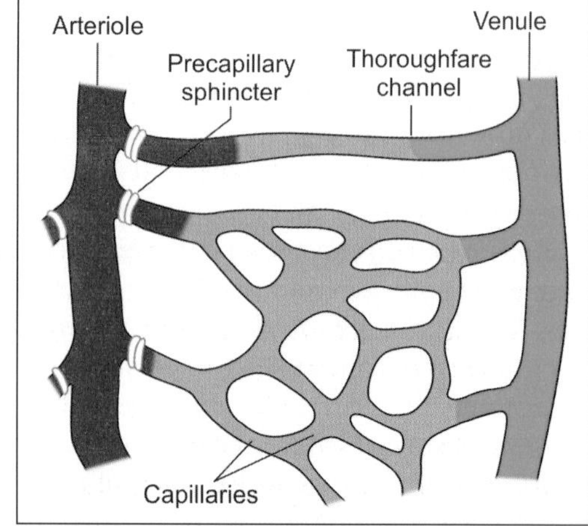

Fig. 13.10: Precapillary sphincters and thoroughfare channels (Schematic representation)

THE HEART

The heart is a muscular organ that pumps blood throughout the blood vessels to various parts of the body by repeated rhythmic contractions.

Structure

There are three layers in the wall of the heart:

- The innermost layer is called the ***endocardium***. It corresponds to the tunica intima of blood vessels. It consists of a layer of endothelium that rests on a thin layer of delicate connective tissue. Outside this there is a thicker ***subendocardial layer*** of connective tissue.
- The main thickness of the wall of the heart is formed by a thick layer of cardiac muscle. This is the ***myocardium***.
- The external surface of the myocardium is covered by the ***epicardium*** (or ***visceral layer of serous pericardium***). It consists of a layer of connective tissue that is covered, on the free surface, by a layer of flattened mesothelial cells.

Added Information

Atrial myocardial fibres secrete a ***natriuretic hormone*** when they are excessively stretched (as in some diseases). The hormone increases renal excretion of water, sodium and potassium. It inhibits the secretion of renin (by the kidneys), and of aldosterone (by the adrenal glands) thus reducing blood pressure.

At the junction of the atria and ventricles, and around the openings of large blood vessels there are rings of dense fibrous tissue. Similar dense fibrous tissue is also present in the interventricular septum. These masses of dense fibrous tissue constitute the 'skeleton' of the heart. They give attachment to fasciculi of heart muscle.

The ***valves of the heart*** are folds of endocardium that enclose a plate like layer of dense fibrous tissue.

Conducting System of the Heart

Conducting system of the heart is made up of a special kind of cardiac muscle. The ***Purkinje fibres*** of this system are chains of cells. The cells are united by desmosomes. Intercalated discs are absent. These cells have a larger diameter, and are shorter, than typical cardiac myocytes. Typically each cell making up a Purkinje fibre has a central nucleus surrounded by clear cytoplasm containing abundant glycogen. Myofibrils are inconspicuous and are confined to the periphery of the fibres. Mitochondria are numerous and the sarcoplasmic reticulum is prominent. ***Nodal myocytes*** [present in the atrioventricular (AV) node and the sinoatrial (SA) node] are narrow, rounded, cylindrical or polygonal cells with single nuclei. They are responsible for pace-maker functions. ***Transitional myocytes*** are present in the nodes, and in the stem and main branches of the AV bundle. They are similar to cardiac myocytes except that they are narrower. Conduction through them is slow.

In the SA node and the AV node the muscle fibres are embedded in a prominent stroma of connective tissue. This tissue contains many blood vessels and nerve fibres.

Chapter 14

The Respiratory System

The respiratory system consists of:

- Respiratory part that includes the lungs
- Conducting part that includes the nasal cavities, the pharynx, the trachea, the bronchi and their intrapulmonary continuations.

The conducting part is responsible for providing passage of air and conditioning the inspired air. The respiratory part is involved in the exchange of oxygen and carbon dioxide between blood and inspired air.

COMMON FEATURES OF AIR PASSAGES

The passages in the conducting part have some features in common. Their walls have a skeletal basis made up variably of bone, cartilage, and connective tissue. The skeletal basis keeps the passages always patent. Smooth muscle present in the walls of the trachea and bronchi enables some alterations in the size of the lumen. The interior of the passages is lined over most of its extent by pseudostratified, ciliated and columnar epithelium. The epithelium is kept moist by the secretions of numerous serous glands. Numerous goblet cells and mucous glands cover the epithelium with a protective mucoid secretion that serves to trap dust particles present in inhaled air. This mucous (along with the dust particles in it) is constantly moved towards the pharynx by action of cilia. When excessive mucous accumulates it is brought out by coughing, or is swallowed. Deep to the mucosa there are numerous blood vessels that serve to warm the inspired air.

THE NASAL CAVITIES

The nasal cavity is the beginning of the respiratory system. These are paired chambers separated by septum. It extends from the nostrils in front to the posterior nasal apertures behind. Each nasal cavity is a hollow organ composed of bone, cartilage and connective tissue covered by mucous membrane.

Histologically, the wall of each half of the nasal cavity is divisible into three distinct regions.

- Vestibule
- Olfactory mucosa
- Respiratory mucosa

Vestibule

It is the anterior dilated part of the nasal cavity. The ***vestibule*** is lined by skin continuous with that on the exterior of the nose. Hair and sebaceous glands are present.

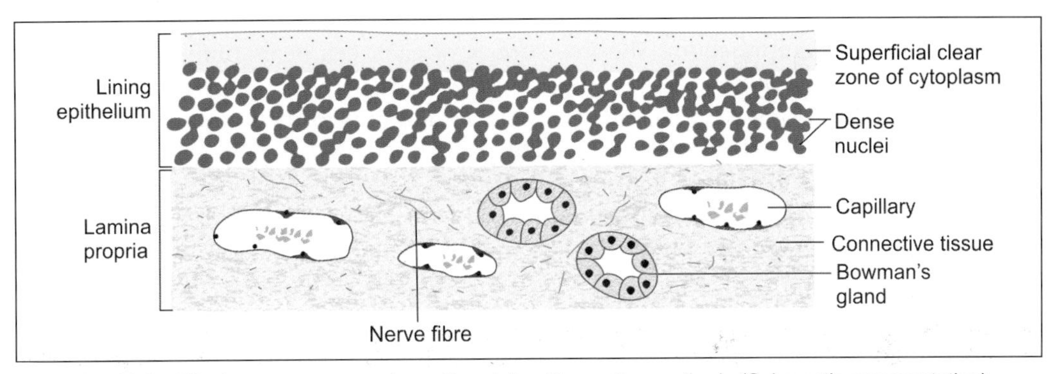

Fig. 14.1: Olfactory mucosa seen in section stained by routine methods (Schematic representation)

Olfactory Mucosa

Apart from their respiratory function the nasal cavities serve as end organs for smell. Receptors for smell are located in the ***olfactory mucosa*** which is confined to a relatively small area on the superior nasal concha, and on the adjoining part of the nasal septum.

Olfactory mucosa is yellow in colour, in contrast to the pink colour of the respiratory mucosa. It is responsible for the sense of smell. It consists of a lining epithelium and a lamina propria.

Olfactory Epithelium

The ***olfactory epithelium*** is pseudostratified. It is much thicker than the epithelium lining the respiratory mucosa (about 100 μm). Within the epithelium there is a superficial zone of clear cytoplasm below which there are several rows of nuclei (Fig. 14.1). Using special methods three types of cells can be recognised in the epithelium (Fig. 14.2).

- The ***olfactory cells*** are modified neurons. Each cell has a central part containing a rounded nucleus. Two processes, distal and proximal, arise from this central part. The distal process (representing the dendrite) passes towards the surface of the olfactory epithelium. It ends in a thickening (called the ***rod*** or ***knob***) from which a number of non-motile olfactory cilia arise and project into a layer of fluid covering the epithelium. (Some of them pass laterally in between the microvilli of adjacent sustentacular cells). The proximal process of each olfactory cell represents the axon. It passes into the subjacent connective tissue where it forms one fibre of the olfactory nerve. The nuclei of olfactory cells lie at various levels in the basal two-third of the epithelium.

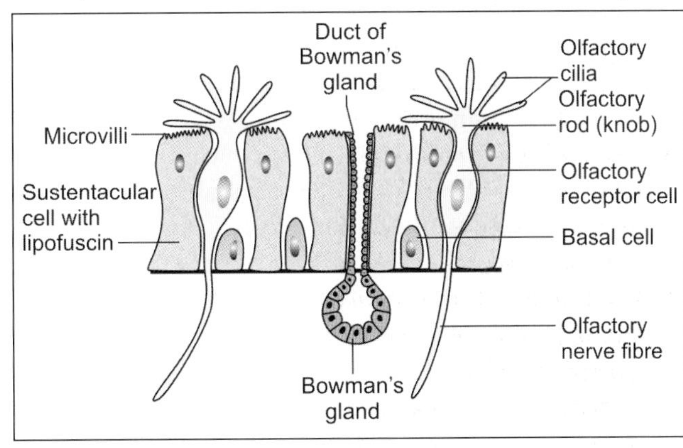

Fig. 14.2: Cells to be seen in olfactory epithelium (Schematic representation)

- The ***sustentacular cells*** support the olfactory cells. Their nuclei are oval, and lie near the free surface of the epithelium. The free surface of each cell bears numerous microvilli (embedded in overlying mucous). The cytoplasm contains yellow pigment (lipofuscin) that gives olfactory mucosa its yellow colour. In addition to their supporting function sustentacular cells may be phagocytic, and the pigment in them may represent remnants of phagocytosed olfactory cells.
- The ***basal cells*** lie deep in the epithelium and do not reach the luminal surface. They divide to form new olfactory cells to replace those that die. Some basal cells have a supporting function.

Added Information

- In vertebrates, olfactory cells are unique in being the only neurons that have cell bodies located in an epithelium.
- Olfactory cells are believed to have a short life. Dead olfactory cells are replaced by new cells produced by division of basal cells. This is the only example of regeneration of neurons in mammals.

Lamina Propria

The lamina propria, lying deep to the olfactory epithelium consists of connective tissue within which blood capillaries, lymphatic capillaries and olfactory nerve bundles are present. It also contains serous glands (of Bowman) the secretions of which constantly 'wash' the surface of the olfactory epithelium. This fluid may help in transferring smell carrying substances from air to receptors on olfactory cells. The fluid may also offer protection against bacteria.

Respiratory Mucosa

The rest of the wall of each half of the nasal cavity is covered by ***respiratory mucosa*** lined by pseudostratified ciliated columnar epithelium.

This mucosa is lined by a pseudostratified ciliated columnar epithelium resting on a basal lamina. In the epithelium, the following cells are present (Fig. 14.3):

- ***Ciliated cells*** are the columnar cells with cilia on their free surfaces and are the most abundant cell type.
- ***Goblet cells*** (flask-shaped cells) scattered in the epithelium produce mucous.
- ***Non-ciliated columnar cells*** with microvilli on the free surface probably secrete a serous fluid that keeps the mucosa moist.
- ***Basal cells*** lying near the basal lamina probably give rise to ciliated cells to replace those lost.

At places the respiratory mucosa may be lined by a simple ciliated columnar epithelium, or even a cuboidal epithelium.

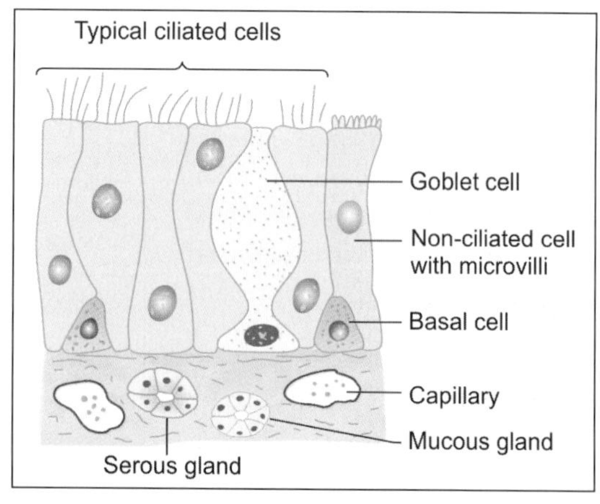

Fig. 14.3: Structure of respiratory part of nasal mucosa (Schematic representation)

Deep to the basal lamina supporting the epithelium lining, the mucosa contains a layer of fibrous tissue, through which the mucosa is firmly connected to underlying periosteum or perichondrium. The fibrous tissue may contain numerous lymphocytes. It also contains mucous and serous glands that open on to the mucosal surface. Some serous cells contain basophilic granules, and probably secrete amylase. Others with eosinophilic granules produce lysozyme.

The deeper parts of the mucosa contain a rich capillary network that constitutes a ***cavernous tissue***. Blood flowing through the network warms inspired air. Variations in blood flow can cause swelling or shrinkage of the mucosa.

Respiratory mucosa also lines the paranasal air sinuses. Here it is closely bound to underlying periosteum forming a ***mucoperiosteum***.

Lamina Propria

The lamina propria of nasal mucosa contains lymphocytes, plasma cells, macrophages, a few neutrophils and eosinophils. Eosinophils increase greatly in number in persons suffering from allergic rhinitis.

Clinical Correlation

- **Acute Rhinitis (Common Cold):** Acute rhinitis or common cold is the common inflammatory disorder of the nasal cavities that may extend into the nasal sinuses. It begins with rhinorrhoea, nasal obstruction and sneezing. Initially, the nasal discharge is watery, but later it becomes thick and purulent.
- **Nasal Polyps:** Nasal polyps are common and are pedunculated grape-like masses of tissue. They are the end-result of prolonged chronic inflammation causing polypoid thickening of the mucosa. They may be allergic or inflammatory. They are frequently bilateral and the middle turbinate is the common site.

THE PHARYNX

The pharynx consists of nasal, oral and laryngeal parts. The nasal part is purely respiratory in function, but the oral and laryngeal parts are more intimately concerned with the alimentary system. The wall of the pharynx is fibromuscular.

Epithelium

In the nasopharynx the epithelial lining is ciliated columnar, or pseudostratified ciliated columnar. Over the inferior surface of the soft palate, and over the oropharynx and laryngopharynx the epithelium is stratified squamous (as these parts come in contact with food during swallowing).

Lymphoid Tissue

Subepithelial aggregations of lymphoid tissue are present specially on the posterior wall of the nasopharynx, and around the orifices of the auditory tubes, forming the nasopharyngeal and tubal tonsils. The palatine tonsils are present in relation to the oropharynx.

Submucosa

Numerous mucous glands are present in the submucosa, including that of the soft palate.

Clinical Correlation

- **Ludwig's Angina:** This is a severe, acute streptococcal cellulitis involving the neck, tongue and back of the throat. The condition was more common in the pre-antibiotic era as a complication of compound fracture of the mandible and periapical infection of the molars. The condition often proves fatal due to glottic oedema, asphyxia and severe toxaemia.
- **Diphtheria:** Diphtheria is an acute communicable disease caused by *Corynebacterium diphtheriae*. It usually occurs in children and results in the formation of a yellowish-grey pseudomembrane in the mucosa of nasopharynx, oropharynx, tonsils, larynx and trachea.
- **Tonsillitis:** Tonsillitis caused by staphylococci or streptococci may be acute or chronic. Acute tonsillitis is characterised by enlargement, redness and inflammation. Acute tonsillitis may progress to acute follicular tonsillitis in which crypts are filled with debris and pus giving it follicular appearance. Chronic tonsillitis is caused by repeated attacks of acute tonsillitis in which case the tonsils are small and fibrosed. Acute tonsillitis may pass on to tissues adjacent to tonsils to form peritonsillar abscess or quinsy.

THE LARYNX

Larynx is a specialised organ responsible for production of voice. It houses the vocal cords. The wall of the larynx has a complex structure made up of a number of cartilages, membranes and muscles.

Mucous Membrane

The epithelium lining the mucous membrane of the larynx is predominantly pseudostratified ciliated columnar. However, over some parts that come in contact with swallowed food the epithelium is stratified squamous. These parts include the epiglottis (anterior surface and upper part of the posterior surface), and the upper parts of the aryepiglottic folds. The vocal folds do not come in contact with swallowed food, but their lining epithelium is exposed to considerable stress during vibration of the folds. These folds are also covered with stratified squamous epithelium.

Numerous goblet cells and subepithelial mucous glands provide a mucous covering to the epithelium. Mucous glands are specially numerous over the epiglottis; in the lower part of the aryepiglottic folds (where they are called ***arytenoid glands***); and in the saccule. The glands in the saccule provide lubrication to the vocal folds. Serous glands and lymphoid tissue are also present.

EM studies have shown that epithelial cells lining the vocal folds bear microvilli and ridge-like foldings of the surface plasma membrane (called ***microplicae***). It is believed that these help to retain fluid on the surface of the cells keeping them moist.

Added Information

The connective tissue subjacent to the epithelial lining of vocal folds is devoid of lymph vessels. This factor slows down lymphatic spread of cancer arising in the epithelium of the vocal folds.

Cartilages of the Larynx

The larynx has a cartilaginous framework which is made of nine cartilages (3 paired and 3 unpaired) that are connected to each other by membranes and ligaments (Fig. 14.4). The cartilages are either hyaline or elastic in nature. These are:

- ***Hyaline cartilages***
  - Thyroid (unpaired)
  - Cricoid (unpaired)
  - Arytenoid (paired)
- ***Elastic cartilages***
  - Epiglottis (unpaired)
  - Cuneiform (paired)
  - Corniculate (paired)

With advancing age, calcification may occur in hyaline cartilage, but not in elastic cartilage.

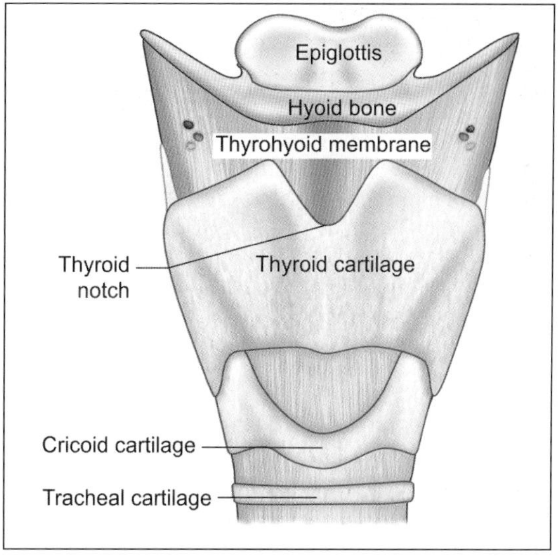

Fig. 14.4: Anterior view of the larynx (Schematic representation)

The Epiglottis

The epiglottis is considered separately because sections through it are usually included in sets of class slides. The epiglottis has a central core of elastic cartilage. Overlying the cartilage there is mucous membrane. The greater part of the mucous membrane is lined by stratified squamous epithelium (non-keratinising). The mucous membrane over the lower part of the posterior surface of the epiglottis is lined by pseudostratified ciliated columnar epithelium (Plate 14.1). This part of the epiglottis does not come in contact with swallowed food as it is overlapped by the aryepiglottic folds. Some taste buds are present in the epithelium of the epiglottis. (A few taste buds may be seen in the epithelium elsewhere in the larynx).

Numerous glands, predominantly mucous, are present in the mucosa deep to the epithelium. Some of them lie in depressions present on the epiglottic cartilage.

Clinical Correlation

- **Acute Laryngitis:** This may occur as a part of the upper or lower respiratory tract infection. Atmospheric pollutants like cigarette smoke, exhaust fumes, industrial and domestic smoke, etc, predispose the larynx to acute bacterial and viral infections. Streptococci and *H. influenzae* cause acute epiglottitis which may be life-threatening.
- **Chronic Laryngitis:** Chronic laryngitis may occur from repeated attacks of acute inflammation, excessive smoking, chronic alcoholism or vocal abuse. The surface is granular due to swollen mucous glands. There may be extensive squamous metaplasia due to heavy smoking, chronic bronchitis and atmospheric pollution.

PLATE 14.1: **Epiglottis**

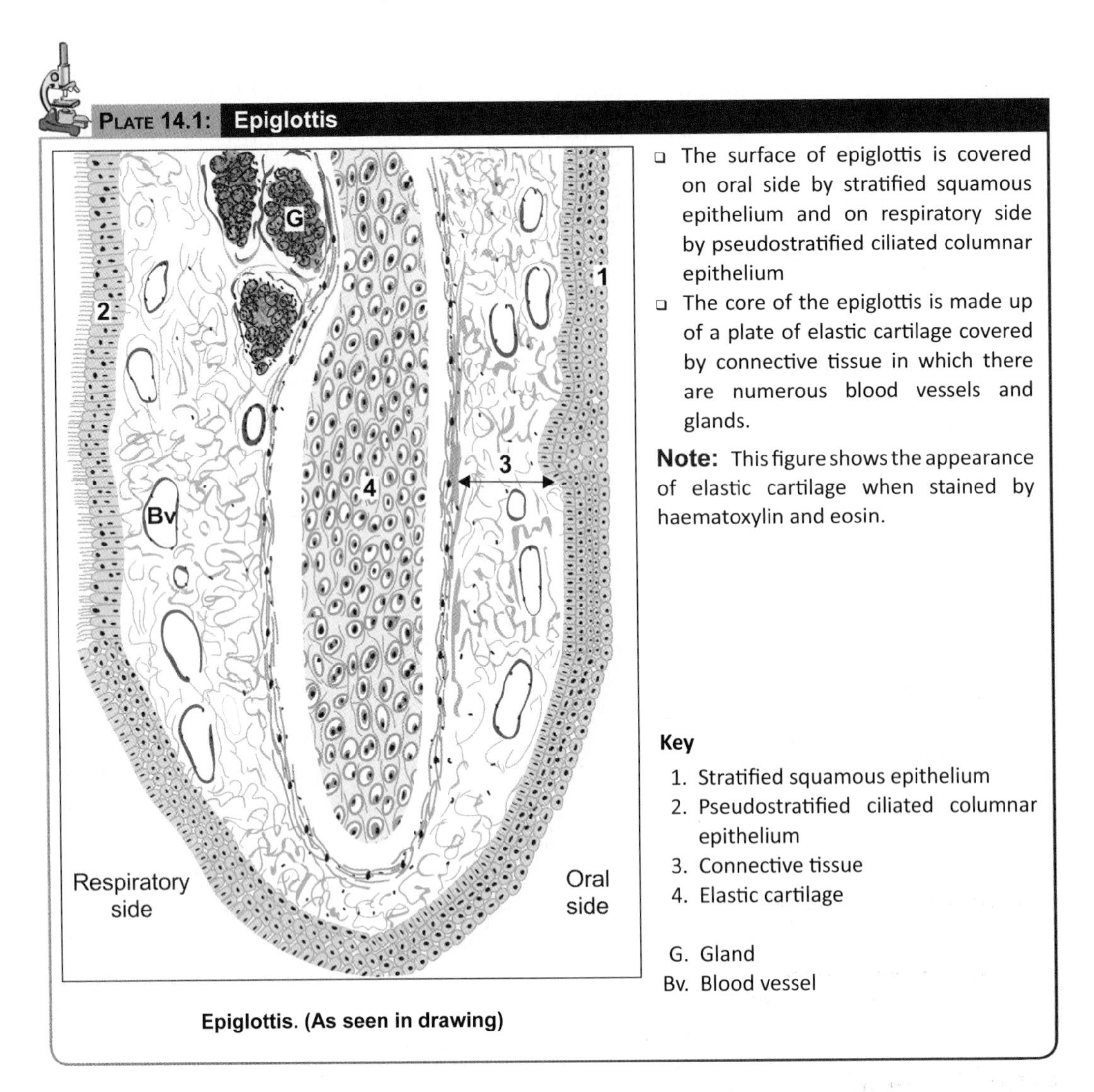

Epiglottis. (As seen in drawing)

- The surface of epiglottis is covered on oral side by stratified squamous epithelium and on respiratory side by pseudostratified ciliated columnar epithelium
- The core of the epiglottis is made up of a plate of elastic cartilage covered by connective tissue in which there are numerous blood vessels and glands.

Note: This figure shows the appearance of elastic cartilage when stained by haematoxylin and eosin.

Key

1. Stratified squamous epithelium
2. Pseudostratified ciliated columnar epithelium
3. Connective tissue
4. Elastic cartilage

G. Gland
Bv. Blood vessel

THE TRACHEA AND PRINCIPAL BRONCHI

Trachea

The trachea is a fibroelastic cartilaginous tube. It extends from the lower border of cricoid cartilage (C_6) to its level of bifurcation (T_4) into right and left bronchi. The trachea consists of four layers (Plate 14.2).

Mucosa

The lumen of the trachea is lined by mucous membrane that consists of a lining epithelium and an underlying layer of connective tissue. The lining epithelium is pseudostratified ciliated columnar. It contains numerous goblet cells, and basal cells that lie next to the basement membrane. Numerous lymphocytes are seen in deeper parts of the epithelium.

PLATE 14.2: Trachea

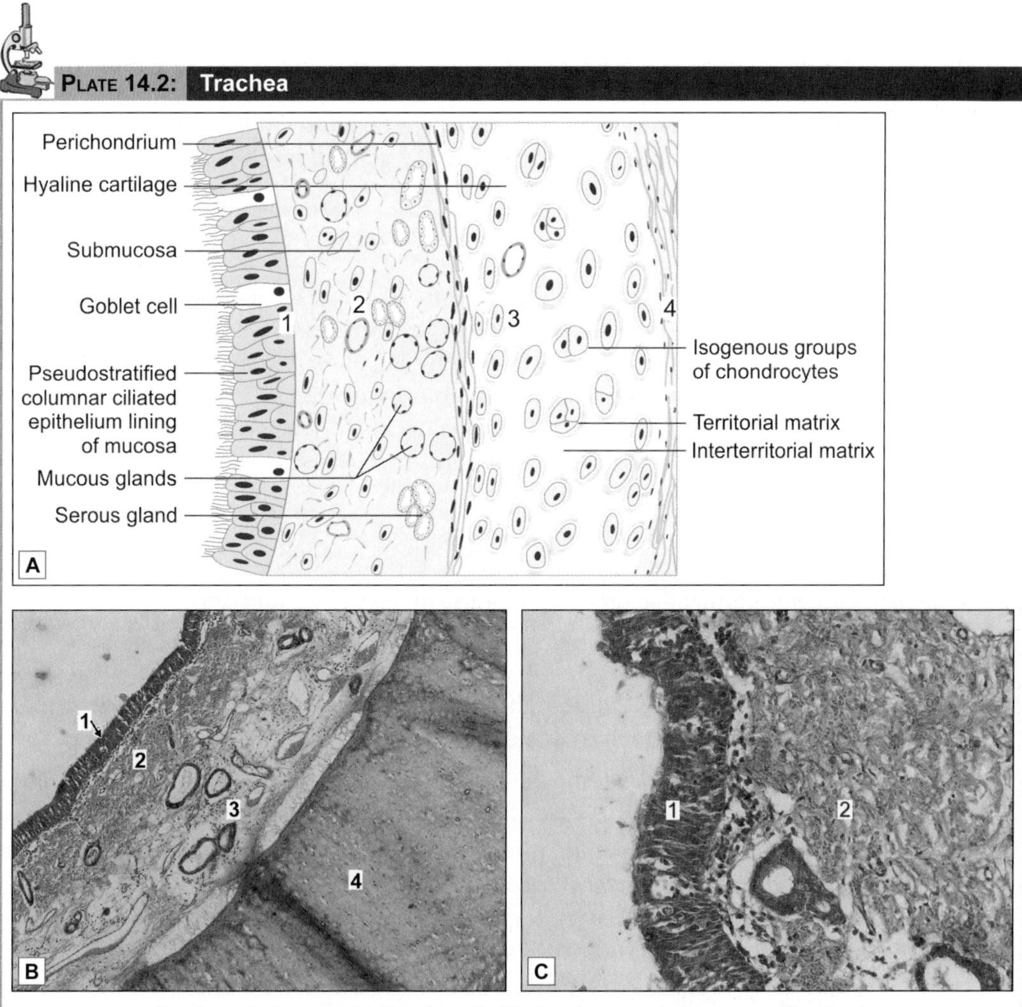

Trachea. A. As seen in drawing; B. Photomicrograph (low magnification); C. Photomicrograph (high magnification)

From within outwards the wall of trachea consists of:

- **Mucosa** formed by pseudostratified ciliated columnar epithelium with goblet cells and the underlying lamina propria
- **Submucosa** made up of loose connective tissue containing mucous glands and serous glands, blood vessels and ducts
- A 'C' shaped plate of **hyaline cartilage**. Perichondrium has outer fibrous and inner chondrogenic layers. Chondrocytes increase in size from periphery to centre. They may appear as isogenous groups surrounded by darkly stained territorial matrix
- Adventitia consisting of collagen fibres (Not shown here).

Key

1. Pseudostratified ciliated columnar epithelium with goblet cells } Mucosa
2. Lamina propria } Mucosa
3. Submucosa
4. Hyaline cartilage

Submucosa

The subepithelial connective tissue contains numerous elastic fibres. It contains serous glands that keep the epithelium moist; and mucous glands that provide a covering of mucous in which dust particles get caught. The mucous is continuously moved towards the larynx by ciliary action. Numerous aggregations of lymphoid tissue are present in the subepithelial connective tissue. Eosinophil leucocytes are also present.

Cartilage and Smooth Muscle Layer

The skeletal basis of the trachea is made up of 16 to 20 tracheal cartilages. Each of these is a C-shaped mass of hyaline cartilage. The open end of the 'C' is directed posteriorly. Occasionally, adjoining cartilages may partly fuse with each other or may have Y-shaped ends. The intervals between the cartilages are filled by fibrous tissue that becomes continuous with the perichondrium covering the cartilages. The gaps between the cartilage ends, present on the posterior aspect, are filled in by smooth muscle and fibrous tissue. The connective tissue in the wall of the trachea contains many elastic fibres.

Adventitia

It is made of fibroelastic connective tissue containing blood vessels and nerves.

Principal Bronchi

The trachea divides at the level of T_4 into right and left principal bronchi (primary or main bronchi). They have a structure similar to that of the trachea.

THE LUNGS

The lungs are the principal respiratory organs that are situated one on either side of mediastinum in the thoracic cavity. They are covered by visceral pleura (Plate 14.3).

The structure of the lungs has to be understood keeping in mind their function of oxygenation of blood. The following features are essential for this purpose.

- A surface at which air (containing oxygen) can be brought into close contact with circulating blood. The barrier between air and blood has to be very thin to allow oxygen (and carbon dioxide) to pass through it. The surface has to be extensive enough to meet the oxygen requirements of the body.
- A system of tubes to convey air to and away from the surface at which exchanges take place.
- A rich network of blood capillaries present in intimate relationship to the surface at which exchanges take place.

Intrapulmonary Passages

On entering the lung the principal bronchus divides into secondary, or ***lobar bronchi*** (one for each lobe). Each lobar bronchus divides into tertiary, or ***segmental bronchi*** (one for each segment of the lobe). The segmental bronchi divide into smaller and smaller bronchi, which ultimately end in ***bronchioles***.

The lung substance is divided into numerous lobules each of which receives a ***lobular bronchiole***. The lobular bronchiole gives off a number of ***terminal bronchioles*** (Fig. 14.5).

As indicated by their name the terminal bronchioles represent the most distal parts of the conducting passage.

Each terminal bronchiole ends by dividing into ***respiratory bronchioles***. These are so called because they are partly respiratory in function as some air sacs (see below) arise from them.

Each respiratory bronchiole ends by dividing into a few ***alveolar ducts***. Each alveolar duct ends in a passage, the ***atrium***, which leads into a number of rounded ***alveolar sacs***. Each alveolar sac is studded with a number of air sacs or ***alveoli***.

The alveoli are blind sacs having very thin walls through which oxygen passes from air into blood, and carbon dioxide passes from blood into air.

The structure of the larger intrapulmonary bronchi is similar to that of the trachea. As these bronchi divide into smaller ones the following changes in structure are observed.

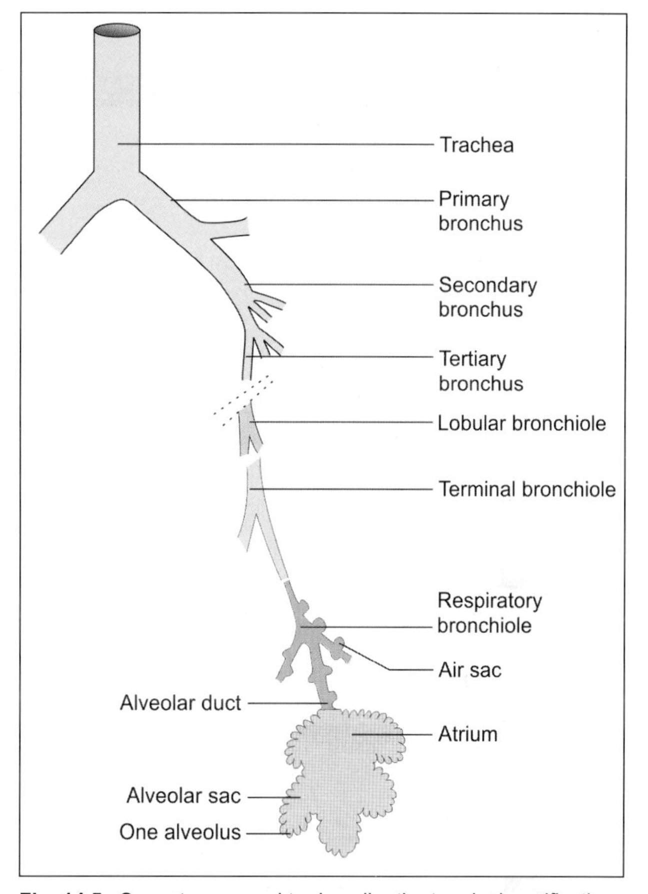

Fig. 14.5: Some terms used to describe the terminal ramifications of the bronchial tree (Schematic representation)

- The cartilages in the walls of the bronchi become irregular in shape, and are progressively smaller. Cartilage is absent in the walls of bronchioles: this is the criterion that distinguishes a bronchiole from a bronchus.
- The amount of muscle in the bronchial wall increases as the bronchi become smaller. The presence of muscle in the walls of bronchi is of considerable clinical significance. Spasm of this muscle constricts the bronchi and can cause difficulty in breathing.
- Subepithelial lymphoid tissue increases in quantity as bronchi become smaller. Glands become fewer, and are absent in the walls of bronchioles.
- The trachea and larger bronchi are lined by pseudostratified ciliated columnar epithelium. As the bronchi become smaller the epithelium first becomes simple ciliated columnar, then non-ciliated columnar, and finally cuboidal (in respiratory bronchioles). The cells contain lysosomes and numerous mitochondria. Plate 14.3 illustrates the salient microscopic features of the lung parenchyma.

EM studies have shown that apart from typical ciliated columnar cells, various other types of cells are to be seen in the epithelium lining the air passages. Some of the cells encountered are as follows (Fig. 14.6):

Plate 14.3: Lung

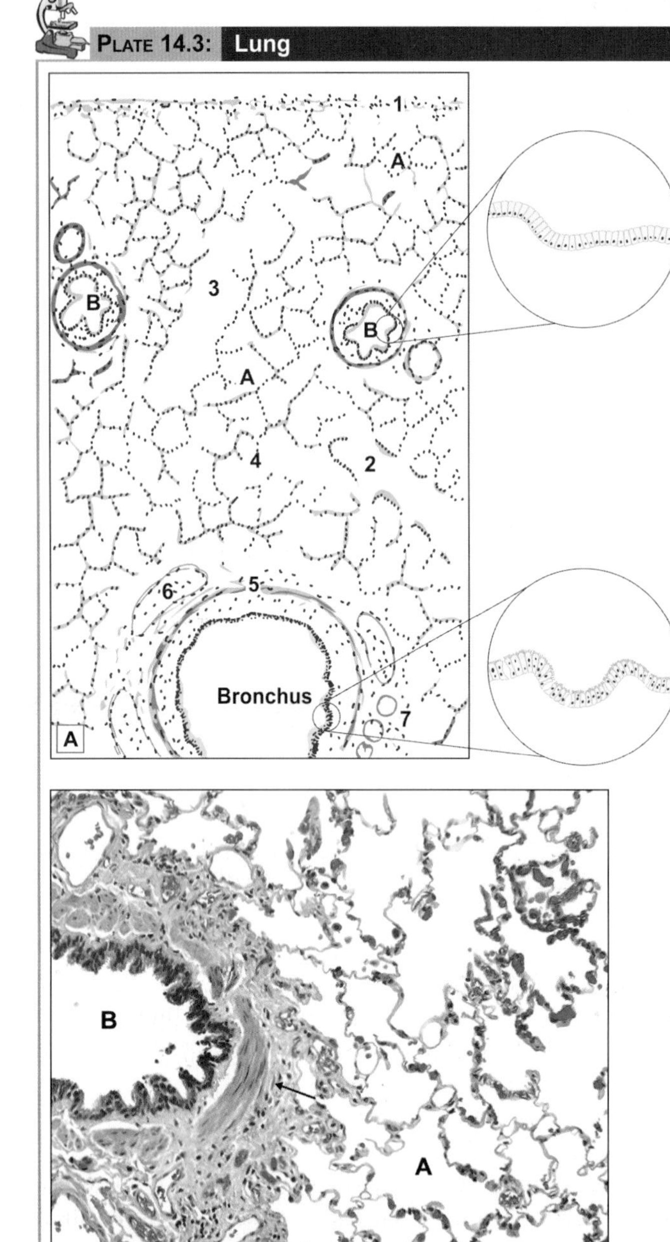

Lung. A. As seen in drawing; B. Photomicrograph.

Courtesy: Atlas of Histopathology, 1st Edition. Ivan Damjanov. Jaypee Brothers. 2012. p37

- The lung surface is covered by pleura. It consists of a lining of mesothelium resting on a layer of connective tissue
- The lung parenchyma is made up of numerous thin-walled spaces or alveoli
- The alveoli give a honey comb appearance and are lined by flattened squamous cells. They are filled with air
- The intrapulmonary bronchus is lined by pseudostratified ciliated columnar epithelium with few goblet cells. Its structure is similar to trachea i.e. it has smooth muscles, cartilage and glands present in its wall
- The bronchiole is lined by simple columnar or cuboidal epithelium surrounded by bundles of smooth muscle cells (see arrow in photomicrograph)
- Bronchioles subdivide and when their diameter is approximately 1 mm or less, they are called terminal bronchiole.
- Arteries are seen near the bronchioles
- Respiratory bronchiole, alveolar duct and atrium are also present
- This slide shows a medium size bronchiole surrounded by alveoli

Key

1. Mesothelium resting on connective tissue
2. Respiratory bronchiole
3. Alveolar duct
4. Atrium
5. Smooth muscle
6. Plates of cartilage
7. Glands

A. Alveoli
B. Bronchus bronchiole

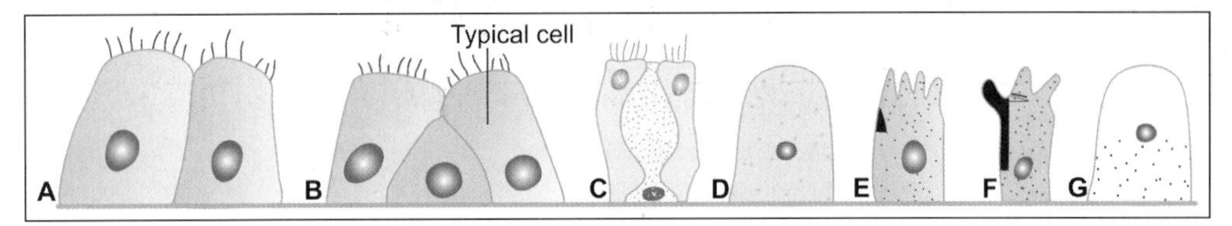

Fig. 14.6: Various types of cells to be seen lining the respiratory passages. A-Typical ciliated columnar, B-Basal, C-Goblet, D-Serous, E-Brush, F-Clara, G-Argyrophil (Schematic representation)

- Goblet cells are numerous. They provide mucous which helps to trap dust entering the passages and is moved by ciliary action towards the larynx and pharynx.
- Non-ciliated serous cells secrete fluid that keeps the epithelium moist.
- Basal cells multiply and transform into other cell types to replace those that are lost.
- Some non-ciliated cells present predominantly in terminal bronchioles (see below) produce a secretion that spreads over the alveolar cells forming a film that reduces surface tension. These include the ***cells of Clara***.
- Cells similar to diffuse endocrine cells of the gut, and containing argyrophil granules are present. They secrete hormones and active peptides including serotonin and bombesin.
- Lymphocytes and other leucocytes may be present in the epithelium. They migrate into the epithelium from surrounding tissues.

The differences between bronchus and bronchioles are given in Table 14.1.

Added Information

Some other functions attributed to cells of Clara include:
- Protection against harmful substances that are inhaled.
- Protection against development of emphysema by opposing the action of substances (proteases) that tend to destroy walls of lung alveoli.
- Stem cell function.

Alveoli

There are about 200 million alveoli in a normal lung. The total area of the alveolar surface of each lung is extensive. It has been estimated to be about 75 square meters. The total capillary surface area available for gaseous exchanges is about 125 square meters.

Structure of Alveolar Wall

Each alveolus has a very thin wall. The wall is lined by an epithelium consisting mainly of flattened squamous cells. The epithelium rests on a basement membrane. Deep to the basement membrane there is a layer of delicate connective tissue through which pulmonary capillaries run. These capillaries have the usual endothelial lining that rests on a basement membrane.

The barrier between air and blood is made up of the epithelial cells and their basement membrane; by endothelial cells and their basement membrane; and by intervening connective tissue. At many places the two basement membranes fuse greatly reducing the thickness of the barrier.

The endothelial cells lining the alveolar capillaries are remarkable for their extreme thinness. With the EM they are seen to have numerous projections extending into the capillary lumen.

Table 14.1: Differences between Bronchus and Bronchiole		
Characteristics	**Bronchus**	**Bronchiole**
Diameter	Larger diameter (more than 1 mm)	Smaller diameter (less than 1 mm)
Lining epithelium	Pseudostratified ciliated columnar epithelium with goblet cells	• Large size bronchioles: simple columnar cells with few cilia and few goblet cells • Small size bronchioles: simple columnar or simple cuboidal cells with no cilia or no goblet cells
Smooth muscle layer	Present between mucosa and cartilage layer	Smooth muscles and elastic fibres form a well-defined layer beneath mucosa
Cartilage	Present in irregular patches	Absent
Glands in submucosa	Both serous and mucous acini present between cartilage and muscle layer	Absent

These projections greatly increase the surface of the cell membrane that is exposed to blood and is, therefore, available for exchange of gases. At many places the basement membrane of the endothelium fuses with that of the alveolar epithelium greatly reducing the thickness of the barrier between blood and air in alveoli.

Pneumocytes

EM studies have shown that the cells forming the lining epithelium of alveoli (***pneumocytes***) are of various types (Fig. 14.7).

- The most numerous cells are the squamous cells already referred to. They are called ***type I alveolar epithelial cells***. Except in the region of the nucleus, these cells are reduced to a very thin layer (0.05 to 0.2 μm). The edges of adjoining cells overlap and are united by tight junctions (preventing leakage of blood from capillaries into the alveolar lumen). They form the lining of 90% of the alveolar surface.
- Scattered in the epithelial lining there are rounded secretory cells bearing microvilli on their free surfaces. These are designated ***type II alveolar epithelial cells*** (Figs 14.7 and 14.8). Their cytoplasm contains secretory granules that appear to be made up of several layers (and are, therefore, called ***multilamellar bodies***). These cells are believed to produce a secretion that forms a film over the alveolar epithelium. This film or ***pulmonary surfactant*** reduces surface tension

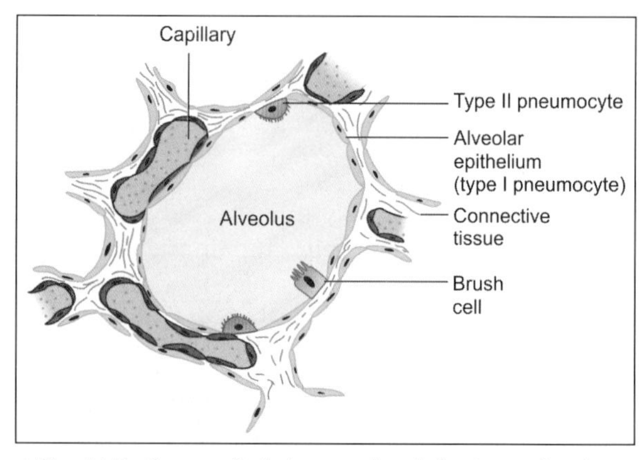

Fig. 14.7: Some cells to be seen in relation to an alveolus (Schematic representation)

Flow chart 14.1: Different types of cells of respiratory system

Different types of cells of respiratory system

- In the epithelium of trachea, bronchus and bronchiole
  - Ciliated cells
  - Goblet cells
  - Basal cells
  - Granule cells
  - Brush cells
  - Clara cells
- In the epithelium lining of lung alveoli
  - Type I alveolar cells (pneumocytes I / squamous cells)
  - Type II alveolar cells (pneumocytes II)
  - Brush cells
- In relation to the lumen of lung alveoli
  - Macrophages

and prevents collapse of the alveolus during expiration.

Surfactant contains phospholipids, proteins and glycosaminoglycans produced in type II cells (A similar fluid is believed to be produced by the cells of Clara present in bronchial passages).

Type II cells may multiply to replace damaged type I cells.

- ***Type III alveolar cells***, or ***brush cells***, of doubtful function, have also been described.

Different types of cells present in the respiratory system are summarised in Flow chart 14.1.

Connective Tissue

The connective tissue in the wall of the alveolus contains collagen fibres and numerous elastic fibres continuous with those of bronchioles. Fibroblasts, histiocytes, mast cells, lymphocytes and plasma cells may be present. Pericytes are present in relation to capillaries.

Some macrophages enter the connective tissue from blood and pass through the alveolar epithelium to reach its luminal surface. Dust particles phagocytosed by them are seen in their cytoplasm. They are therefore called ***dust cells***. These dust cells are expelled to the outside through the respiratory passages. In congestive heart failure (in which pulmonary

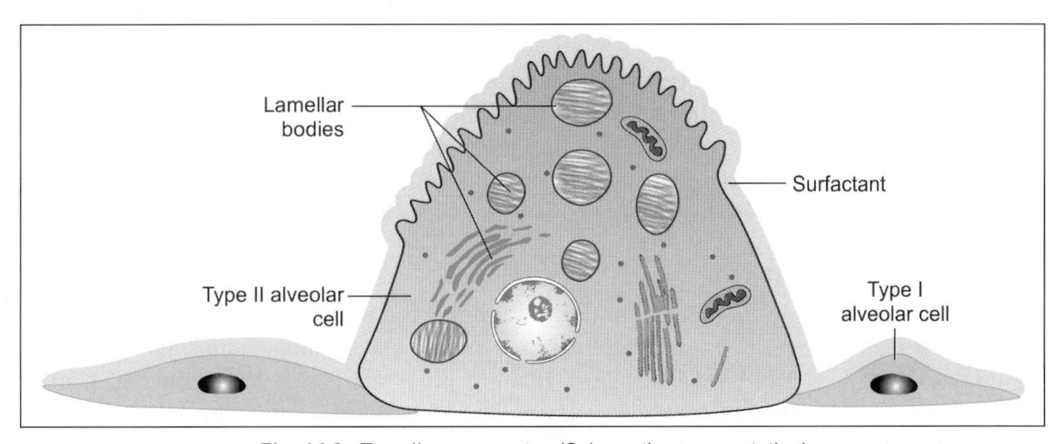

Fig. 14.8: Type II pneumocytes (Schematic representation)

capillaries become overloaded with blood) these macrophages phagocytose erythrocytes that escape from capillaries. The cells, therefore, acquire a brick red colour and are then called ***heart failure cells***. Macrophages also remove excessive surfactant, and secrete several enzymes.

Connective Tissue Basis of the Lung

The greater part of the surface of the lung is covered by a serous membrane, the visceral pleura. This membrane consists of a layer of flattened mesothelial cells, supported on a layer of connective tissue.

Deep to the pleura there is a layer of subserous connective tissue. This connective tissue extends into the lung substance along bronchi and their accompanying blood vessels, and divides the lung into lobules. Each lobule has a lobular bronchiole and its ramifications, blood vessels, lymphatics and nerves.

The epithelial lining of air passages is supported by a basal lamina deep to which there is the connective tissue of the lamina propria. Both in the basal lamina and in the lamina propria there are numerous elastic fibres. These fibres run along the length of respiratory passages and ultimately become continuous with elastic fibres present in the walls of air sacs. This elastic tissue plays a very important role by providing the physical basis for elastic recoil of lung tissue. This recoil is an important factor in expelling air from the lungs during expiration. Elastic fibres passing between lung parenchyma and pleura prevent collapse of alveoli and small bronchi during expiration.

Pleura

The pleura is lined by flat mesothelial cells that are supported by loose connective tissue rich in elastic fibres, blood vessels, nerves and lymphatics. There is considerable adipose tissue under parietal pleura.

Blood Supply of Lungs

The lungs receive deoxygenated blood from the right ventricle of the heart through pulmonary arteries. Within the lung the arteries end in an extensive capillary network in the walls of alveoli. Blood oxygenated here is returned to the left atrium of the heart through pulmonary veins.

Oxygenated blood required for nutrition of the lung itself reaches the lungs through bronchial arteries. They are distributed to the walls of bronchi as far as the respiratory bronchioles. Blood reaching the lung through these arteries is returned to the heart partly through bronchial veins, and partly through the pulmonary veins.

Plexuses of lymph vessels are present just deep to the pleura and in the walls of bronchi.

Nerve Supply of Lungs

The lungs receive autonomic nerves, both sympathetic and parasympathetic, and including both afferent and efferent fibres. Efferent fibres supply the bronchial musculature. Vagal stimulation produces bronchoconstriction. Efferent fibres also innervate bronchial glands. Afferent fibres are distributed to the walls of bronchi and of alveoli. Afferent impulses from the lungs play an important role in control of respiration through respiratory reflexes.

Clinical Correlation

- **Acute respiratory distress syndrome (ARDS)** is a severe, at times life-threatening, form of progressive respiratory insufficiency which involves pulmonary tissues diffusely, i.e. involvement of the alveolar epithelium, alveolar lumina and interstitial tissue. ARDS exists in 2 forms: neonatal and adult type. Both have the common morphological feature of formation of hyaline membrane in the alveoli, and hence, is also termed as hyaline membrane disease (HMD).
- **Bacterial pneumonia:** Bacterial infection of the lung parenchyma is the most common cause of pneumonia or consolidation of one or both the lungs. Two types of acute bacterial pneumonias are distinguished—lobar pneumonia and broncho-(lobular-) pneumonia, each with distinct aetiologic agent and morphologic changes.
- **Chronic bronchitis** is a common condition defined clinically as persistent cough with expectoration on most days for at least three months of the year for two or more consecutive years. The cough is caused by oversecretion of mucous. In spite of its name, chronic inflammation of the bronchi is not a prominent feature. The condition is more common in middle-aged males than females.
- **Asthma** is a disease of airways that is characterised by increased responsiveness of the tracheobronchial tree to a variety of stimuli resulting in widespread spasmodic narrowing of the air passages which may be relieved spontaneously or by therapy. Asthma is an episodic disease manifested clinically by paroxysms of dyspnoea, cough and wheezing. However, a severe and unremitting form of the disease termed status asthmaticus may prove fatal.
- **Immotile cilia syndrome** that includes Kartagener's syndrome (bronchiectasis, situs inversus and sinusitis) is characterised by ultrastructural changes in the microtubules causing immotility of cilia of the respiratory tract epithelium, sperms and other cells. Males in this syndrome are often infertile.

Chapter 15

Digestive System: Oral Cavity and Related Structures

The abdominal part of the alimentary canal (consisting of the stomach and intestines) is often referred to as the ***gastrointestinal tract***. Closely related to the alimentary canal there are several accessory organs that form part of the alimentary system. These include the structures of the oral cavity (lips, teeth, tongue, salivary glands) and the liver and pancreas.

ORAL CAVITY

The wall of the oral cavity is made up partly of bone (jaws and hard palate), and partly of muscle and connective tissue (lips, cheeks, soft palate, and floor of mouth). These structures are lined by mucous membrane which is lined by stratified squamous epithelium that rests on connective tissue, similar to that of the dermis.

The epithelium differs from that on the skin in that it is not keratinised. Papillae of connective tissue (similar to dermal papillae) extend into the epithelium. The size of these papillae varies considerably from region to region. Over the alveolar processes (where the mucosa forms the gums), and over the hard palate, the mucous membrane is closely adherent to underlying periosteum. Elsewhere it is connected to underlying structures by loose connective tissue. In the cheeks, this connective tissue contains many elastic fibres and much fat (specially in children).

THE LIPS

Lips are fleshy folds which on the 'external' surface are lined by skin, and on 'internal' surface are lined by mucous membrane. It must be noted, however, that part of the mucosal surface is 'free' and constitutes the region the lay person thinks of as the lip. The substance of each lip (upper or lower) is predominantly muscular (skeletal muscle) (Fig. 15.1).

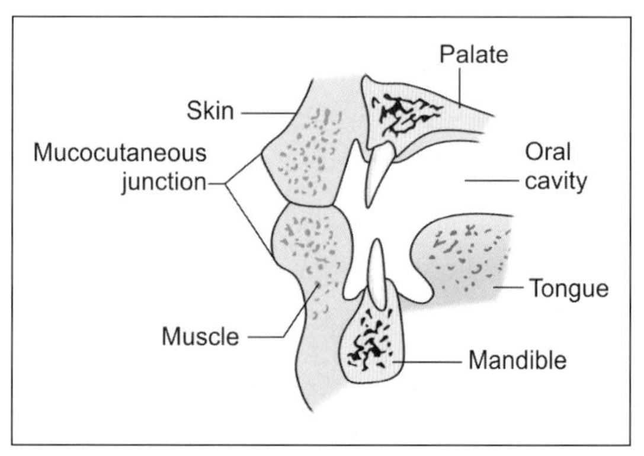

Fig. 15.1: Some relationships of the lips (Schematic representation)

The upper and lower lips close along the red margin which represents the ***mucocutaneous junction***. There is a transitional zone between the skin and mucous

PLATE 15.1: Longitudinal Section through the Lip

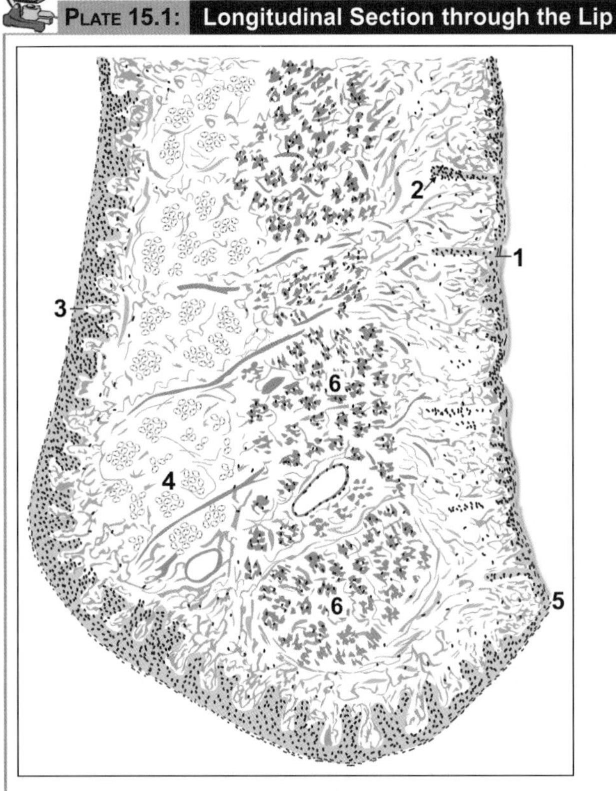

Longitudinal Section through the Lip
as seen in drawing

- The substance of the lip is formed by a mass of muscle.
- Each lip has an 'external' surface covered by skin and an 'internal' surface lined by mucous membrane.
- There is a transitional zone between the skin and mucous membrane known as vermilion border.

Key

1. Outer surface covered by skin (keratinised epithelium)
2. Hair follicle
3. Inner surface (mucosa) covered by stratified squamous non-keratinised epithelium
4. Glands
5. Junction of skin and mucosa
6. Bundles of skeletal muscle

membrane which is sometimes referred to as the ***vermilion***, because of its pink colour in fair skinned individuals. This part meets the skin along a distinct edge.

The 'external' surface of the lip is lined by true skin in which hair follicles and sebaceous glands can be seen.

The mucous membrane is lined by stratified squamous nonkeratinised epithelium (Plate 15.1). This epithelium is much thicker than that lining the skin (specially in infants). The epithelium has a well marked ***rete ridge system***. The term rete ridges is applied to finger-like projections of epithelium that extend into underlying connective tissue, just like the epidermal papillae. This arrangement anchors the epithelium to underlying connective tissue, and enables it to withstand friction. A similar arrangement is seen in the mucosa over the palate.

Subjacent to the epithelium the mucosa has a layer of connective tissue (corresponding to the dermis), and a deeper layer of loose connective tissue. The latter contains numerous mucous glands. Sebaceous glands, not associated with hair follicles, may be present. Their secretions prevent dryness and cracking of the exposed part of the mucosa.

Clinical Correlation

- ***Fordyce's granules***: Fordyce's granules are symmetric, small, light yellow macular spots on the lips and buccal mucosa and represent collections of sebaceous glands.
- ***Pyogenic granuloma***: This is an elevated, bright red swelling of variable size occurring on the lips, tongue, buccal mucosa and gingiva. It is a vasoproliferative inflammatory lesion. ***Pregnancy tumour*** is a variant of pyogenic granuloma.

THE TEETH

General Structure

A tooth consists of an 'upper' part, the ***crown***, which is seen in the mouth; and of one or more ***roots*** which are embedded in sockets in the jaw bone (mandible or maxilla). It is composed of 3 calcified tissues, namely: ***enamel, dentine*** and the ***pulp*** (Fig. 15.2).

The greater part of the tooth is formed by a bone-like material called ***dentine***. In the region of the crown the dentine is covered by a much harder white material called the ***enamel***. Over the root the dentine is covered by a thin layer of ***cementum***. The cementum is united to the wall of the bony socket in the jaw by a layer of fibrous tissue that is called the ***periodontal ligament***.

Within the dentine there is a ***pulp canal*** (or ***pulp cavity***) that contains a mass of cells, blood vessels, and nerves that constitute the ***pulp***. The blood vessels and nerves enter the pulp canal through the ***apical foramen*** which is located at the apex of the root.

The Enamel

The enamel is the hardest material in the body. It is made up almost entirely (96%) of inorganic salts. These salts are mainly in the form of complex crystals of hydroxyapatite (as in bone). The crystals contain calcium phosphate and calcium carbonate. Some salts are also present in amorphous form. The crystals of hydroxyapatite are arranged in the form of rod-shaped ***prisms***, which run from the deep surface of the enamel (in contact with dentine) to its superficial (or free) surface.

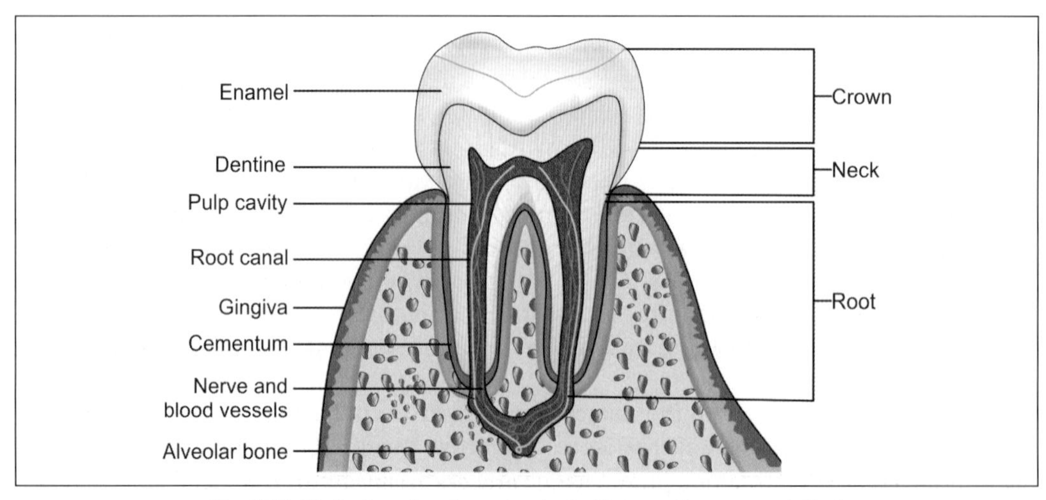

Fig. 15.2: Vertical section through a tooth (Schematic representation)

Prisms are separated by ***interprismatic material***. There is no essential difference in the structure of prisms and of interprismatic material, the two appearing different only because of differing orientation of the hydroxyapatite crystals in them. The most superficial part of the enamel is devoid of prisms.

During development, enamel is laid down in the form of layers. When seen in section these layers can be distinguished as they are separated by lines running more or less parallel to the surface of the enamel. These lines are called the ***incremental lines*** or the ***lines of Retzius*** (Fig. 15.3). In some teeth, in which enamel formation takes place partly before birth and partly after birth (e.g., in milk teeth), one of the incremental lines is particularly marked. It represents the junction of enamel formed before birth with that formed after birth, and is called the ***neonatal line***.

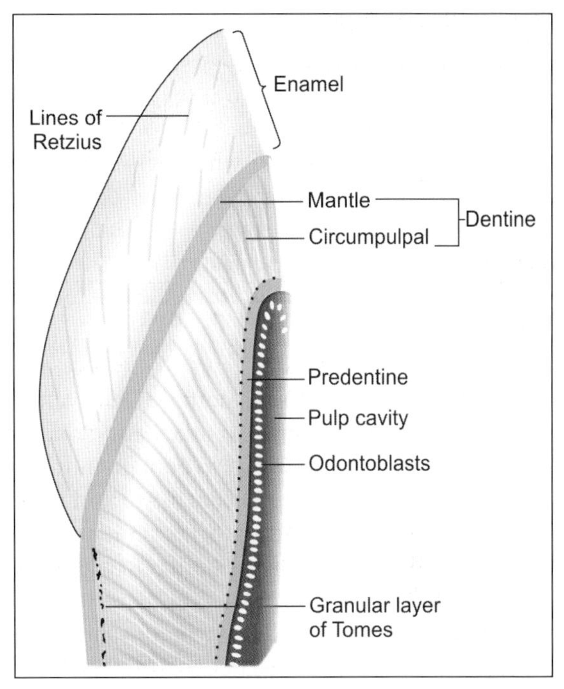

Fig. 15.3: Part of a tooth to show some features of the structure of enamel and dentine (Schematic representation)

At places, the enamel is penetrated by extraneous material. Projections entering the enamel from the dentine-enamel junction are called ***enamel tufts***; and projections entering it from the free surface are called ***enamel lamellae*** (because of their shape). Dentineal tubules (see below) may extend into the enamel forming ***enamel spindles***.

The Dentine

Structure and Composition

Dentine is a hard material having several similarities to bone. It is made up basically of calcified ground substance (glycosaminoglycans) in which there are numerous collagen fibres (type 1). The calcium salts are mainly in the form of hydroxyapatite. Amorphous salts are also present. The inorganic salts account for 70% of the weight of dentine. Like bone, dentine is laid down in layers that are parallel to the pulp cavity. The layers may be separated by less mineralized tissue that forms the incremental lines of Von Ebner. Dentine is permeated by numerous fine canaliculi that pass radially from the pulp cavity towards the enamel (or towards cement). These are the dentinal tubules. The tubules may branch specially near the enamel-dentine junction. We have seen (above) that some dentinal tubules extend into the enamel as enamel spindles.

The ground substance of dentine is more dense (than elsewhere) immediately around the dentinal tubules, and forms the ***peritubular dentine*** or the ***dentineal sheath*** (of Newmann). Each dentineal tubule contains a protoplasmic process arising from cells called ***odontoblasts***, that line the pulp cavity. These protoplasmic processes are called the ***fibres of Tomes***.

Near the surface of the root of the tooth (i.e., just deep to the cementum) the dentine contains minute spaces that give a granular appearance. This is the ***granular layer of Tomes***.

Types of Dentin

At the junction of dentine with the pulp cavity there is a layer of ***predentine*** that is not mineralised. Dentine near the enamel-dentine junction is less mineralised (than elsewhere) and is called the ***mantle dentine***. The main part of dentine (between predentine and mantle dentine) is called the ***circumpulpal dentine*** (Fig. 15.3). Dentine formed before eruption of the tooth is called ***primary dentine***, while that formed after eruption is called ***secondary dentine.***

The Cementum

Cementum is the portion of tooth which covers the dentine at the root of tooth and is the site where periodontal ligament is attached. In old people, cementum may be lost, exposing the dentine. Cementum is similar to bone in morphology and composition and may be regarded as a layer of true bone that covers the root of the tooth.

Towards the apex of the tooth, the cementum contains lacunae and canaliculi as in bone. The lacunae are occupied by cells similar to osteocytes (***cementocytes***). Some parts of cementum are acellular.

Cementum is covered by a fibrous membrane called the ***periodontal membrane*** (or ligament). This membrane may be regarded as the periosteum of the cementum. Collagen fibres from this membrane extend into the cementum, and also into the alveolar bone (forming the socket in which the root lies) as fibres of Sharpey. The periodontal membrane fixes the tooth in its socket. It contains numerous nerve endings that provide sensory information.

The Pulp

Dental pulp lies inner to dentine and occupies the pulp cavity and root canal. The dental pulp is made up of very loose connective tissue resembling embryonic mesenchyme (mucoid tissue). The ground substance is gelatinous and abundant. In it there are many spindle-shaped and star-shaped cells. Delicate collagen fibres, numerous blood vessels, lymphatics and nerve fibres are present. The nerve fibres are partly sensory and partly sympathetic.

Clinical Correlation

Dental caries

- Dental caries is the most common disease of dental tissues, causing destruction of the calcified tissues of the teeth.
- Caries occurs chiefly in the areas of pits and fissures, mainly of the molars and premolars, where food retention occurs, and in the cervical part of the tooth.
- The earliest change is the appearance of a small, chalky-white spot on the enamel which subsequently enlarges and often becomes yellow or brown and breaks down to form carious cavity. Eventually, the cavity becomes larger due to fractures of enamel. Once the lesion reaches enamel-dentine junction, destruction of dentine also begins.

Pulpitis

- It refers to the inflammation of the pulp. It may be acute, which is accompanied by severe pain and requires immediate intervention or it may be chronic where the pulp is exposed widely. It may protrude through the cavity forming polyp of the pulp.

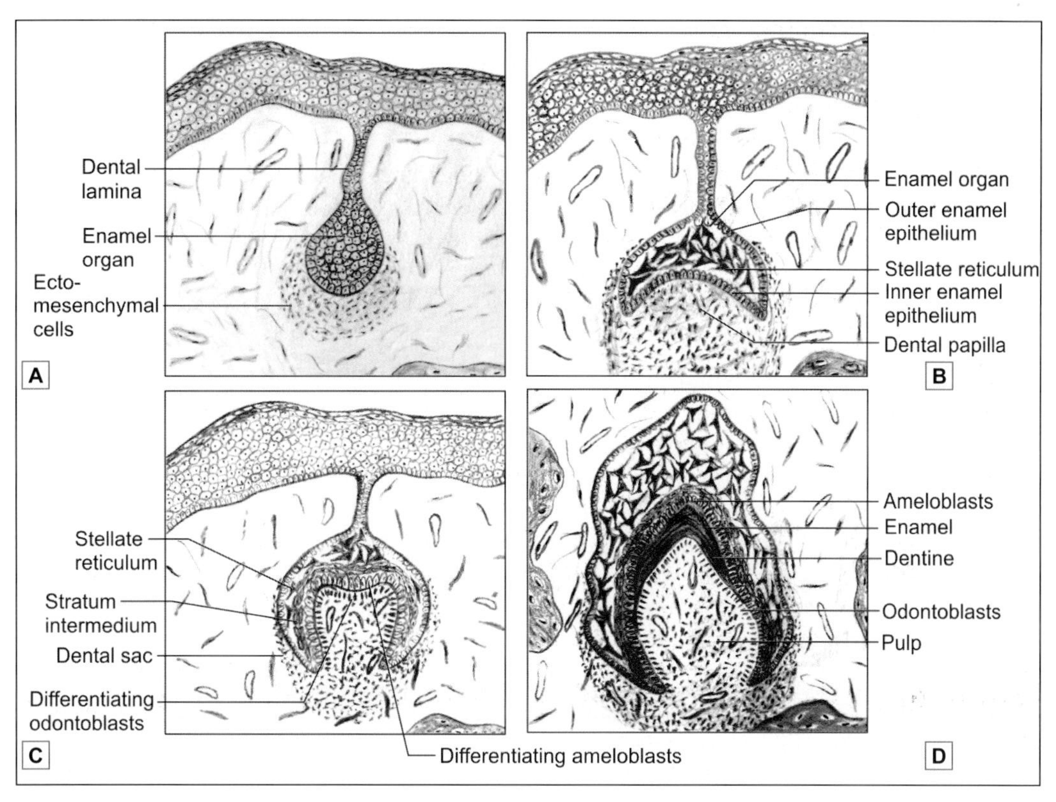

Fig. 15.4: Stages of tooth development. **A.** Bud stage; **B.** Cap stage; **C.** Early bell stage; **D.** Advance bell stage (Schematic representation)

Stages in Tooth Development

Each tooth may be regarded as a highly modified form of the stratified squamous epithelium covering the developing jaw (alveolar process). A thickening of epithelium grows downwards into the underlying connective tissue and enlarges to form an ***enamel organ*** (Fig. 15.4).

Bud Stage (Fig. 15.4A)

The enamel organ resembles a small bud, which is surrounded by the condensation of ectomesenchymal cells. In this stage, the enamel organ is made up of peripherally located low columnar cells and centrally located polygonal cells.

Cap Stage (Fig. 15.4B)

The enamel organ then proliferates to form a cap over the central condensation of ectomesenchymal cells which is called the dental papilla. The dental papilla and the dental sac become well-defined. Three layers are differentiated from the enamel organ.

- Inner dental/inner enamel epithelium
- Stellate reticulum
- Outer dental/outer enamel epithelium

Early Bell Stage (Fig. 15.4C)

The enamel organ resembles bell shape as a result of deepening of the undersurface of the epithelial cap. A cell layer forms in between the inner dental epithelium and stellate reticulum, called the ***stratum intermedium***.

The inner dental epithelium differentiates into tall columnar cells called the ***ameloblasts***. The peripheral cells of the dental papilla differentiate into ***odontoblasts*** under the organizing influence of inner dental epithelium. Ameloblasts form enamel and odontoblasts produce dentine.

Advance Bell Stage (Fig. 15.4D)

In this stage, apposition of dental hard tissues occurs. Odontoblasts are dentine forming cells; ameloblasts are enamel forming cells. A layer of predentine is secreted by the odontoblasts. Both ameloblasts and odontoblasts behave in a way very similar to that of osteoblasts and lay down layer upon layer of enamel, or of dentine. The deposition of enamel and dentine continues until the crown formation is complete.

The formation of layers of enamel and dentine results in separation of ameloblasts and odontoblasts. The original line at which enamel and dentine formation begins is known as the enamel-dentine junction. At last, ameloblasts come to line the external aspect of the enamel. The odontoblasts persist as a lining for the pulp cavity.

Added Information

When examined by EM both ameloblasts and odontoblasts show the features typical of actively secreting cells. They have prominent Golgi complexes and abundant rough ER. The apical part of each cell is prolonged into a process. In the case of odontoblasts, the process runs into the proximal part of a dentineal tubule. In ameloblasts, the projection is called ***Tomes process***. This process contains numerous microtubules, and many secretory vesicles. Other smaller processes are present near the base of Tomes process. The organic matrix of enamel is released mainly by Tomes process, which also appears to be responsible for forming prisms of enamel.

Root Formation

Root formation is carried out by ***Hertwig's epithelial root sheath*** which is formed by the cervical portion of the enamel organ. It molds the shape of the roots and initiates radicular dentine formation.

The root sheath encloses the dental pulp except at apical portion. The rim of root sheath, the ***epithelial diaphragm*** surrounds the ***primary apical foramen***. The root apex remains wide open until about 2 to 3 years after the eruption of the tooth, when the root development is completed.

THE TONGUE

The tongue lies in the floor of the oral cavity. It has a dorsal surface that is free; and a ventral surface that is free anteriorly, but is attached to the floor of the oral cavity posteriorly. The dorsal and ventral surfaces become continuous at the lateral margins, and at the tip (or apex) of the tongue.

Near its posterior end the dorsum of the tongue is marked by a V-shaped groove called the ***sulcus terminalis*** (Fig. 15.5). The apex of the 'V' points backwards and is marked by a depression called the ***foramen caecum***. The limbs of the sulcus terminalis run forwards and laterally. The sulcus terminalis divides the tongue into a larger (2/3) anterior, or oral part; and a smaller (1/3) posterior, or pharyngeal part.

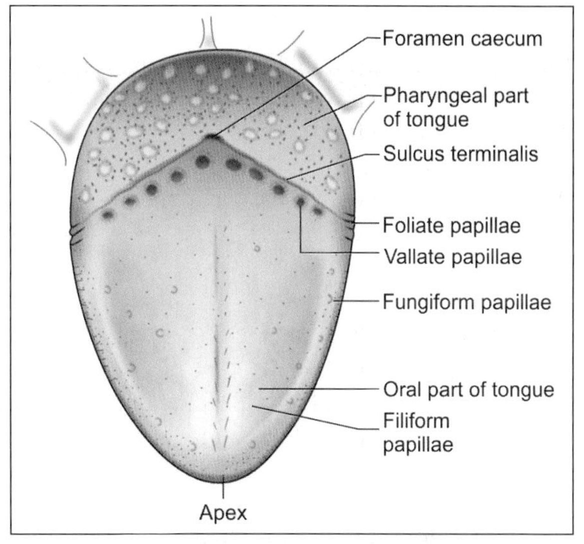

Fig. 15.5: Dorsal surface of the tongue (Schematic representation)

The substance of the tongue is made up chiefly of skeletal muscle supported by connective tissue (Plate 15.2). The muscle is arranged in bundles that run in vertical, transverse and longitudinal directions. This arrangement of muscle permits intricate movements of the tongue associated with the chewing and swallowing of food, and those necessary for speech. The substance of the tongue is divided into right and left halves by a connective tissue septum.

The surface of the tongue is covered by mucous membrane lined by stratified squamous epithelium. The epithelium is supported on a layer of connective tissue. On the undersurface (ventral surface) of the tongue the mucous membrane resembles that lining the rest of the oral cavity, and the epithelium is not keratinised.

The mucous membrane covering the dorsum of the tongue is different over the anterior and posterior parts. Over the part lying in front of the sulcus terminalis the mucosa bears numerous projections or ***papillae***.

Papillae

Each papilla consists of a lining of epithelium and a core of connective tissue. The epithelium over the papillae is partially keratinised (parakeratinised).

The papillae are of various types (Fig. 15.6):

- ***Filiform papillae:*** They are the most numerous papillae. They are small and conical in shape. The epithelium at the tips of these papillae is keratinised. It may project in the form of threads.
- ***Fungiform papillae:*** At the apex of the tongue and along its lateral margins there are larger fungiform papillae with rounded summits and narrower bases. Fungiform papillae bear taste buds (described below). In contrast to the filiform papillae the epithelium on fungiform papillae is not keratinised.
- ***Circumvallate papillae:*** These are the largest papillae of the tongue. They are arranged in a row just anterior to the sulcus terminalis. When viewed from the surface each papilla is seen to have a circular top demarcated from the rest of the mucosa by a groove. In sections through the papilla it is seen that the papilla has a circumferential 'lateral wall' that lies in

PLATE 15.2: Tongue: Anterior Part

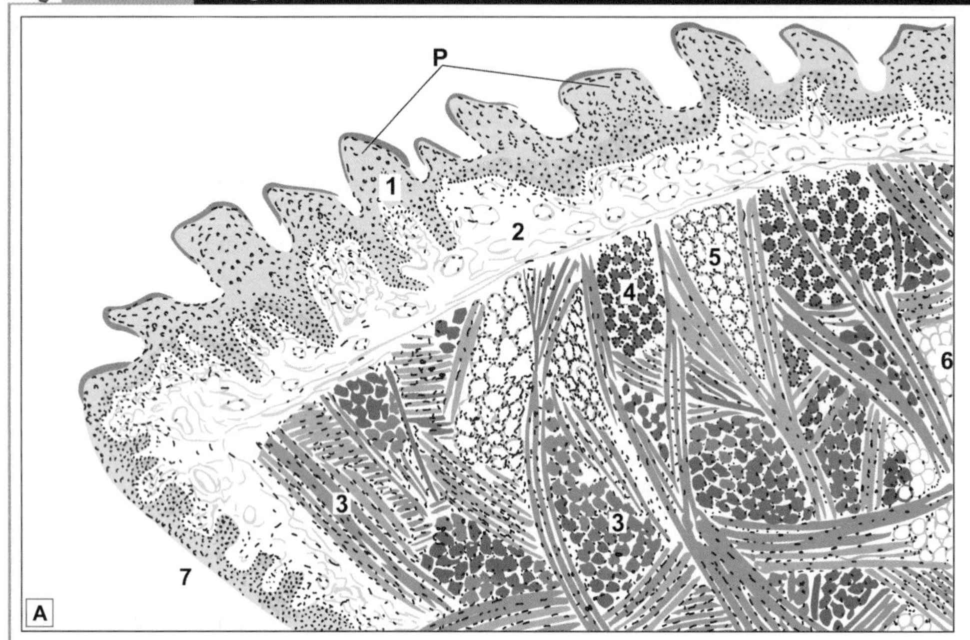

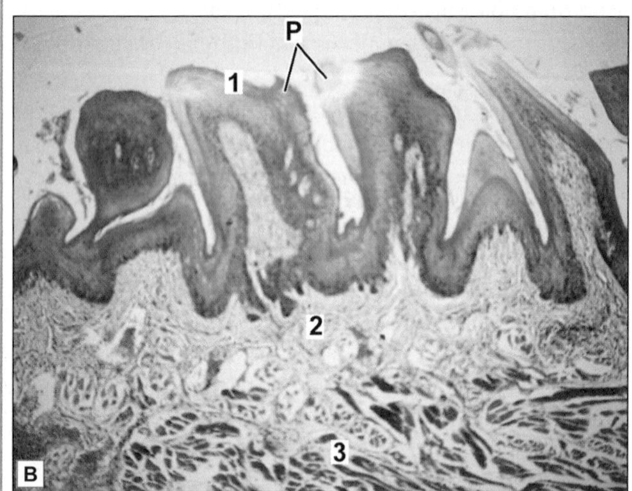

Tongue. A. As seen in drawing; B. Photomicrograph

Key

1. Stratified squamous epithelium
2. Lamina propria
3. Skeletal muscle
4. Serous gland
5. Mucous gland
6. Adipose tissue
7. Smooth ventral surface of tongue

P. Papillae

- The tongue is covered on both surfaces by stratified squamous epithelium (non-keratinised)
- The ventral surface of the tongue is smooth, but on the dorsum the surface shows numerous projections or papillae
- Each papilla has a core of connective tissue covered by epithelium. Some papillae are pointed (filiform), while others are broad at the top (fungiform). A third type of papilla is circumvallate, the top of this papilla is broad and lies at the same level as the surrounding mucosa
- The main mass of the tongue is formed by skeletal muscle seen below the lamina propria. Muscle fibres run in various directions so that some are cut longitudinally and some transversely.
- Numerous serous and mucous glands are present amongst the muscle fibres.

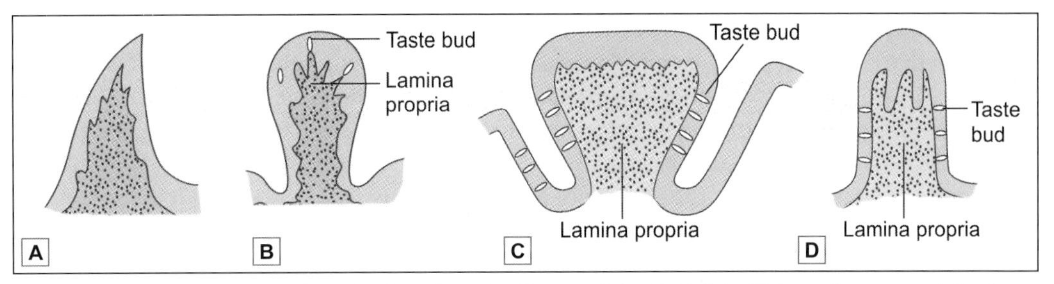

Fig. 15.6: Papillae, **A.** Filliform; **B.** Fungiform; **C.** Circumvallate; **D.** Foliate (Schematic representation)

the depth of the groove (Fig. 15.6 and Plate 15.3). Taste buds are present on this wall, and also on the 'outer' wall of the groove.

- ***Foliate papillae***: These are rudimentary in man. They can be seen along the posterior part of lateral margin of tongue.

Added Information

Another variety of papilla sometimes mentioned in relation to the tongue is the ***papilla simplex***. Unlike the other papillae which can be seen by naked eye, the papillae simplex are microscopic and are quite distinct from other papillae. They are not surface projections at all, but are projections of subjacent connective tissue into the epithelium. In other words these papillae are equivalent to dermal papillae of the skin.

Note: The mucous membrane of the posterior (pharyngeal) part of the dorsum of the tongue bears numerous rounded elevations that are quite different from the papillae described above. These elevations are produced by collections of lymphoid tissue present deep to the epithelium which are collectively called the ***lingual tonsil***.

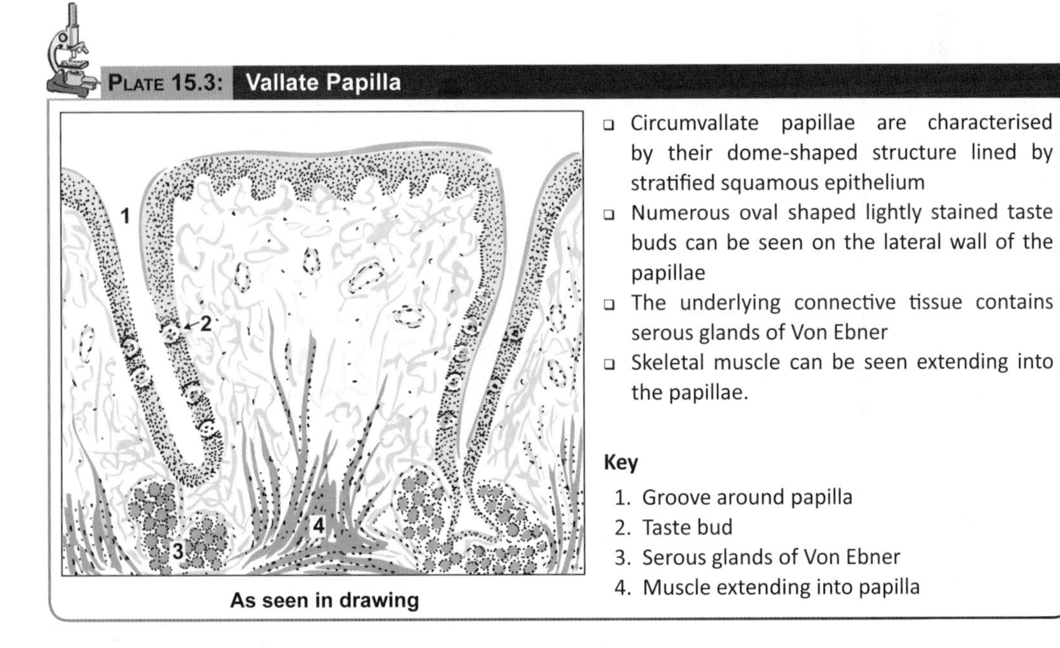

Plate 15.3: Vallate Papilla

- Circumvallate papillae are characterised by their dome-shaped structure lined by stratified squamous epithelium
- Numerous oval shaped lightly stained taste buds can be seen on the lateral wall of the papillae
- The underlying connective tissue contains serous glands of Von Ebner
- Skeletal muscle can be seen extending into the papillae.

Key

1. Groove around papilla
2. Taste bud
3. Serous glands of Von Ebner
4. Muscle extending into papilla

As seen in drawing

- ***Mucous and serous glands:*** Numerous mucous and serous glands are present in the connective tissue deep to the epithelium of the tongue. Mucous glands are most numerous in the pharyngeal part, in relation to the masses of lymphoid tissue. They open into recesses of mucosa that dip into the masses of lymphoid tissue. The serous glands are present mainly in relation to circumvallate papillae, and open into the furrows surrounding the papillae. Serous glands also open in the vicinity of other taste buds. It is believed that the secretions of these glands dissolve the substance to be tasted and spread it over the taste bud; and wash it away after it has been tasted.

The largest glands in the tongue are present on the ventral aspect of the apex. They contain both mucous and serous acini and are referred to as the ***anterior lingual glands***.

Taste Buds

Taste buds are present in relation to circumvallate papillae, to fungiform papillae, and to leaf-like folds of mucosa (***folia linguae***) present on the posterolateral part of the tongue. Taste buds are also present on the soft palate, the epiglottis, the palatoglossal arches, and the posterior wall of the oropharynx.

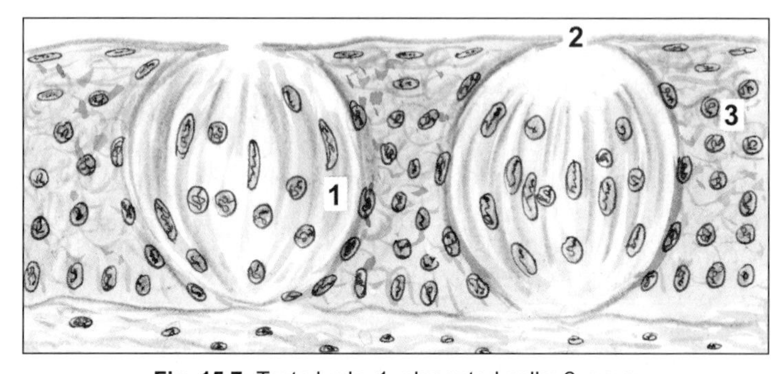

Fig. 15.7: Taste buds. 1–elongated cells; 2–pore; 3–stratified squamous epithelium; (Schematic representation)

Each taste bud is a piriform structure made up of modified epithelial cells (Fig. 15.7). It extends through the entire thickness of the epithelium. Each bud has a small cavity that opens to the surface through a ***gustatory pore***. The cavity is filled by a material rich in polysaccharides.

The cells present in taste buds are elongated and are vertically orientated, those towards the periphery being curved like crescents (Fig. 15.8). Each cell has a central broader part containing the nucleus, and tapering ends. The cells are of two basic types. Some of them are ***receptor cells*** or ***gustatory cells***. Endings of afferent nerves end in relation to them. Other cells perform a supporting function and are called ***supporting cells***.

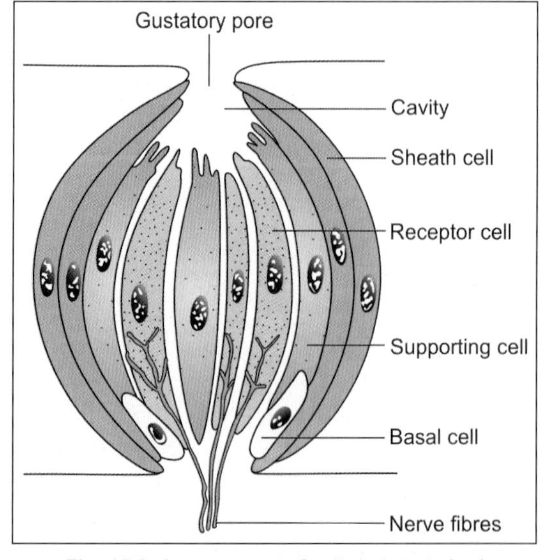

Fig. 15.8: Arrangement of cells in a taste bud (Schematic representation)

Note: However, it is by no means easy to distinguish between receptor and supporter cells, the essential difference being the presence of innervation. Early observers using the light microscope found hairs at the tips of some cells and concluded that these were the receptor cells. However, this has not been confirmed by EM studies. The latter have shown that the 'hair' seen with the light microscope are microvilli that are more common on supporting cells rather on receptors. Two types of receptor cells can be distinguished on the basis of the vacuoles present in them.

Added Information

Supporting cells are probably of three types. Some of them that lie at the periphery of the taste bud form a sheath for it. Those near the centre of the bud are truly supporting. They probably secrete a material that fills the cavity at the apex of the taste bud. Microvilli are often present at the tips of these cells. A third variety of supporting cell is seen in the basal part of the bud. These basal cells multiply and produce new supporting and receptor cells to replace those that are worn out. This may be correlated with the fact that cells of taste buds have a short life and are continuously being replaced.

Recognition of Various Tastes by Tongue

It has been held that taste buds in different parts of the tongue may respond best to particular modalities of taste. However, it is now known that the same taste bud can respond to different types of taste (sweet, sour, salt and bitter) and that taste is a complicated sensation depending upon the overall pattern of responses from taste buds all over the tongue. With this reservation in mind, we may note that sweet taste is best appreciated at the tip of the tongue, salt by the area just behind the tip and along the lateral border, and bitter taste by circumvallate papillae.

Clinical Correlation

Fissured Tongue

It is a genetically-determined condition characterised by numerous small furrows or grooves on the dorsum of the tongue.

Hairy Tongue

In this condition, the filiform papillae are hypertrophied and elongated. These 'hairs' (papillae) are stained black, brown or yellowish-white by food, tobacco, oxidising agents or by oral flora.

SALIVARY GLANDS

There are two main groups of salivary glands—major and minor. The major salivary glands are the three paired glands: parotid, submandibular and sublingual. The parotid glands are located laterally to the mandibular ramus and its main duct drains into the oral cavity opposite the second maxillary molar. The submandibular glands are present in the floor of the mouth, superior to digastric muscles. The sublingual gland lie anterior to submandibular glands. The ducts of submandibular and sublingual glands empty in the floor of the mouth.

The minor salivary glands are numerous and are widely distributed in the mucosa of oral cavity. Some of the minor salivary glands are the Von Ebner's gland in tongue, buccal glands in cheeks and labial glands in lips.

The secretions of these glands help to keep the mouth moist, and provide a protective and lubricant coat of mucous. Some enzymes (amylase, lysozyme), and immunoglobulin IgA are also present in the secretions.

Structural Organisation

Basically, a salivary gland consists of stroma, parenchyma and a duct system which carries the secretions into the oral cavity.

Stroma

- The stroma consists of connective tissue capsule and septa.
- Numerous septa arise from the capsule and enter the parenchyma of the gland, dividing the gland into numerous lobules.
- These septa bring the blood vessels and nerves into the gland. Large ducts of the glands are also present in it.

Parenchyma

- Parenchyma has two components: the secretory part and conducting part.

Secretory Part

Salivary glands are compound tubuloalveolar glands (racemose glands). Their secretory elements (also referred to as ***end pieces*** or as the ***portio terminalis***) may be rounded (acini), pear shaped (alveoli), tubular, or a mixture of these (tubuloacinar, tubuloalveolar). The secretory elements lead into a series of ducts through which their secretions are poured into the oral cavity.

In sections through salivary glands we see a large number of closely packed aciniwith ducts scattered between them (Plates 15.3, 15.4 and 15.6). These elements are supported by connective tissue that also divides the glands into lobules, and forms capsules around them. Blood vessels, lymphatics and nerves run in the connective tissue that may at places contain some adipose tissue.

The acini are made up of either ***serous*** or ***mucous*** cells.

A salivary gland may have only one type of acini or there may be a mix of both serous and mucous acini, these are called mixed acini. A secretory unit, or gland, with only one type of cell (serous or mucous) is said to be ***homocrine.*** If it contains more than one variety of cells it is said to be ***heterocrine.***

The structure of mucous and serous acini has already been described and compared in chapter 4. It would be worth while if your revise that chapter now.

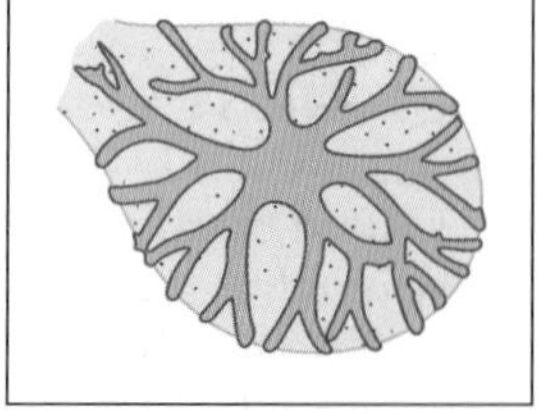

Fig. 15.9: Surface view of an acinus showing myoepithelial cell. Its processes form a basket around the acinus (Schematic representation)

Myoepithelial Cells

Myoepithelial cells are present in relation to acini and intercalated ducts of salivary glands. They may also be seen in relation to larger ducts (intralobular and extralobular). These cells lie between the epithelial cells and their basement membrane. The myoepithelial cells located on acini are often branched (stellate) and may form 'baskets' around the acini. Those located on the ducts are fusiform and run longitudinally along them.

Conducting part:

Duct System

Secretions produced in acini pass along a system of ducts, different parts of which have differing structure. The smallest ducts are called ***intercalated ducts***. These are lined by cuboidal or flattened cells. Intercalated ducts open into ***striated ducts*** lined by columnar cells. They are so called because the basal parts of the cells show vertical striations. Both intercalated and striated ducts are intralobular ducts. Striated ducts open into ***excretory ducts*** (interlobular) that are lined by simple columnar epithelium.

Clinical Correlation

Sialorrhoea (Ptyalism) : Increased flow of saliva is termed sialorrhoea or ptyalism.

Xerostomia : Decreased salivary flow is termed xerostomia.

Sialadenitis : Inflammation of salivary glands is called as sialadenitis.

Tumours of Salivary Glands

- **Pleomorphic adenoma (mixed salivary tumour)**: It is the most common tumour of major and minor salivary gland. It is characterised by pleomorphic or mixed appearance in which there are epithelial elements present in a matrix of mucoid, myxoid and chondroid tissue.
- **Mucoepidermoid carcinoma**: It is the most common malignant salivary gland tumour. The tumour is composed of combination of 4 types of cells: mucin-producing, squamous, intermediate and clear cells. Well-differentiated tumours have predominance of mucinous cells, while poorly differentiated have more solid and infiltrative pattern.

Added Information

Cells of Salivary Glands

Serous Cells

- Serous cells are usually arranged in the form of rounded acini. As a result each cell is roughly pyramidal having a broad base (towards the basement membrane) and a narrow apex (towards the lumen) (Fig. 15.10). Some microvilli and pinocytotic vesicles are seen at the apex of the cell. The lumen of the acinus often extends for some distance between adjacent cells: these extensions are called ***intercellular secretory canaliculi***. Deep to these canaliculi the cell membranes of adjoining cells are united by tight junctions. Deep to these junctions, the lateral cell margins show folds that interdigitate

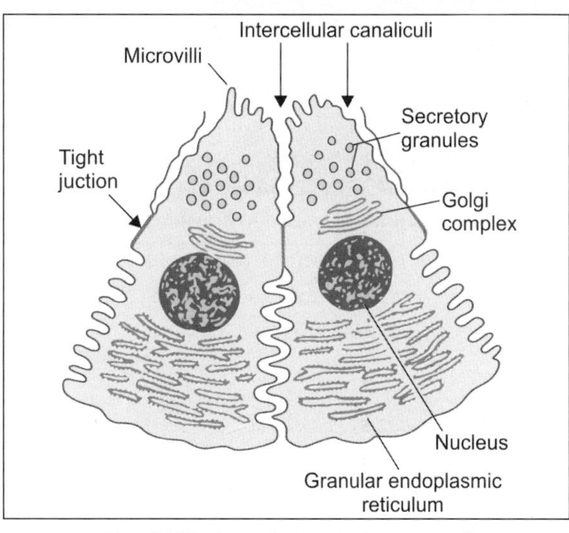

Fig. 15.10: Some features of serous cell in a salivary gland (Schematic representation)

PLATE 15.4: Parotid Gland

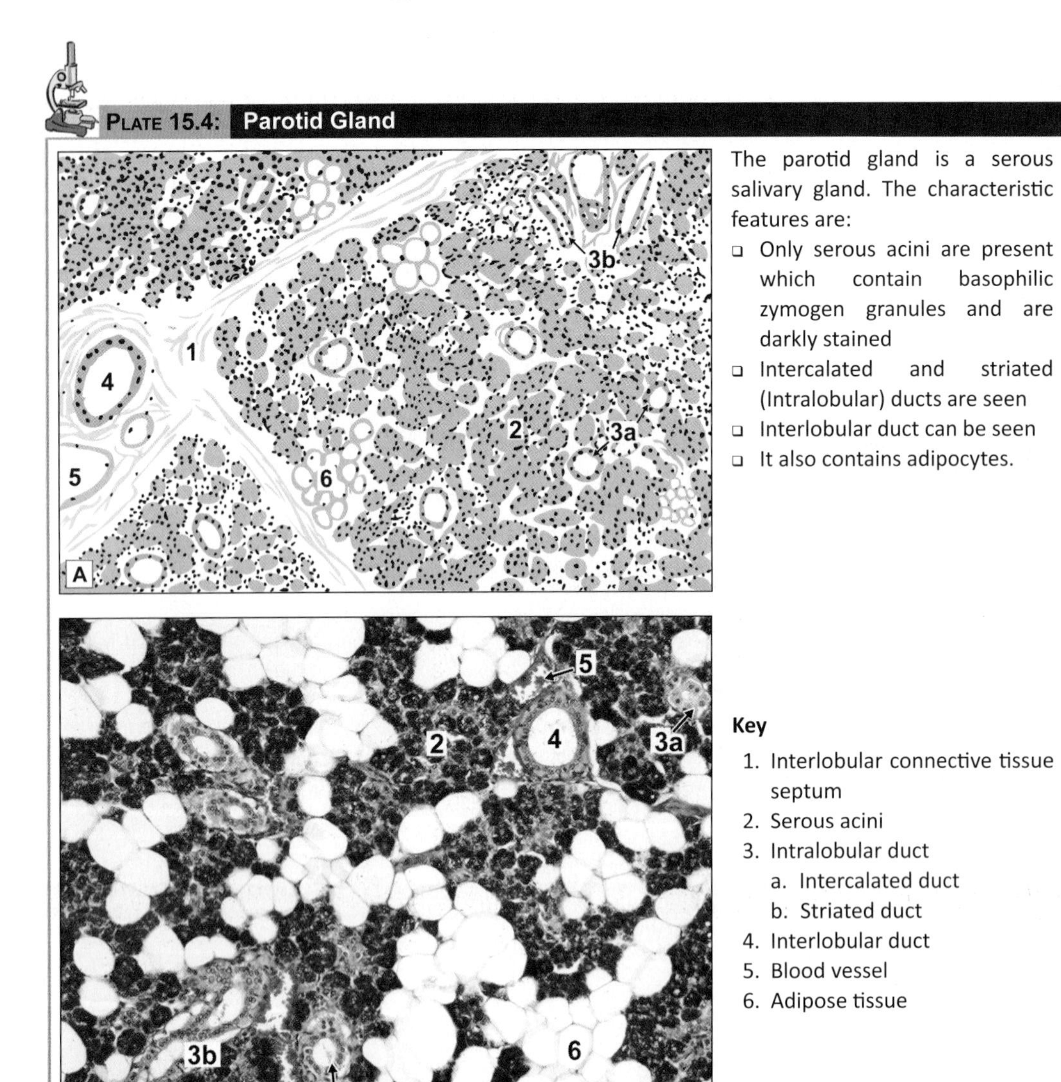

The parotid gland is a serous salivary gland. The characteristic features are:

- Only serous acini are present which contain basophilic zymogen granules and are darkly stained
- Intercalated and striated (Intralobular) ducts are seen
- Interlobular duct can be seen
- It also contains adipocytes.

Key

1. Interlobular connective tissue septum
2. Serous acini
3. Intralobular duct
   a. Intercalated duct
   b. Striated duct
4. Interlobular duct
5. Blood vessel
6. Adipose tissue

Parotid gland. A. As seen in drawing; B. Photomicrograph

PLATE 15.5: Submandibular Gland

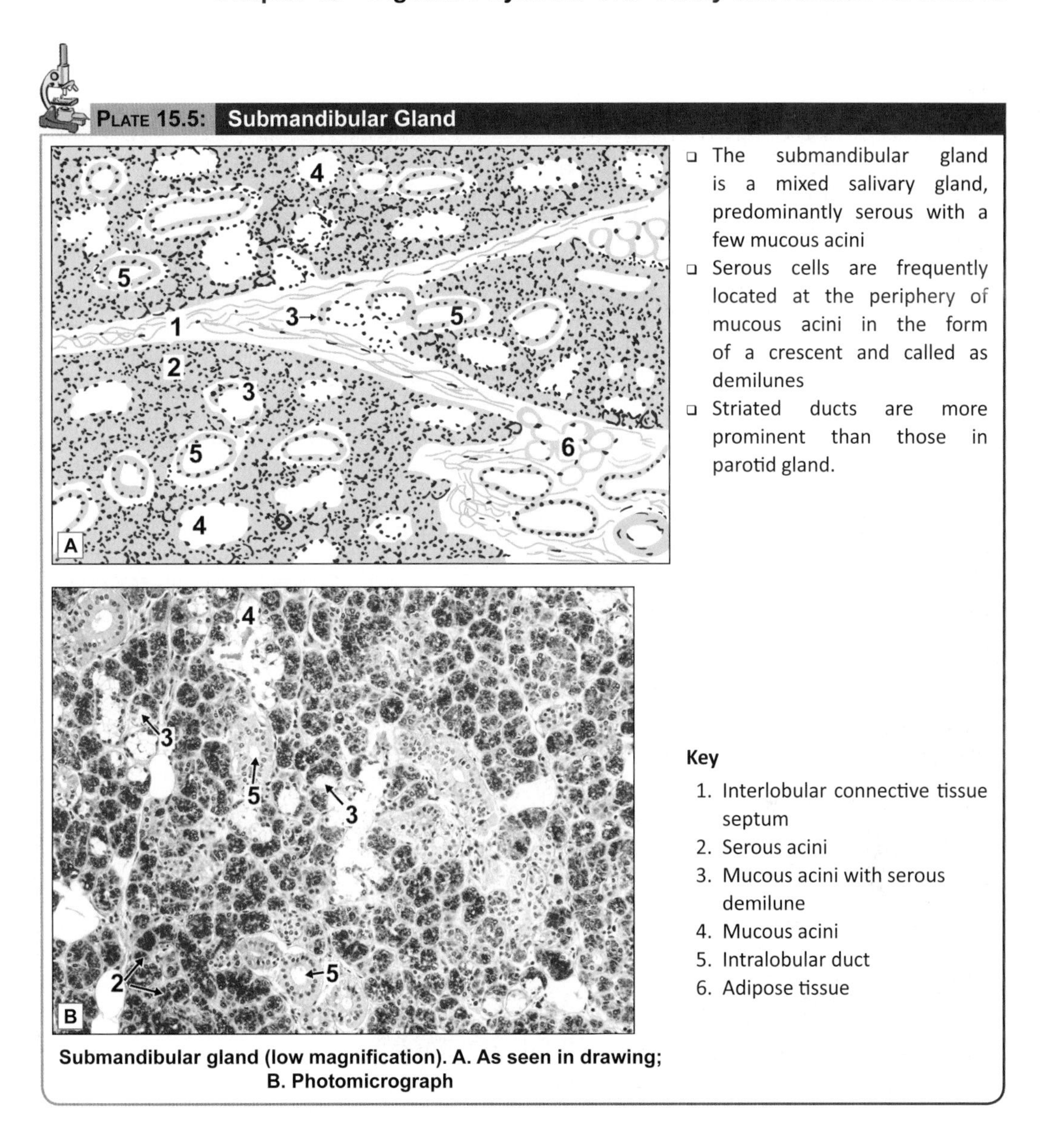

Submandibular gland (low magnification). A. As seen in drawing; B. Photomicrograph

- The submandibular gland is a mixed salivary gland, predominantly serous with a few mucous acini
- Serous cells are frequently located at the periphery of mucous acini in the form of a crescent and called as demilunes
- Striated ducts are more prominent than those in parotid gland.

Key

1. Interlobular connective tissue septum
2. Serous acini
3. Mucous acini with serous demilune
4. Mucous acini
5. Intralobular duct
6. Adipose tissue

Plate 15.6: Submandibular Gland (High Magnification)

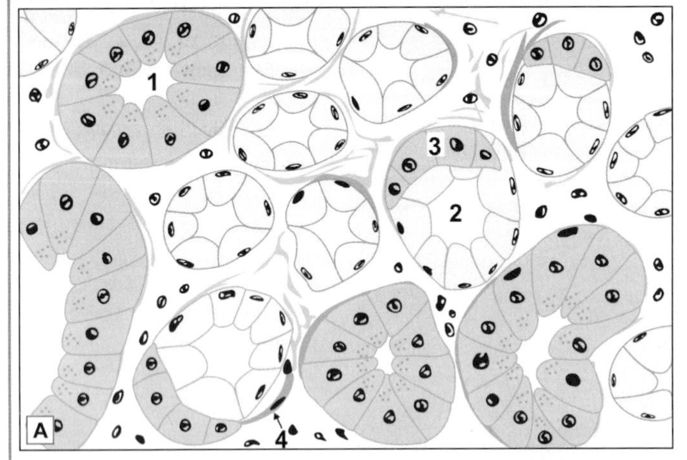

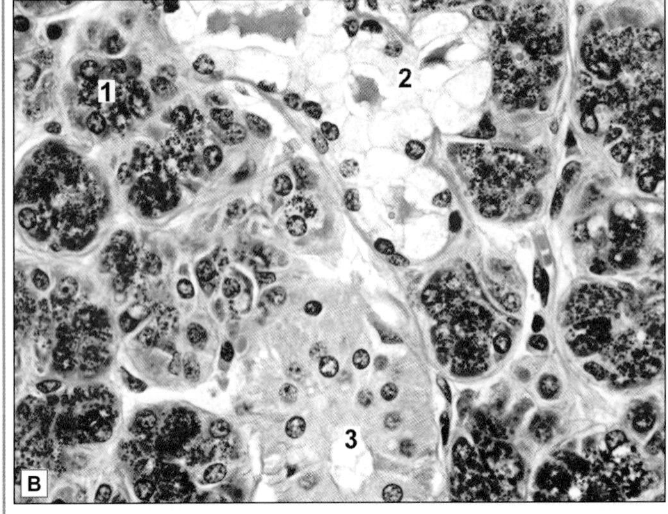

Submandibular gland (high magnification). A. As seen in drawing; B. Photomicrograph

- In the high power view the serous and mucous acini can be identified by their staining reaction, and shape and position of nucleus
- The serous acini are darkly stained and have rounded nucleus placed near the centre of the cell
- The mucous acini are lightly stained with flat nucleus placed towards the basement membrane
- The mucous acini are often associated with darkly stained crescentic patch of serous cells called serous demilune.

Key

1. Serous acini
2. Mucous acini
3. Serous demilune
4. Myoepithelial cells

Plate 15.7: Sublingual Gland

- The sublingual gland is predominantly a mucous gland but few serous acini may also be seen
- Serous demilunes may be present

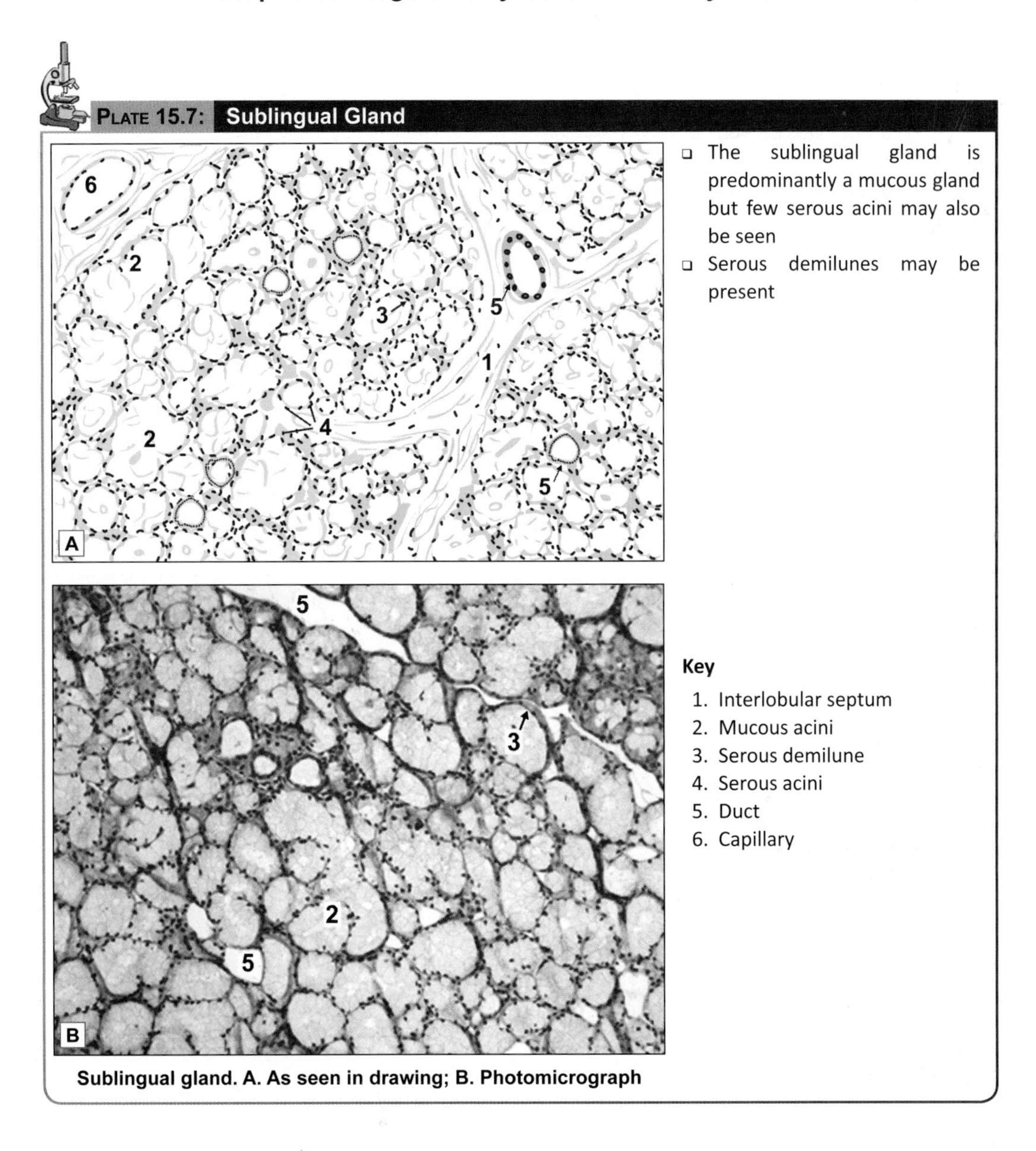

Key

1. Interlobular septum
2. Mucous acini
3. Serous demilune
4. Serous acini
5. Duct
6. Capillary

Sublingual gland. A. As seen in drawing; B. Photomicrograph

with those of adjoining cells. The apical cytoplasm contains secretory granules that are small, homogeneous, and electron dense. The cytoplasm also contains a prominent Golgi complex and abundant rough endoplasmic reticulum, both features indicating considerable synthetic activity. Mitochondria, lysosomes, and microfilaments are also present.

Mucous Cells

Mucous cells are usually arranged in the form of tubular secretory elements (Fig. 15.11). Crescents present in relation to them are located at the ends of the tubules. The cells lining mucous cells tend to be columnar rather than pyramidal. Their secretory granules are large and ill defined. Rough endoplasmic reticulum and Golgi complex are similar to those in serous cells, but microvilli, foldings of plasma membrane, and intercellular canaliculi are not usually seen.

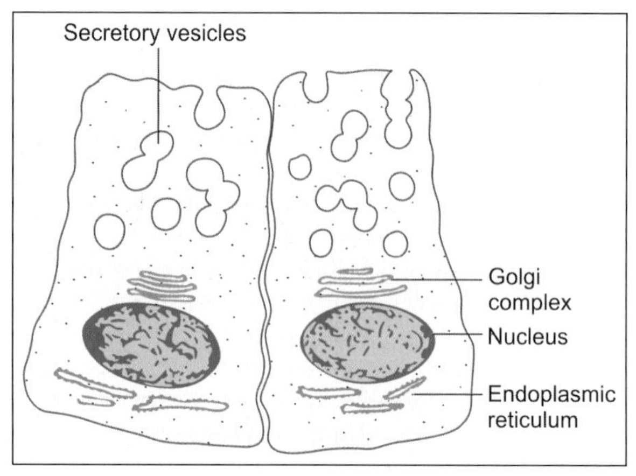

Fig. 15.11: Some features of mucous cells in salivary glands (Schematic representation)

Seromucous Cells

From the point of view of ultrastructure many cells of salivary glands are intermediate between serous and mucous cells. They are referred to as ***seromucous cells***. Most of the cells identified as serous with light microscopy in the parotid and submandibular glands are really seromucous.

The secretions of all types of salivary secretory cells contain protein-carbohydrate complexes. Their concentration is lowest in cases of serous cells, very high in mucous cells, and with widely differing concentrations in seromucous cells.

We have seen that in the submandibular glands mucous acini are often capped by serous demilunes. The serous cells of a demilune drain into the lumen of the acinus through fine canaliculi passing through the intervals between mucous cells.

With the EM myoepithelial cells are seen to contain the usual organelles. In addition they have conspicuous filaments that resemble myofilaments of smooth muscle cells. These filaments are numerous in processes arising from the cell. Cilia are present on some myoepithelial cells. It has been suggested that the cilia may subserve a sensory or chemoreceptor function.

Myoepithelial cells are contractile, their contraction helping to squeeze out secretion from acini.

The cells lining the striated ducts show an interesting ultrastructure (Fig. 15.12). Their basal striations are seen to be due to the presence of numerous deep infoldings of the basal parts of the cell membranes. Numerous elongated mitochondria are present in the intervals between the folds. Similar cells are also present scattered in the epithelium of the excretory ducts. These cells are believed to play a

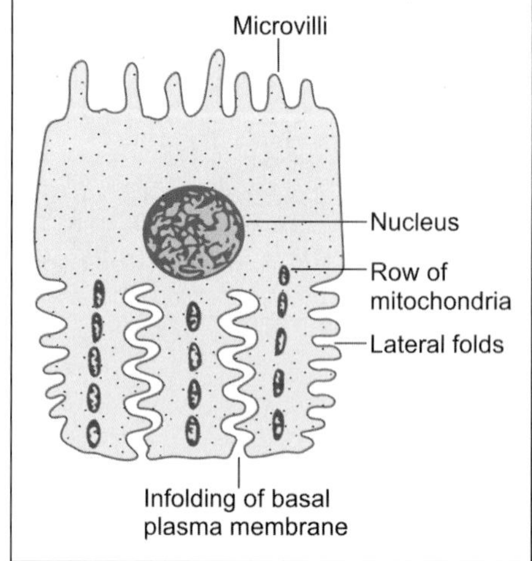

Fig. 15.12: Electron microscopic structure of a cell from striated duct (Schematic representation)

contd...

role in regulating the water and electrolyte content of saliva to make it hypotonic. Immunoglobulin A, produced by plasma cells lying subjacent to the epithelium, passes into saliva through the cells lining the striated ducts.

Innervation of Salivary Glands

Secretion by salivary glands is under hormonal as well as neural control. A local hormone ***plasmakinin*** formed by secretory cells influences vasodilation. Salivary glands are innervated by autonomic nerves, both parasympathetic (cholinergic) and sympathetic (adrenergic). Parasympathetic nerves travel to secretory elements along ducts, while sympathetic nerves travel along arteries. Synaptic contacts between nerve terminals and effector cells form ***neuro-effector junctions***.

Two types of junction, ***epilemmal*** and ***hypolemmal***, are present. At epilemmal junctions the nerve terminal is separated from the secretory or effector cell by the basal lamina. At hypolemmal junctions the nerve terminal pierces the basal lamina and comes into direct contact with the effector cell. Nerve impulses reaching one effector cell spread to others through intercellular contacts. Classically, salivary secretion has been attributed to parasympathetic stimulation. While this is true, it is believed that sympathetic nerves can also excite secretion either directly, or by vasodilation. Autonomic nerves not only stimulate secretion, but also appear to determine its viscosity and other characteristics. Autonomic nerve terminals are also seen on myoepithelial cells and on cells lining the ducts of salivary glands. The latter probably influence reabsorption of sodium by cells lining the ducts. Salivary glands are sensitive to pain, and must therefore have a sensory innervation as well.

Chapter 16

Digestive System: Oesophagus, Stomach and Intestines

The gastrointestinal tract (GIT) or alimentary canal is a long muscular tube that begins at the oral cavity and ends in the anus. Different parts of the tract are specialised to perform different functions, and hence structural modifications are seen in various parts of the GIT.

The oesophagus and anal canal are merely transport passages. The part of the alimentary canal from the stomach to the rectum is the proper digestive tract, responsible for digestion and absorption of food. Reabsorption of secreted fluids is an important function of the large intestine.

GENERAL STRUCTURE OF GIT

The structure of the alimentary canal, from the oesophagus up to the anal canal, shows several features that are common to all these parts. We shall consider these common features before examining the structure of individual parts of the canal.

The walls of the oral cavity and pharynx are partly bony, and partly muscular. From the upper end of the oesophagus up to the lower end of the anal canal the alimentary canal has the form of a fibromuscular tube. The wall of the tube is made up of the following layers (from inner to outer side) (Fig. 16.1).

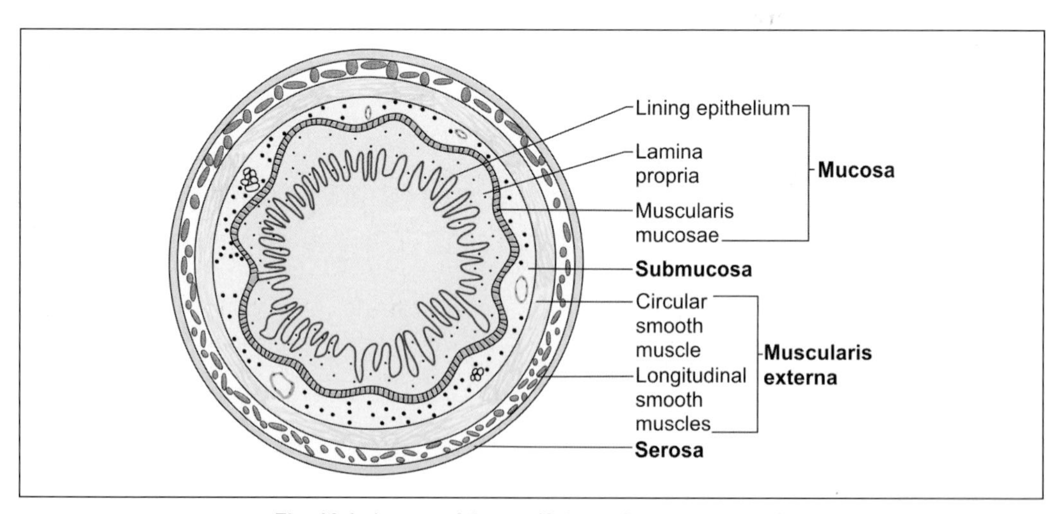

Fig. 16.1: Layers of the gut (Schematic representation)

- The innermost layer is the ***mucous membrane*** that is made up of:
  - A lining epithelium
  - A layer of connective tissue, the ***lamina propria***, that supports the epithelium
  - A thin layer of smooth muscle called the ***muscularis mucosae***.
- The mucous membrane rests on a layer of loose areolar tissue called the ***submucosa***.
- The gut wall derives its main strength and form because of a thick layer of muscle (***muscularis externa***) that surrounds the submucosa.
- Covering the muscularis externa there is a ***serous layer*** or (alternatively) an ***adventitial layer***.

Primarily, it is the mucosa in which changes are seen in the alimentary tract; the other layers remain almost the same.

The Mucosa

The Lining Epithelium

The lining epithelium is columnar all over the gut; except in the oesophagus, and in the lower part of the anal canal, where it is stratified squamous. This stratified squamous epithelium has a protective function in these situations. The cells of the more typical columnar epithelium are either absorptive or secretory.

The epithelium of the gut presents an extensive absorptive surface. The factors contributing to the extent of the surface are as follows:

- The ***considerable length*** of the alimentary canal, and specially that of the small intestine.
- The presence of ***numerous folds*** involving the entire thickness of the mucous membrane. These folds can be seen by naked eye. The submucosa extends into the folds.
- At numerous places the epithelium dips into the lamina propria forming ***crypts*** (see below).
- In the ***small intestine*** the mucosa bears numerous finger-like processes that project into the lumen. These processes are called ***villi***. Each villus has a surface lining of epithelium and a core formed by an extension of the connective tissue of the lamina propria. The luminal surfaces of the epithelial cells bear numerous microvilli.

The epithelium of the gut also performs a very important secretory function. The secretory cells are arranged in the form of numerous glands as follows:

- Some glands are unicellular, the secretory cells being scattered among the cells of the lining epithelium.
- In many situations, the epithelium dips into the lamina propria forming simple tubular glands (These are the crypts referred to above).
- In other situations (e.g., in the oesophagus, duodenum) there are compound tubuloalveolar glands lying in the submucosa. They open into the lumen of the gut through ducts traversing the mucosa.
- Finally, there are the pancreas and the liver that form distinct organs lying outside the gut wall. They pour their secretions into the lumen of the gut through large ducts (In this respect, these glands are similar to the salivary glands). The liver and pancreas, and most of the salivary glands, are derivatives of the epithelial lining of the gut. Embryologically, this epithelium is derived from endoderm.

The Lamina Propria

The lamina propria is made up of collagen and reticular fibres embedded in a glycosaminoglycan matrix. Some fibroblasts, blood capillaries, lymph vessels, and nerves are seen in this layer. In the small intestine the lamina propria forms the core of each villus. It surrounds and supports glandular elements and the overlying epithelium.

Prominent aggregations of lymphoid tissue (as well as scattered lymphocytes) are present in the lamina propria. Some of this lymphoid tissue extends into the submucosa and is called as gut associated lymphoid tissue (GALT).

The Muscularis Mucosae

This is a thin layer of smooth muscle that separates the connective tissue of the lamina propria from the submucosa. It consists of an ***inner layer*** in which the muscle fibres are arranged ***circularly*** (around the lumen) and an ***outer layer*** in which the muscle fibres run ***longitudinally***. The muscularis mucosae extends into mucosal folds, but not into villi. Contractions of the muscularis mucosae are important for the local mixing of intestinal contents.

The Submucosa

This layer of loose areolar tissue connects the mucosa to the muscularis externa. Its looseness permits some mobility of the mucosa over the muscle. Numerous blood vessels, lymphatics and nerve fibres traverse the submucosa. Smaller branches arising from them enter the mucous membrane.

The Muscularis Externa

Over the greater part of the gut the muscularis externa consists of smooth muscle. ***The only exception is the upper part of the oesophagus where this layer contains striated muscle fibres.*** Some striated muscle fibres are also closely associated with the wall of the anal canal.

The muscle layer consists (typically) of an ***inner layer of circularly arranged muscle fibres, and an outer longitudinal layer.***

Both layers really consist of spirally arranged fasciculi, the turns of the spiral being compact in the circular layer, and elongated in the longitudinal layer.

The arrangement of muscle fibres shows some variation from region to region. In the stomach an additional oblique layer is present. In the colon the longitudinal fibres are gathered to form prominent bundles called the ***taenia coli***.

Localised thickenings of circular muscle fibres form ***sphincters*** that can occlude the lumen of the gut. For example, the ***pyloric sphincter*** is present around the pyloric end of the stomach, and the ***internal anal sphincter*** surrounds the anal canal. A functional sphincter is seen at the junction of the oesophagus with the stomach. A valvular arrangement at the ileocaecal junction (ileocaecal valve) prevents regurgitation of caecal contents into the ileum.

The Serous and Adventitial Layers

Covering the muscle coat, there is the serous layer which is the outermost layer of the alimentary canal. This layer is merely the visceral peritoneum that covers most parts of the gastrointestinal tract. In some places where a peritoneal covering is absent (e.g., over part of the oesophagus) the muscle coat is covered by an adventitia made up of connective tissue.

Nerve Plexuses

The gut is richly supplied with nerves. A number of nerve plexuses are present as follows:

- The ***myenteric plexus*** (***of Auerbach***) lies between the circular and longitudinal coats of muscularis externa.
- The ***submucosal plexus*** (***of Meissner***) lies in the submucosa (near its junction with the circular muscle layer).
- A third plexus is present near the muscularis mucosae.

The nerve fibres in these plexuses are both afferent and efferent. The efferent fibres supply smooth muscle and glands. Postganglionic neurons meant for these structures lie amongst the nerve fibres forming these plexuses.

THE OESOPHAGUS

The oesophagus is a long muscular tube beginning at the end of cricoid cartilage and opens into the cardiac end of stomach. It conducts chewed food (bolus) and liquids to stomach.

MICROSCOPIC FEATURES

The wall of oesophagus has the usual four layers viz., mucosa, submucosa, muscularis externa and an external adventitia (Fig. 16.2 and Plate 16.1). The oesophagus does not have a serous covering except over a short length near its lower end.

The Mucosa

- The mucous membrane of the oesophagus shows several longitudinal folds that disappear when the tube is distended.
- The mucosa is lined by stratified squamous epithelium, which is normally not keratinised.
- Occasional melanocytes and endocrine cells are present. A columnar epithelium, similar to that lining the cardiac end of the stomach, may extend for some distance into the abdominal part of the oesophagus.
- Finger-like processes (or papillae) of the connective tissue of the lamina propria project into the epithelial layer (just like dermal papillae). This helps to prevent separation of epithelium from underlying connective tissue.

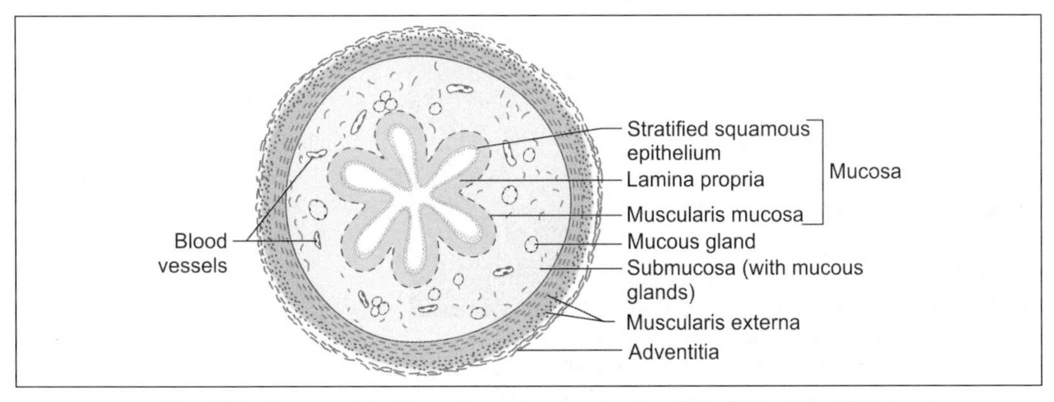

Fig. 16.2: Transverse section of oesophagus showing all the four layers of wall. The lumen of oesophagus is star shaped (Schematic representation)

PLATE 16.1: Oesophagus

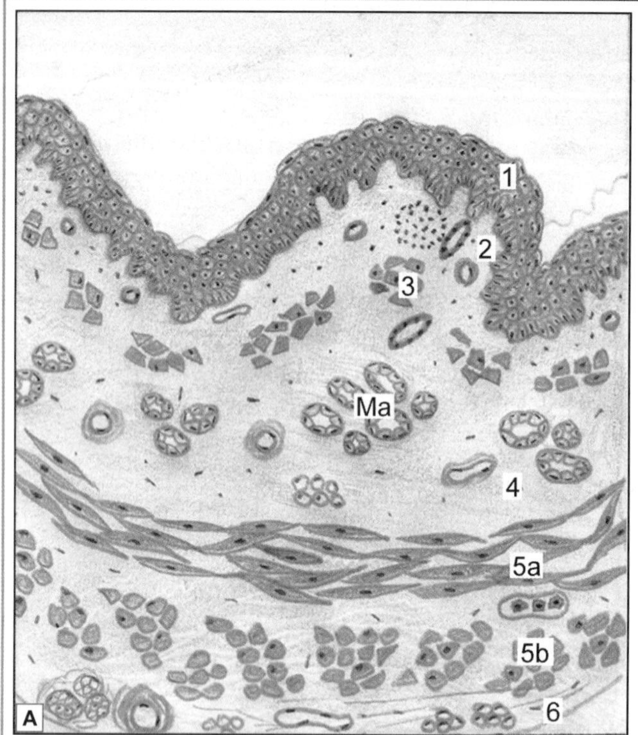

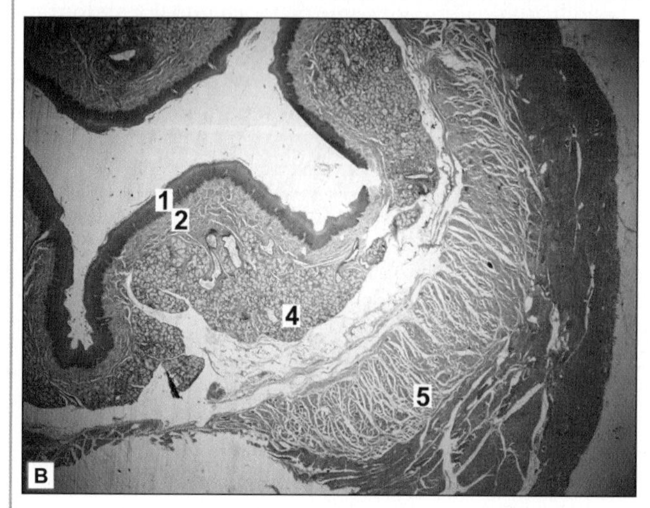

Oesophagus. A. As seen in drawing [to be provided by author]; B. Photomicrograph

In transverse section the oesophagus shows the following layers:

- Lining of non-keratinised stratified squamous epithelium
- The underlying connective tissue of the lamina propria
- The muscularis mucosae in which the muscle fibres are cut transversely
- The lining epithelium, lamina propria and muscularis mucosa collectively constitute the mucosa
- The submucosa having esophageal glands (mucous acini)
- The layer of circular muscle, and the layer of longitudinal muscle constituting the muscularis externa. In muscularis externa the muscle is of the striated variety in the upper one-third of the oesophagus, mixed in the middle one-third, and smooth in the lower one third.

Note: In the photomicrograph muscularis mucosa cannot be differentiated.

Key

1. Muscosa lined by stratified squamous epithelium
2. Lamina propria
3. Muscularis mucosa
4. Submucosa displaying mucous acini.
5. Muscularis externa
   a. Inner circular layer
   b. Outer circular layer
6. Adventitia

Ma. Mucous acini.

- At the upper and lower ends of the oesophagus some tubuloalveolar mucous glands are present in the lamina propria.
- The muscularis mucosae is absent or poorly developed in the upper part of the oesophagus. It is distinct in the lower part of the oesophagus, and is thickest near the eosophagogastric junction. It consists chiefly of longitudinal muscular fibres, but a few circular fibres are also present.

The Submucosa

The only special feature of the submucosa is the presence of compound tubuloalveolar mucous glands. Small aggregations of lymphoid tissue may be present in the submucosa, specially near the lower end. Some plasma cells and macrophages are also present.

The Muscularis Externa

The muscle layer consists of the usual circular and longitudinal layers. However, it is unusual in that the muscle fibres are partly striated and partly smooth. In the upper one-third (or so) of the oesophagus the muscle fibres are entirely of the striated variety, while in the lower one-third all the fibres are of the smooth variety. Both types of fibres are present in the middle one-third of the oesophagus.

Added Information

The circular muscle fibres present at the lower end of the oesophagus could possibly act as a sphincter guarding the cardiooesophageal junction. However, the circular muscle is not thicker here than elsewhere in the oesophagus, and its role as a sphincter is not generally accepted. However, a ***physiological sphincter*** does appear to exist. The anatomical factors that could account for this sphincteric action are not agreed upon.

The Adventitia

The muscle layer of the oesophagus is surrounded by dense fibrous tissue that forms an adventitial coat for the oesophagus. The lowest part of the oesophagus is intra-abdominal and has a covering of peritoneum.

Pathological Correlation

- **Achalasia (Cardiospasm):** Achalasia of the oesophagus is a neuromuscular dysfunction due to which the cardiac sphincter fails to relax during swallowing and results in progressive dysphagia and dilatation of the oesophagus (mega-oesophagus).
- **Barrett's Oesophagus:** This is a condition in which, following reflux oesophagitis, stratified squamous epithelium of the lower oesophagus is replaced by columnar epithelium (columnar metaplasia). The condition is seen more commonly in later age and is caused by factors producing gastro-oesophageal reflux disease.

THE STOMACH

Stomach is a muscular bag that receives food bolus from oesophagus. The food passes through the oesophagus and enters the stomach where it is converted into a thick paste known as ***chyme***. Anatomaically, stomach is divided into four regions: ***Cardia, fundus, body and pylorus*** (Fig. 16.3).

Histologically fundus and body of stomach have a similar structure.

MICROSCOPIC FEATURES

The wall of the stomach has the four basic layers a mucous membrane, a submucosa, a muscularis externa, and a serous layer. The mucous membrane and the muscularis externa have some special features that are described below.

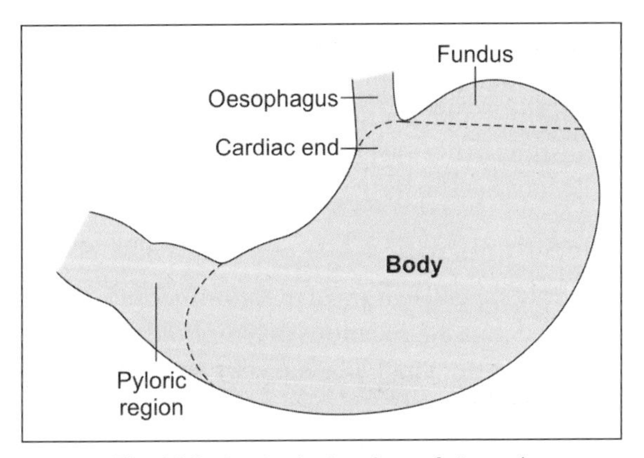

Fig. 16.3: Anatomical regions of stomach (Schematic representation)

The Mucous Membrane

As seen with the naked eye the mucous membrane shows numerous folds (or ***rugae***) that disappear when the stomach is distended.

Lining Epithelium

The lining epithelium is columnar and mucous secreting. The apical parts of the lining cells are filled by mucin that is usually removed during processing of tissues so that the cells look empty (or vacuolated). Mucous secreted by cells of the lining epithelium protects the gastric mucosa against acid and enzymes produced by the mucosa itself. (The mucous cells lining the surface are also believed to produce blood group factors).

At numerous places the lining epithelium dips into the lamina propria to form the walls of depressions called ***gastric pits*** (Fig. 16.4). These pits extend for a variable distance into the thickness of the mucosa. Deep to the gastric pits the mucous membrane is packed with numerous ***gastric glands***. These glands are of three types: main gastric, cardiac and pyloric (depending on their presence in different regions of stomach).

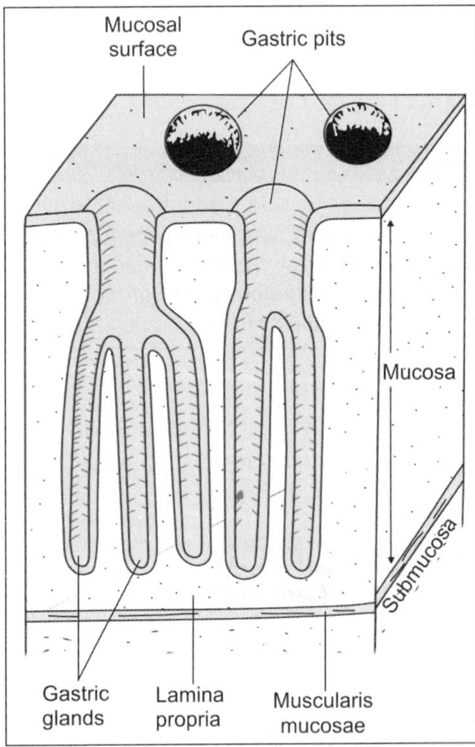

Fig. 16.4: Basic structure of the mucous membrane of the stomach (Schematic representation)

The Lamina Propria

As the mucous membrane of the stomach is packed with glands, the connective tissue of the lamina propria is, therefore, scanty. It contains the usual connective tissue cells. Occasional aggregations of lymphoid tissue are present in it.

The Muscularis Mucosae

The muscularis mucosae of the stomach is well developed. Apart from the usual circular (inner) and longitudinal (outer) layers an additional circular layer may be present outside the longitudinal layer.

The Muscularis Externa

The muscularis externa of the stomach is well developed. Three layers, oblique, circular and longitudinal (from inside out) are usually described. The appearance of the layers in sections is, however, highly variable depending upon the part of the stomach sectioned. The circular fibres are greatly thickened at the pylorus where they form the pyloric sphincter. There is no corresponding thickening at the cardiac end. Salient features of cardiac, fundus and pyloric part of stomach have been summarised in Table 16.1 and discussed in detail in Plates 16.2, 16.3 and 16.4.

Table 16.1: Salient features of each region of stomach

Cardia	Fundus and body	Pylorus
Presence of cardiac glands (mucous secreting glands) in lamina propria of mucosa. Cardiac glands are either simple tubular, or compound tubulo alveolar	Presence of gastric glands in the lamina propria of mucosa. Gastric glands are simple or branched tubular glands. They secrete enzymes and hydrochloric acid	Presence of pyloric glands in the lamina propria of mucosa. Pyloric glands (mucous glands) are simple or branched tubular glands that are coiled.
Shallow gastric pits	Shallow gastric pits occupying superficial 1/4th or less of the mucosa	Deep gastric pits occupying 2/3rd of the depth of the mucosa
Change of epithelium from stratified squamous of the oesophagus to simple columnar epithelium in stomach	Epithelium is simple columnar.	Epithelium is simple columnar. Circular muscle layer is thick and is called as pyloric sphincter.

GASTRIC GLANDS

The Cardiac Glands

These are confined to a small area (cardia) near the opening of the oesophagus. In this region the mucosa is relatively thin (Plate 16.2). Gastric pits are shallow (as in the body of the stomach). The cardiac glands are either simple tubular, or compound tubulo-alveolar. They are mucous secreting. An occasional oxyntic or peptic cell may be present.

The Main Gastric Glands

The main gastric glands are present over most of the stomach, but not in the pyloric region and in a small area near the cardiac end. In other words they are present in the body of the stomach, and in the fundus (Plate 16.3 and Fig. 16.6).

Note: These glands are often inappropriately called fundic glands in many books of histology: they are not confined to the fundus.

The main gastric glands are simple or branched tubular glands that lie at right angles to the mucosal surface. The glands open into gastric pits, each pit receiving the openings of several glands. Here the gastric pits occupy the superficial one-fourth or less of the mucosa, the remaining thickness being closely packed with gastric glands. The following varieties of cells are present in the epithelium lining the glands.

PLATE 16.2: Stomach (Cardia)

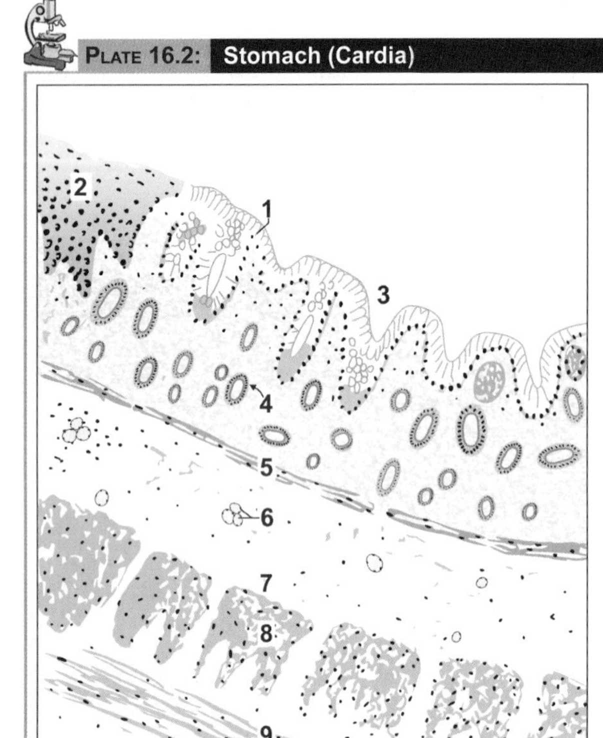

Stomach (cardia). As seen in drawing

- At low magnification, the cardiac end of stomach shows all the four layers seen in stomach:
  - Mucosa
  - Submucosa
  - Muscularis externa
  - Serosa
- At its cardiac end the stomach is lined by simple columnar cells. The epithelium is sharply demarcated from the stratified squamous epithelium lining the lower end of the oesophagus
- Important distinguishing points of cardiac end of stomach are the columnar epithelium lining, the absence of goblet cells, and the simple tubular nature of cardiac glands. If the lower end of the oesophagus is included in the section, the diagnosis becomes obvious (as seen in the drawing).

Key

1. Columnar epithelium
2. Stratified squamous lining of lower end of oesophagus
3. Gastric pit
4. Cardiac gland in mucosa
5. Muscularis mucosae
6. Oesophageal gland in submucosa
7. Submucosa
8. Circular muscle
9. Longitudinal muscle

- ***Chief cells:*** The most numerous cells are called ***chief cells***, ***peptic cells***, or ***zymogen cells***. They are particularly numerous in the basal parts of the glands. The cells are cuboidal or low columnar. Their cytoplasm is basophilic. With special methods the chief cells are seen to contain prominent secretory granules in the apical parts of their cytoplasm. The granules contain pepsinogen that is a precursor of pepsin. With the EM the cytoplasm is seen to contain abundant rough endoplasmic reticulum and a prominent Golgi complex. The luminal surfaces of the cells bear small irregular microvilli.

Note: Chief cells secrete the digestive enzymes of the stomach including pepsin. Pepsin is produced by action of gastric acid on pepsinogen. Pepsin breaks down proteins into small peptides. It is mainly through the action of pepsin that solid food is liquefied.

- ***Oxyntic cells:*** The ***oxyntic*** or ***parietal cells*** are large, ovoid or polyhedral, with a large central nucleus (Fig. 16.5). They are present singly, amongst the peptic cells. They are more

PLATE 16.3: Stomach (Body/Fundus)

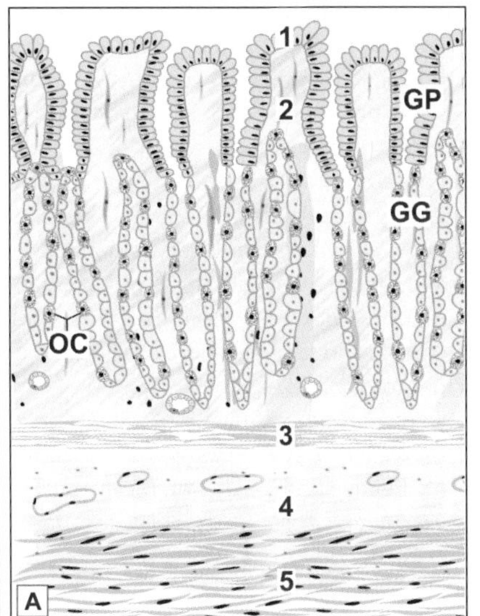

The basic structure of stomach is similar to oesophagus i.e. it is composed of:

- Mucosa
- Submucosa
- Muscularis externa
- Serosa

- Mucosa is lined by simple tall columnar epithelium. It shows invaginations called gastric pits that occupy the superficial one fourth of the mucosa
- The area between the pits and the muscularis mucosae is packed with tubular gastric glands. The glands are lined mainly by blue staining chief cells or peptic cells. Amongst these there are pink staining oxyntic cells. These are large cells that are placed peripherally in the wall of the gland. They are more numerous in the upper parts of the gastric glands. The main point to note is that in the region of the glands we see different types of cells that appear to be closely packed together
- Muscularis externa is composed of three layers of smooth muscle–inner oblique, middle circular and outer longitudinal.

A photomicrograph of the body of the stomach is shown in Plate 16.3B. The gastric pits and gastric glands can be distinguished. Observe that the gastric pits occupy the upper one fourth of the lamina propria of mucosa.

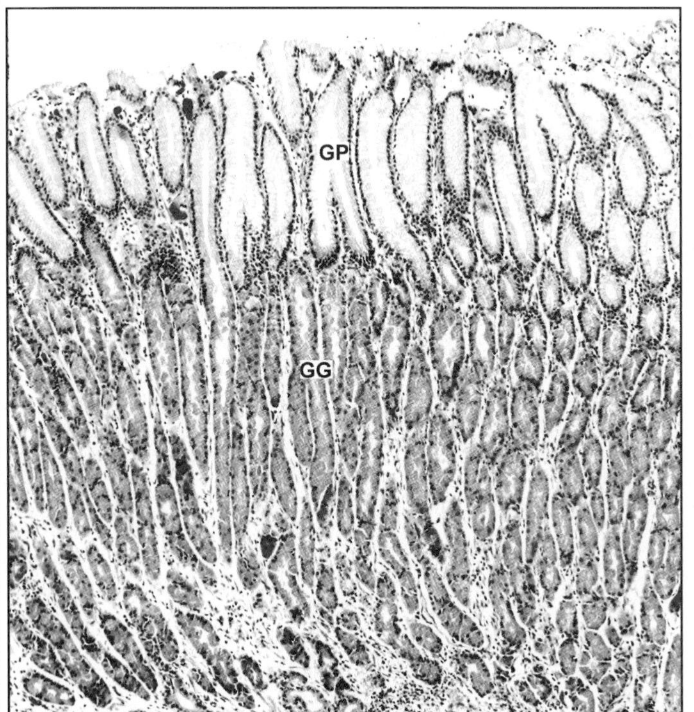

Stomach (body/fundus). A. As seen in drawing; B. Photomicrograph

Key

1. Columnar epithelium lining
2. Lamina propria
3. Muscularis mucosa
4. Submucosa
5. Muscularis externa

GP. Gastric pit
GG. Gastric gland
OC. Oxyntic cells

PLATE 16.4: Stomach (Pylorus)

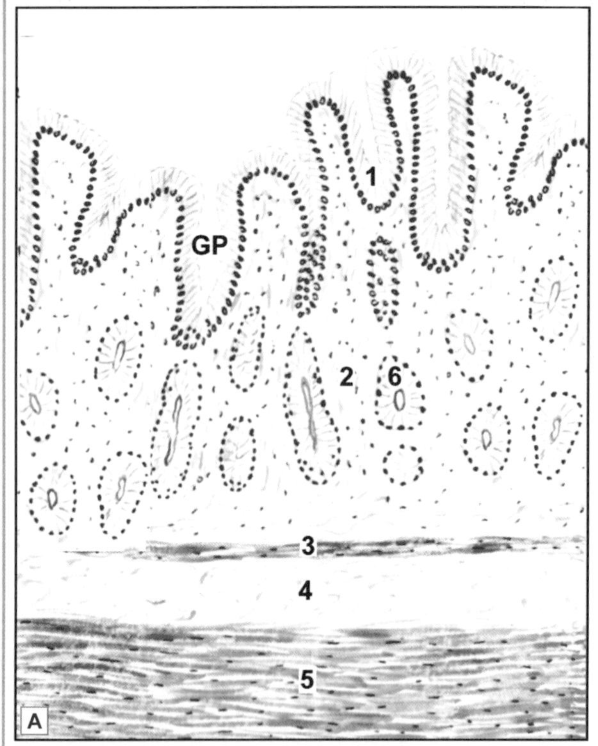

- In the pyloric part of the stomach the gastric pits are much deeper than in the body of the stomach (occupying up to two-third of the depth of the mucosa)
- Deep to the pits there are pyloric glands that are lined by mucous secreting cells. These are pale staining
- The muscularis mucosae, submucosa, and part of the muscle coat are also seen
- It is important to note that the stomach does not have villi. In the photomicrograph folds of epithelial lining may be confused with villi. Observe that each fold merges with underlying connective tissue completely. When true villi are present (as in the small intestine), small parts of them appear as circular or oval masses not attached to a villus or to the submucosa. These are villi that have been cut transversely or obliquely
- Another important feature to note is that the lining epithelium does not have typical goblet cells, but some epithelial cells are mucous secreting.

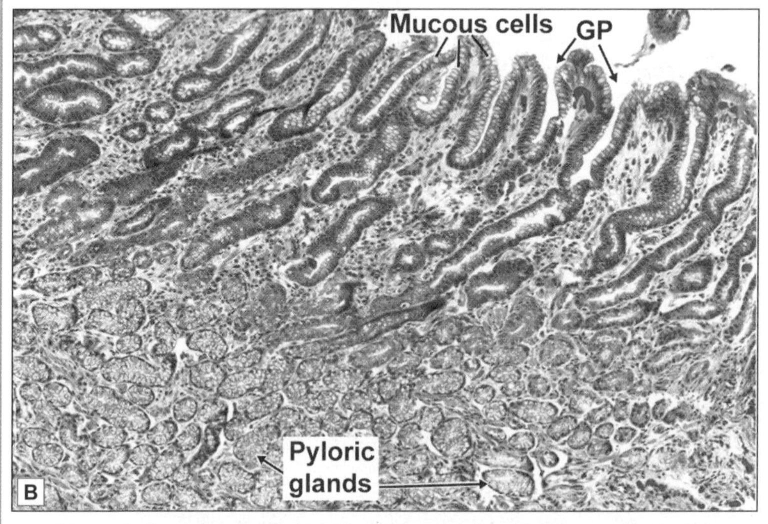

Key

1. Columnar epithelium lining
2. Lamina propria
3. Muscularis mucosa
4. Submucosa
5. Muscularis externa
6. Pyloric gland

GP. Gastric pit

Stomach (body/fundus). A. As seen in drawing; B. Photomicrograph

Courtesy: **Atlas of Histopathology, Ist Edition. Ivan Damjanov. Jaypee Brothers. 2012. p 104**

numerous in the upper half of the gland than in its lower half. They are called oxyntic cells because they stain strongly with eosin. They are called parietal cells as they lie against the basement membrane, and often bulge outwards (into the lamina propria) creating a beaded appearance. With the light microscope they appear to be buried amongst the chief cells.

The EM shows, however, that each parietal cell has a narrow apical part that reaches the lumen of the gland. The cell membrane of this apical region shows several invaginations into the cytoplasm, producing tortuous intracellular canaliculi that communicate with the glandular lumen. The walls of the canaliculi bear microvilli that project into the canaliculi. The cytoplasm (in the intervals between the canaliculi) is packed with mitochondria. The mitochondria are responsible for the granular appearance and eosinophilia of the cytoplasm (seen with the light microscope). Secretory granules are not present.

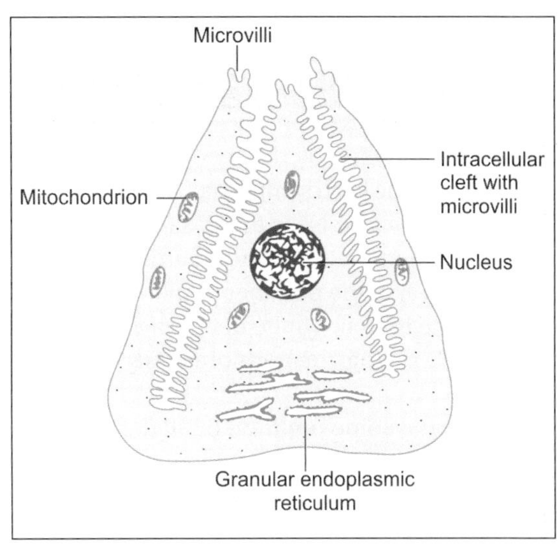

Fig. 16.5: Some features of the electrone microscope structure of an oxyntic cell (schematic representation)

Note: Oxyntic cells are responsible for the secretion of hydrochloric acid. They also produce an ***intrinsic factor*** (a glucoprotein) that combines with vitamin B_{12} (present in ingested food and constituting an ***extrinsic factor***) to form a complex necessary for normal formation of erythrocytes.

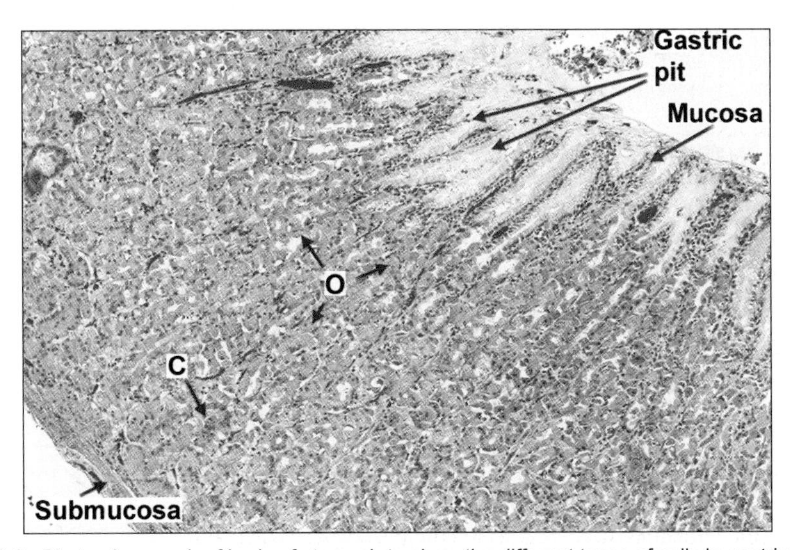

Fig. 16.6: Photomicrograph of body of stomach to show the different types of cells in gastric glands O–oxyntic cells; C–chief cells

Courtesy: Atlas of Histopathology. Ist Edition. Ivan Damjanov. Jaypee Brothers. 2012. p 104

- ***Mucous neck cells:*** Near the upper end (or 'neck') of the glands there are mucous secreting cells that are called ***mucous neck cells***. These are large cells with a clear cytoplasm. The nucleus is flattened and is pushed to the base of the cell by accumulated mucous. The supranuclear part of the cell contains prominent granules. The chemical structure of the mucous secreted by these cells is different from that secreted by mucous cells lining the surface of the gastric mucosa.
- ***Endocrine cells:*** Near the basal parts of the gastric glands there are ***endocrine cells*** that contain membrane bound neurosecretory granules. As the granules stain with silver salts these have, in the past, been called ***argentaffin cells.*** These cells are flattened. They do not reach the lumen, but lie between the chief cells and the basement membrane. These cells probably secrete the hormone ***gastrin***. Some of the cells can be shown to contain serotonin (5HT).
- ***Stem cells:*** Some undifferentiated cells (stem cells) that multiply to replace other cells are also present. They increase in number when the gastric epithelium is damaged (for example when there is a gastric ulcer), and play an important role in healing.

The Pyloric Glands

In the pyloric region of the stomach the gastric pits are deep and occupy two-third of the depth of the mucosa. The pyloric glands that open into these pits are short and occupy the deeper one-third of the mucosa. They are simple or branched tubular glands that are coiled. The glands are lined by mucous secreting cells (Plate 16.4). Occasional oxyntic and argentaffin cells may be present. In addition to other substances, pyloric glands secrete the hormone gastrin.

Clinical Correlation

- **Gastritis:** The term 'gastritis' is commonly employed for any clinical condition with upper abdominal discomfort like indigestion or dyspepsia in which the specific clinical signs and radiological abnormalities are absent. The condition is of great importance due to its relationship with peptic ulcer and gastric cancer.
- **Gastric ulcer:** Gastric ulcer may occur due to damage to the gastric mucosa barrier. It is most common along the lesser curvature and pyloric antrum. Food-pain pattern, vomiting, significant weight loss and deep tenderness in the midline in epigastrium are the main presentations

THE SMALL INTESTINE

The small intestine is a tube about five meters long. It is divided into three parts. These are (in craniocaudal sequence) the ***duodenum*** (about 25 cm long); the ***jejunum*** (about 2 meters long); and the ***ileum*** (about 3 meters long). (Gut length is shorter in the living person than in a cadaver, because of muscle tone). It is the principal site for absorption of products of digestion. It also secretes some hormones through enteroendocrine cells. Digestion is completed in small intestine.

MICROSCOPIC FEATURES

The wall of the small intestine is made up of the four layers: mucous, submucous, muscularis serous. The serous and muscular layers correspond exactly to the general structure of alimentary canal. The submucosa is also typical except in the duodenum, where it contains the ***glands ofBrunner***. The mucous membrane exhibits several special features that are described below.

The Mucous Membrane

The surface area of the mucous membrane of the small intestine is extensive (to allow adequate absorption of food). This is achieved by virtue of the following.

- The considerable length of the intestine.
- The presence of numerous circular folds in the mucosa.
- The presence of numerous finger-like processes, or ***villi***, that project from the surface of the mucosa into the lumen.
- The presence of numerous depressions or ***crypts*** that invade the lamina propria.
- The presence of microvilli on the luminal surfaces of the cells lining the mucosa (Fig. 16.7).

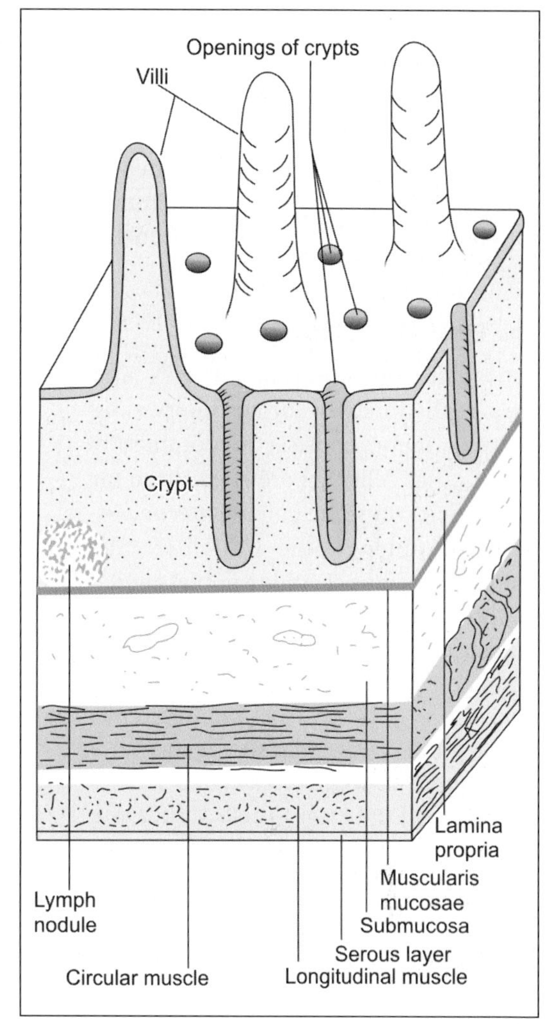

Fig. 16.7: Basic structure of the small intestine (Schematic representation)

Circular Folds

The circular folds are also called the ***valves of Kerkring***. Each fold is made up of all layers of the mucosa (lining epithelium, lamina propria and muscularis mucosae). The submucosa also extends into the folds. The folds are large and readily seen with the naked eye. They are absent in the first one or two inches of the duodenum. They are prominent in the rest of the duodenum, and in the whole of the jejunum. The folds gradually become fewer and less marked in the ileum. The terminal parts of the ileum have no such folds.

Apart from adding considerably to the surface area of the mucous membrane, the circular folds tend to slow down the passage of contents through the small intestine thus facilitating absorption.

The Villi

The villi are, typically, finger-like projections consisting of a core of reticular tissue covered by a surface epithelium (Fig. 16.8). The connective tissue core contains numerous blood capillaries forming a plexus. The endothelium lining the capillaries is fenestrated thus allowing rapid absorption of nutrients into the blood. Each villus contains a central lymphatic vessel called a lacteal. Distally, the lacteal ends blindly near the tip of the villus; and proximally it ends in a plexus of lymphatic vessels present in the lamina propria. Occasionally, the lacteal may be double. Some muscle fibres derived from the muscularis mucosae extend into the villus core.

The Crypts

The crypts (of Lieberkuhn) are tubular invaginations of the epithelium into the lamina propria. They are really simple tubular ***intestinal glands*** that are lined by epithelium (Plate 16.4). The epithelium is supported on the outside by a basement membrane.

The Epithelium Lining

The epithelium covering the villi, and areas of the mucosal surface intervening between them, consists predominantly of columnar cells that are specialised for absorption. These are called ***enterocytes***. Scattered amongst the columnar cells there are mucous secreting goblet cells.

The cells lining the crypts (intestinal glands) are predominantly undifferentiated. These cells multiply to give rise to absorptive columnar cells and to goblet cells. Near the bases of the crypts there are ***Paneth cells*** that secrete enzymes. Endocrine cells (bearing membrane bound granules filled with various neuroactive peptides) are also present.

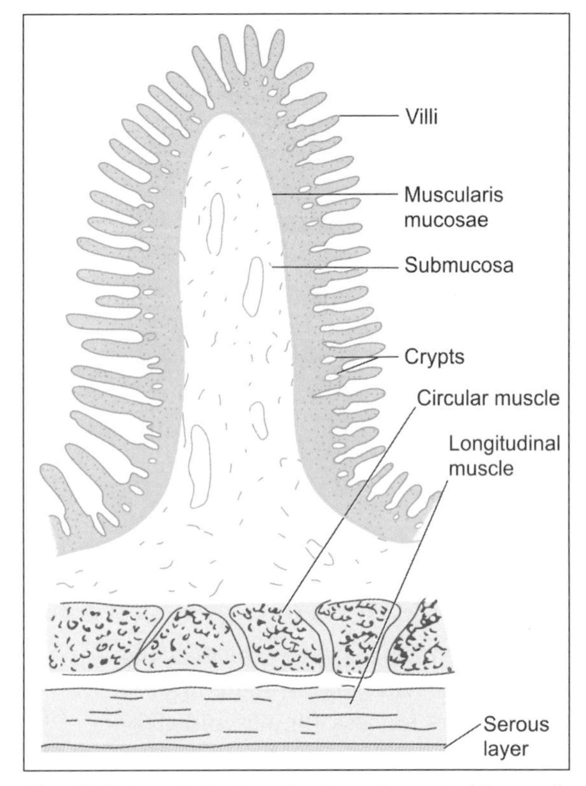

Fig. 16.8: Longitudinal section through a part of the small intestine seen at a very low magnification to show a circular mucosal fold (Schematic representation) Note the submucosa extending into the mucosal fold.

Added Information

In some situations the villi, instead of being finger-like, are flattened and leaf-like, while in some other situations they are in the form of ridges. The villi are greatest and most numerous (for a given area) in the duodenum. They progressively decrease in size, and in number, in proceeding caudally along the small intestine. It has been estimated that the presence of villi increases the surface area of the epithelial lining of the small intestine about eight times.

Cells of Small Intestine

Absorptive Columnar Cells

The general characteristics of columnar epithelium apply to the columnar cells lining the mucous membrane of the small intestine. Each cell has an oval nucleus located in its lower part. When seen with the light microscope the luminal surface of the cell appears to be thickened and to have striations in it, perpendicular to the surface (Fig. 16.9). With the EM this ***striated border*** is seen to be made up of microvilli arranged in a very regular manner. The presence of microvilli greatly increases the absorptive surface of the cell.

Each microvillus has a wall of plasma membrane within which there are fine filaments. These filaments extend into the apical part of the cell. Here they become continuous with a plexus of similar filaments that form the ***terminal web***. The surface of each microvillus is covered by a layer of fine fibrils and mucous (***glycocalyx***).

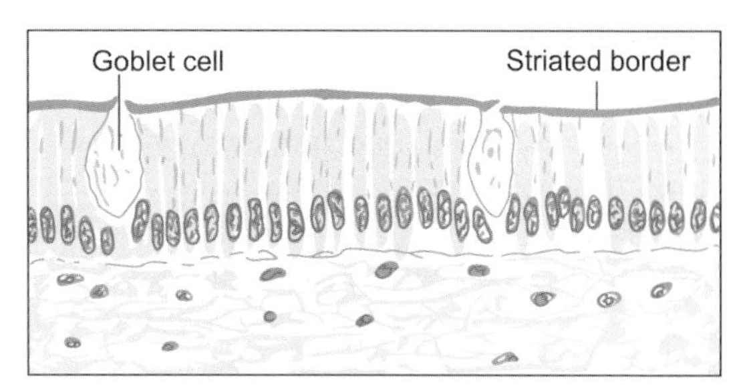

Fig. 16.9: Columnar epithelium lining the small intestine. Note the striated border and some goblet cells (Schematic representation)

The plasma membrane on the lateral sides of absorptive cells shows folds that interdigitate with those of adjoining cells. Adjacent cells are united by typical junctional complexes and by scattered desmosomes. Intercellular canals may be present between adjacent cells. The cytoplasm of absorptive cells contains the usual organelles, including lysosomes and smooth ER. These cells are responsible for absorption of amino acids, carbohydrates, and lipids present in digested food.

Goblet Cells

A goblet is literally a drinking glass that is broad above, and has a narrow stem attached to a base. Goblet cells are so named because of a similar shape.

Each goblet cell has an expanded upper part that is distended with mucin granules (Figs. 16.9 and 16.10). The nucleus is flattened and is situated near the base of the cell. Goblet cells are mucous secreting cells. In consonance with their secretory function these cells have a prominent Golgi complex and abundant rough endoplasmic reticulum. The luminal surface of the cell bears some irregular microvilli. In haematoxylin and eosin stained preparations, the mucin content of goblet cells appears to be unstained. It stains brightly with the PAS technique. Mucous cells increase in number as we pass down the small intestine, being few in the duodenum and most numerous in the terminal ileum.

Undifferentiated Cells

These are columnar cells present in the walls of intestinal crypts. They are similar to absorptive

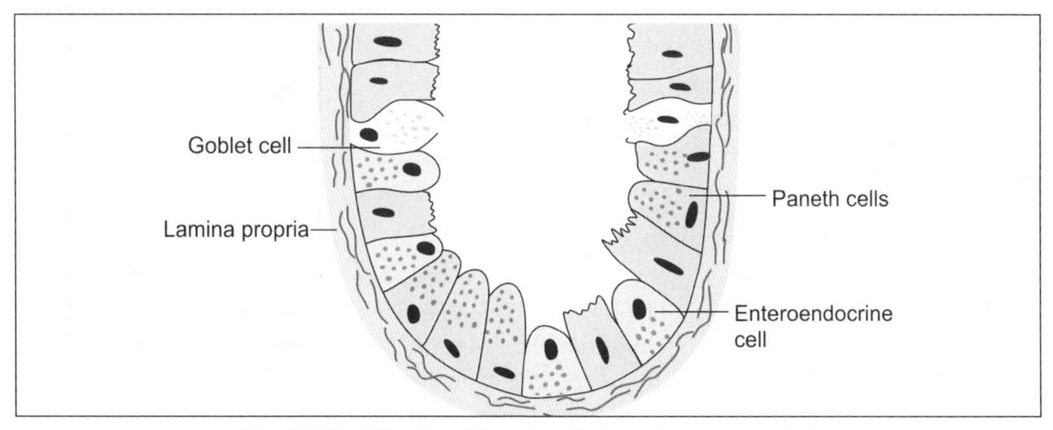

Fig. 16.10: Cells of small intestine (Schematic representation)

cells, but their microvilli and terminal webs are not so well developed. The cytoplasm contains secretory granules.

Undifferentiated cells proliferate actively by mitosis. The newly formed cells migrate upwards from the crypt to reach the walls of villi. Here they differentiate either into typical absorptive cells, or into goblet cells. These cells migrate towards the tips of the villi where they are shed off. In this way, the epithelial lining is being constantly replaced, each cell having a life of only a few days. The term ***intermediate cells*** has been applied to differentiating stem cells that show features intermediate between those of stem cells and fully differentiated cells.

Paneth Cells (Zymogen Cells)

These cells are found only in the deeper parts of intestinal crypts. They contain prominent eosinophilic secretory granules (Figs 16.10 and 16.11). With the EM Paneth cells are seen to contain considerable rough endoplasmic reticulum. Other organelles and some irregular microvilli are present. The cells are rich in zinc.

The function of zymogen cells is not well known. They are known to produce lysozyme that destroys bacteria. They may also produce other enzymes.

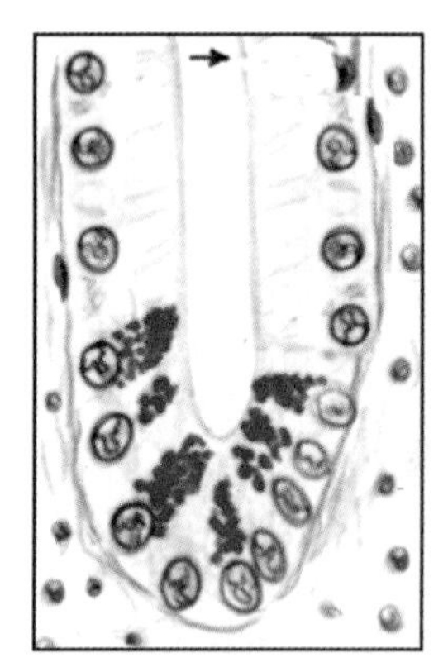

Fig. 16.11: High power view of an intestinal crypt showing Paneth cells with red staining supranuclear granules (Schematic representation)

Endocrine Cells

Cells containing membrane bound vesicles filled with neuroactive substances are present in the epithelial lining of the small intestine. They are most numerous near the lower ends of crypts. As the granules in them stain with silver salts these cells have, in the past, been termed argentaffin cells (Figs 16.11 and 16.12). Some of them also give a positive chromaffin reaction. They are, therefore, also called ***enterochromaffin cells***. With the introduction of immunohistochemical techniques it has now been demonstrated that these cells are of various functional types, and contain many amines having an endocrine function.

Other Cells in the Lamina Propria

Apart from connective tissue associated fibroblasts, and lymphocytes (mentioned above) the lamina propria of the small intestine contains eosinophil leucocytes, macrophages, and mast cells. Plasma cells are present in relation to aggregations of lymphoid tissue.

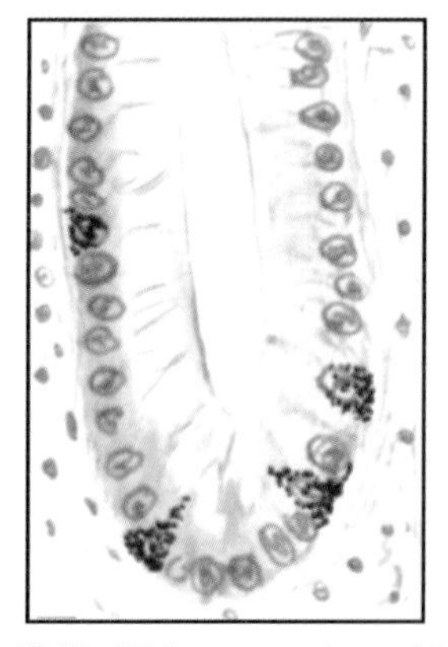

Fig. 16.12: High power view of intestinal crypt showing argentaffin cell with black staining granules that are mostly infranuclear (Schematic representation)

Lymphoid Tissue of the Small Intestine

Solitary and aggregated lymphatic follicles (Peyer's patches) are present in the lamina propria of the small intestine. The solitary follicles become more numerous, and the aggregated follicles larger, in proceeding caudally along the small intestine. They are most prominent in the terminal ileum (Plate 16.5). Their lymphoid tissue may occasionally extend into the submucosa. Villi are few or missing in the mucosa overlying aggregated follicles.

The epithelium overlying lymphatic follicles contains special ***follicle-associated epithelial cells*** or ***M-cells*** (M for 'microfold' or 'membrane'). These cells are columnar. They are believed to take up antigens present in the lumen of the intestine and to transport them to subjacent lymphoid tissue, which can then produce antibodies against the antigens. The lateral borders of these cells are deeply indented by small lymphocytes (lying within the thickness of the epithelium).

Peyer's Patches

Small collections of lymphoid tissue, similar in structure to the follicles of lymph nodes, may be present anywhere along the length of the gut. They are called ***solitary lymphatic follicles***. Larger aggregations of lymphoid tissue, each consisting of 10 to 200 follicles are also present in the small intestine. They are called ***aggregated lymphatic follicles*** or ***Peyer's patches*** (Plate 16.7). These patches can be seen by naked eye, and about 200 of them can be counted in the human gut. The mucosa overlying them may be devoid of villi or may have rudimentary villi. Peyer's patches always lie along the antemesenteric border of the intestine and measure 2 cm to 10 cm. Both solitary and aggregated follicles increase in number and size in proceeding caudally along the small intestine, being most numerous and largest in the terminal ileum. In addition to lymphoid follicles, a large number of lymphocytes and plasma cells are present in the connective tissue of the gut wall.

It has been held that gut associated lymphoid tissue may possibly have a role to play in the processing of B-lymphocytes (similar to that of T-lymphocytes in the thymus), but at present there is not much evidence to support this view.

DISTINGUISHING FEATURES OF DUODENUM, JEJUNUM, AND ILEUM

Sections through the small intestine are readily distinguished from those of other parts of the gut because of the presence of villi.

Duodenum (Plate 16.6)

The duodenum is easily distinguished from the jejunum or ileum because of the presence of glands in the submucosa. (No glands are present in the submucosa of the jejunum or ileum). These ***duodenal glands*** (of Brunner) are compound tubulo-alveolar glands (Plate 16.6). Their ducts pass through the muscularis mucosae to open into the intestinal crypts (of Lieberkuhn).

The cells lining the alveoli of duodenal glands are predominantly mucous secreting columnar cells having flattened basal nuclei. Some endocrine cells are also present. The duodenal glands are most numerous in the proximal part of the duodenum. They are few (or missing) in the distal part. The secretions of the duodenal glands contain mucous, bicarbonate ions (to neutralise gastric acid entering the duodenum) and an enzyme that activates trypsinogen produced by the pancreas.

Jejunum (Plate 16.5)

The proximal part of the jejunum shows significant differences in structure from the terminal part of the ileum. The changes take place gradually in proceeding caudally along the small intestine, there being no hard and fast line of distinction between the jejunum and the ileum. As compared to the ileum the jejunum has the following features:

- A larger diameter
- A thicker wall
- Larger and more numerous circular folds
- Larger villi
- Fewer solitary lymphatic follicles. Aggreated lymphatic follicles are absent in the proximal jejunum, and small in the distal jejunum
- Greater vascularity

Ileum

The villi are thin and slender in the region of ileum. The submucosa contains the Peyer's patches. M cells are found overlying the lymphoid follicles.

Pathological Correlation

- **Crohn's Disease or Regional Enteritis** is an idiopathic chronic ulcerative inflammatory bowel disease, characterised by transmural, non-caseating granulomatous inflammation, affecting most commonly the segment of terminal ileum and/or colon, though any part of the gastrointestinal tract may be involved.
- **Coeliac Sprue** is the most important cause of primary malabsorption occurring in temperate climates. The condition is characterised by significant loss of villi in the small intestine and thence diminished absorptive surface area. The condition occurs in 2 forms:
  - Childhood form, seen in infants and children and is commonly referred to as coeliac disease.
  - Adult form, seen in adolescents and early adult life and used to be called idiopathic steatorrhoea.

THE LARGE INTESTINE (COLON)

It consists of the caecum, appendix, colon, rectum and anal canal. The main functions of the large intestine are absorption of water and conversion of the liquid, undigested material into solid faeces. It harbours some nonpathogenic bacteria that produce vitamin B_{12} and vitamin K. The former is necessary for haemopoiesis and the latter for coagulation of blood.

THE COLON

The structure of the colon conforms to the general description of the structure of the gut. The following additional points may be noted (Fig. 16.13 and Plate 16.7).

Mucous Membrane

The mucous membrane of the colon shows numerous crescent-shaped folds. There are no villi. The mucosa shows numerous closely arranged tubular glands or crypts similar to those in the small intestine. The mucosal surface, and the glands, are lined by an epithelium made up predominantly of columnar cells. Their main function is to absorb excess water and electrolytes from intestinal contents.

Plate 16.5: Jejunum

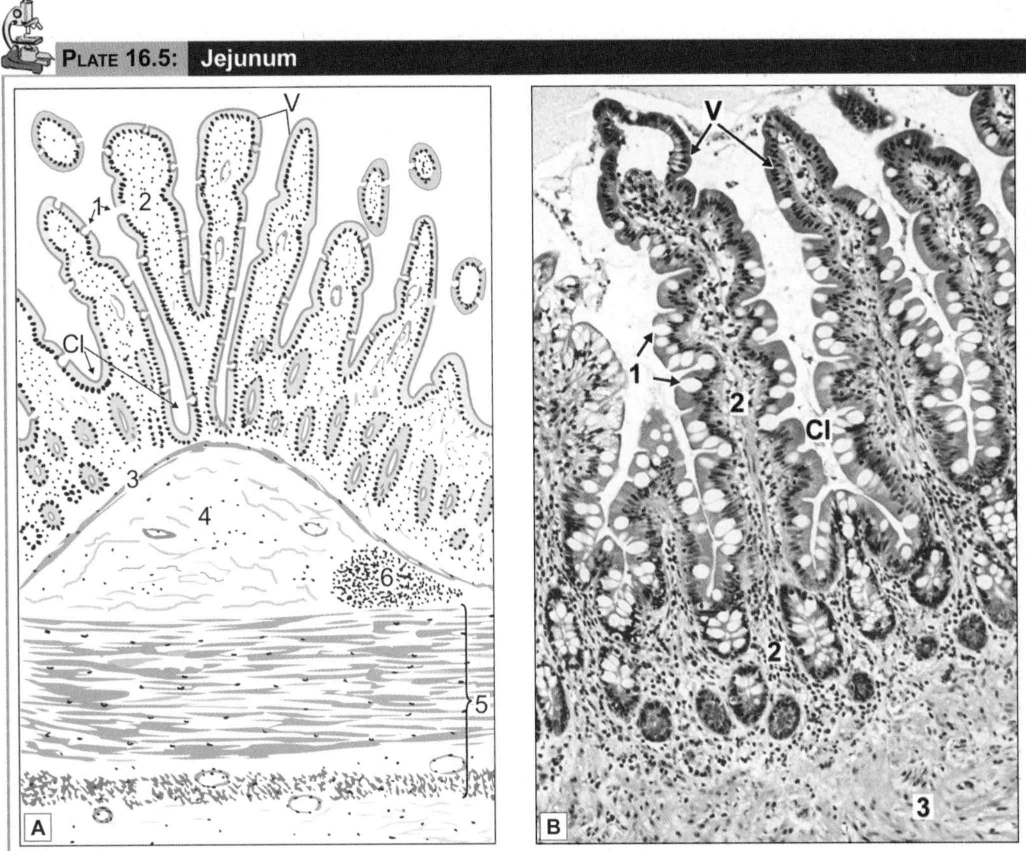

Jejunum. A. As seen in drawing; B. Photomicrograph

***Courtesy:* Atlas of Histopathology. Ist Edition. Ivan Damjanov. Jaypee Brothers. 2012. p 117**

In this figure we see features of the typical structure of the small intestine. The mucosa shows numerous finger-like projections or villi. Each villus has a covering of columnar epithelium that covers a core of delicate connective tissue. Some goblet cells are also seen. Numerous tubular depressions, or crypts dip into the lamina propria. These crypts are also lined by columnar cells. The mucosa is separated from the submucosa by the muscularis mucosae. The intestine is surrounded by circular and longitudinal layers of smooth muscle which constitutue muscularis externa.

Note: A solitary lymph nodule is present in the submucosa in the drawing.

The photomicrograph shows the finger-like villi lined by columnar epithelial cells and goblet cells. The villi dip down into the crypts of Lieberkuhn. At the bases of these are Paneth cells with eosinophilic supranuclear granules.

Key

1. Columnar epithelial lining with goblet cells
2. Lamina propria
3. Muscularis mucosa
4. Submucosa
5. Muscularis externa
6. Lymph nodule

V. Villi

Cl. Crypts of Lieberkuhn

PLATE 16.6: Duodenum

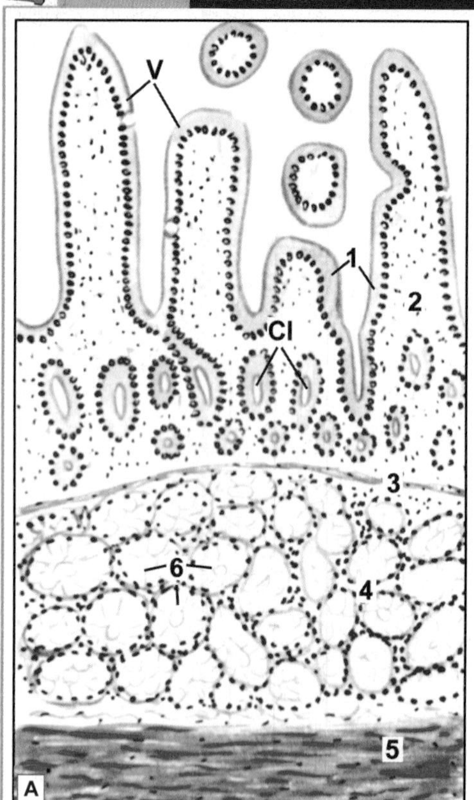

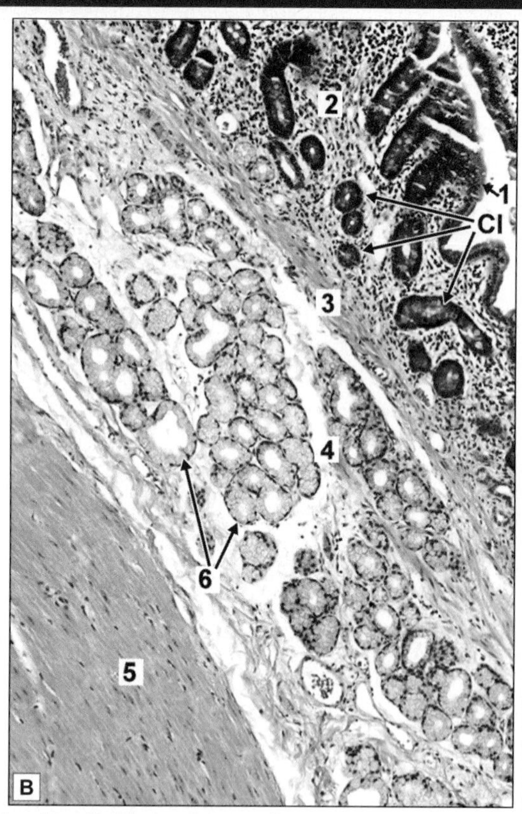

Duodenum. A. As seen in drawing; B. Photomicrograph

***Courtesy:* Atlas of Histopathology. Ist Edition. Ivan Damjanov. Jaypee Brothers. 2012. p 117**

The general structure of the duodenum is the same as that described for the jejunum, except that the submucosa is packed with mucous secreting glands of Brunner.

Note: The intestinal crypts lie 'above' the muscularis mucosae while the glands of Brunner lie 'below' it. The presence of the glands of Brunner is a distinctive feature of the duodenum.

The photomicrograph shows Brunner's glands which are mucous glands present in the submucosa of the duodenum.

Key

1. Columnar epithelial lining with goblet cells
2. Lamina propria
3. Muscularis mucosa
4. Submucosa with duodenal glands of Brunner
5. Muscularis externa
6. Glands of Brunner

V. Villi

Cl. Crypts of Lieberkuhn

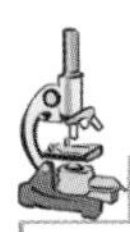

Plate 16.7: Ileum (Peyers Patch)

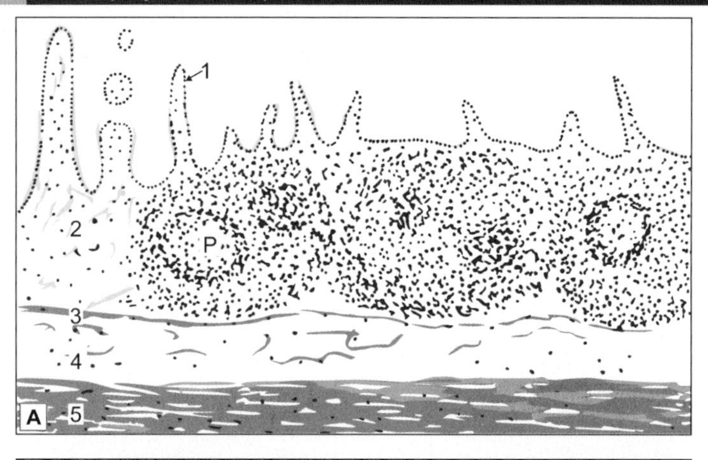

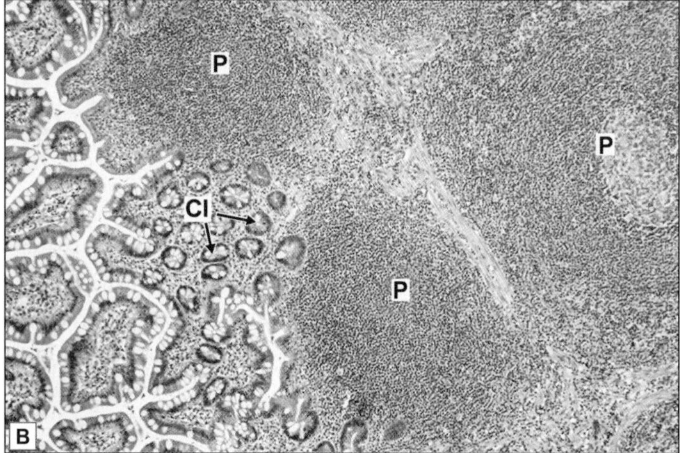

Duodenum. A. As seen in drawing; B. Photomicrograph.

***Courtesy:* Atlas of Histopathology. Ist Edition. Ivan Damjanov. Jaypee Brothers. 2012. p 117**

The general structure of the ileum is similar to that of the jejunum except for

- The entire thickness of the lamina propria is infiltrated with lymphocytes amongst which typical lymphatic follicles can be seen which may extend into the submucosa. These lymphatic follicles are called as Peyer's patches
- In the region overlying the 'Peyer's patch' villi may be rudimentary or absent.

Note: Presence of Peyer's patches is the most distinguishing feature of ileum.

Key

1. Columnar epithelial lining with goblet cells
2. Lamina propria
3. Muscularis mucosa
4. Submucosa
5. Muscularis externa

P. Peyer's patches
Cl. Crypts of Lieberkuhn

Many columnar cells secrete mucous and antibodies (IgA). The antibodies provide protection against pathogenic organisms. Numerous goblet cells are present, their number increasing in proceeding caudally. The mucous secreted by them serves as a lubricant that facilitates the passage of semisolid contents through the colon. Paneth cells are not present. Some endocrine cells, and some stem cells, are seen.

The epithelium overlying solitary lymphatic follicles (present in the lamina propria) contains M-cells similar to those described in the small intestine. Scattered cells bearing tufts of long microvilli are also seen. They are probably sensory cells.

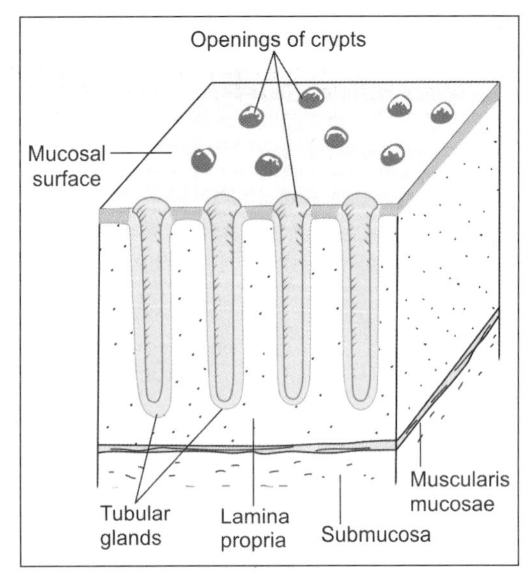

Fig. 16.13: Basic features of the structure of the mucous membrane of the large intestine (Schematic representation)

Submucosa

The submucosa often contains fat cells. Some cells that contain PAS-positive granules, termed ***muciphages***, are also present. These are most numerous in the rectum.

Muscularis Externa

The longitudinal layer of muscle is unusual. Most of the fibres in it are collected to form three thick bands, the ***taenia coli*** (Fig. 16.14). A thin layer of longitudinal fibres is present in the intervals between the taenia. The taenia are shorter in length than other layers of the wall of the colon. This results in the production of ***sacculations*** on the wall of the colon.

Serosa

The serous layer is missing over the posterior aspect of the ascending and descending colon. In many situations the peritoneum forms small pouch-like processes that are filled with fat. These yellow masses are called the ***appendices epiploicae***.

THE VERMIFORM APPENDIX

The appendix is the narrowest part of the gut. The structure of the vermiform appendix resembles that of the colon with the following differences (Plate 16.8):

- The crypts are poorly formed.
- The longitudinal muscle coat is complete and equally thick all round. Taenia coli are not present.
- The submucosa contains abundant lymphoid tissue that may completely fill the submucosa. The lymphoid tissue is not present at birth. It gradually increases and is best seen in children about 10 years old. Subsequently, there is progressive reduction in quantity of lymphoid tissue.

PLATE 16.8: Large Intestine

- The most important feature to note is the absence of villi
- The mucosa shows numerous tubular glands or crypts. The surface of the mucosa, and the crypts, are lined by columnar cells amongst which there are numerous goblet cells
- A section of the large intestine is easily distinguished from that of the small intestine because of the absence of villi; and from the stomach because of the presence of goblet cells (which are absent in the stomach)
- The muscularis mucosae, submucosa and circular muscle coat are similar to those in the small intestine
- However, the longitudinal muscle coat is gathered into three thick bands called taenia coli (as shown in drawing). The longitudinal muscle is thin in the intervals between the taenia.

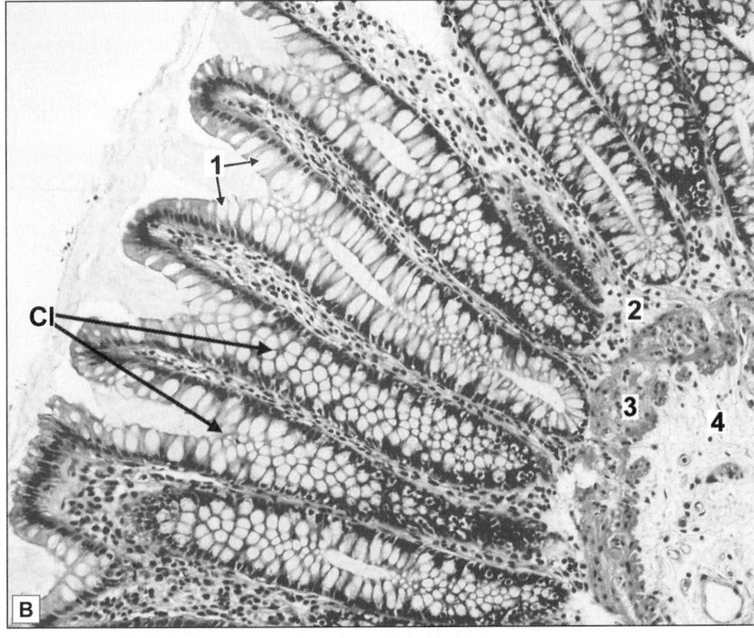

Note: A lymphatic nodule can be seen in the lamina propria (as shown by arrow head) in the drawing.

Key

1. Columnar epithelial lining with goblet cells
2. Lamina propria
3. Muscularis mucosa
4. Submucosa
5. Muscle coat
6. Taenia coli
7. Longitudinal muscle

Cl. Crypts of Lieberkuhn

Large Intestine. A. As seen in drawing; B. Photomicrograph

***Courtesy:* Atlas of Histopathology. Ist Edition. Ivan Damjanov. Jaypee Brothers. 2012. p 122**

THE RECTUM

The structure of the rectum is similar to that of the colon except for the following:

- A continuous coat of longitudinal muscle is present. There are no taenia.
- Peritoneum covers the front and sides of the upper one-third of the rectum; and only the front of the middle third. The rest of the rectum is devoid of a serous covering.
- There are no appendices epiploicae.

Sacculation
Taenia coli
Appendices epiploicae

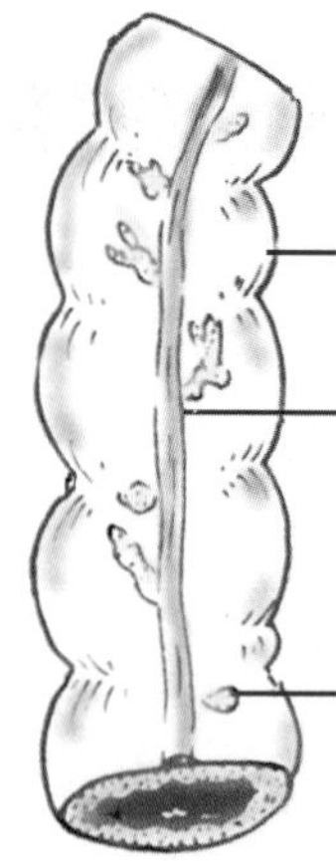

Fig. 16.14: Segment of the colon (Schematic representation)

THE ANAL CANAL

The anal canal is about 4 cm long. The upper 3 cm are lined by mucous membrane, and the lower 1 cm by skin.

Mucosa

The mucous membrane of the upper 15 mm of the canal is lined by columnar epithelium. The mucous membrane of this part shows six to twelve longitudinal folds that are called the ***anal columns***. The lower ends of the anal columns are united to each other by short transverse folds called the ***anal valves***. The anal valves together form a transverse line that runs all round the anal canal, this is the ***pectinate line*** (Fig. 16.15).

The mucous membrane of the next 15 mm of the rectum is lined by non-keratinised stratified squamous epithelium. This region does not have anal columns. The mucosa has a bluish appearance because of the presence of a dense venous plexus between it and the muscle coat. This region is called the ***pecten*** or ***transitional zone***. The lower limit of the pecten forms the ***white line*** (***of Hilton***).

The lowest 8 to 10 mm of the anal canal are lined by true skin in which hair follicles, sebaceous glands and sweat glands are present.

Above each anal valve there is a depression called the anal sinus. Atypical (apocrine) sweat glands open into each sinus. They are called the anal (or circumanal) glands.

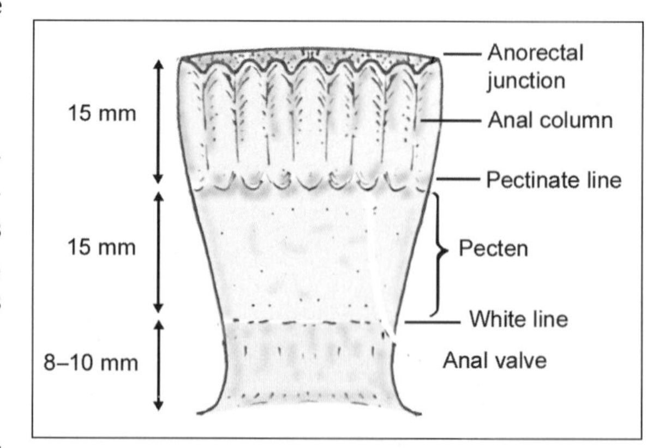

Fig. 16.15: Some features in the interior of the anal canal (Schematic representation)

Submucosa

Prominent venous plexuses are present in the submucosa of the anal canal. The internal haemorrhoidal plexus lies above the level of the pectinate line, while the external haemorrhoidal plexus lies near the lower end of the canal.

Muscularis Externa

The anal canal is surrounded by circular and longitudinal layers of muscle

PLATE 16.9: Vermiform Appendix

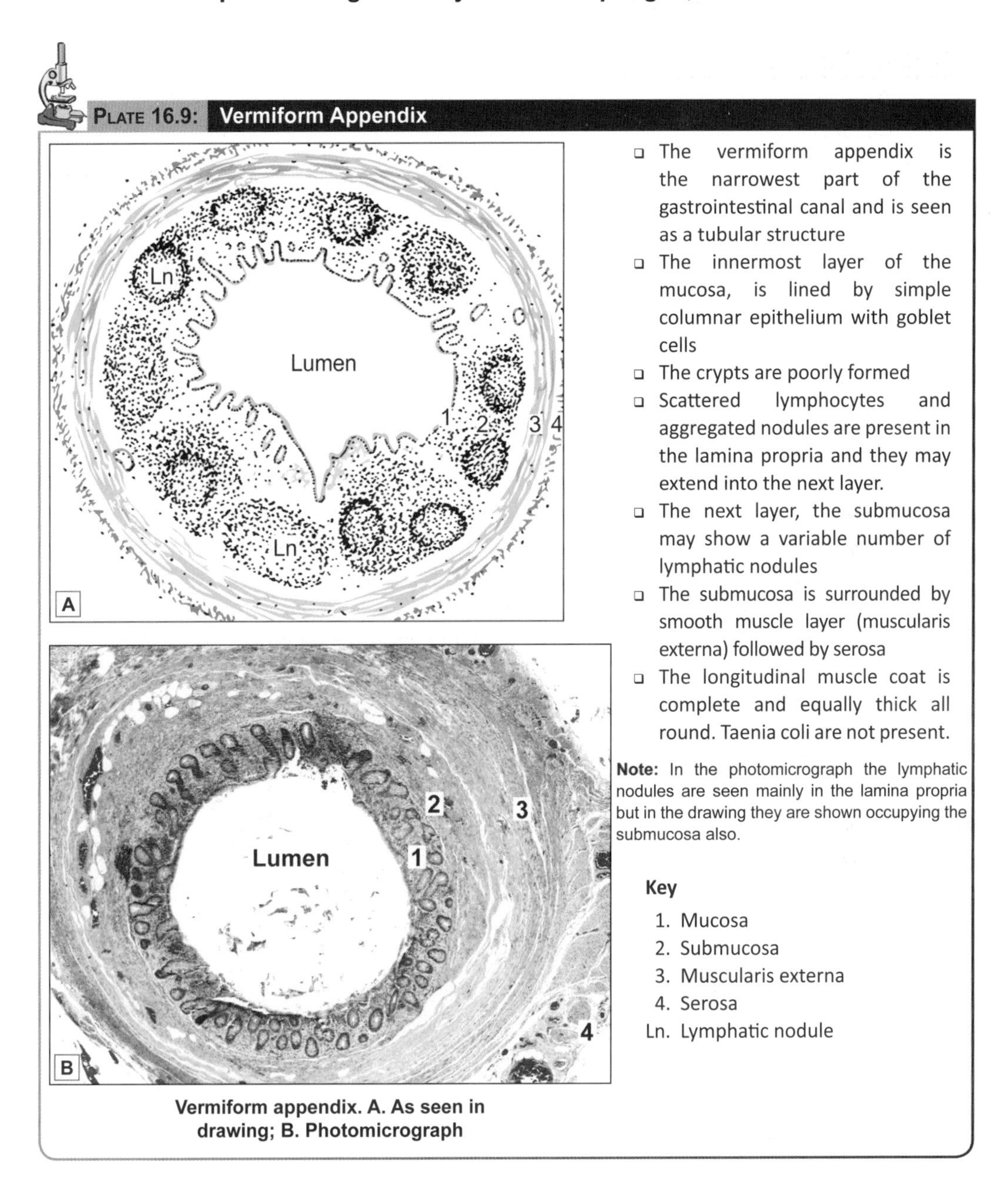

- The vermiform appendix is the narrowest part of the gastrointestinal canal and is seen as a tubular structure
- The innermost layer of the mucosa, is lined by simple columnar epithelium with goblet cells
- The crypts are poorly formed
- Scattered lymphocytes and aggregated nodules are present in the lamina propria and they may extend into the next layer.
- The next layer, the submucosa may show a variable number of lymphatic nodules
- The submucosa is surrounded by smooth muscle layer (muscularis externa) followed by serosa
- The longitudinal muscle coat is complete and equally thick all round. Taenia coli are not present.

Note: In the photomicrograph the lymphatic nodules are seen mainly in the lamina propria but in the drawing they are shown occupying the submucosa also.

Key

1. Mucosa
2. Submucosa
3. Muscularis externa
4. Serosa

Ln. Lymphatic nodule

Vermiform appendix. A. As seen in drawing; B. Photomicrograph

continuous with those of the rectum. The circular muscle is thickened to form the ***internal anal sphincter***. Outside the layer of smooth muscle, there is the ***external anal sphincter*** that is made up of striated muscle. For further details of the anal musculature see a book on gross anatomy.

Clinical Correlation

- **Ulcerative colitis** is an idiopathic form of acute and chronic ulcero-inflammatory colitis affecting chiefly the mucosa and submucosa of the rectum and descending colon, though sometimes it may involve the entire length of the large bowel.
- Acute inflammation of the appendix, **acute appendicitis,** is the most common acute abdominal condition confronting the surgeon. The condition is seen more commonly in older children and young adults, and is uncommon at the extremes of age. The disease is seen more frequently in the West and in affluent societies which may be due to variation in diet—a diet with low bulk or cellulose and high protein intake more often causes appendicitis.
- Haemorrhoids or piles are the varicosities of the haemorrhoidal veins. They are called 'internal piles' if dilatation is of superior haemorrhoidal plexus covered over by mucous membrane, and 'external piles' if they involve inferior haemorrhoidal plexus covered over by the skin. They are common lesions in elderly and pregnant women. They commonly result from increased venous pressure.

Added Information

The Endocrine Cells of the Gut

The lining epithelium of the stomach, and of the small and large intestines, contains scattered cells that have an endocrine function. These were recognised by early workers because of the presence, in them, of infranuclear granules that blackened with silver salts. They were, therefore, termed ***argentaffin*** cells. The granules also show a positive chromaffin reaction and have, therefore, also been called ***enterochromaffin*** cells. More recently, by the use of immunohistochemical methods, several other biologically active substances (amines or polypeptides) have been located in these cells. Many of these substances are also found in the nervous system where they function as neurotransmitters. They also act as hormones. This action is either local on neighbouring cells (paracrine effect); or on cells at distant sites (through the bloodstream). Very similar cells are also to be seen in the pancreas. All these cells are now grouped together under the term ***gastro-entero-pancreatic endocrine system***. Some of the cell types recognised, and their secretory products are given in the Table 16.2.

Some features of this system are similar to those of amine producing cells in other organs. All these are included under the term APUD cell system.

Table 16.2: Different types of endocrine cells in the gut and in the pancreas

Cell type	Secretory products	Distribution (+ = present)			
		Stomach	Small intestine	Large intestine	Pancreas
D_1	Vasoactive intestinal polypeptide	+	+	+	+
D	Somatostatin	+	+		+
EC_1	5HT + Substance P		+	+	
L	Enteroglucagon		+	+	
G	Gastric encephalin	+	+		
EC_2	5HT + Motilin		+		

contd...

S	Secretin		+		
I	Cholecystokinin pancreozymin		+		
K	Gastric inhibitory peptide		+		
N	Neurokinin		+		
ECn	5HT + unknown	+			
B	Insulin				+
A	Glucagon	?	?		+
PP	Pancreatic polypeptide		?		+

Chapter 17

Hepatobiliary System and Pancreas

The hepatobiliary system comprises of liver, gallbladder and extrahepatic ducts.

THE LIVER

Liver is the largest gland of the body situated mainly in the right hypochondrium, below the right dome of diaphragm in the abdomen.

The liver may be regarded as a modified exocrine gland that also has other functions. The liver substance is divisible into a large number of large lobes, each of which consists of numerous lobules (Plate 17.1).

MICROSCOPIC FEATURES

Glisson's Capsule

The liver is covered by a capsule (Glisson's capsule) made up of connective tissue. This connective tissue extends into the liver substance through the portal canals (mentioned above) where it surrounds the portal triads. The sinusoids are surrounded by reticular fibres. Connective tissue does not intervene between adjoining liver cells.

Hepatic Lobules

In sections through the liver, the substance of the organ appears to be made up of hexagonal areas that constitute the hepatic lobules (Fig. 17.1). In some species (e.g., the pig) the lobules are distinctly demarcated by connective tissue septa, but in the human liver the connective tissue is scanty and the lobules often appear to merge with one another.

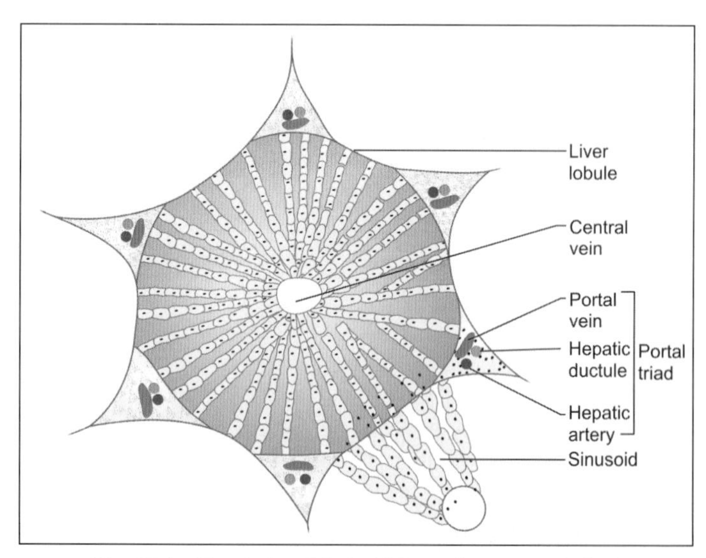

Fig. 17.1: Classic liver lobule (Schematic representation)

In transverse sections, each lobule appears to be made up of cords of liver cells that are separated by sinusoids. However, the cells are really arranged in the form of plates (one cell thick) that branch and anastomose with one another to form a network. Spaces within the network are occupied by sinusoids.

Portal Canal

Along the periphery of each lobule there are angular intervals filled by connective tissue. These intervals are called portal canals, the 'canals' forming a connective tissue network permeating the entire liver substance. Each 'canal' contains:

- A branch of the portal vein
- A branch of the hepatic artery
- An interlobular bile duct

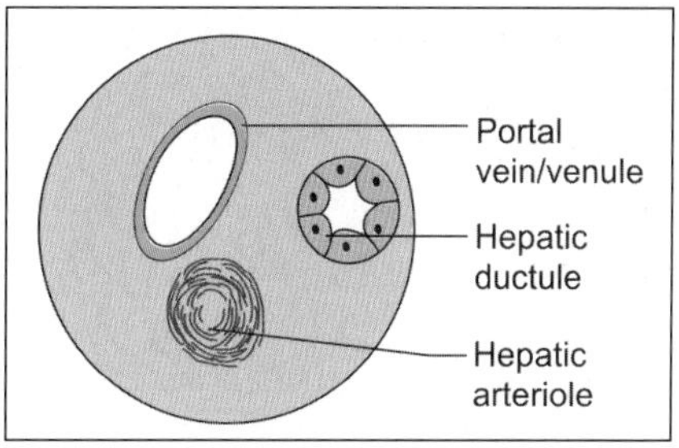

Fig. 17.2: Portal triad (Schematic representation)

These three structures collectively form a portal triad (Fig. 17.2). Blood from the branch of the portal vein, and from the branch of the hepatic artery, enters the sinusoids at the periphery of the lobule and passes towards its centre. Here the sinusoids open into a central vein that occupies the centre of the lobule. The central vein drains into hepatic veins (which leave the liver to end in the inferior vena cava).

Blood vessels and hepatic ducts present in portal canals are surrounded by a narrow interval called the space of Mall.

Portal Lobules

The vessels in a portal triad usually give branches to parts of three adjoining lobules. The area of liver tissue (comprising parts of three hepatic lobules) supplied by one branch of the portal vein is regarded by many authorities as the true functional unit of liver tissue, and is referred to as a portal lobule (Fig. 17.3).

A still smaller unit, the portal acinus has also been described. It consists of a diamond shaped area of liver tissue supplied by one hepatic arteriole (Fig. 17.4) running along the line of junction of two hepatic lobules. Two central veins lie at the ends of the acinus.

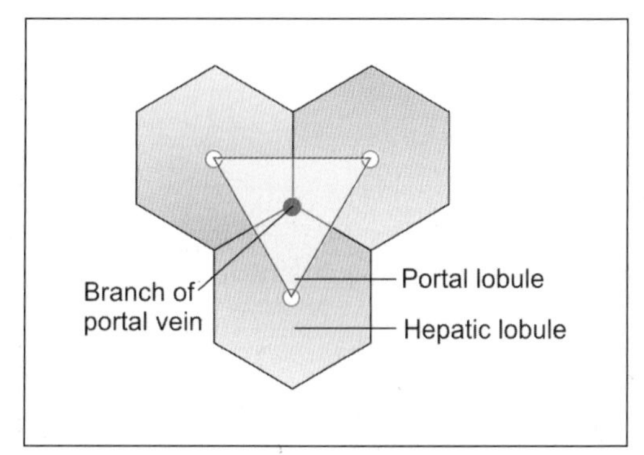

Fig. 17.3: Scheme to show the concept of portal lobules (pink). Hepatic lobules are shaded green. Note that the portal lobule is made up of parts of three hepatic lobules

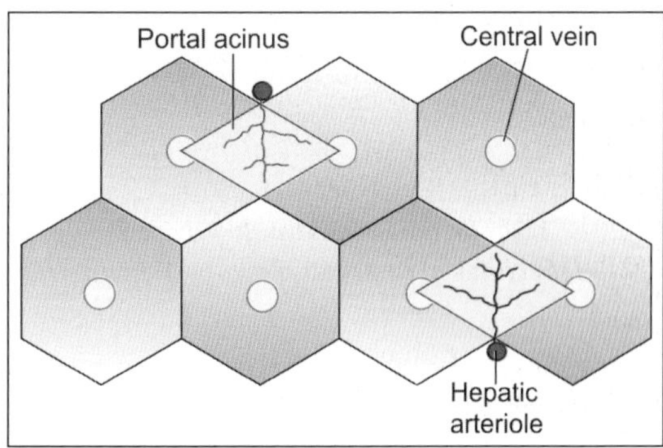

Fig. 17.4: Scheme to show the concept of portal acini (pink)

PLATE 17.1: Liver (Panoramic View)

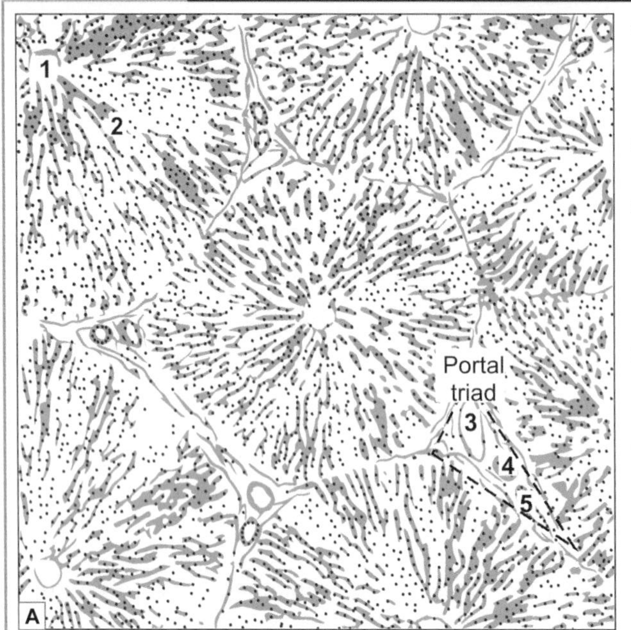

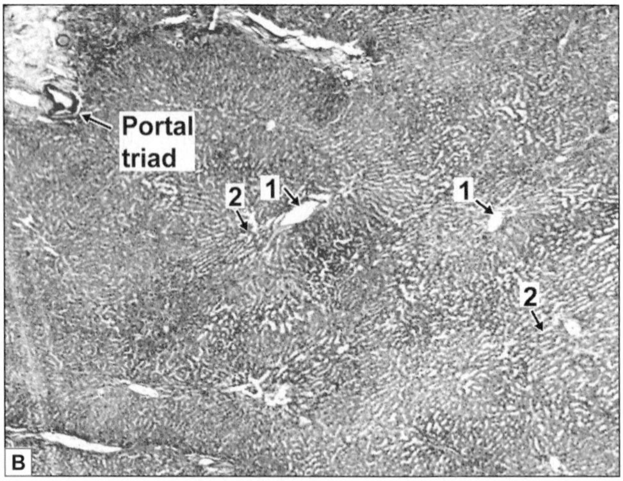

- The panoramic view of liver shows many hexagonal areas called hepatic lobules. The lobules are partially separated by connective tissue
- Each lobule has a small round space in the centre. This is the central vein
- A number of broad irregular cords of cells seem to pass from this vein to the periphery of the lobule. These cords are made up of polygonal liver cells—hepatocytes
- Along the periphery of the lobules there are angular intervals filled by connective tissue
- Each such area contains a branch of the portal vein, a branch of the hepatic artery, and an interlobular bile duct
- These three constitute a portal triad. The identification of hepatic lobules and of portal triads is enough to recognise liver tissue.

Liver (panoramic view). A. As seen in drawing; B. Photomicrograph

***Courtesy:* Atlas of Histopathology. Ist Edition. Ivan Damjanov. Jaypee Brothers. 2012. p142**

Key

1. Central vein
2. Radiating cords of hepatocytes
3. Branch of portal vein
4. Branch of hepatic artery
5. Interlobular duct

(3, 4, 5: Portal Triad)

Plate 17.2: Liver (High Magnification)

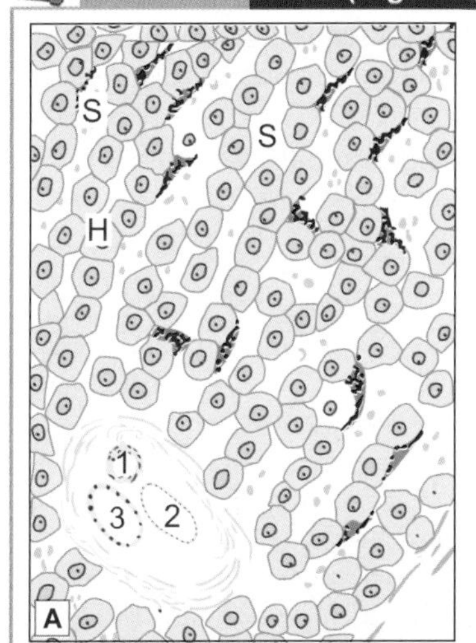

On high magnification:

- The lobule is made up of polygonal liver cells arranged in the form of radiating cords
- The central round nucleus of hepatocyte is surrounded by abundant pink cytoplasm
- The cords are separated from each other by spaces called sinusoids
- The sinusoids are lined by endothelial cells and Kupffer cells (macrophage cells)
- Plate 17.2C shows magnified view of portal tract
- Each portal tract contains a hepatic arteriole, portal venule and one or two interlobular bile ducts. Normally these structures are surrounded by fibroconnective tissue and a few lymphocytes.

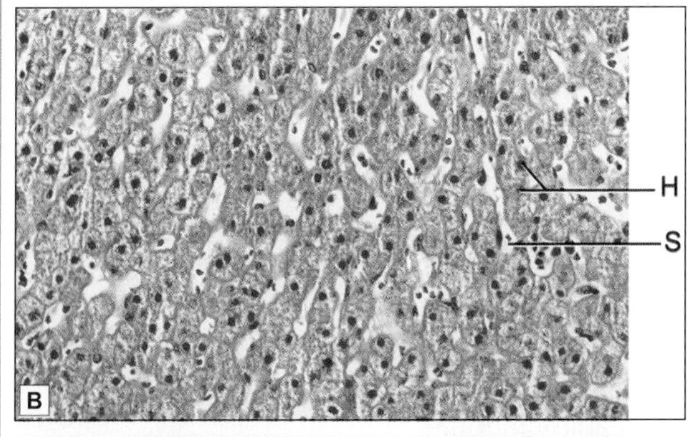

Key

1. Hepatic arteriole
2. Portal venule
3. Interlobular bile duct

S. Hepatic Sinusoids
H. Radiating cords of Hepatocytes

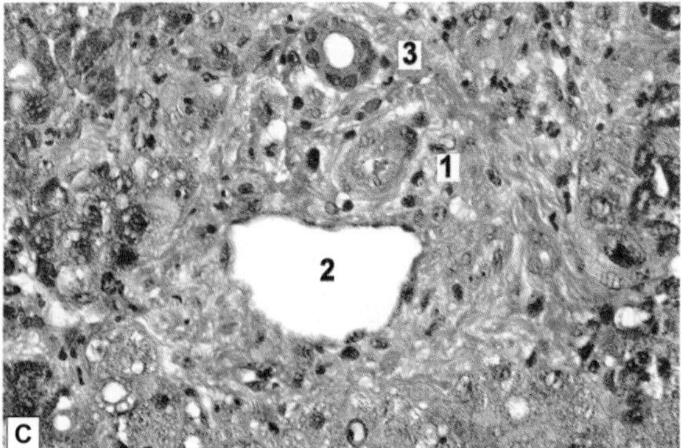

Liver (high magnification). A. As seen in drawing; B&C. Photomicrograph
Courtesy: **Atlas of Histopathology. Ist Edition. Ivan Damjanov. Jaypee Brothers. 2012. p142**

Duct System

Bile secreted by liver cells is poured into bile canaliculi. These canaliculi have no walls of their own. They are merely spaces present between plasma membranes of adjacent liver cells. The canaliculi form hexagonal networks around the liver cells. At the periphery of a lobule the canaliculi become continuous with delicate intralobular ductules, which in turn become continuous with larger interlobular ductules of portal triads.

The interlobular ductules are lined by cuboidal epithelium. Some smooth muscle is present in the walls of larger ducts.

Hepatocytes

Liver is made up, predominantly, of liver cells or hepatocytes. Each hepatocyte is a large cell with a round open-faced nucleus, with prominent nucleoli (Plate 17.2).

The cytoplasm of liver cells contains numerous mitochondria, abundant rough and smooth endoplasmic reticulum, a well developed Golgi complex, lysosomes, and vacuoles containing various enzymes. Numerous free ribosomes are present. These features are to be correlated with the high metabolic activity of liver cells. Stored glycogen, lipids, and iron (as crystals of ferratin and haemosiderin) are usually present. Glycogen is often present in relation to smooth endoplasmic reticulum. Many hepatocytes show two nuclei; or a single polyploid nucleus.

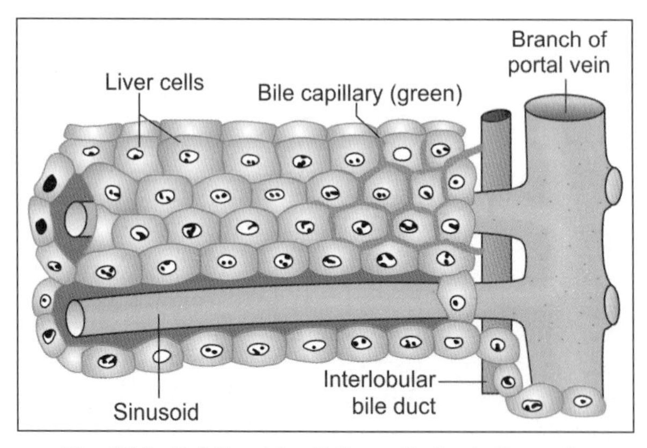

Fig. 17.5: Relationship of bile capillaries to liver cells (Schematic representation)

Liver cells are arranged in the form of anastomosing plates, one cell thick; and that the plates form a network in the spaces of which sinusoids lie (Fig. 17.5). In this way each liver cell has a sinusoid on two sides. The sinusoids are lined by an endothelium in which there are numerous pores (fenestrae). A basement membrane is not seen. Interspersed amongst the endothelial cells there are hepatic macrophages (Kupffer cells).

The surface of a hepatocyte can show three kinds of specialisation (Fig. 17.6):

- ***Sinusoidal surface***: The cell surfaces adjoining sinusoids bears microvilli that project into the space of Disse. The cell surface here also shows many coated pits that are concerned with exocytosis. Both these features are to be associated with active transfer of materials

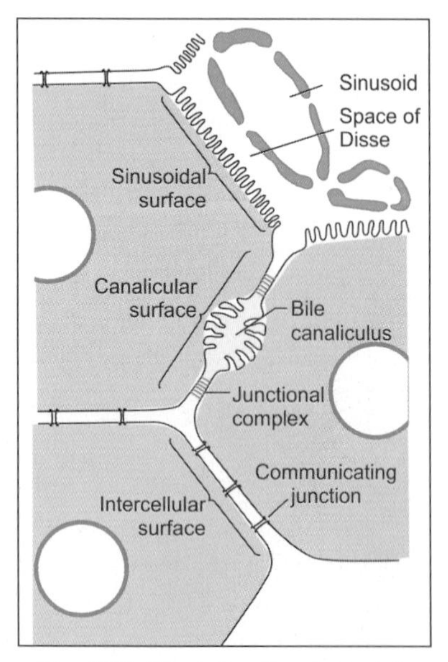

Fig. 17.6: Three functional specialisations of cell surface of a hepatocyte (Schematic representation)

from sinusoids to hepatocytes, and ***vice versa***. About 70% of the surface of hepatocytes is of this type.

- ***Canalicular surface***: Such areas of cell membrane bear longitudinal depressions that are apposed to similar depressions on neighbouring hepatocytes, to form the wall of a bile canaliculus. Irregular microvilli project into the canaliculus. On either side of the canaliculus, the cell membranes of adjoining cells are united by junctional complexes. About 15% of the hepatocyte surface is canalicular.
- ***Intercellular surface***: These are areas of cell surface where adjacent hepatocytes are united to each other just as in typical cells. Communicating junctions allow exchanges between the cells. About 15% of the hepatocyte surface is intercellular.

Space of Disse

The surface of the liver cell is separated from the endothelial lining of the sinusoid by a narrow perisinusoidal space (of Disse) (Fig. 17.7). Microvilli, present on the liver cells, extend into this space. As a result of these factors hepatocytes are brought into a very intimate relationship with the circulating blood. Some fat cells may also be seen in the space of Disse.

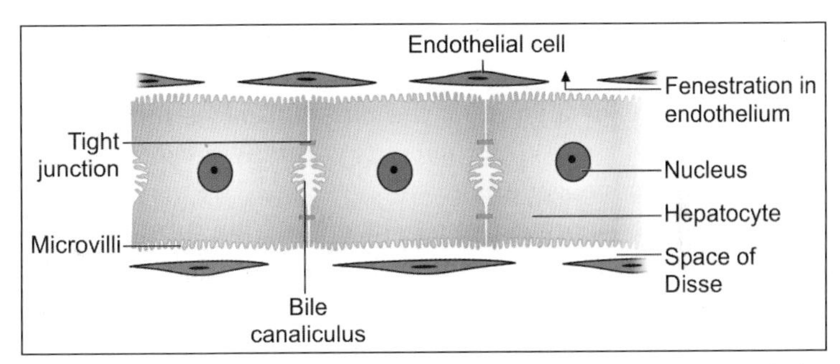

Fig. 17.7: Space of Disse and bile canaliculus (Schematic representation)

BILE

The exocrine secretion of the liver cells is called bile. Bile is poured out from liver cells into very delicate bile canaliculi that are present in intimate relationship to the cells. From the canaliculi bile drains into progressively larger ducts that end in the bile duct. This duct conveys bile into the duodenum where bile plays a role in digestion of fat.

BLOOD SUPPLY OF LIVER

In addition to deoxygenated blood reaching the liver through the portal vein (Fig. 17.8), the organ also receives oxygenated blood through the ***hepatic artery*** and its branches. The blood entering the liver from both these sources passes through the hepatic sinusoids and is collected by

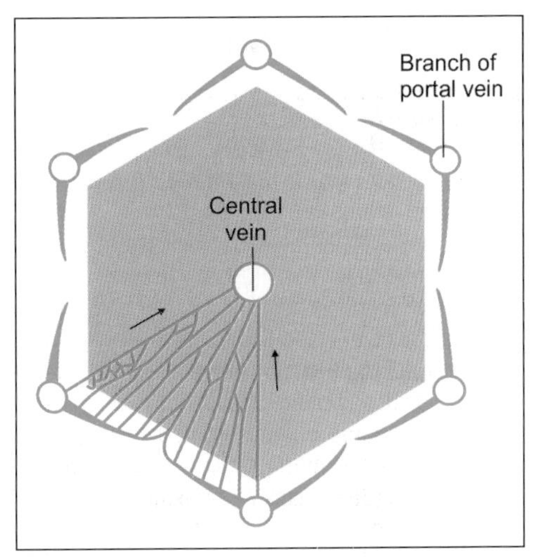

Fig. 17.8: Scheme to show the presence of several branches of the portal vein around a hepatic lobule. The manner in that they open into sinusoids is shown. The intervals between the sinusoids are occupied by liver cells (Schematic representation)

tributaries of hepatic veins. One such tributary runs through the centre of each lobule of the liver where it is called the ***central vein*** (Fig. 17.8).

Branches of the hepatic artery, the portal vein, and the hepatic ducts, travel together through the liver. The tributaries of hepatic veins follow a separate course.

FUNCTIONS OF LIVER

The liver performs numerous functions. Some of these are as follows:

- The liver acts as an exocrine gland for the secretion of bile. However, the architecture of the liver has greater resemblance to that of an endocrine gland, the cells being in intimate relationship to blood in sinusoids. This is to be correlated with the fact that liver cells take up numerous substances from the blood, and also pour many substances back into it.
- The liver plays a prominent role in metabolism of carbohydrates, proteins and fats. Metabolic functions include synthesis of plasma proteins fibrinogen and prothrombin, and the regulation of blood glucose and lipids.
- The liver acts as a store for various substances including glucose (as glycogen), lipids, vitamins and iron. When necessary, the liver can convert lipids and amino acids into glucose (gluconeogenesis).
- The liver plays a protective role by detoxifying substances (including drugs and alcohol). Removal of bile pigments from blood (and their excretion through bile) is part of this process. Amino acids are deaminated to produce urea, which enters the bloodstream to be excreted through the kidneys. The macrophage cells (of Kupffer) lining the sinusoids of the liver have a role similar to that of other cells of the mononuclear phagocyte system. They are of particular importance as they are the first cells of this system that come in contact with materials absorbed through the gut. They also remove damaged erythrocytes from blood.
- During fetal life the liver is a centre for haemopoiesis.

Clinical Correlation

- Inflammation in the liver is called hepatitis. It is frequently caused by viruses (viral hepatitis), and by a protozoan parasite ***Entamoeba histolytica*** (amoebic hepatitis). An abscess may form in the liver as a sequel of amoebic hepatitis.
- Cirrhosis of the liver is a disease in which many hepatocytes are destroyed, the areas being filled by fibrous tissue. This gradually leads to collapse of the normal architecture of the liver.
- One effect of cirrhosis of the liver is to disrupt the flow of blood through the liver. As a result of increased resistance to blood flow there is increased blood pressure in the portal circulation (***portal hypertension***). In portal hypertension anastomoses between the portal and systemic veins dilate to form varices (e.g., at the lower end of the oesophagus). Rupture of these varices can result in fatal bleeding.
- When a large number of hepatocytes are destroyed this leads to liver failure. All the above listed functions are interfered with. Hepatic failure may be acute or chronic. Accumulation of waste products in blood (due to lack of detoxification by the liver) ultimately leads to unconsciousness (***hepatic coma***) and death.

EXTRAHEPATIC BILIARY APPARATUS

The extrahepatic biliary apparatus consists of the gallbladder and the extrahepatic bile ducts.

THE GALLBLADDER

Gallbladder is a muscular sac situated on the visceral surface of liver in the fossa for gallbladder. The gallbladder stores and concentrates bile. This bile is discharged into the duodenum when required.
The wall of the gallbladder is made up of:
- A mucous membrane
- A fibromuscular coat
- A serous layer that covers part of the organ (Plate 17.3)

Mucous Membrane

The mucous membrane of the gall bladder is lined by a tall columnar epithelium with a striated border. The mucosa is highly folded. In sections, the folds may look like villi.

Note: Because of the resemblance to villi, students sometimes mistake sections of the gallbladder for those of the intestines. The two are easily distinguished if it is remembered that there are no goblet cells in the epithelium of the gallbladder.

Fibromuscular Coat

The fibromuscular coat is made up mainly of connective tissue containing the usual elements. Smooth muscle fibres are present and run in various directions.

Serosa

The serous layer has a lining of mesothelium resting on connective tissue. The fundus and lower surface of body of gallbladder is covered by serosa, whereas the upper surface is attached to the fossa for gallbladder by means of connective tissue (adventitia).

Cells of GallBladder

With the EM the lining cells of the gallbladder are seen to have irregular microvilli on their luminal surfaces. Near the lumen the lateral margins of the cells are united by prominent junctional complexes. More basally the lateral margins are separated by enlarged intercellular spaces into which complex folds of plasma membrane extend. Numerous blood capillaries are present near the bases of the cells.

These features indicate that bile is concentrated by absorption of water at the luminal surface of the cell. This water is poured out of the cell into basal intercellular spaces from where it passes into blood. Absorption of salt and water from bile into blood is facilitated by presence of Na+ and K+ ATPases in cell membranes of cells lining the gallbladder.

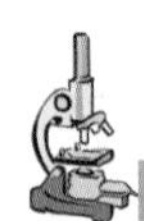

PLATE 17.3: Gallbladder

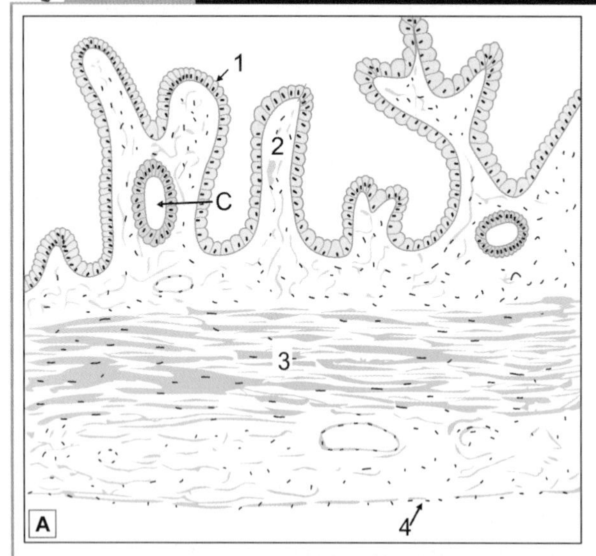

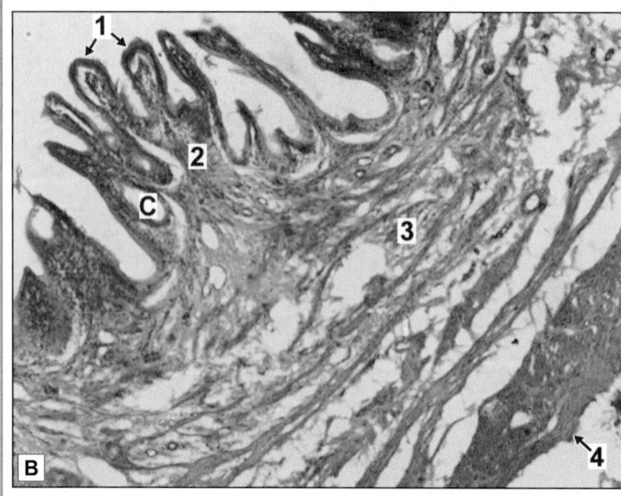

Gallbladder. A. As seen in drawing; B. Photomicrograph

- The mucous membrane is lined by tall columnar cells with striated border
- The mucosa is highly folded and some of the folds might look like villi
- Crypts may be found in lamina propria
- Submucosa is absent.
- The muscle coat is poorly developed there being numerous connective tissue fibres amongst the muscle fibres. This is called as fibromuscular coat
- A serous covering lined by flattened mesothelium is seen.

Note: Gallbladder can be differentiated from small intestine by –
- Absence of villi
- Absence of goblet cells
- Absence of submucosa
- Absence of proper muscularis externa

Key

1. Mucous membrane lined by tall columnar cells with striated border
2. Lamina propria
3. Fibromuscular coat
4. Serosa

C. Crypt in lamina propria

Clinical Correlation

Inflammation of the gallbladder is called cholecystitis. Stones may form in the gallbladder (gallstones; cholelithiasis). In such cases surgical removal of the gallbladder may be necessary (cholecystectomy).

THE EXTRAHEPATIC DUCTS

These are the right, left and common hepatic ducts; the cystic duct; and the bile duct. All of them have a common structure. They have a mucosa surrounded by a wall made up of connective tissue, in which some smooth muscle may be present.

The mucosa is lined by a tall columnar epithelium with a striated border.

Hepato-pancreatic Duct

At its lower end the bile duct is joined by the main pancreatic duct, the two usually forming a common hepato-pancreatic duct (or ampulla) that opens into the duodenum at the summit of the major duodenal papilla.

The mucosa of the hepato-pancreatic duct is highly folded. These folds are believed to constitute a valvular mechanism that prevents duodenal contents from entering the bile and pancreatic ducts.

Sphincter of Oddi

Well developed smooth muscle is present in the region of the lower end of the bile duct. This muscle forms the ***sphincter of Oddi***.

From a functional point of view this sphincter consists of three separate parts. The ***sphincter choledochus*** surrounds the lower end of the bile duct. It is always present, and its contraction is responsible for filling of the gallbladder. A less developed ***sphincter pancreaticus*** surrounds the terminal part of the main pancreatic duct (Fig. 17.9).

A third sphincter surrounds the hepato-pancreatic duct (or ampulla) and often forms a ring round the lower ends of both the bile and pancreatic ducts. This is the ***sphincter ampullae***. The sphincter ampullae and the sphincter pancreaticus are often missing.

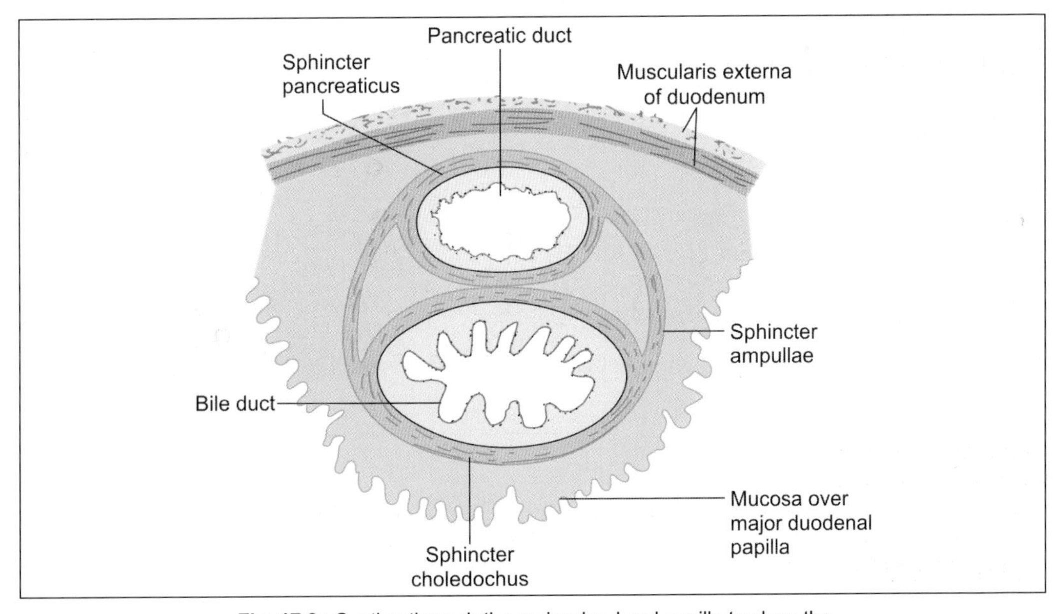

Fig. 17.9: Section through the major duodenal papilla to show the components of the sphincter of Oddi (Schematic representation)

Clinical Correlation

Blockage of the bile duct (by inflammation, by a gallstone, or by carcinoma) leads to accumulation of bile in the biliary duct system, and within the bile capillaries. As pressure in the passages increases bile passes into blood leading to jaundice. The sclera, the skin, and the nails appear to be yellow in colour, and bile salts and pigments are excreted in urine. Jaundice occurring as a result of such obstruction is called obstructive jaundice. Jaundice is seen in the absence of obstruction in cases of hepatitis.

A gallstone passing through the bile duct can cause severe pain. This pain is ***biliary colic***.

THE PANCREAS

The pancreas extends from the concavity of the duodenum, on the right to the spleen and on the left in the posterior abdominal wall retroperitoneally. The pancreas is covered by connective tissue that forms a capsule for it. Septa arising from the capsule extend into the gland dividing it into lobules. Each pancreatic islet is surrounded by a network of reticular fibres.

It is a gland that is partly exocrine, and partly endocrine, the main bulk of the gland being constituted by its exocrine part (Plate 17.4).

- The exocrine pancreas secretes enzymes that play a very important role in the digestion of carbohydrates, proteins and fats. After digestion, and absorption through the gut, these products are carried to the liver through the portal vein.
- The endocrine pancreas produces two very important hormones, ***insulin*** and ***glucagon***. These two hormones are also carried through the portal vein to the liver where they have a profound influence on the metabolism of carbohydrates, proteins and fats.

The functions of the exocrine and endocrine parts of the pancreas are thus linked. The linkage between the two parts is also seen in their common embryonic derivation from the endodermal lining of the gut.

THE EXOCRINE PANCREAS

The exocrine pancreas is in the form of compound tubuloalveolar serous gland.

Note: Its general structure is very similar to that of the parotid gland, but the two are easily distinguished because of the presence in the pancreas of endocrine elements.

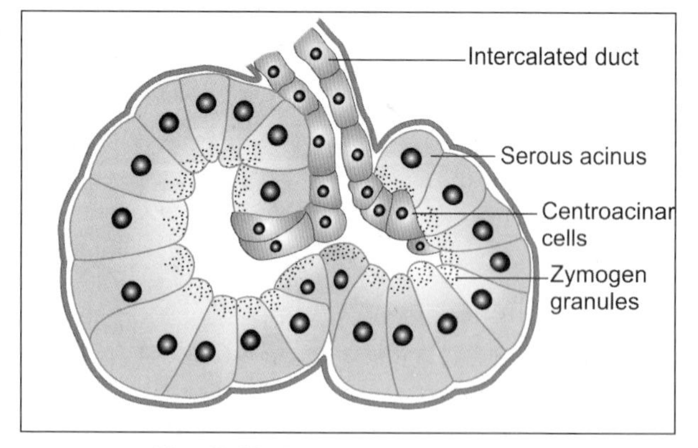

Fig. 17.10: Pancreatic serous acinus

Capsule

A delicate capsule surrounds the pancreas. Septa extend from the capsule into the gland and divide it into lobules.

Pancreatic Acini

The secretory elements of the exocrine pancreas are long and tubular (but they are usually described as acini as they appear rounded or oval in sections). Their lumen is small (Fig. 17.10).

PLATE 17.4: Pancreas

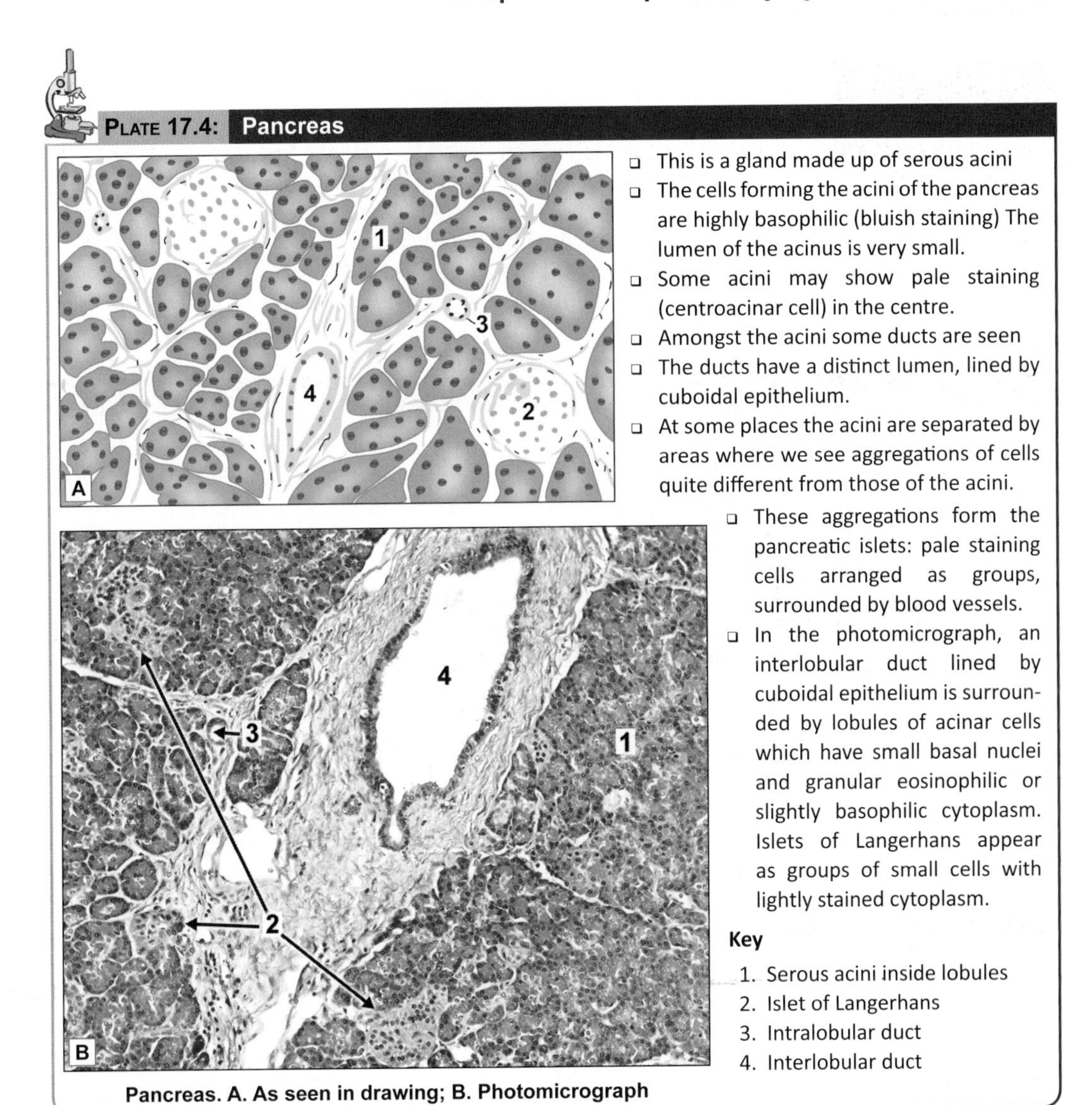

- This is a gland made up of serous acini
- The cells forming the acini of the pancreas are highly basophilic (bluish staining) The lumen of the acinus is very small.
- Some acini may show pale staining (centroacinar cell) in the centre.
- Amongst the acini some ducts are seen
- The ducts have a distinct lumen, lined by cuboidal epithelium.
- At some places the acini are separated by areas where we see aggregations of cells quite different from those of the acini.
- These aggregations form the pancreatic islets: pale staining cells arranged as groups, surrounded by blood vessels.
- In the photomicrograph, an interlobular duct lined by cuboidal epithelium is surrounded by lobules of acinar cells which have small basal nuclei and granular eosinophilic or slightly basophilic cytoplasm. Islets of Langerhans appear as groups of small cells with lightly stained cytoplasm.

Key

1. Serous acini inside lobules
2. Islet of Langerhans
3. Intralobular duct
4. Interlobular duct

Pancreas. A. As seen in drawing; B. Photomicrograph

Secretory Cells

The cells lining the alveoli appear triangular in section, and have spherical nuclei located basally. In sections stained with haematoxylin and eosin the cytoplasm is highly basophilic (blue) particularly in the basal part. With suitable fixation and staining numerous secretory (or zymogen) granules can be demonstrated in the cytoplasm, specially in the apical part of the cell. These granules are eosinophilic. They decrease considerably after the cell has poured out its secretion.

With the EM the cells lining the alveoli show features that are typical of secretory cells. Their basal cytoplasm is packed with rough endoplasmic reticulum (this being responsible for the basophilia of this region). A well developed Golgi complex is present in the supranuclear part of the cell. Numerous secretory granules (membrane bound, and filled with enzymes) occupy the greater part of the cytoplasm (except the most basal part).

Centroacinar Cells

In addition to secretory cells, the alveoli of the exocrine pancreas contain ***centroacinar cells*** that are so called because they appear to be located near the centre of the acinus. These cells really belong to the intercalated ducts (see below) that are invaginated into the secretory elements. Some cell bodies of autonomic neurons, and undifferentiated cells are also present in relation to the secretory elements.

Added Information

The secretory cells produce two types of secretion:

- One of these is watery and rich in bicarbonate. Bicarbonate is probably added to pancreatic secretion by cells lining the ducts. It helps to neutralise the acid contents entering the duodenum from the stomach. Production of this secretion is stimulated mainly by the hormone ***secretin*** liberated by the duodenal mucosa.
- The other secretion is thicker and contains numerous enzymes (including trypsinogen, chymotrypsinogen, amylase, lipases, etc.). The production of this secretion is stimulated mainly by the hormone cholecystokinin (pancreozymin) liberated by endocrine cells in the duodenal mucosa.

Secretion by cells of the exocrine pancreas, and the composition of the secretion, is influenced by several other amines produced either in the gastrointestinal mucosa or in pancreatic islets. (These include gastrin, vasoactive intestinal polypeptide, and pancreatic polypeptide). Secretion is also influenced by autonomic nerves, mainly parasympathetic.

The enzymes are synthesised in the rough endoplasmic reticulum. From here they pass to the Golgi complex where they are surrounded by membranes, and are released into the cytoplasm as secretory granules. The granules move to the luminal surface of the cell where the secretions are poured out by exocytosis. Within the cell the enzymes are in an inactive form. They become active only after mixing with duodenal contents. Activation is influenced by enzymes present in the epithelium lining the duodenum.

Duct System

Secretions produced in the acini are poured into ***intercalated ducts*** (also called ***intralobular ducts***). These ducts are invaginated deeply into the secretory elements (Fig. 17.10). As a result of this invagination, the interacalated ducts are not conspicuous in sections.

From the intercalated ducts the secretions pass into larger, ***interlobular ducts***. They finally pass into the duodenum through the ***main pancreatic duct*** and the ***accessory pancreatic duct.*** The cells lining the pancreatic ducts control the bicarbonate and water content of pancreatic secretion. These actions are under hormonal and neural control. The walls of the larger ducts are formed mainly of fibrous tissue. They are lined by a columnar epithelium.

The terminal part of the main pancreatic duct is surrounded by a sphincter. A similar sphincter may also be present around the terminal part of the accessory pancreatic duct.

THE ENDOCRINE PANCREAS

The endocrine pancreas is in the form of numerous rounded collections of cells that are embedded within the exocrine part. These collections of cells are called the pancreatic islets, or the islets of Langerhans. The human pancreas has about one million islets. They are most numerous in the tail of the pancreas.

Each islet is separated from the surrounding alveoli by a thin layer of reticular tissue. The islets are very richly supplied with blood through a dense capillary plexus. The intervals between the capillaries are occupied by cells arranged in groups or as cords. In ordinary preparations stained with haematoxylin and eosin, all the cells appear similar, but with the use of special procedures three main types of cells can be distinguished.

Alpha Cells (A-Cells)

In islets of the human pancreas the ***alpha cells*** (or ***A-cells***) tend to be arranged towards the periphery (or cortex) of the islets. They form about 20% of the islet cells. They contain smaller granules that stain brightly with acid fuchsin. They do not stain with aldehyde fuchsin.

When seen with electron microscopy, the granules of alpha cells (A2) appear to be round or ovoid with high electron density.

The alpha cells secrete the hormone ***glucagon***.

Beta Cells (B-Cells)

The ***beta cells*** (or ***B-cells***) tend to lie near the centre (or medulla) of the islet. About 70% of the cells of islet are of this type. The beta cells contain granules (larger than alpha cells) that can be stained with aldehyde fuchsin. When seen with electron microscopy, the granules of beta cells are fewer, larger, and of less electron density than those of alpha cells.

The beta cells secrete the hormone ***insulin***.

Delta Cells (D-Cells)

Delta cells (or ***D-cells***), like alpha cells, are also peripherally placed. The delta cells (also called type III cells) stain black with silver salts (i.e., they are argyrophile). They resemble alpha cells in having granules that stain with acid fuchsin; and are, therefore, sometimes called A1 cells in distinction to the glucagon producing cells that are designated A2. The two can be distinguished by the fact that A2 cells are not argyrophile.

When seen with electron microscopy, the granules of delta cells (A1) appear to be round or ovoid with low electron density.

The delta cells probably produce the hormones gastrin and somatostatin. Somatostatin inhibits the secretion of glucagon by alpha cells, and (to a lesser extent) that of insulin by beta cells.

Added Information

Apart from the three main types of cells described above some other types are also present in the islets of Langerhans. These are the PP cells containing pancreatic polypeptide (and located mainly in the head and neck of the pancreas), and D1 cells (or type IV cells) probably containing vasoactive intestinal polypeptide (or a similar amine). A few cells secreting serotonin, motilin and substance P are also present.

BLOOD SUPPLY

The gland is richly supplied with blood vessels that run through the connective tissue. The capillary network is most dense in the islets. Here the endothelial lining is fenestrated providing intimate contact of islet cells and circulating blood.

Lymphatics are also present in the pancreas.

NERVE SUPPLY

The connective tissue of the pancreas also serves as a pathway for nerve fibres, both myelinated and unmyelinated. Groups of neurons are also present.

Pancreatic islets are richly innervated by autonomic nerves. Noradrenaline and acetyl choline released at nerve endings influence secretion by islet cells.

Clinical Correlation

- **Acute pancreatitis:** Acute pancreatitis is an acute inflammation of the pancreas presenting clinically with 'acute abdomen'. The condition occurs in adults between the age of 40 and 70 years and is more common in females than in males. The onset of acute pancreatitis is sudden, occurring after a bout of alcohol or a heavy meal. Characteristically, there is elevation of ***serum amylase*** level within the first 24 hours and elevatd serum lipase level after 3–4 days.
- **Chronic pancreatitis:** Chronic pancreatitis or ***chronic relapsing pancreatitis*** is the progressive destruction of the pancreas due to repeated mild and subclinical attacks of acute pancreatitis. Most patients present with recurrent attacks of severe abdominal pain at intervals of months to years. Weight loss and jaundice are often associated.

Chapter 18

The Urinary System

The urinary organs are:
- A pair of kidneys
- A pair of ureters
- The urinary bladder
- The urethra.

Urine production, and the control of its composition, is exclusively the function of the kidneys. The urinary bladder is responsible for storage of urine until it is voided. The ureter and urethra are simple passages for transport of urine.

Functions

- The urinary organs are responsible for the production, storage, and passing of urine. Many harmful waste products (that result from metabolism) are removed from blood through urine. These include urea and creatinine that are end products of protein metabolism.
- Many drugs, or their breakdown products, are also excreted in urine.
- Considerable amount of water is excreted through urine. The quantity is strictly controlled being greatest when there is heavy intake of water, and least when intake is low or when there is substantial water loss in some other way (for example by perspiration in hot weather). This enables the water content of plasma and tissues to remain fairly constant.

Note: In diseased conditions urine can contain glucose (as in diabetes mellitus), or proteins (in kidney disease), the excretion of which is normally prevented.

THE KIDNEYS

Each kidney has a characteristic bean-like shape. A thin layer of fibrous tissue, which constitutes the capsule, intimately covers kidney tissue. The capsule of a healthy kidney can be easily stripped off, but it becomes adherent in some diseases.

The kidney has a convex lateral margin; and a concavity on the medial side that is called the ***hilum***. The hilum leads into a space called the ***renal sinus***. The renal sinus is occupied by the upper expanded part of the ureter called the ***renal pelvis***.

Within the renal sinus the pelvis divides into two (or three) parts called ***major calyces***. Each major calyx divides into a number of ***minor calyces*** (Fig. 18.1). The end of each minor calyx is shaped like a cup. A projection of kidney tissue, called a ***papilla*** fits into the cup.

Kidney tissue consists of an outer part called the ***cortex***, and an inner part called the ***medulla*** (Fig. 18.1 and Plate 18.1).

Medulla

The medulla is made up of triangular areas of renal tissue that are called the ***renal pyramids*** (Fig. 18.1). Each pyramid has a base directed towards the cortex; and an apex (or papilla) that is directed towards the renal pelvis, and fits into a minor calyx. Pyramids show striations that pass radially towards the apex.

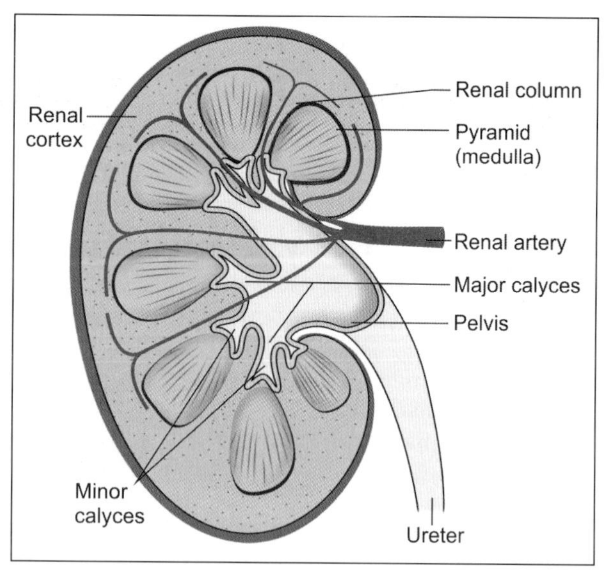

Fig. 18.1: Some features to be seen in a coronal section through the kidney (Schematic representation)

Cortex

The renal cortex consists of the following:

- Tissue lying between the bases of the pyramids and the surface of the kidney, forming the ***cortical arches*** or ***cortical lobules***. This part of the cortex shows light and dark striations. The light lines are called ***medullary rays*** (Plate 18.1).
- Tissue lying between adjacent pyramids is also a part of the cortex. This part constitutes the ***renal columns***.
- In this way each pyramid comes to be surrounded by a 'shell' of cortex. The pyramid and the cortex around it constitute a lobe of the kidney. This lobulation is obvious in the fetal kidney.

Added Information

Interstitial Tissue of the Kidney

Most of the interstitial space in the renal cortex is occupied by blood vessels and lymphatics. In the medulla the interstitium is composed mainly of a matrix containing proteins and glycosaminoglycans. Collagen fibres and interstitial cells are present.

It has been held that interstitial cells produce prostaglandins, but it now appears that prostaglandins are produced by epithelial cells of collecting ducts.

The Uriniferous Tubules

From a functional point of view the kidney may be regarded as a collection of numerous ***uriniferous tubules*** that are specialised for the excretion of urine. Each uriniferous tubule consists of an excretory part called the ***nephron***, and of a ***collecting tubule***. The collecting tubules draining different nephrons join to form larger tubules called the ***papillary ducts*** (of Bellini), each of which opens into a minor calyx at the apex of a renal papilla. Each kidney contains one to two million nephrons.

Urinary tubules are held together by scanty connective tissue. Blood vessels, lymphatics and nerves lie in this connective tissue.

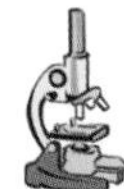

PLATE 18.1: Kidney (Low Magnification)

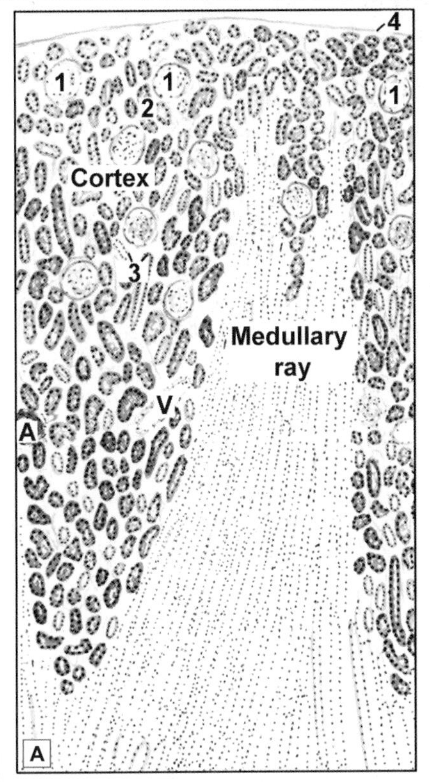

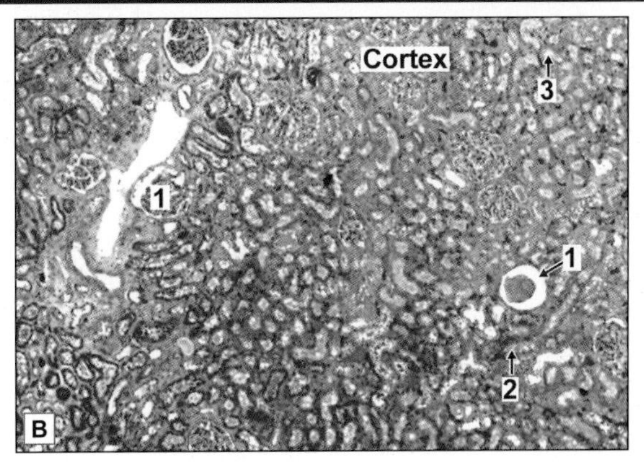

Key

1. Renal corpuscle
2. Proximal convoluted tubule
3. Distal convoluted tubule
4. Capsule

A. Artery

V. Vein

Kidney (low magnification). A. As seen in drawing; B. Photomicrograph of cortex

- The kidney is covered by a capsule
- Deep to the capsule there is the cortex
- Deep to the cortex there is the medulla of the kidney
- In the cortex we see circular structures called renal corpuscles surrounding which there are tubules cut in various shapes
- The dark pink stained tubules are parts of the proximal convoluted tubules (PCT): their lumen is small and indistinct. It is lined by cuboidal epithelium with brush border
- Lighter staining tubules, each with a distinct lumen, are the distal convoluted tubules (DCT). They are lined by simple cuboidal epithelium
- PCT are more in number than DCT
- In the medulla we see very light staining, elongated, parallel running tubules. These are collecting ducts and loop of Henle. Some of them extend into the cortex forming a medullary ray. The collecting ducts are lined by simple cuboidal epithelium and loop of Henle (thin segments) are lined by simple squamous epithelium
- Cut sections of blood vessels are seen both in the cortex and medulla.

Note: When we look at a section of the kidney we see that most of the area is filled with a very large number of tubules. These are of various shapes and have different types of epithelial lining. This fact by itself suggests that the tissue is the kidney.

Nephron

Nephron is the structural and functional unit of kidney and there are about 1–4 million nephrons in each kidney. The nephron consists of a ***renal corpuscle*** (or ***Malpighian corpuscle***), and a long complicated ***renal tubule***. Renal tubule is made up of three parts:

- The proximal convoluted tubule
- Loop of Henle
- The distal convoluted tubule (Fig. 18.2)

Renal corpuscle is situated in the cortex of the kidney either near the periphery or near the medulla. Based on the situation of renal corpuscle, the nephrons are classified into two types:

- Cortical nephrons or superficial nephrons (which have their corpuscles in the outer cortex).
- Juxtamedullary nephrons (which have their corpuscles in the inner cortex near medulla or corticomedullary junction).

Renal corpuscles, and (the greater parts of) the proximal and distal convoluted tubules are located in the cortex of the kidney. The loops of Henle and the collecting ducts lie in the medullary rays and in the substance of the pyramids.

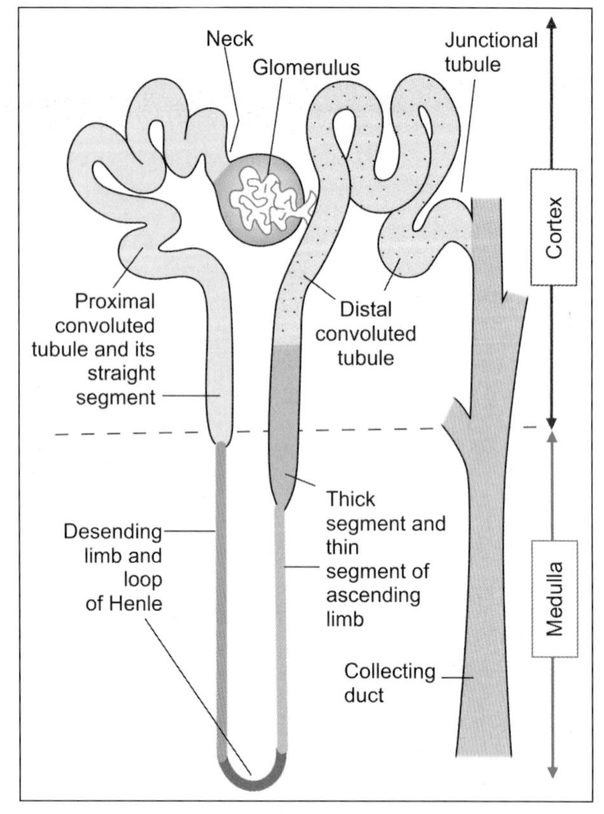

Fig. 18.2: Parts of a nephron. A collecting duct is also shown (Schematic representation)

The Renal Corpuscle

The renal corpuscle is a rounded structure consisting of (a) a rounded tuft of blood capillaries called the ***glomerulus***; and (b) a cup-like, double layered covering for the glomerulus called the ***glomerular capsule*** (or ***Bowman's capsule***) (Fig. 18.3). The glomerular capsule represents the cup-shaped blind beginning of the renal tubule. Between the two layers of the capsule there is a ***urinary space*** that is continuous with the lumen of the renal tubule.

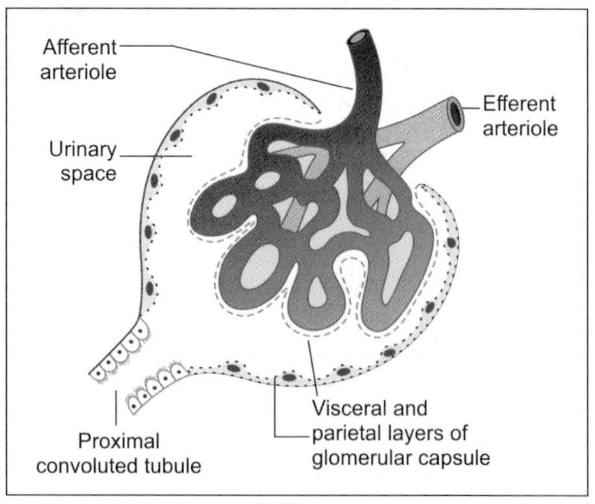

Fig. 18.3: Basic structure of a renal corpuscle (Schematic representation)

Plate 18.2: Renal Cortex (High Magnification)

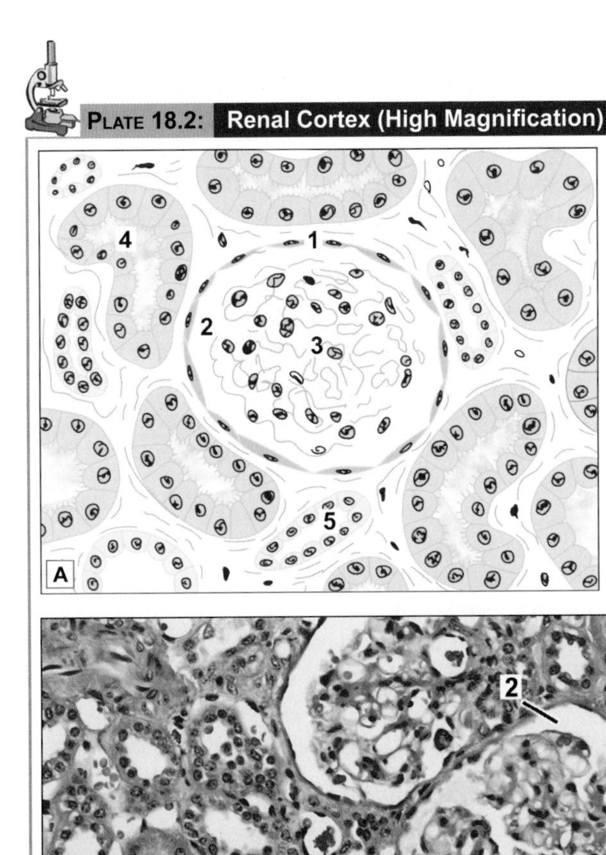

Renal cortex (high magnification). A. As seen in drawing; B. Photomicrograph

- In the high power view of renal cortex large renal corpuscles can be identified
- The renal corpuscle consists of a tuft of capillaries that form a rounded glomerulus, and an outer wall, the glomerular capsule (Bowman's capsule)
- A urinary space between the glomerulus and the capsule is seen
- Proximal convoluted tubules are dark staining. They are lined by cuboidal cells with a prominent brush border. Their lumen is indistinct
- Distal convoluted tubules are lighter staining. The cuboidal cells lining them do not have a brush border. Their lumen is distinct.

Key

1. Bowman's capsule
2. Urinary space
3. Glomerulus
4. Proximal convoluted tubule
5. Distal convoluted tubule

Glomerulus

The glomerulus is a rounded tuft of anastomosing capillaries (Plate 18.2). Blood enters the tuft through an afferent arteriole and leaves it through an efferent arteriole (Note that the efferent vessel is an arteriole, and not a venule. It again breaks up into capillaries). The afferent and efferent arterioles lie close together at a point that is referred to as the ***vascular pole*** of the renal corpuscle.

The Mesangium

On entering the glomerulus the afferent arteriole divides (usually) into five branches, each branch leading into an independent capillary network. The glomerular circulation can, therefore, be divided into a number of lobules or segments.

Glomerular capillaries are supported by the ***mesangium*** that is made up of mesangial cells surrounded by a non-cellular mesangial matrix. The mesangium forms a mesentery-like fold over the capillary loop. Mesangial cells give off processes that run through the matrix. Mesangium intervenes between the capillaries of the glomerular segments.

Mesangial cells contain filaments similar to myosin. They bear angiotensin II receptors. It is believed that stimulation by angiotensin causes the fibrils to contract. In this way mesangial cells may play a role in controlling blood flow through the glomerulus. Other functions attributed to mesangial cells include phagocytosis, and maintenance of glomerular basement membrane. The mesangium becomes prominent in a disease called glomerulonephritis.

Glomerular Capsule

The glomerular capsule is a double-layered cup, the two layers of which are separated by the urinary space. The outer layer is lined by squamous cells. With the light microscope the inner wall also appears to be lined by squamous cells, but the EM shows that these cells, called ***podocytes***, have a highly specialised structure. The urinary space becomes continuous with the lumen of the renal tubule at the ***urinary pole*** of the renal corpuscle. This pole lies diametrically opposite the vascular pole.

Podocytes

The ***podocytes*** are so called because they possess foot-like processes. Each podocyte has a few ***primary processes*** that give the cell a star-shaped appearance (Fig. 18.4). These processes are wrapped around glomerular capillaries and interdigitate with those of neighbouring podocytes. Each primary process terminates in numerous ***secondary processes*** also called ***pedicels*** (or end feet) that rest on the basal lamina. The cell body of the podocyte comes in contact with the basal lamina only through the pedicels.

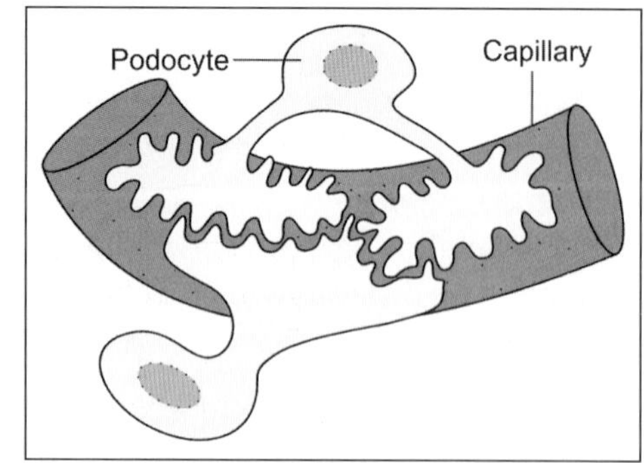

Fig. 18.4: Relationship of podocytes to a glomerular capillary. Note that the entire surface of the capillary is covered by processes of podocytes, the bare areas being shown only for sake of clarity (Schematic representation)

Glomerular Basement Membrane

As compared to typical membranes the glomerular basement membrane is very thick (about 300 nm). It is made up of three layers. There is a central electron dense layer (***lamina densa***), and inner and outer electron lucent layers (***lamina rara interna*** and ***externa***). The lamina densa contains a network of collagen (type IV) fibrils, and thus acts as a physical barrier. The electron lucent layers contain the glycosaminoglycan heparan sulphate. This bears the negative charges referred to above. The glomerular basement membrane is, therefore, both a physical barrier and an electrical barrier to the passage of large molecules.

As shown in Figure 18.5, the glomerular basement membrane (and overlying podocyte cytoplasm) does not go all round a glomerular capillary. The gap is filled in by mesangium.

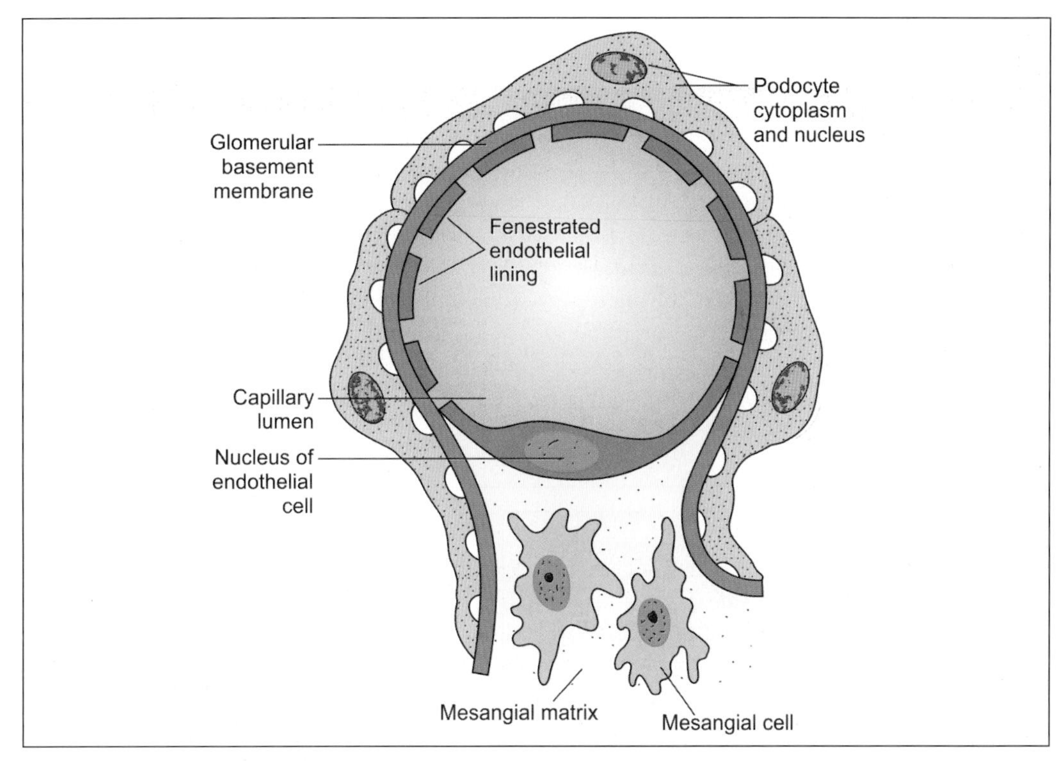

Fig. 18.5: Relationship of a glomerular capillary to podocytes, basement membrane and mesangium (Schematic representation)

A thin membrane continuous with the lamina rara interna may separate endothelial cytoplasm from mesangium. In the interval between capillaries the basement membrane is in contact with mesangium.

The Renal Tubule

The renal tubule is divisible into several parts that are shown in Figure 18.2 and Plate 18.3. Starting from the glomerular capsule (in proximodistal sequence). They are: (**a**) the ***proximal convoluted tubule***; (**b**) the ***loop of Henle*** consisting of a ***descending limb***, a ***loop***, and an ***ascending limb***; and (**c**) the ***distal convoluted tubule***, which ends by joining a collecting tubule. Along its entire length the renal tubule is lined by a single layer of epithelial cells that are supported on a basal lamina.

Proximal Convoluted Tubule

The junction of the proximal convoluted tubule with the glomerular capsule is narrow and is referred to as the ***neck***. The proximal convoluted tubule is made up of an initial part having many convolutions (lying in the cortex), and of a terminal straight part that descends into the medulla to become continuous with the descending limb of the loop of Henle.

The ***neck*** is lined by simple squamous epithelium continuous with that of the glomerular capsule (Some texts refer to the neck as part of the glomerulus).

PLATE 18.3: Renal Medulla (High Magnification)

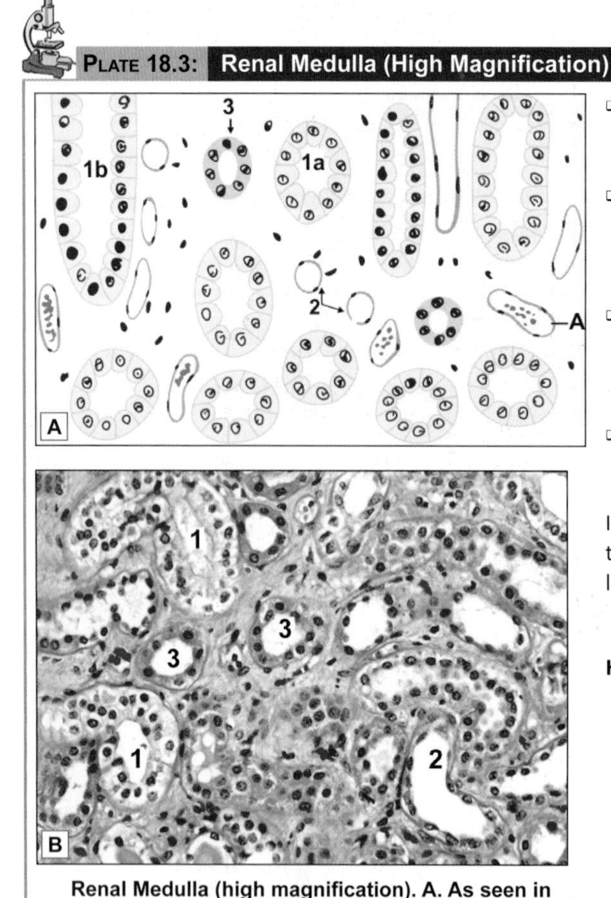

Renal Medulla (high magnification). A. As seen in drawing; B. Photomicrograph

- A high power view of a part of the renal medulla shows a number of collecting ducts cut longitudinally or transversely
- They are lined by a cuboidal epithelium, the cells of which stain lightly. Cell boundaries are usually distinct. The lumen of the tubule is also distinct
- Sections of the thin segment of the loop of Henle are seen. They are lined by flattened cells, the walls being very similar in appearance to those of blood capillaries
- Sections through the thick segments of loops of Henle are seen. They are lined by cuboidal epithelium.

In the photomicrograph, collecting ducts and thick segments and thin segments of the loops of Henle can be identified.

Key

1. Collecting duct
   1a. As seen in transverse section
   1b. As seen in longitudinal section
2. Loop of Henle - thin
3. Loop of Henle - thick

A. Artery

The ***proximal convoluted tubules*** are 40–60 µm in diameter. They have a relatively small lumen. They are lined by cuboidal (or columnar) cells having a prominent brush border. The nuclei are central and euchromatic. The cytoplasm stains pink (with haematoxylin and eosin). The basal part of the cell shows a vertical striation.

With the EM the lining cells of the proximal convoluted tubules show microvilli on their luminal surfaces. The striae, seen with the light microscope near the base of each cell, are shown by EM to be produced by infoldings of the basal plasma membrane, and by numerous mitochondria that lie longitudinally in the cytoplasm intervening between the folds (Fig. 18.6). The presence of microvilli, and of the basal infoldings greatly increases the surface area available for transport. Adjacent cells show some lateral interdigitations. Numerous enzymes associated with ionic transport are present in the cytoplasm.

Loop of Henle

The descending limb, the loop itself, and part of the ascending limb of the loop of Henle are narrow and thin walled. They constitute the ***thin segment*** of the loop. The upper part of the ascending limb has a larger diameter and thicker wall and is called the ***thick segment***.

The thin segment of the loop of Henle is about 15–30 μm in diameter. It is lined by low cuboidal or squamous cells. The thick segment of the loop is lined by cuboidal cells.

The loop of Henle is also called the ***ansa nephroni***. With the EM the flat cells lining the thin segment of the loop of Henle show very few organelles indicating that the cells play only a passive role in ionic movements across them. In some areas the lining epithelium may show short microvilli, and some basal and lateral infoldings.

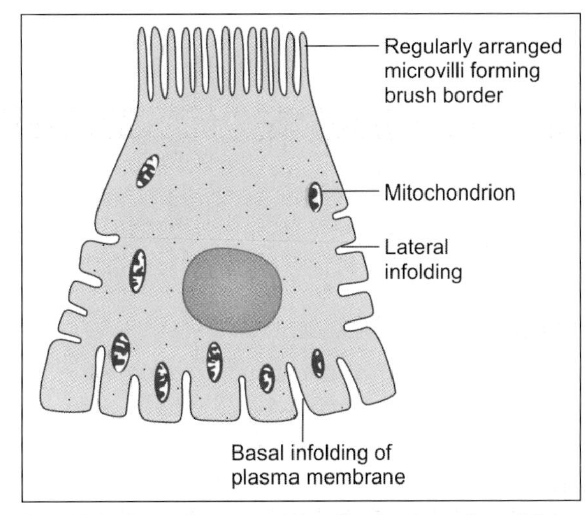

Fig. 18.6: Some features of the ultrastructure of a cell lining a proximal convoluted tubule (Schematic representation)

The length of the thin segment of the loop of Henle is variable. The loops of nephrons having glomeruli lying deep in the cortex (juxtamedullary glomeruli) pass deep into the medulla. Those associated with glomeruli lying in the middle of the cortical thickness extend into the medulla to a lesser degree, so that part of the loop of Henle lies in the cortex. Some loops (associated with glomeruli placed in the superficial part of the cortex) may lie entirely within the cortex.

Distal Convoluted Tubule

The distal convoluted tubule has a straight part continuous with the ascending limb of the loop of Henle, and a convoluted part lying in the cortex. At the junction between the two parts, the distal tubule lies very close to the renal corpuscle of the nephron to which it belongs.

The distal convoluted tubules are 20–50 μm in diameter.

They can be distinguished (in sections) from the proximal tubules as

- They have a much larger lumen
- The cuboidal cells lining them do not have a brush border
- They stain less intensely pink (with eosin).

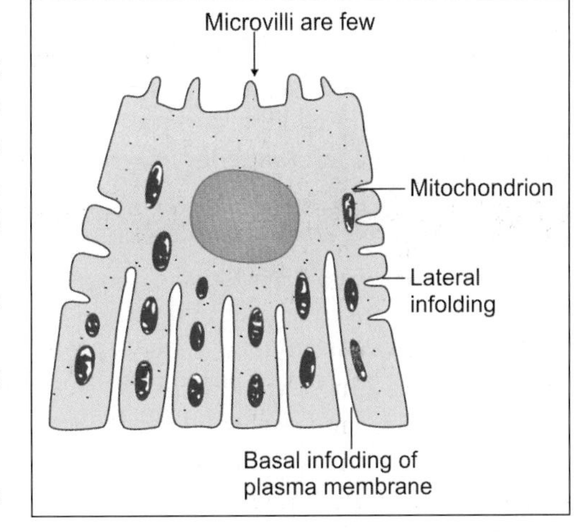

Fig. 18.7: Some features of ultrastructure of a cell lining a distal convoluted tubule (Schematic representation)

The structure of the (ascending) thick segment of the loop of Henle is similar to that of distal convoluted tubules. The cells of the distal convoluted tubules resemble those of the proximal convoluted tubules with the following differences.

- They have only a few small microvilli.
- The basal infoldings of plasma membrane are very prominent and reach almost to the luminal surface of the cell. This feature is characteristic of cells involved in the active transport of ions. Enzymes concerned with active transport of ions are present in the cells.

At the junction of the straight and convoluted parts of the distal convoluted tubules, the cells show specialisations that are described below in connection with the juxtaglomerular apparatus.

Note: Students may be confused by somewhat different terminology used in some books. The straight part of the proximal convoluted tubule is sometimes described as part of the loop of Henle, and is termed the ***descending thick segment,*** in distinction to the (ascending) thick segment. Some workers regard the thin segment alone to be the loop of Henle. They include the descending and ascending thick segments with the proximal and distal convoluted tubules respectively.

Collecting Tubule

The terminal part of the distal convoluted tubule is again straight. This part is called the ***junctional tubule*** or ***connecting tubule,*** and ends by joining a collecting duct.

The smallest collecting tubules are 40-50 μm in diameter, and the largest as much as 200 μm. They are lined by a simple cuboidal, or columnar, epithelium. Collecting tubules can be easily distinguished from convoluted tubules as follows.

- Collecting tubules have larger lumina. In transverse sections their profiles are circular in contrast to the irregular shapes of convoluted tubules.
- The lining cells have clear, lightly staining cytoplasm, and the cell outlines are usually distinct. They do not have a brush border.

The walls of collecting tubules (in the proximal part of the collecting system) are lined by two types of cells. The majority of cells (called ***clear cells***) have very few organelles, a few microvilli and some basal infoldings. The lining epithelium also contains some ***dark cells*** (or ***intercalated cells***). These have microvilli, but basal infoldings are not seen. They contain numerous mitochondria.

The cells of the collecting ducts do not have microvilli, or lateral infoldings of plasma membrane. Very few organelles are present in the cytoplasm.

Added Information

Glomerular Filtration Barrier

In the renal corpuscle water and various small molecules pass, by filtration, from blood (in the glomerular capillaries) to the urinary space of the glomerular capsule. Theoretically the barrier across which the filtration would have to occur is constituted by (a) the capillary endothelium, (b) by the cells (podocytes) forming the glomerular (or visceral) layer of the glomerular capsule, and (c) by a ***glomerular basement membrane*** that intervenes between the two layers of cells named above, and represents the fused basal laminae of the two layers. In fact, however, the barrier is modified as follows.

- Firstly, the endothelial cells show numerous fenestrae or pores that are larger than pores in many other situations. The fenestrae are not closed by membrane. As a result filtrate passes easily through the pores, and the endothelial cells do not form an effective barrier.
- Between the areas of attachment of individual pedicels there are gaps in which the basal lamina is not covered by podocyte cytoplasm. Filtration takes place through the basal lamina at these gaps that are, therefore, called ***filtration slits*** or ***slit pores*** (Fig. 18.8). These slits are covered by a layer of fine filaments that constitute the ***glomerular slit diaphragm***. From what has been said above it will be clear that the filtrate does not have to pass through podocyte cytoplasm.

Contd...

It follows, therefore, that the only real barrier across which filtration occurs is the basal lamina (or the glomerular basement membrane) that is thickened at the filtration slits by the glomerular slit diaphragm. The efficacy of the barrier is greatly enhanced by the presence of a high negative charge in the basement membrane and in podocyte processes. (Loss of this charge, in some diseases, leads to excessive leakage of protein through the barrier).

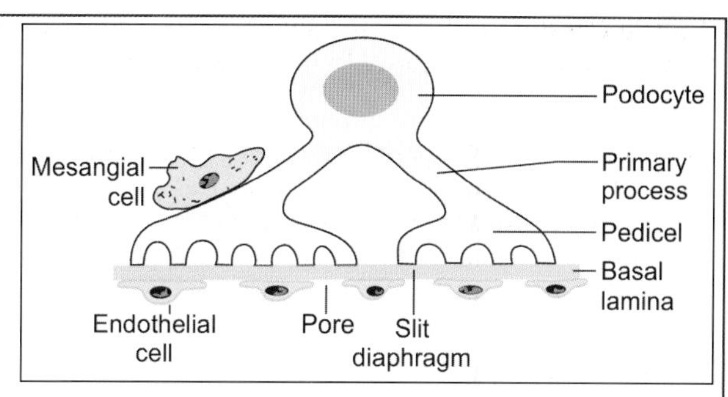

Fig. 18.8: Filtration slits (Schematic representation)

Defects in the glomerular basement membrane are responsible for the nephrotic syndrome in which large amounts of protein are lost through urine. The regular arrangement of podocyte processes is also disorganised in this condition.

Juxtaglomerular Apparatus

Juxtaglomerular apparatus is a specialised organ situated near the glomerulus of each nephron (juxta = near). The juxtaglomerular apparatus is formed by three different structures.

- Juxtaglomerular cells
- Macula densa
- Lacis cells

Juxtaglomerular Cells

A part of the distal convoluted tubule (at the junction of its straight and convoluted parts) lies close to the vascular pole of the renal corpuscle, between the afferent and efferent arterioles. In this region the muscle cells in the wall of the afferent arteriole are modified. They are large and rounded (epithelioid) and have spherical nuclei. Their cytoplasm contains granules that can be stained with special methods. These are ***juxtaglomerular cells***. They are innervated by unmyelinated adrenergic nerve fibres. Juxtamedullary cells are regarded, by some, as highly modified myoepithelial cells as they contain contractile filaments in the cytoplasm.

The granules of the juxtaglomerular cells are seen by EM to be membrane bound secretory granules. They contain an enzyme called ***renin***.

The juxtaglomerular cells also probably act as baroreceptors reacting to a fall in blood pressure by release of renin. Secretion of renin is also stimulated by low sodium blood levels and by sympathetic stimulation.

Note: In addition to renin the kidney produces the hormone erythropoietin (which stimulates erythrocyte production). Some workers have claimed that erythropoietin is produced by juxtaglomerular cells, but the site of production of the hormone is uncertain.

Macula Densa

The wall of the distal convoluted tubule is also modified at the site of contact with the arteriole. Here the cells lining it are densely packed together, and are columnar (rather than cuboidal as in the rest of the tubule). These cells form the ***macula densa***. The cells of the macula densa lie in close contact with the juxtaglomerular cells.

Lacis Cells

In addition to the renin producing cells, and the macula densa, the juxtaglomerular apparatus has a third component: these are ***lacis cells***. These cells are so called as they bear processes that form a lace-like network. They are located in the interval between the macula densa and the afferent and efferent arterioles. The function of lacis cells is unknown.

Mode of Action of Juxtaglomerular Apparatus

The juxtaglomerular apparatus is a mechanism that controls the degree of resorption of ions by the renal tubule. It appears likely that cells of the macula densa monitor the ionic constitution of the fluid passing across them (within the tubule). The cells of the macula densa appear to influence the release of renin by the juxtaglomerular cells.

Renin acts on a substance called ***angiotensinogen*** present in blood and converts it into ***angiotensin I***. Another enzyme (present mainly in the lungs) converts angiotensin I into ***angiotensin II***. Angiotensin II increases blood pressure. It also stimulates the secretion of aldosterone by the adrenal cortex, thus influencing the reabsorption of sodium ions by the distal convoluted tubules, and that of water through the collecting ducts. In this way it helps to regulate plasma volume and blood pressure.

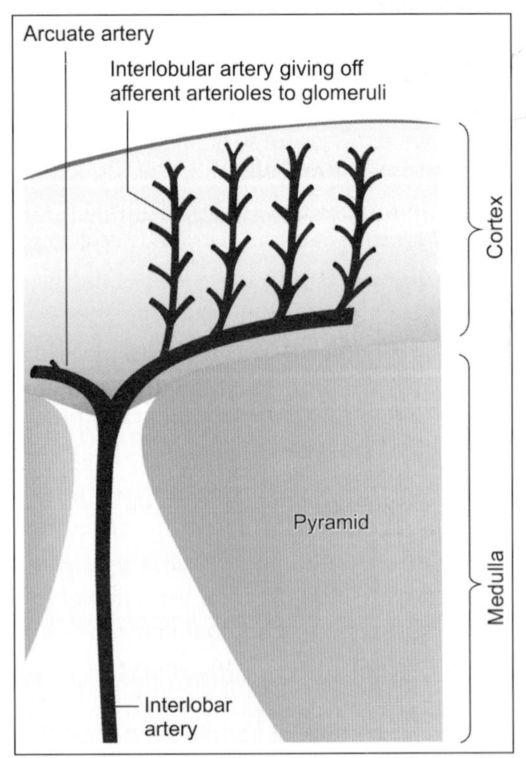

Fig. 18.9: Arrangement of arteries within the kidney (Schematic representation)

Renal Blood Vessels

At the hilum of the kidney each renal artery divides into a number of ***lobar arteries*** (one for each pyramid). Each lobar artery divides into two (or more) ***interlobar arteries*** that enter the tissue of the renal columns and run towards the surface of the kidney. Reaching the level of the bases of the pyramids, the interlobar arteries divide into ***arcuate arteries*** (Fig. 18.9). The arcuate arteries run at right angles to the parent interlobar arteries.

They lie parallel to the renal surface at the junction of the pyramid and the cortex. They give off a series of ***interlobular arteries*** that run through the cortex at right angles to the renal surface to end in a subcapsular plexus. Each interlobular artery gives off a series of arterioles that enter glomeruli as ***afferent arterioles***. Blood from these arterioles circulates through glomerular capillaries that join to form ***efferent arterioles*** that emerge from glomeruli.

The behaviour of efferent arterioles leaving the glomeruli differs in the case of glomeruli located more superficially in the cortex, and those lying near the pyramids. Efferent arterioles arising from the majority of glomeruli (superficial) divide into capillaries that surround the proximal and distal convoluted tubules. These capillaries drain into ***interlobular veins***, and through them into ***arcuate veins*** and ***interlobar veins***. Efferent arterioles arising from glomeruli nearer the medulla (***juxtamedullary glomeruli***) divide into 12 to 25 straight vessels that descend into the medulla. These are the ***descending vasa recta*** (Fig. 18.10). Side branches arising from the vasa recta join a capillary plexus that surrounds the descending and ascending limbs of the loop of Henle (and also the collecting tubules). The capillary plexus consists predominantly of vessels running longitudinally along the tubules. It is drained by ***ascending vasa recta*** that run upwards parallel to the descending vasa recta to reach the cortex. Here they drain into interlobular or arcuate veins. The parallel orientation of the ascending and descending vasa recta and their close relationship to the ascending and descending limbs of the loops of Henle is of considerable physiological importance.

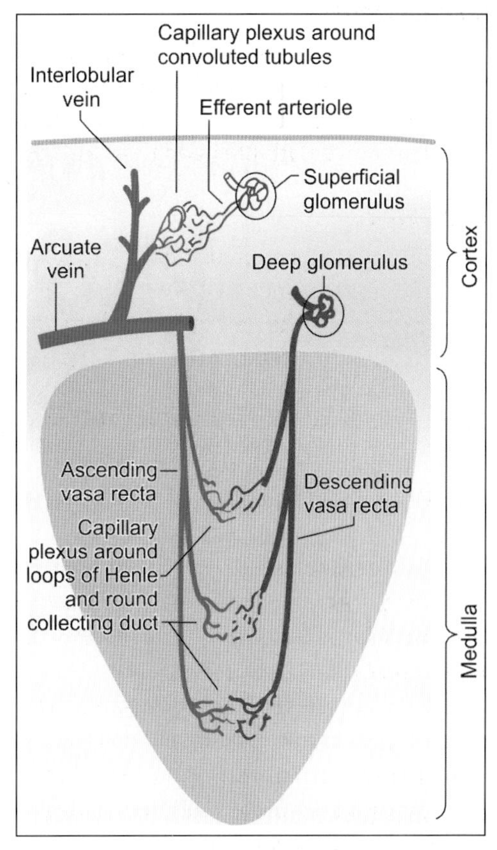

Fig. 18.10: Behaviour of efferent arterioles of glomeruli in the superficial and deeper parts of the renal cortex (Schematic representation)

From this account of the renal blood vessels it will be clear that two sets of arterioles and capillaries intervene between the renal artery and vein. The first capillary system, present in glomeruli, is concerned exclusively with the removal of waste products from blood. It does not supply oxygen to renal tissues. Exchanges of gases (oxygen, carbon dioxide) between blood and renal tissue is entirely through the second capillary system (present around tubules).

It has been said that in most tissues the blood supply exists to provide service to the parenchyma (in the form of the supply of oxygen and nutrients, and the removal of carbon dioxide and other waste products). In the kidney, on the other hand, the parenchyma exists to provide service to blood (by removal of waste products in it).

It has also been held that interlobular arteries divide the renal cortex into small lobules. Each lobule is defined as the region of cortex lying between two adjacent interlobular arteries. A medullary ray, containing a collecting duct, runs vertically through the middle of the lobule (midway between the two arteries) (Fig. 18.11). Glomeruli lie in a zone adjacent to the arteries while other parts of the nephron lie nearer the centre of the lobule.

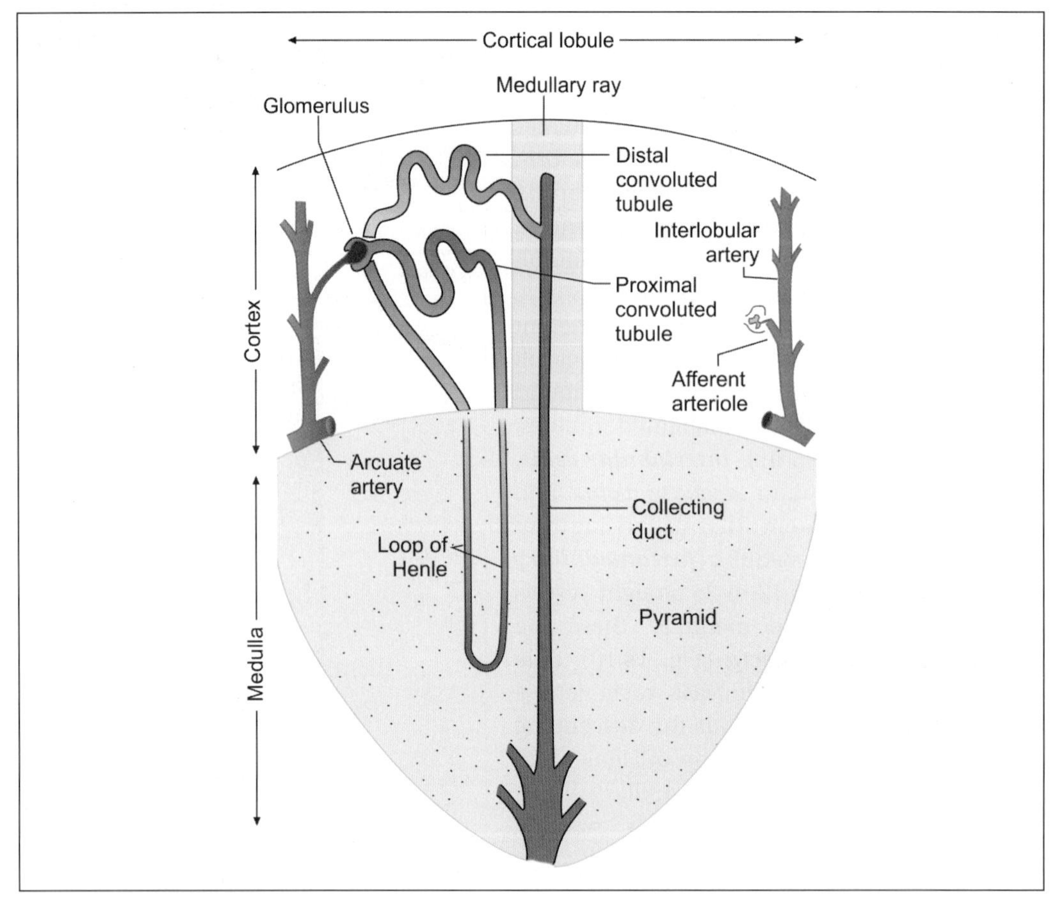

Fig. 18.11: The concept of cortical lobules (Schematic representation)

Clinical Correlation

- **Glomerulonephritis or Bright's disease** is the term used for diseases that primarily involve the renal glomeruli. It is convenient to classify glomerular diseases into 2 broad groups:
  - *Primary glomerulonephritis* in which the glomeruli are the predominant site of involvement.
  - *Secondary glomerular diseases* include certain systemic and hereditary diseases which secondarily affect the glomeruli.
- **Nephrotic syndrome** is a constellation of features in different diseases having varying pathogenesis; it is characterised by findings of massive proteinuria, hypoalbuminaemia, oedema, hyperlipidaemia, lipiduria, and hypercoagulability.
- **Hydronephrosis** is the term used for dilatation of renal pelvis and calyces due to partial or intermittent obstruction to the outflow of urine. Hydronephrosis develops if one or both the pelviureteric sphincters are incompetent, as otherwise there will be dilatation and hypertrophy of the urinary bladder but no hydronephrosis. Hydroureter nearly always accompanies hydronephrosis. Hydronephrosis may be unilateral or bilateral.

THE URETERS

Ureters are muscular tubes that conduct urine from renal pelvis to the urinary bladder. The wall of the ureter has three layers:

- An inner lining of mucous membrane
- A middle layer of smooth muscle
- An outer fibrous coat: adventitia (Fig. 18.12 and Plate 18.4).

Mucous Membrane

The mucous membrane has a lining of ***transitional epithelium*** that is 4 to 5 cells thick and an underlying connective tissue, ***lamina propria.***

The mucosa shows a number of longitudinal folds that give the lumen a star-shaped appearance in transverse section. The folds disappear when the ureter is distended.

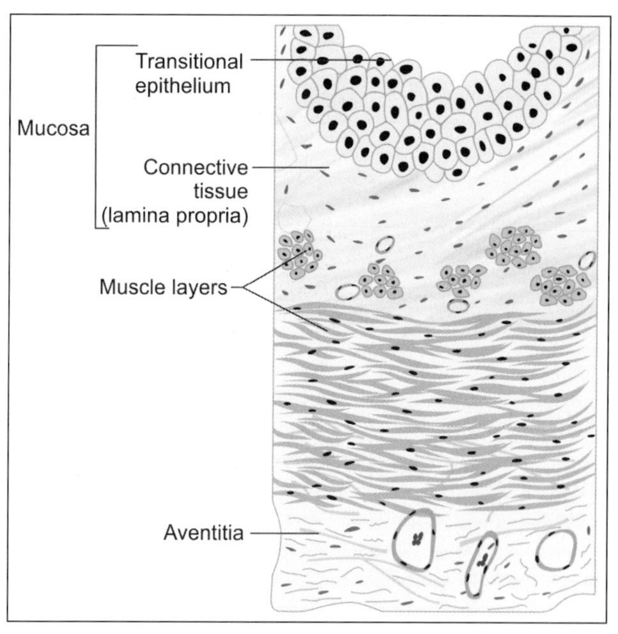

Fig. 18.12: Layers of ureter (Schematic representation)

Muscle Coat

The muscle coat is usually described as having an inner longitudinal layer and an outer circular layer of smooth muscle. A third layer of longitudinal fibres is present outside the circular coat in the middle and lower parts of the ureter. The layers are not distinctly marked off from each other. Some workers have reported that the musculature of the ureter is really in the form of a meshwork formed by branching and anastomosing bundles of muscle fibres.

Adventitia

Adventitia is the outer fibrous coat consisting of loose connective tissue. It contains numerous blood vessels, nerves, lymphatics and some fat cells.

Added Information

Reflux of urine from the urinary bladder into the ureters is prevented by the oblique path followed by the terminal part of the ureter, through the bladder wall. When the musculature of the bladder contracts this part of the ureter is compressed. This mechanism constitutes a physiological sphincter.

Clinical Correlation

Ureterocele is cystic dilatation of the terminal part of the ureter which lies within the bladder wall. The cystic dilatation lies beneath the bladder mucosa and can be visualised by cystoscopy.

PLATE 18.4: Ureter

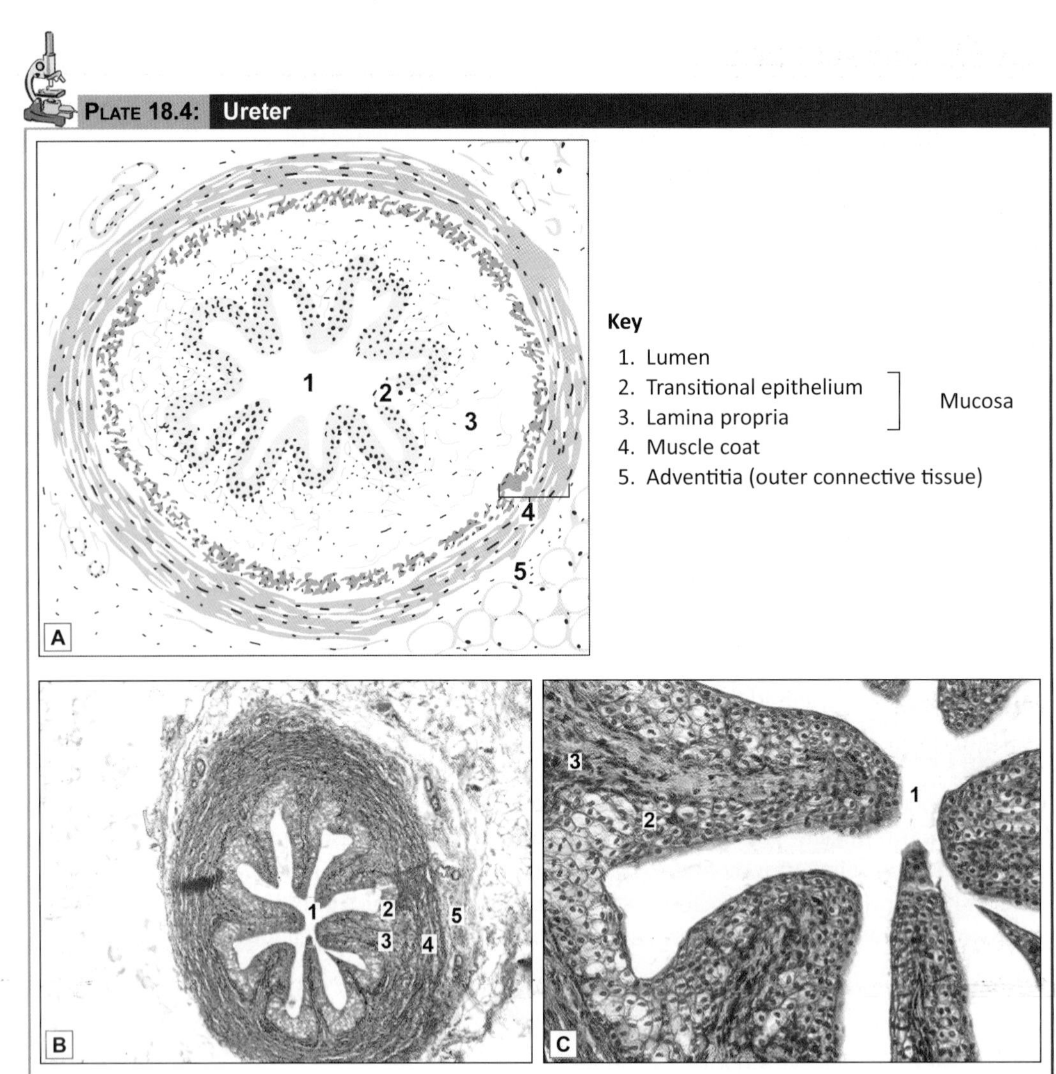

Ureter. A. As seen in drawing; B. Photomicrograph (low magnification); C. Photomicrograph (high magnification)

- The ureter can be recognised because it is tubular and its mucous membrane is lined by transitional epithelium
- The epithelium rests on a layer of connective tissue (lamina propria)
- The mucosa shows folds that give the lumen a star-shaped appearance
- The muscle coat has an inner layer of longitudinal fibres and an outer layer of circular fibres. This arrangement is the reverse of that in the gut
- The muscle coat is surrounded by connective tissue–adventitia in which blood vessels and fat cells are present.

THE URINARY BLADDER

Urinary bladder is a muscular bag, where urine is stored temporarily and is discharged periodically via urethra during micturition.

The wall of the urinary bladder consists of three layers:

- An inner mucous membrane
- A thick coat of smooth muscle
- An outer serous layer (Plate 18.5).

Mucous Membrane

The mucous membrane is lined by transitional epithelium. There is no muscularis mucosae.

In the empty bladder the mucous membrane is thrown into numerous folds that disappear when the bladder is distended. Some mucous glands may be present in the mucosa specially near the internal urethral orifice.

When the bladder is distended (with urine) the lining epithelium becomes thinner. This results from the ability of the epithelial cells to change shape and shift over one another.

Note: The transitional epithelium lining the urinary bladder (and the rest of the urinary passages) is capable of withstanding osmotic changes caused by variations in concentrations of urine. It is also resistant to toxic substances present in urine.

Muscle Coat

The muscle layer is thick. The smooth muscle in it forms a meshwork. Internally and externally the fibres tend to be longitudinal. In between them there is a thicker layer of circular (or oblique) fibres. Contraction of this muscle coat is responsible for emptying of the bladder. That is why it is called the ***detrusor muscle***. Just above the junction of the bladder with the urethra the circular fibres are thickened to form the ***sphincter vesicae***.

Serous Layer

The superior surface of the bladder is covered by mesothelium of peritoneum, forming serous layer. The inferior part of the bladder is covered with adventitia which is made of fibroelastic connective tissue carrying blood vessels, nerves and lymphatics.

Clinical Correlation

- **Ectopia vesicae** is a rare condition owing to congenital developmental deficiency of anterior wall of the bladder and is associated with splitting of the overlying anterior abdominal wall. This results in exposed interior of the bladder. There may be prolapse of the posterior wall of the bladder through the defect in the anterior bladder and abdominal wall. The condition in males is often associated with epispadias in which the urethra opens on the dorsal aspect of penis.
- **Cystitis** is the inflammation of the urinary bladder. Primary cystitis is rare since the normal bladder epithelium is quite resistant to infection. Cystitis may occur by spread of infection from upper urinary tract as seen following renal tuberculosis, or may spread from the urethra such as in instrumentation. Cystitis is caused by a variety of bacterial and fungal infections. Cystitis, like UTI, is more common in females because of the shortness of urethra which is liable to faecal contamination and due to mechanical trauma during sexual intercourse.

PLATE 18.5: Urinary Bladder

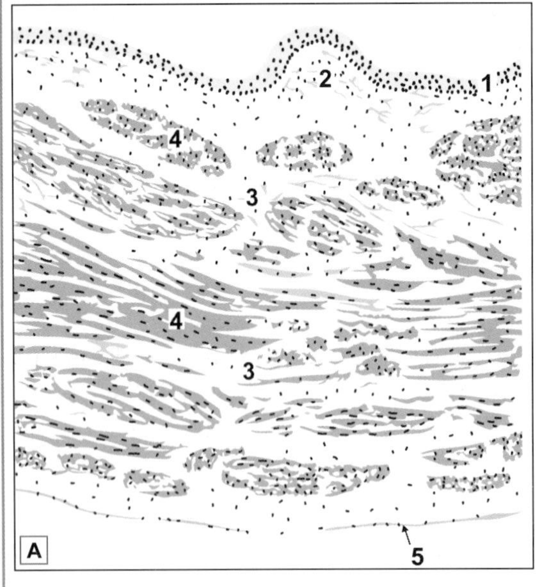

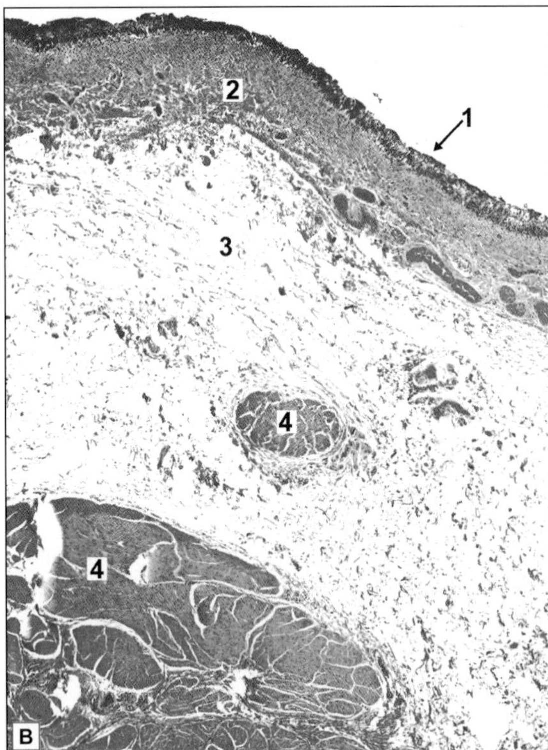

- The urinary bladder is easily recognised because the mucous membrane is lined by transitional epithelium
- The epithelium rests on lamina propria
- The muscle layer is thick. It has inner and outer longitudinal layers between which there is a layer of circular or oblique fibres. The distinct muscle layers may not be distinguishable
- The outer surface is lined in parts by peritoneum (serosa) (not seen in the photomicrograph).

Key

1. Transitional epithelium
2. Lamina propria
3. Interstitial connective tissue
4. Smooth muscle bundles
5. Serous layer

Urinary bladder. A. As seen in drawing; B. Photomicrograph

The Urethra

Urethra is a tube that carries urine from bladder to the exterior. In males, semen also passes through the urethra.

Although the male urethra is much longer than the female urethra, the structure of the two is the same. The wall of the urethra is composed of:

- Mucous membrane
- Submucosa
- Muscle layer.

In the case of the male, the prostatic urethra is surrounded by prostatic tissue; and the penile urethra by erectile tissue of the corpus spongiosum.

Mucous Membrane

The mucous membrane consists of a lining epithelium that rests on connective tissue. The epithelium varies in different parts of the urethra. Both in the male and female, the greater part of the urethra is lined by pseudostratified columnar epithelium. A short part adjoining the urinary bladder is lined by transitional epithelium, while the part near the external orifice is lined by stratified squamous epithelium.

The mucosa shows invaginations or recesses into which mucous glands open.

Submucosa

The submucosa consists of loose connective tissue.

Muscle Coat

The muscle coat consists of an inner longitudinal layer and an outer circular layer of smooth muscle. This coat is better defined in the female urethra. In the male urethra it is well defined only in the membranous and prostatic parts, the penile part being surrounded by occasional fibres only.

In addition to this smooth muscle the membranous part of the male urethra, and the corresponding part of the female urethra are surrounded by striated muscle that forms the ***external urethral sphincter***.

Chapter 19

Male Reproductive System

The male reproductive system consists of (Fig. 19.1)

- A pair of testis
- Genital ducts: the epididymis, the ductus deferens and ejaculatory ducts
- Accessory sex glands: a pair of seminal vesicle, a single prostate and a pair of bulbourethral glands
- Male urethra and
- Penis.

TESTIS

General Structure of Testis

The adult testes are paired ovoid organs placed outside the body in scrotum. This is ideal for normal spermatogenesis as the temperature is 2 to 3°C less than the body temperature. The right and left testes produce the male gametes or ***spermatozoa***. Each testis is about 4 cm long. The testis lies within a double layered serous sac called the ***tunica vaginalis***. The outermost layer of the organ is formed by a dense fibrous membrane called the ***tunica albuginea*** (Fig. 19.2 and Plate 19.1).

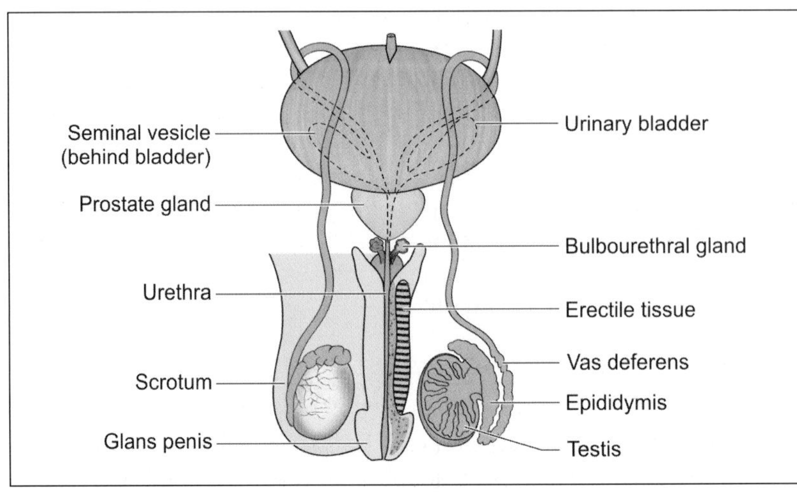

Fig. 19.1: Male reproductive system (Schematic representation)

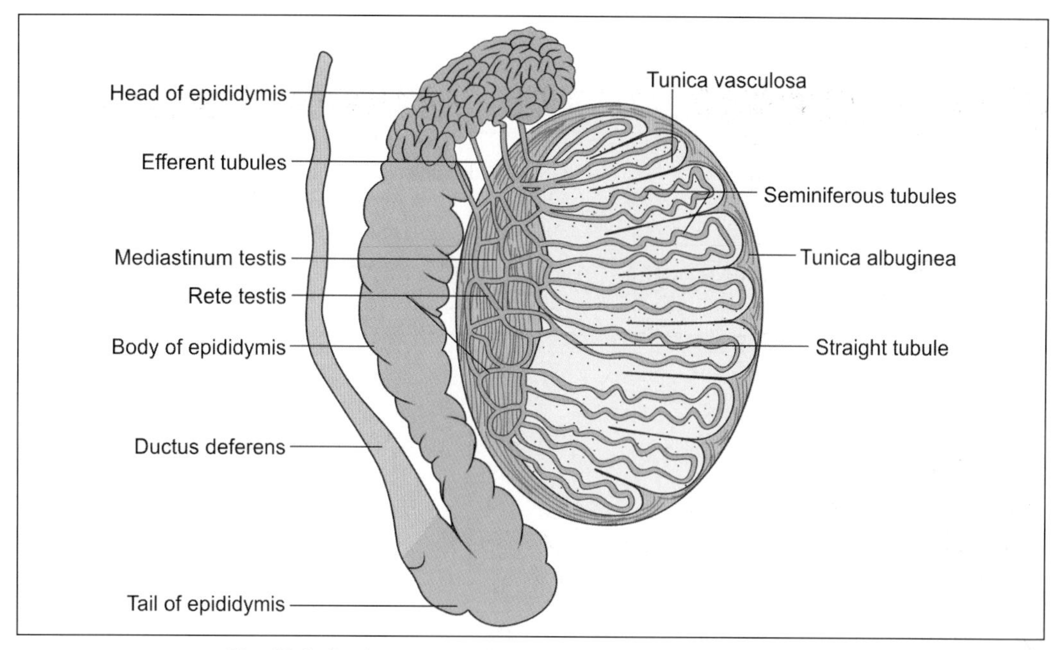

Fig. 19.2: Basic structure of the testis (Schematic representation)

The tunica albuginea consists of closely packed collagen fibres amongst which there are many elastic fibres. In the posterior part of the testis the connective tissue of the tunica albuginea expands into a thick mass that projects into the substance of the testis. This projection is called the ***mediastinum testis***. The visceral layer of tunica vaginalis covers the tunica albuginea, except in the region of the mediastinum testis.

Numerous septa pass from the mediastinum testis to the tunica albuginea, and divide the substance of the testis into a number of lobules. Each lobule is roughly conical, the apex of the cone being directed towards the mediastinum testis. Each lobule contains one or more highly convoluted ***seminiferous tubules***.

Near the apex of the lobule the seminiferous tubules lose their convolutions and join one another to form about 20 to 30 larger, ***straight tubules*** (or ***tubuli recti***) (Fig. 19.2). These straight tubules enter the fibrous tissue of the mediastinum testis and unite to form a network called the ***rete testis***. At its upper end the retetestis gives off 12 to 20 ***efferent ductules***. These ductules pass from the upper part of the testis into the ***epididymis***. This duct is highly coiled on itself and forms the body and tail of the epididymis. The epididymis has a head, a body and a tail. The head of the epididymis is made up of highly convoluted continuations of the efferent ductules. At the lower end of the head of the epididymis these tubules join to form a single tube called the duct of the epididymis. At the lower end of the tail of the epididymis the duct becomes continuous with the ***ductus deferens***.

General Structure of Seminiferous Tubules

Seminiferous tubules are highly convoluted structures present in each lobule of the testes. It has been estimated that each testis has about 200 lobules, and that each lobule has one to three seminiferous tubules. The total number of tubules is between 400 and 600.

PLATE 19.1: Testis (Low Magnification)

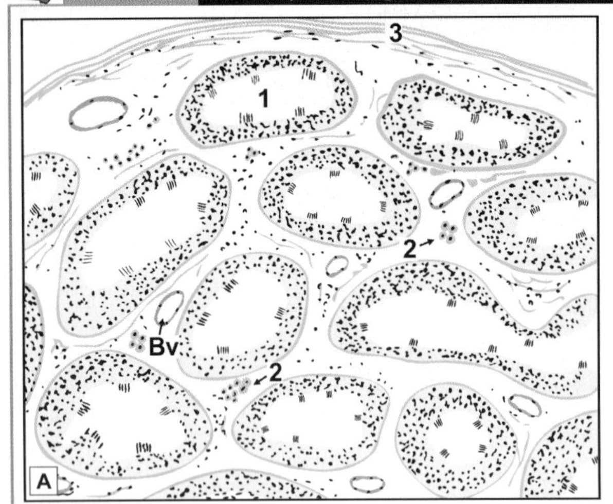

The testis has an outer fibrous layer, the tunica albuginea deep to which:

- A number of seminiferous tubules cut in various directions are seen
- The tubules are separated by connective tissue, containing blood vessels and groups of interstitial cells of Leydig
- Each seminiferous tubule is lined by several layers of cells
- Cells are of two types:
  - Spermatogenic cells which produce spermatozoa
  - Sustentacular (Sertoli) cells which have a supportive function.

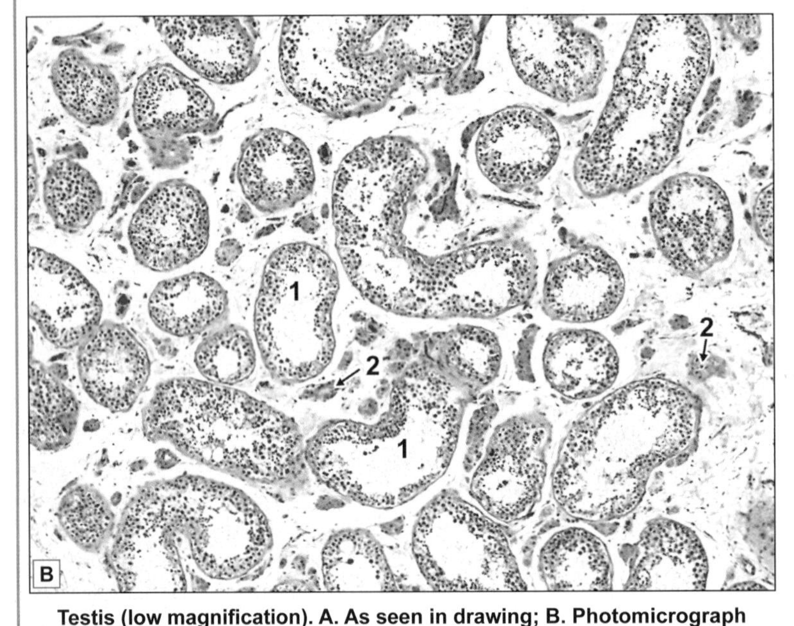

Key

1. Seminiferous tubule
2. Interstitial cells
3. Tunica albuginea

Bv. Blood vessel

Testis (low magnification). A. As seen in drawing; B. Photomicrograph

When stretched out each tubule is 70–80 cm in length. It has a diameter of about 150 μm. The combined length of all seminiferous tubules in one testis is between 300 and 900 metres. These tubules are lined by cells that are concerned with the production of spermatozoa.

Within a lobule, the spaces between seminiferous tubules are filled by very loose connective tissue, containing blood vessels and lymphatics. Interstitial cells of Leydig are also present here. The wall of each tubule is made up of an outer layer of fibrous tissue that also contains muscle-like (myoid) cells. Contractions of these cells probably help to move spermatozoa along the tubule.

Between this connective tissue and the lumen of the tubule there are several layers of cells. The cells rest on a basal lamina. They are of various shapes and sizes. Most of the cells represent stages in the formation of spermatozoa: they are referred to as ***germ cells***. Other cells that have a supporting function are called ***sustentacular cells*** or the ***cells of Sertoli*** (Plate 19.2).

Note: The appearance of the cellular lining of the seminiferous tubules is characteristic, and a student who has studied sections through them carefully (even at low magnification) is not likely to mistake the seminiferous tubules for anything else. The points to note are (a) the many layers of cells; (b) the great variety in size and shape of the cells and of their nuclei; (c) the lack of a well defined margin of the lumen; and (d) inconspicuous cell boundaries.

Spermatogenic Cells (Germ Cells)

Spermatogenic cells represent the various stages in the formation of spermatozoa. They are arranged in developmentally higher order from the basal lamina to the lumen, namely spermatogonia, spermatocytes, spermatids and spermatozoa (Plate 19.2).

However, all types of cells are not seen in any one part of the seminiferous tubule at a given time.

In a given segment of the tubule there is a gradual change in the type of cells encountered (with passage of time). Six phases have been recognised. At a given point of time different segments of a seminiferous tubule show cell patterns corresponding to the six phases.

It has, therefore, been suggested that over a period of time waves of maturation (of germ cells) pass along the length of a seminiferous tubule, this is referred to as spermatogenic cycle.

Sustentacular Cells or Cells of Sertoli (Fig. 19.3)

These are tall, slender cells having an irregularly pyramidal or columnar shape. The nucleus lies near the base of the cell. It is light staining and is of irregular shape. There is a prominent nucleolus. The base of each sustentacular cell rests on the basement membrane, spermatogonia being interposed amongst the sustentacular cells. The apex of the sustentacular cell reaches the lumen of the seminiferous tubule. Numerous spermatids, at various stages of differentiation into spermatozoa, appear to be embedded in the apical part of the cytoplasm (Fig. 19.3 and Plate 19.2). Nearer the basement membrane spermatocytes and spermatogonia indent the sustentacular cell cytoplasm.

With the EM it is seen that the sides and apices of these cells are marked by recesses that are occupied by spermatogonia, spermatocytes, and spermatids. However, there is no cytoplasmic continuity between these cells and the sustentacular cell. On the basis of light microscopic studies some workers were of the view that the sustentacular cells formed a syncytium. However, EM studies have shown that the cells are distinct.

PLATE 19.2: Testis (High Magnification)

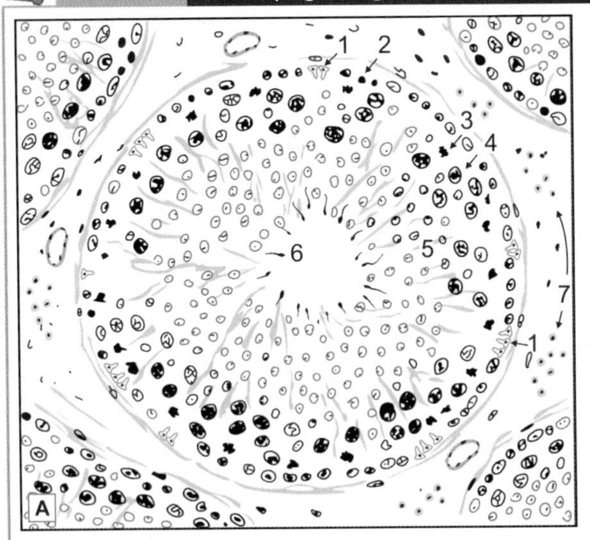

Details of cells lining a seminiferous tubule seen at a high magnification. Note that the cell boundaries are indistinct, and nuclei are prominent.

- The outermost row of nuclei belongs to sustentacular cells (Sertoli) and to spermatogonia, some of which are undergoing mitosis (note very dense nucleus of irregular shape)
- Passing inwards towards the centre of the tubule we have large darkly staining nuclei of spermatocytes, and many smaller nuclei of spermatids
- Towards the centre of the tubule a number of developing spermatozoa are seen. The sperms are often found in clusters embedded in the cytoplasm of Sertoli cells

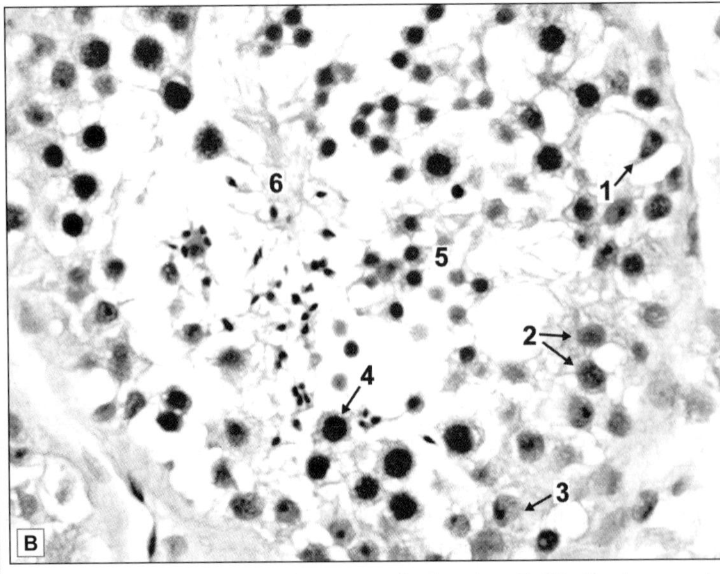

Testis (high magnification). A. As seen in drawing; B. Photomicrograph

Key

1. Sustentacular cells (Sertoli)
2 & 3. Spermatogonia
4. Spermatocytes
5. Spermatids
6. Spermatozoa
7. Interstitial cells of Leydig

- In the adult testis, sustentacular cells are less prominent than germ cells. They are more prominent than germ cells before puberty and in old age
- In the drawing, groups of interstitial cells can be seen in the connective tissue between the seminiferous tubules (not seen in photomicrograh).

Note: In the practical class you may not be able to recognise these cells. Observe that the presence of many cells located at different levels gives the appearance of a stratified epithelium which are actually the spermatogonia at different stages of maturation.

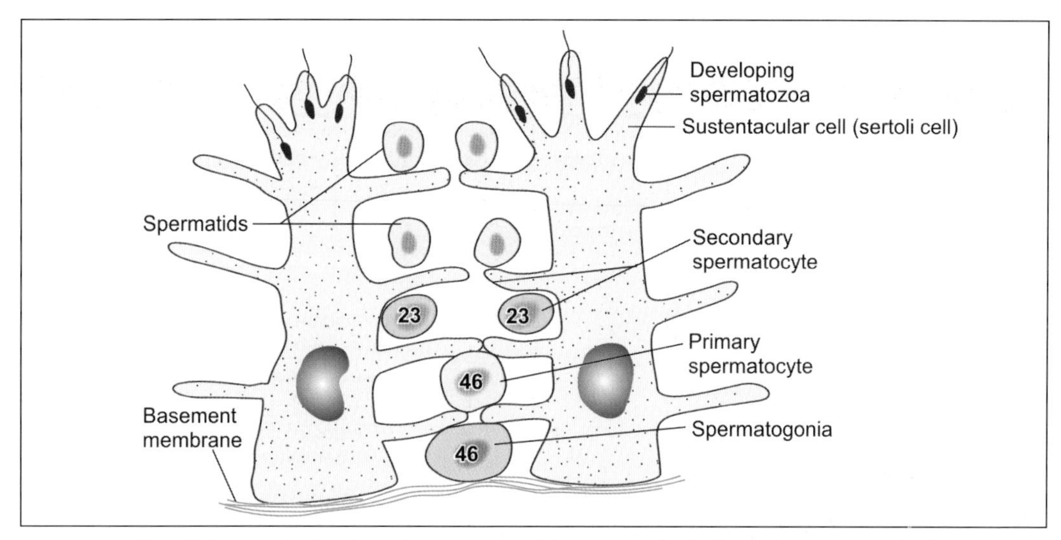

Fig. 19.3: A sustentacular cell and some related germ cells (Schematic representation)

The plasma membranes of adjoining sustentacular cells are connected by tight junctions that divide the wall of the seminiferous tubule into two compartments, ***superficial*** (or ***adluminal***) and ***deep*** (***abluminal***). The deep compartment contains spermatogonia (and preleptotene spermatocytes) and the superficial compartment contains other stages of spermatogenesis.

The two compartments are believed to be separated by a ***blood-testis barrier***. Sustentacular cells contain abundant mitochondria, endoplasmic reticulum, and other organelles. Microfilaments and microtubules form a cytoskeleton that appears to be important in cohesive functions of these cells.

In the adult testis sustentacular cells are less prominent than germ cells. They are more prominent than germ cells before puberty, and in old age.

Functions

- They provide physical support to germ cells, and provide them with nutrients. Waste products from germ cells are transferred to blood or lymph through them.
- They phagocytose residual cytoplasm that remains after conversion of spermatids to spermatozoa.
- They probably secrete fluid that helps to move spermatozoa along the seminiferous tubules. This fluid is rich in testosterone that may stimulate activity of cells lining the epididymis.
- Sertoli cells produce a number of hormones. In the eighth month of fetal life they secrete a hormone (***Mullerian inhibitory substance*** or ***MIS***) that suppresses development of the paramesonephric (Mullerian) ducts (in male fetuses). They are also believed to produce a substance that inhibits spermatogenesis before puberty.
- In the male adult, Sertoli cells produce ***androgen binding protein (ABP)*** which binds to testosterone (and hydroxytestosterone) making these available in high concentration to germ cells within seminiferous tubules. Secretion of ABP is influenced by FSH. A hormone ***inhibin*** produced by Sertoli cells inhibits production of FSH (Flow chart 17.1).

Flow chart 19.1: Scheme to show control of male genital system

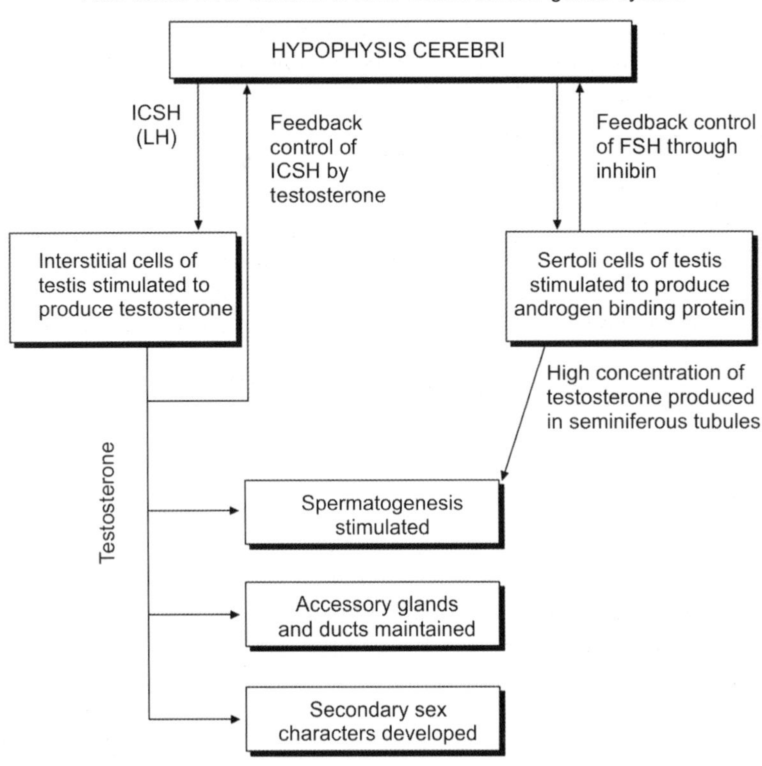

Interstitial Cells

The interstitial cells (of Leydig) are large, round or polyhedral cells lying in the connective tissue that intervenes between the coils of seminiferous tubules (Plate 19.2A). Their nuclei are eccentric. The cytoplasm stains lightly and often has a foamy appearance (because of the removal of lipids during processing of tissues). It contains yellow granules that are seen by EM to be vacuoles containing various enzymes. Rod-shaped crystalloids (Reinke's crystalloids) are also present in the cytoplasm. Much agranular endoplasmic reticulum is present. Yellow-brown pigment (lipofuscin) is seen in some cells.

Interstitial cells secrete male sex hormone (testicular androgens). Secretion is stimulated by the interstitial cell stimulating hormone of the hypophysis cerebri. (This hormone is identical with the luteinising hormone present in the female).

Some interstitial cells may be present in the mediastinum testis, in the epididymis, or even in the spermatic cord.

Apart from interstitial cells, the interstitial tissue contains collagen fibres, fibroblasts, macrophages, mast cells, blood vessels and lymphatics.

Structure of Rete Testis and Efferent Ductules

The rete testis consists of anastomosing tubules that are lined by flattened or cuboidal cells. They bear microvilli. The epithelium is surrounded by connective tissue of the mediastinum testis.

The efferent ductules are lined by ciliated columnar epithelium. Some non-ciliated cells bearing microvilli are also present. The tubules have some smooth muscle in their walls. Movement of spermatozoa through the tubules is facilitated by ciliary action, and by peristaltic contraction of smooth muscle.

Pathological Correlation

Tumours of Testis

Testicular tumours can arise from germ cells (Germ cell tumours); stroma (sex cord stromal tumours) or both (combined germ cell-sex cord stromal tumours). Examples of germ cell tumours are seminoma, embryonal carcinoma, teratoma and choriocarcinoma; sex cord stromal tumours are Leydig cell tumour, Sertoli cell tumour and granulosa cell tumour. Gonadoblastoma is an example of combined germ cell-sex cord-stromal tumours.

Germ cell tumours comprise approximately 95% of all testicular tumours and are more frequent before the age of 45 years. Testicular germ cell tumours are almost always malignant.

SPERMATOGENESIS

The process of the formation of spermatozoa is called ***spermatogenesis*** (Flow chart 17.2). This process occurs in waves along the length of seminiferous tubules taking about 64 days to complete. It consists of several stages as described below.

Flowchart 19.2: Spermatogenesis

- Germ cells
- ↓ *Mitosis*
- Spermatogonia
- ↓
- Primary spermatocyte (46, XY)
- ↓
- First meiotic division
- ↓
- Secondary spermatocytes
- ↓
- 23, X | 23, Y
- ↓
- Second meiotic division
- ↓
- 23, X | 23, X | 23, Y | 23, Y
- ↓
- Spermatids
- ↓
- Morphological change to spermatozoa (Spermiogenesis)

Stages of Spermatogenesis

- The stem cells from which all stages in the development of spermatozoa are derived are called ***spermatogonia***. These cells lie near the basal lamina. Spermatogonia undergo several mitotic divisions and, because of this, spermatogonia of varied structure are seen in the wall of a seminiferous tubule. These mitoses give rise to more spermatogonia, and to primary spermatocytes.

Note: Three main types of spermatogonia are described: ***dark type-A (AD)***, ***light*** (or ***pale***) ***type-A (AP)***, and ***type-B*** (Fig. 19.4). In type A spermatogonia (dark and light) the nuclei are oval and possess nucleoli that are eccentric, and are attached to the nuclear membrane. The terms dark and light, applied to these cells, refer to the intensity of staining of the nuclei. In type B spermatogonia the nuclei are spherical. Each nucleus has a spherical nucleolus.

Dark type A spermatogonia (also called type A_1) represent a reserve of resting stem cells. They divide to form more dark type A cells and also some light type A cells (or A_2 cells). Light type A spermatogonia divide to form more light type A spermatogonia, and also some spermatogonia

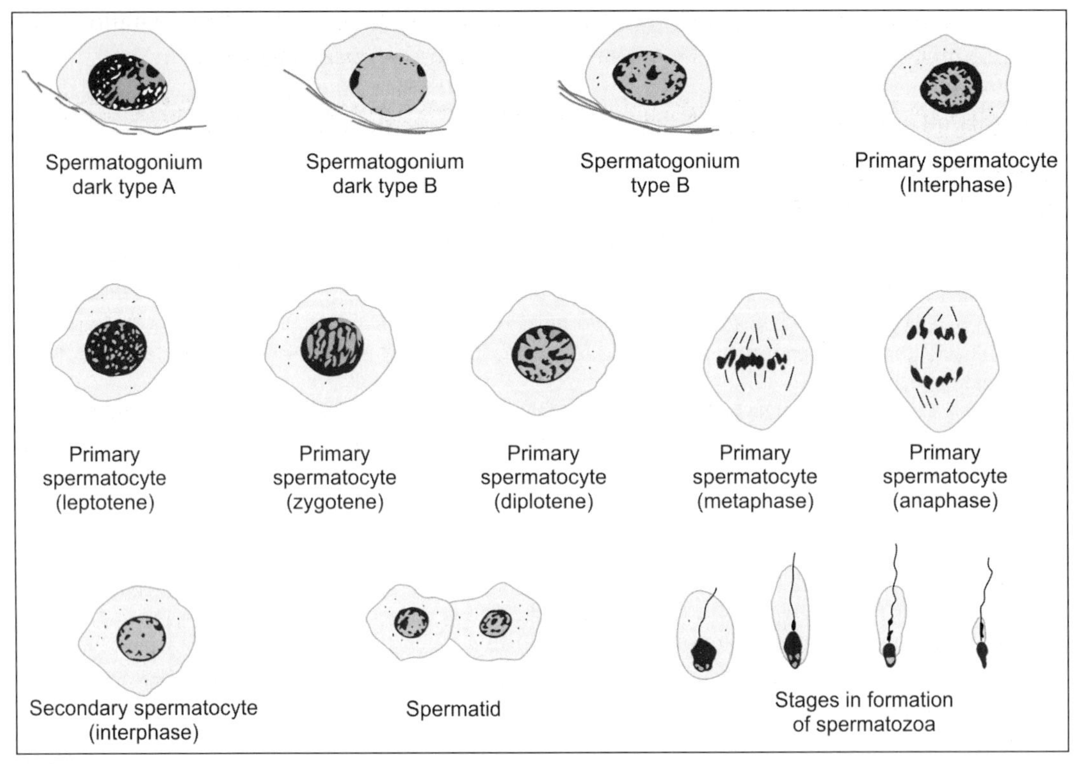

Fig. 19.4: Some stages in spermatogenesis as seen in the walls of seminiferous tubules (Schematic representation)

of type B. Each type B spermatogonium divides several times (probably four times in man). The resulting cells are designated B_1, B_2, B_3 etc. Each of the resulting cells divides to form two primary spermatocytes.

- ***Primary spermatocytes*** are formed by mitotic division of spermatogonia. These are largest cells with large spherical nuclei occupying the middle region of the seminiferous epithelium. Each primary spermatocyte undergoes meiosis to give rise to two secondary spermatocytes (see below). This is the first meiotic division in which the number of chromosomes is reduced to half. (Each primary spermatocyte has 46 chromosomes, whereas each secondary spermatocyte has only 23).
- The prophase of the first meiotic division is prolonged and passes through several stages (leptotene, zygotene, pachytene, diplotene) in which the appearance of the nucleus shows considerable alterations. As a result, primary spermatocytes at various stages of prophase (and subsequent stages of division) can be recognised in the wall of a seminiferous tubule (Fig. 19.4).
- ***Secondary spermatocytes*** are smaller than primary spermatocytes, and so are their nuclei. Each secondary spermatocyte has the haploid number of chromosomes (i.e., 23). It divides to form two spermatids (see below). This is the second meiotic division and this time there is no further reduction in chromosome number. They are scarcely seen in sections as they undergo second meiotic division as soon as they are formed.
- Each ***spermatid*** is a rounded cell with a spherical nucleus. Both cell and nucleus are much smaller than in the case of spermatogonia or spermatocytes. They lie in groups in association

with Sertoli cells. The spermatid undergoes changes in shape, and in the orientation of its organelles, to form a spermatozoon. This process is called ***spermiogenesis***.

Added Information

Diploid and Haploid Chromosome Number and DNA content

We have seen that a typical cell contains 46 chromosomes, this being referred to as the diploid number. Spermatozoa (or ova) have only half this number i.e., 23 which is the haploid number. At fertilisation, the diploid number is restored, the zygote receiving 23 chromosomes from the ovum, and 23 from the sperm.

A primary spermatocyte contains the diploid number of chromosomes (46). During the first meiotic division the number is halved so that secondary spermatocytes have the haploid number (23).

Now let us look at these facts in relation to DNA content, rather than chromosome number. Let us designate the DNA content of a gamete (sperm or ovum) as ***n***. The DNA content of a zygote formed as a result of fertilisation is, therefore, ***n + n = 2n***. Before the zygote can undergo division its DNA has to be replicated. In other words it has to become ***4n,*** of which ***2n*** goes to each daughter cell.

When first formed primary spermatocytes (or oocytes) have ***2n*** DNA. After replication this becomes ***4n***. At the first meiotic division ***2n*** goes to each secondary spermatocyte. The point to note is that although the chromosome number in a secondary spermatocyte is haploid, DNA is ***2n***. There is no replication of DNA in secondary spermatocytes. As a result the DNA content of spermatids formed during the second meiotic division is ***n***. Therefore, note that although there is no reduction in chromosome number during the second meiotic division, DNA content is reduced from ***2n to n***.

Spermiogenesis

The process by which a spermatid becomes a spermatozoon is called ***spermiogenesis*** (or ***spermateleosis***). The spermatid is a more or less circular cell containing a nucleus, Golgi complex, centriole and mitochondria. All these components take part in forming the spermatozoon.

The nucleus undergoes condensation and changes shape to form the head. The Golgi complex is transformed into the acrosomic cap that comes to lie over one side of the nucleus. The acrosome marks the future anterior pole of the spermatozoon. The centriole divides into two parts that are at first close together. They migrate to the pole of the cell that is away from the acrosome. The axial filament grows out of the distal centriole. The region occupied by the two centrioles later becomes the neck of the spermatozoon. The proximal centriole probably forms the basal body.

The part of the axial filament between the head and the annulus becomes surrounded by mitochondria, and together with them forms the middle piece. Most of the cytoplasm of the spermatid is shed, and is phagocytosed by Sertoli cells. The cell membrane persists as a covering for the spermatozoon.

Structure of a Mature Spermatozoon

The spermatozoa are motile male gametes. With their tails projecting into the lumen of seminiferous tubules, they are found in close association with the sertoli cells.

The spermatozoon has a ***head***, a ***neck***, and a ***principal piece*** or ***tail***. The tail is made of three pieces i: e middle piece, principal piece and end piece.

The ***head*** is covered by a cap called t͘ .e ***acrosomic cap***, ***anterior nuclear cap***, or ***galea capitis*** (Fig. 19.5). It is flattened from before backwards so that it is oval when seen from the front, but appears to be pointed (somewhat like a spear-head) when seen from one side, or in section. It consists of chromatin (mostly DNA) that is extremely condensed and, therefore, appears to have a homogeneous structure even when examined by EM. This condensation makes it highly resistant to various physical stresses.

Note: The acrosome is made up of glycoprotein. It can be regarded as a large lysosome containing numerous enzymes (proteases, acid phosphatase, neuraminidase, hyaluronidase).

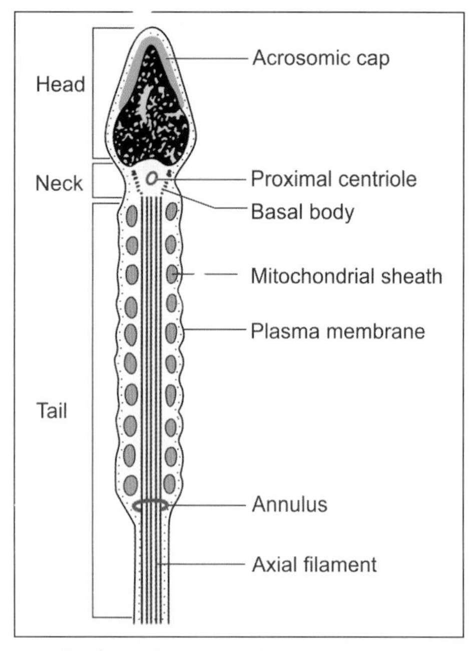

Fig. 19.5: Structure of spermatozoon as seen by electron microscope (Schematic representation)

The ***neck*** of the spermatozoon is narrow. It contains a funnel-shaped ***basal body*** and a spherical ***centriole***.

The chief structure to be seen in the neck is the ***basal body***. It is also called the ***connecting piece*** because it helps to establish an intimate union between the head and the remainder of the spermatozoon. The basal body is made up of nine segmented rod-like structures each of which is continuous distally with one coarse fibre of the axial filament. On its proximal side (i.e., towards the head of the spermatozoon) the basal body has a convex ***articular surface*** that fits into a depression (called the ***implantation fossa***) present in the head.

An ***axial filament*** (or ***axoneme***) begins just behind the centriole. It passes through the middle piece and extends into the tail. At the point where the middle piece joins the tail, this axial filament passes through a ring-like ***annulus***. That part of the axial filament that lies in the middle piece is surrounded by a ***spiral sheath*** made up of mitochondria.

The ***axial filament*** is really composed of several fibrils arranged as illustrated in Fig. 19.6. There is a pair of central fibrils, surrounded by nine pairs (or ***doublets***) arranged in a circle around the central pair. (This arrangement of one central pair of fibrils surrounded by nine doublets is similar to that seen in cilia).

In addition to these doublets there are nine coarser petal-shaped fibrils of unequal size, one such fibril lying just outside each doublet. These coarse fibrils are present in the middle piece and most of the tail, but do not extend into the terminal part of the tail. The whole system of fibrils is kept in position by a series of coverings. Immediately outside the fibrils there is a fibrous sheath. In the region of the middle piece the fibrous sheath is surrounded by spirally arranged mitochondria. Finally, the entire sperm is enclosed in a plasma membrane.

From Fig. 19.6 it will be seen that one of the coarse fibrils is larger than the others. This is called fibril 1, the others being numbered in a clockwise direction from it. The fibrous sheath is adherent to fibrils 3 and 8. The line joining fibrils 3 and 8 divides the tail into a major compartment containing 4 fibrils and a minor compartment containing 3 fibrils. This line also passes through both the central fibrils and provides an axis in reference to which sperm movements can be analysed. The part of tail connected to neck is middle piece. The middle piece contains the mitochondrial sheath and provides energy for sperm maturation.

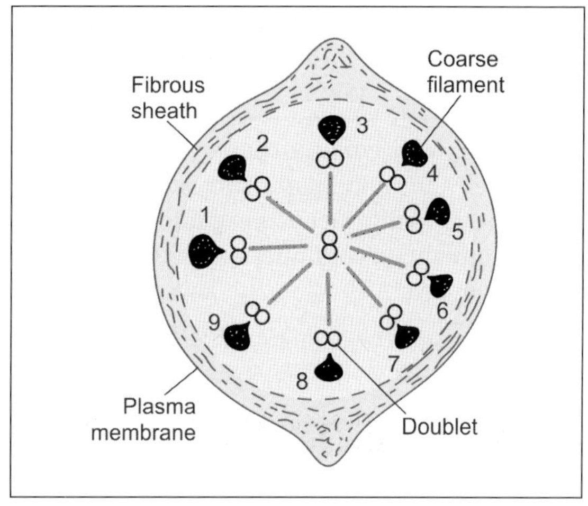

Fig. 19.6: Transverse section across the tail of a spermatozoon to show the arrangement of fibrils (Schematic representation)

Principal piece contains the 9 + 2 pattern of microtubules in the central core surrounded by 9 coarse fibres enclosed in a fibrous sheath (Fig. 19.6). End piece contains 9 + 2 axonema enclosed by plasma membrane

Maturation and Capacitation of Spermatozoa

Maturation

As fully formed spermatozoa pass through the male genital passages they undergo a process of ***maturation***. Spermatozoa acquire some motility only after passing through the epididymis. The secretions of the epididymis, seminal vesicles and the prostate have a stimulating effect on sperm motility, but spermatozoa become fully motile only after ejaculation.

When introduced into the vagina, spermatozoa reach the uterine tubes much sooner than their own motility would allow, suggesting that contractions of uterine and tubal musculature exert a sucking effect.

Capacitation

Spermatozoa acquire the ability to fertilise the ovum only after they have been in the female genital tract for some time. This final step in their maturation is called ***capacitation***. During capacitation some proteins and glycoproteins are removed from the plasma membrane overlying the acrosome.

When the sperm reaches near the ovum, changes take place in membranes over the acrosome and enable release of lysosomal enzymes present within the acrosome. This is called the ***acrosome reaction***. The substances released include hyaluronidase that helps in separating corona radiata cells present over the ovum. ***Trypsin-like substances*** and a substance called ***acrosin***, help in digesting the zona pellucida and penetration of the sperm through it. Changes in the properties of the zona pellucida constitute the ***zona reaction***.

ACCESSORY UROGENITAL ORGANS

EPIDIDYMIS

Epididymis is a comma shaped structure present on the postero-lateral aspect of testis. Structurally, the epididymis consists of a head and a duct. The body and tail of the epididymis are made up of the duct of the epididymis, that is greatly coiled on itself. The head is formed by highly convoluted continuations of the efferent ductules. These are lined by ciliated columnar epithelium (Plate 19.3). At the lower end of the head of the epididymis these tubules join to form a single tube called the ***duct of the epididymis***.

PLATE 19.3: The Epididymis

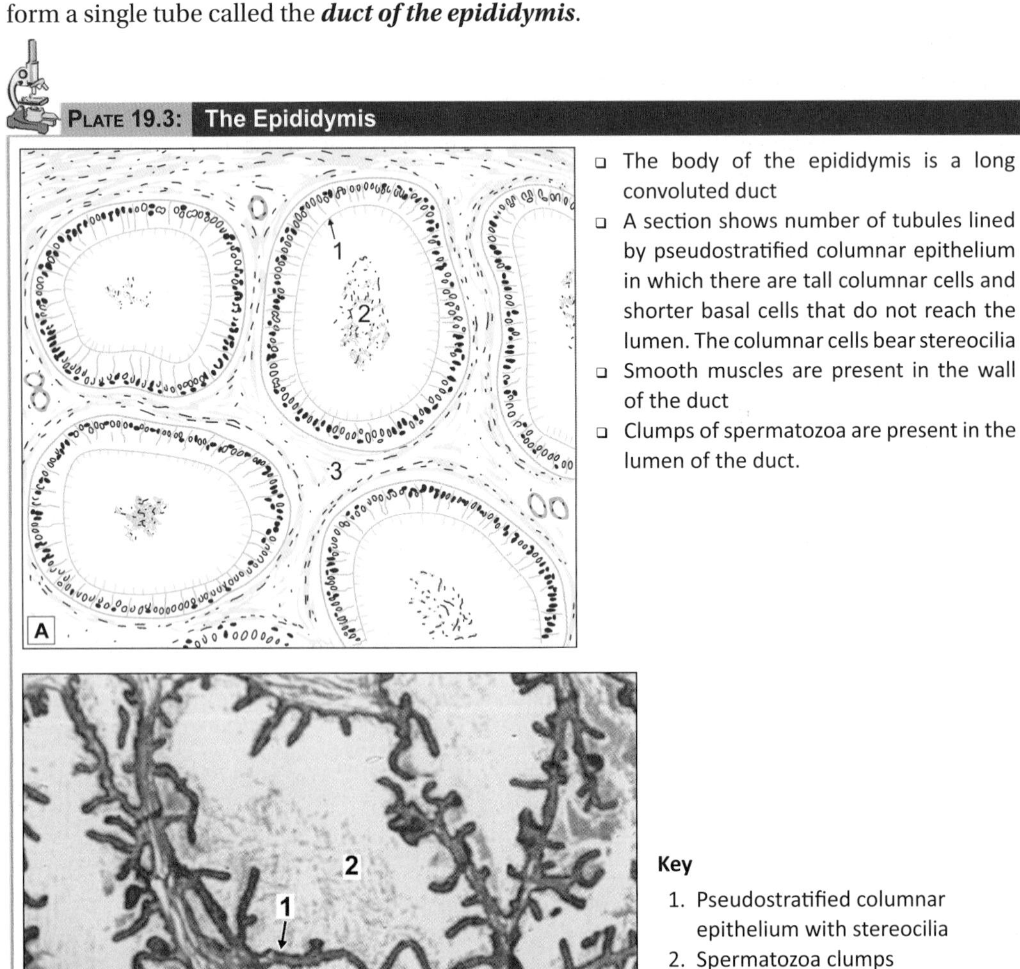

- The body of the epididymis is a long convoluted duct
- A section shows number of tubules lined by pseudostratified columnar epithelium in which there are tall columnar cells and shorter basal cells that do not reach the lumen. The columnar cells bear stereocilia
- Smooth muscles are present in the wall of the duct
- Clumps of spermatozoa are present in the lumen of the duct.

Key

1. Pseudostratified columnar epithelium with stereocilia
2. Spermatozoa clumps
3. Smooth muscle fibres

The epididymis. A. As seen in drawing; B. Photomicrograph

The duct is lined by pseudostratified columnar epithelium which is made of 2 types of cells tall columnar cells, and shorter basal cells that do not reach the lumen.

The luminal surface of each columnar cell bears non-motile projections that resemble cilia. These stereocilia are seen by EM to be thick microvilli. They do not have the structure of true cilia. The EM also shows the presence of agranular endoplasmic reticulum, lysosomes and a prominent Golgi complex in these cells. The basal cells are precursors of the tall cells. Beneath the epithelium there is a layer of circularly arranged smooth muscle fibres. This muscle layer increases in thickness gradually from head to tail and may be organised into inner circular and outer longitudinal layers in the tail region.

Functions

- Phagocytosis of defective spermatozoa.
- Absorption of excess fluid.
- Secretion of substances (sialic acid, glyceryl-phosphoryl-choline) that play a role in maturation of spermatozoa.

DUCTUS DEFERENS

The ductus deferens (deferent duct or vas deferens) is a muscular tube extending from the lower end of epididymis to the prostatic urethra. The wall of the ductus deferens consists (from inside out) of:

- Mucous membrane
- Muscle
- Connective tissue (Plate 19.4).

Mucous Membrane

The mucous membrane shows a number of longitudinal folds so that the lumen appears to be stellate in section. The lining epithelium is simple columnar, but becomes pseudostratified columnar in the distal part of the duct. The cells are ciliated in the extra-abdominal part of the duct. The epithelium is supported by a lamina propria in which there are many elastic fibres.

Muscle

The muscle coat is very thick and consists of smooth muscle. It is arranged in the form of an inner circular layer and an outer longitudinal layer. An inner longitudinal layer is present in the proximal part of the duct.

Connective Tissue

The fibroelastic connective tissue forms the adventitial layer containing blood vessels and nerves.

The terminal dilated part of the ductus deferens is called the ***ampulla***, which joints the duct of seminal vesicles to form ejaculatory duct. It has the same structure as that of the seminal vesicle.

PLATE 19.4: The Ductus Deferens

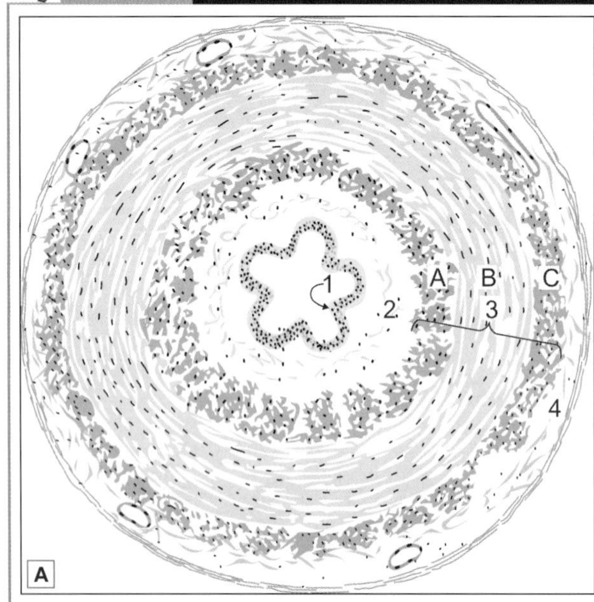

This tubular structure displays:

- A small irregular lumen
- Mucous membrane lined by pseudostratified columnar epithelium with underlying lamina propria
- The muscle coat is very thick. Three layers, inner longitudinal, middle circular and outer longitudinal are seen
- Outer most layer is adventitia composed of collagen fibres and containing blood vessels.

Key

1. Small irregular lumen lined by pseudostratified columnar epithelium
2. Lamina propria
3. Muscle layer
   A. Inner longitudinal muscle layer
   B. Middle circular muscle layer
   C. Outer longitudinal muscle layer.
4. Adventitia

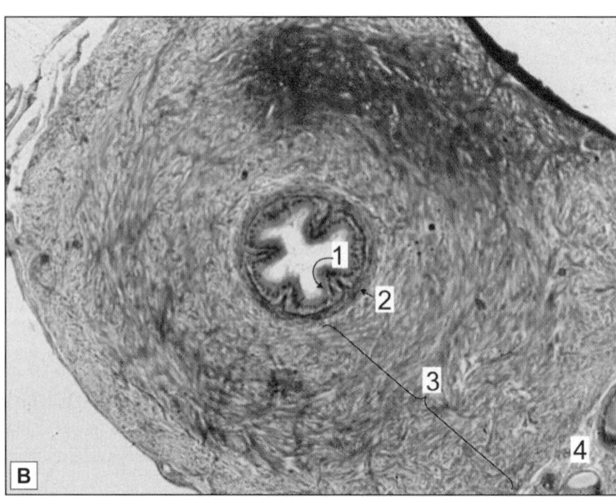

The Ductus Deferens. A. As seen in drawing; B. Photomicrograph.

THE SEMINAL VESICLE

The seminal vesicle is a sac-like mass that is really a convoluted tube. The tube is cut several times in any section (Plate 19.5). The wall of the seminal vesicle (from inside out) consists of:
- Mucous membrane
- Muscle
- Connective tissue.

Mucous Membrane

The mucosal lining is thrown into numerous thin folds that branch and anastomose thus forming a network. The lining epithelium is simple columnar, or pseudostratified. Goblet cells are present in the epithelium.

Muscle

The seminal vesicles consist of a thin intermediate layer of smooth muscles. The muscle layer contains outer longitudinal and inner circular fibres.

Connective Tissue

The outer covering of loose connective tissue forms the adventitial layer containing blood vessels and nerves.

Function

The seminal vesicles produce a thick secretion that forms the bulk of semen. The secretion contains fructose which provides nutrition to spermatozoa. It also contains amino acids, proteins, prostaglandins, ascorbic acid and citric acid. This secretion is expelled during ejaculation by contraction of the smooth muscle of the vesicle.

Clinical Correlation

Fructose content in seminal fluid: In cases of male infertility, on semen analysis absence of frustose suggests congenital absence of seminal vesicle or portion of the ductal system or both.

PROSTATE

The prostate is the largest accessory sex gland. It surrounds the beginning of urethra and is the shape of a chestnut. It is made up of 30 to 50 compound tubulo-alveolar glands that are embedded in a framework of fibromuscular tissue. The glandular part of the prostate is poorly developed at birth. It undergoes considerable proliferation at puberty, and degenerates in old age.

In sections, the glandular tissue is seen in the form of numerous follicles that are lined by columnar epithelium (Plate 19.6). The epithelium is thrown into numerous folds (along with some underlying connective tissue). The follicles drain into 12 to 20 excretory ducts that open into the prostatic urethra. The ducts are lined by a double layered epithelium. The superficial (luminal) layer is columnar, and the deeper layer is cuboidal.

Small rounded masses of uniform or lamellated structure are found within the lumen of the follicles. They are called ***amyloid bodies*** or ***corpora amylacea***. These are more abundant in older individuals. These consist of condensed glycoprotein. They are often calcified.

PLATE 19.5: Seminal Vesicle (Low Magnification)

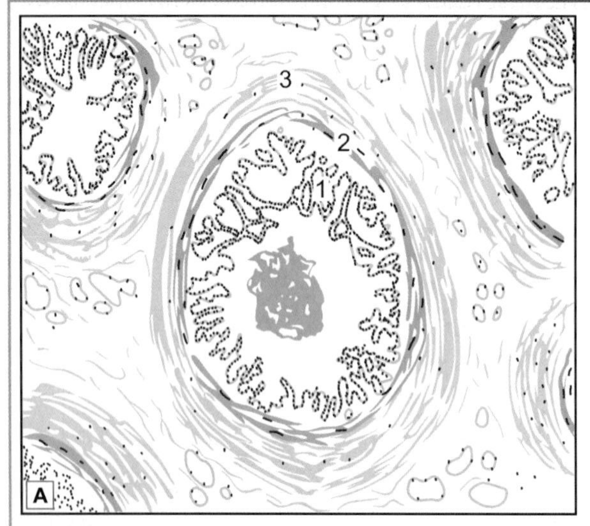

- The seminal vesicle is made up of a convoluted tubule
- The tube has an outer covering of connective tissue, a thin layer of smooth muscle and an inner mucosa
- The mucosal lining is thrown into numerous folds that branch and anastomose to form a network
- The lining epithelium is usually simple columnar or pseudostratified.

Key

1. Folds of mucosa
2. Muscle wall
3. Adventitial layer

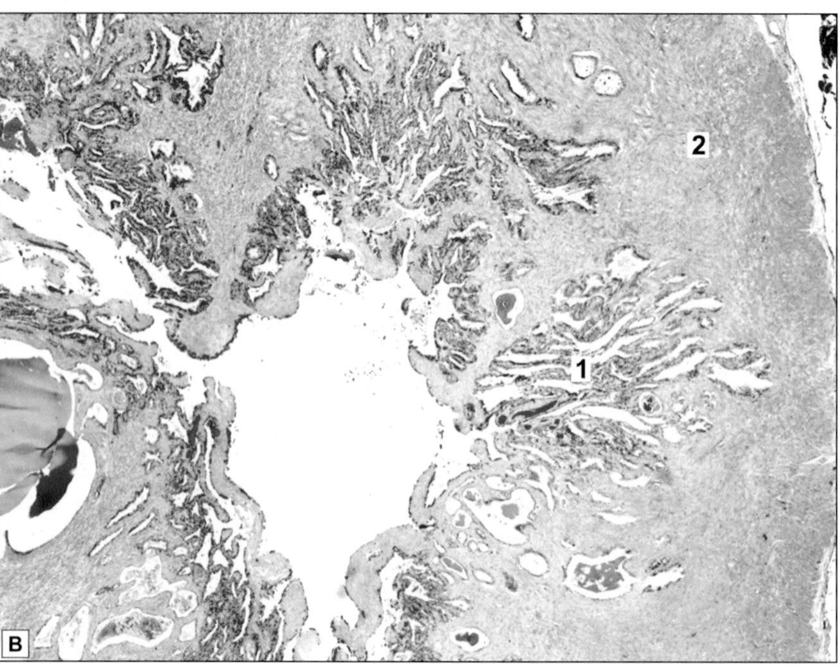

Seminal vesicle (low magnification). A. As seen in drawing; B. Photomicrograph

Plate 19.6: Prostate (Low Magnification)

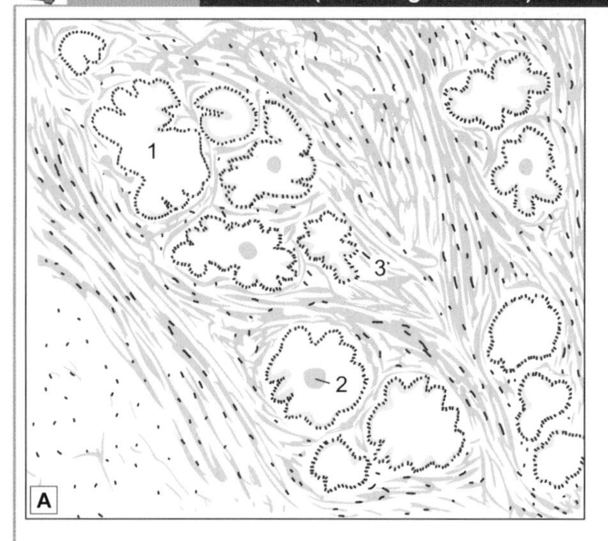

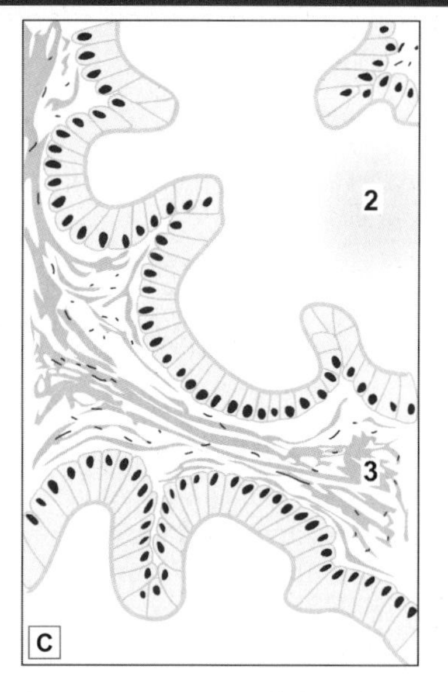

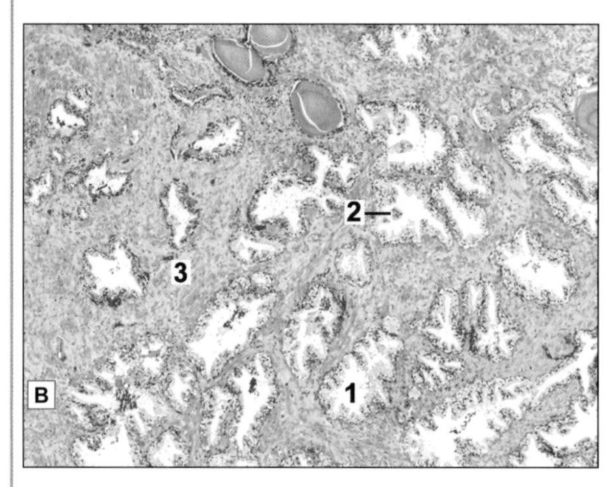

Prostate. A. As seen in drawing (low magnification); B. Photomicrograph (low magnification); C. Prostate as seen in drawing (high magnification).

- The prostate consists of glandular tissue embedded in prominent fibromuscular stroma.
- The glandular tissue is in the form of follicles with serrated edges. They are lined by columnar epithelium. The lumen may contain amyloid bodies.
- The follicles are separated by broad bands of fibromuscular tissue.

Key

1. Follicles lined by columnar epithelium
2. Amyloid bodies
3. Fibromuscular tissue

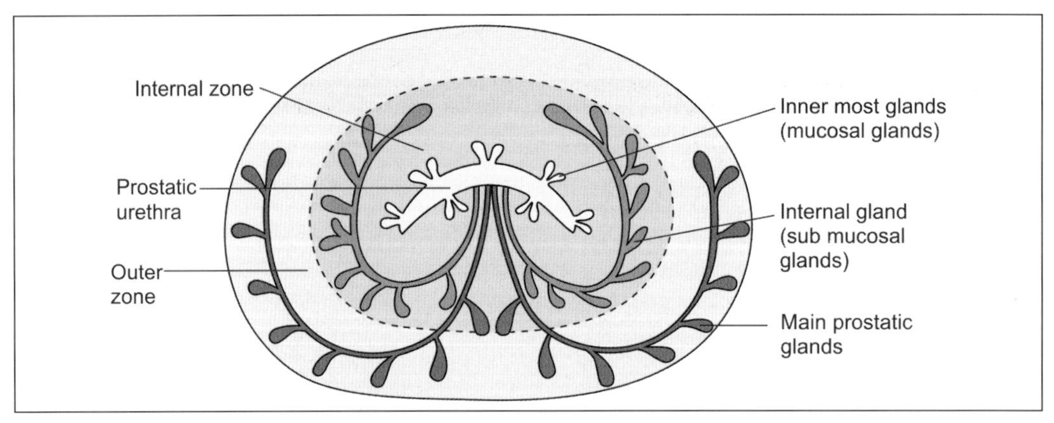

Fig. 19.7: Arrangement of glandular tissue in prostate (Schematic representation)

The fibromuscular tissue forms a conspicuous feature of sections of the prostate. It contains collagen fibres and smooth muscle. Within the gland the fibromuscular tissue forms septa that separate the glandular elements. These septa are continuous with a fibrous capsule that surrounds the prostate. The capsule contains numerous veins and parasympathetic ganglion cells.

On the basis of differences in the size and nature of the glands the prostate can be divided into an ***outer*** (or ***peripheral***) ***zone***, and an ***internal zone***. An innermost zone lying immediately around the prostatic urethra is also described (Fig. 19.7).

The glands in the outer zone are the main prostatic glands. They open into long ducts that join the urethra. The internal (or submucous) glands have short ducts. The innermost (or mucous) glands open directly into the urethra. The internal and innermost zones together form the central zone.

The prostate is traversed by the prostatic urethra. The gland is also traversed by the ejaculatory ducts.

Function

- The prostate produces a secretion that forms a considerable part of semen. The secretion is rich in enzymes (acid phosphatase, amylase, protease) and in citric acid.
- The prostate also produces substances called ***prostaglandins*** that have numerous actions.

Clinical Correlation

- **Benign nodular hyperplasia of prostate:** Non-neoplastic tumour-like enlargement of the prostate, is a very common condition in men and considered by some as normal ageing process. It becomes increasingly more frequent above the age of 50 years and its incidence approaches 75–80% in men above 80 years. The central zone commonly undergoes benign hypertrophy in old persons. Enlargement of the prostate can compress the urethra leading to problems in passing urine.
- **Carcinoma of prostate:** Cancer of the prostate is the second most common form of cancer in males, followed in frequency by lung cancer. It is a disease of men above the age of 50 years and its prevalence increases with increasing age so that more than 50% of men 80 years old have asymptomatic (latent) carcinoma of the prostate. Many a times, carcinoma of the prostate is small and detected as microscopic foci in a prostate removed for benign enlargement of prostate or found incidentally at autopsy. The peripheral zone is often the site of carcinoma.

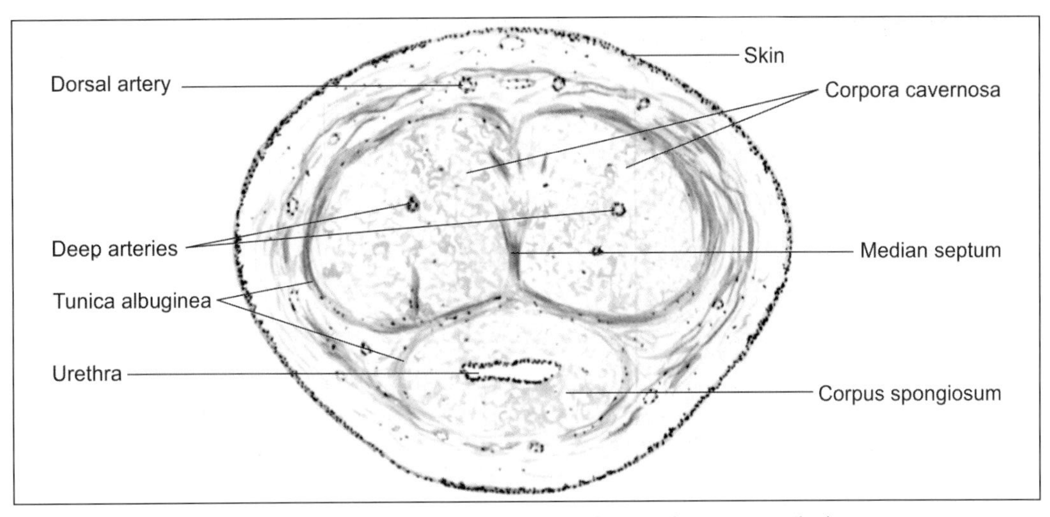

Fig. 19.8: Transverse section of penis (Schematic representation)

PENIS

The penis is the erectile copulatory organ in males. It consists of a ***root*** that is fixed to the perineum, and of a free part that is called the ***body*** or ***corpus***. A transverse section through the free part of the penis is shown in Figure 19.8. The penis is covered all round by thin skin that is attached loosely to underlying tissue.

The substance of the penis is made up of three masses of erectile tissue, two dorsal and one ventral. The dorsal masses are the right and left ***corpora cavernosa***, while the ventral mass is the ***corpus spongiosum***.

Each corpus cavernosum is surrounded by a dense sheath containing collagen fibres, elastic fibres and some smooth muscle. In the midline the sheaths of the right and left corpora cavernosa fuse to form a median septum. The corpora cavernosa lie side by side and are separated only by a median fibrous septum.

The corpus spongiosum is placed in the midline ventral to the corpora cavernosa. The corpus spongiosum is also surrounded by a sheath, but this sheath is much thinner than that around the corpora cavernosa. An additional sheath surrounds both the corpora cavernosa and the corpus spongiosum. The corpus spongiosum is traversed by the penile urethra throughout its length.

The tip of urethra at glans penis is lined by stratified squamous non-keratinised epithelium. Many small mucous glands of Littre are scattered along the length of urethra that secrete mucus and have lubricating functions.

Many sensory nerve endings are present in the penis, particularly on the glans.

Added Information

Erectile Tissue

Numerous septa arising from the connective tissue sheath extend into the corpora cavernosa and into the corpus spongiosum, and form a network. The spaces of the network are lined by endothelium. The spaces are in communication with arteries and veins. They are normally empty.

During erection of the penis they are filled with blood under pressure. This results in enlargement and rigidity of the organ. The process of erection involves the corpora cavernosa more than the corpus spongiosum, rigidity of the former being made possible by the presence of a dense fibrous sheath. The corpus spongiosum does not become so rigid as its sheath is elastic, and the vascular spaces within it are smaller. As a result, the penile urethra remains patent during erection, and semen can flow through it.

Blood to the corpora cavernosa is supplied mainly by the deep arteries of the penis. These arteries give off branches that follow a spiral course before opening into the cavernous spaces. They are called ***helicrine arteries*** and have an unusual structure. The circular muscle in their media is very thick so that the vessels can be completely occluded. The tunica intima shows longitudinal thickenings.

Erection is produced by complete relaxation of smooth muscle both in the walls of arteries and in the septa. Helicrine arteries are connected to veins by arteriovenous anastomoses. Normally, the anastomoses are patent. Their closure (caused by parasympathetic nerves) causes cavernous spaces to fill leading to erection.

As the cavernous spaces fill with blood, increasing pressure in them compresses the veins that lie just deep to the fibrous sheath. In this way blood is prevented from draining out of the spaces. At the end of erection smooth muscle in the walls of arteries contracts stopping inflow of blood. Contraction of muscle in the trabeculae gradually forces blood out of the spaces.

Chapter 20

Female Reproductive System

The female reproductive system includes (Fig. 20.1)

- A pair of ovaries
- A pair of uterine tubes
- Uterus
- Vagina
- External genitalia
- Mammary glands

THE OVARIES

The ovaries are the female gonads, responsible for the formation of ova. They also produce hormones (oestrogen and progesterone) that are responsible for the development of the female secondary sex characters, and produce marked cyclical changes in the uterine endometrium. Each ovary is an oval structure about 3 cm in long diameter.

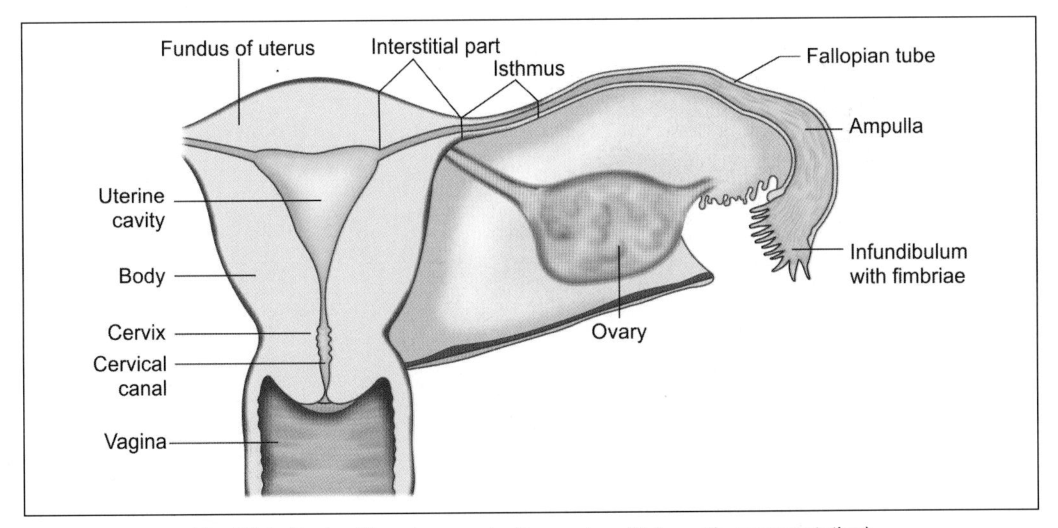

Fig. 20.1: Parts of female reproductive system (Schematic representation)

General Structure

Each ovary consists of the following parts (Fig. 20.2 and Plate 20.1):

- ***Germinal epithelium:*** The free surface of ovary is covered by a single layer of cubical cells that constitute the ***germinal epithelium***. This epithelium is continuous with the mesothelium lining the peritoneum, and represents a modification of the latter.

Note: The term germinal epithelium is a misnomer. The epithelium does not produce germ cells. The cells of this epithelium bear microvilli, and contain numerous mitochondria. They become larger in pregnancy.

- ***Tunica albuginea:*** The germinal epithelium rests on a connective tissue layer called the ***tunica albuginea***. The tunica albuginea of the ovary is much thinner, and less dense, than that of the testis
- ***Cortex:*** Deep to the tunica albuginea the cortex has a stroma made up of reticular fibres and numerous fusiform cells that resemble mesenchymal cells. Scattered in this stroma there are ***ovarian follicles*** at various stages of development. Each follicle contains a developing ovum
- ***Medulla:*** The ***medulla*** consists of connective tissue in which numerous blood vessels (mostly veins) are seen. Elastic fibres and smooth muscle are also present. The hilum of the ovary is the site for entry of blood vessels and lymphatics. It is continuous with the medulla. The hilum also contains some remnants of the mesonephric ducts; and ***hilus cells*** that are similar to interstitial cells of the testis.

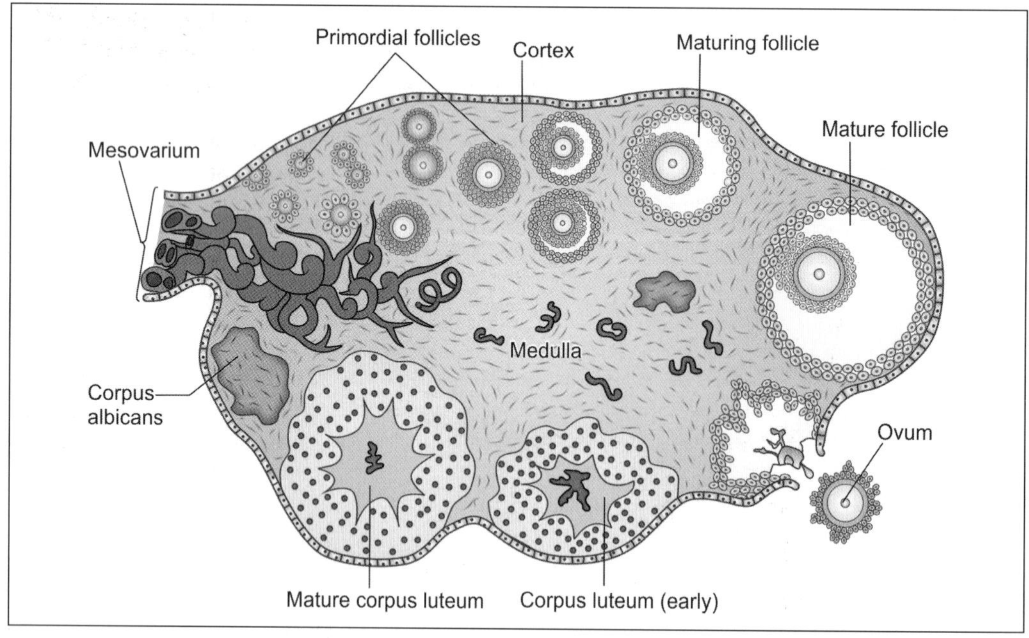

Fig. 20.2: Histological structure of ovary showing follicles at various stages of development (Schematic representation)

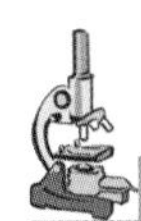

PLATE 20.1: Ovary

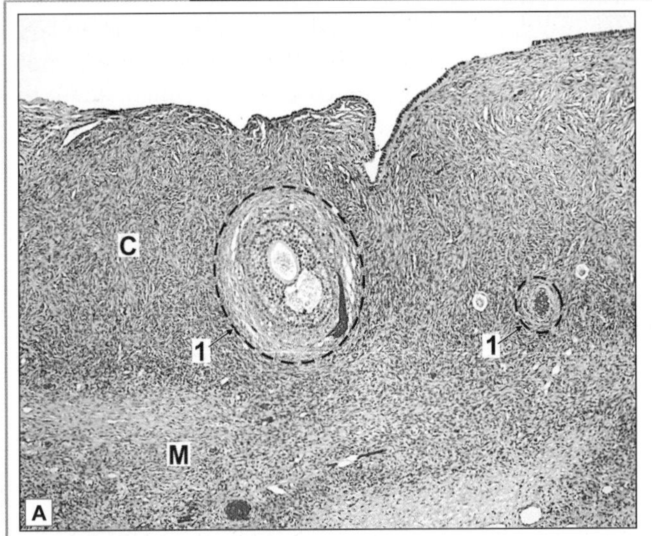

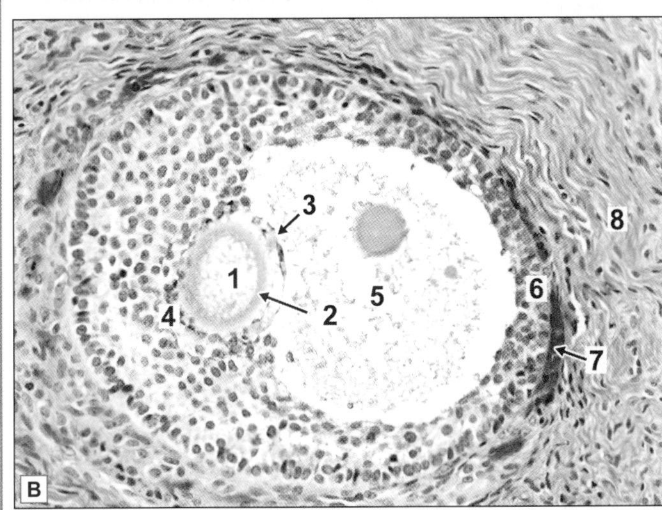

A. Ovary (photomicrograph); B. Graafian follicle (photomicrograph)

- The surface is covered by a cuboidal epithelium. Deep to the epithelium there is a layer of connective tissue that constitutes the tunica albuginea
- The substance of the ovary has an outer cortex in which follicles of various sizes are present; and an inner medulla consisting of connective tissue containing numerous blood vessels
- Just deep to the tunica albuginea many primordial follicles each of which contains a developing ovum surrounded by flattened follicular cells are present.
- Large follicles have a follicular cavity surrounded by several layers of follicular cells
- The cells surrounding the ovum constitute the cumulus oophoricus
- The follicle is surrounded by a condensation of connective tissue which forms a capsule for it
- The capsule consists of an inner cellular part (the theca interna), and an outer fibrous part (the theca externa) collectively called as Theca folliculi. The follicle is surrounded by a stroma made up of reticular fibres and fusiform cells.

Key

C. Cortex
M. Medulla
1. Ovarian follicle
2. Zona pellucida
3. Cumulus oophoricus
4. Discus proligerus
5. Antrum folliculi
6. Membrana granulosa
7. Capsule of follicle
8. Stroma

OOGENESIS

The process of formation of ovum from the stem cells is called oogenesis (Flowchart 20.1). The process of oogenesis consists of following stages:

- ***Oogonia:*** The stem cells from which ova are derived are called ***oogonia***. These are large round cells present in the cortex of the ovary. Oogonia are derived (in fetal life) from ***primordial germ cells*** that are formed in the region of the yolk sac, and migrate into the developing ovary. They increase in number by mitosis.

  All oogonia to be used throughout the life of a woman are produced at a very early stage (before birth) and do not multiply thereafter. At birth the number of oogonia in an ovary is about one million. Many oogonia formed in this way degenerate, the process starting before birth and progressing throughout life, so that the number of oogonia becomes less and less with increasing age.
- ***Primary oocyte:*** An oogonium enlarges to form a ***primary oocyte***. The primary oocyte contains the diploid number of chromosomes i.e., 46. It undergoes the first meiotic division to form two daughter cells each of which has 23 chromosomes.
- ***Secondary oocyte:*** The cytoplasm of the primary oocyte is not equally divided. Most of it goes to one daughter cell that is large and is called the ***secondary oocyte***. The second daughter cell has hardly any cytoplasm, and forms the ***first polar body***.

Flow chart 20.1: Stages of oogenesis.

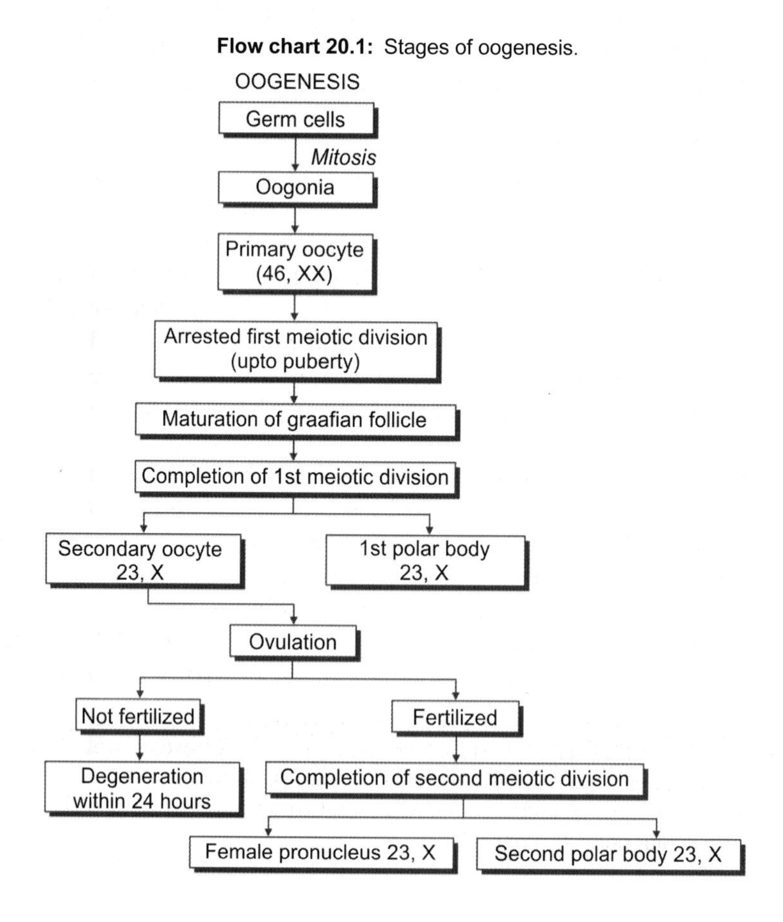

Embryological Considerations

- At the time of birth all primary oocytes are in the prophase of first meiotic division. Their number is about 40,000.
- The primary oocytes remain in prophase and do not complete their first meiotic division until they begin to mature and are ready to ovulate.
- The reproductive period of a female is between 12 and 50 years of age. With each menstrual cycle, a few primary oocytes (about 5 to 30) begin to mature and complete the first meiotic division shortly before ovulation.
- The secondary oocyte immediately enters the second meiotic cell division. Ovulation takes place while the oocyte is in metaphase. The secondary oocyte remains arrested in metaphase till fertilization occurs.
- The second meiotic division is completed only if fertilization occurs.
- If fertilization does not occur, the secondary oocyte fails to complete the second meiotic division and degenerates about 24 hours after ovulation.
- In each menstrual cycle, 5 to 30 primary oocytes start maturing, but only one of them reaches maturity and is ovulated. The remaining degenerate.
- During the entire reproductive life of a female, only around 400 ova are discharged (out of 40,000 primary oocytes available).

- ***Ovum:*** The secondary oocyte now undergoes the second meiotic division, the daughter cells being again unequal in size. The larger daughter cell produced as a result of this division is the ***mature ovum***. The smaller daughter cell (which has hardly any cytoplasm) is the ***second polar body***. Thus, one primary oocyte ultimately gives rise to only one ovum.

Formation of Ovarian Follicles

Ovarian follicles (or ***Graafian follicles***) consists of a developing ova surrounded by follicular (granulosa) cells. The development and maturation of an ovarian follicle passes through four stages—the process is called folliculogenesis (Figs 20.2 and 20.3).

- ***Primordial follicle:*** Some cells of the stroma become flattened and surround an oocyte (Figs 20.3 and 20.4A). These stromal cells are now called ***follicular cells***. The oocyte (20–25 μm) and the flat surrounding cells form a ***primordial follicle***. Primordial follicles are the smallest and simplest in structure located at the periphery of the cortex. Numerous primordial follicles are present in the ovary at birth. They undergo further development only at puberty.
- ***Primary follicle:*** The first indication that a primordial follicle is beginning to undergo further development is that the flattened follicular cells become columnar (Figs 20.3 and 20.4B). Follicles at this stage of development are called ***primary follicles***. The outermost layer of the follicular cells rest on a well defined basement membrane which separates it from the ovarian stroma. A homogeneous membrane, the ***zona pellucida***, appears between the follicular cells and the oocyte (which enlarges in size 50–80 μm in diameter. With the appearance of the zona pellucida the follicle is now referred to as a ***multilaminar primary follicle***.

Added Information

The origin of the zona pellucida is controversial. It consists of glycoprotein that is eosinophilic and PAS positive. It is traversed by microvilli projecting outwards from the oocyte. Thin cytoplasmic processes from follicular cells also enter it.

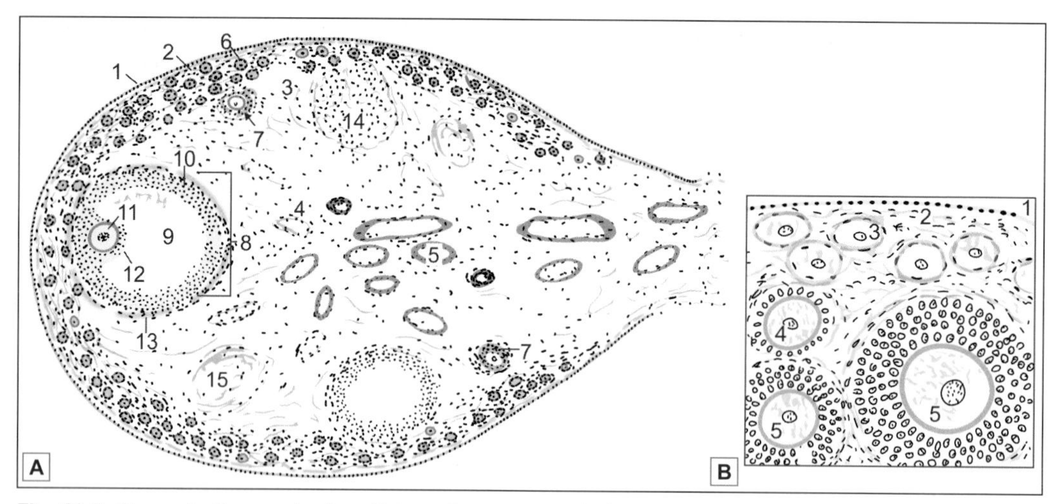

Fig. 20.3: Ovary, **A.** Panoramic view (Schematic representation). 1–cuboidal epithelium over surface; 2–tunica albuginea; 3–cortex; 4–medulla; 5–blood vessels; 6–primordial follicle; 7–secondary follicle; 8–graffian follicle; 9–follicular cavity; 10–granulosa cells; 11–ovum; 12-cumulus oophoricus; 13–capsule of follicle; 14–corpus luteum; 15-atretic follicle. **B.** Ovary (high magnification drawing) showing early stages in formation of follicles. 1–germinal epithelium; 2–tunica albuginea; 3–primordial follicle; 4–primary follicle; 5–secondary follicles (Schematic representation)

- ***Secondary follicle:*** The follicular cells proliferate to form several layers of cells that constitute the ***membrana granulosa***. The cells are now called ***granulosa cells***. The oocyte enlarges and reaches its maximum size (125 μm). This is a ***secondary follicle*** (Fig. 20.3).

  So far the granulosa cells are in the form of a compact mass. However, the cells to one side of the ovum soon partially separate from one another so that a ***follicular cavity*** (or ***antrum folliculi***) appears between them. It is with the appearance of this cavity that a true follicle (= small sac) can be said to have been formed. The follicular cavity is filled by a fluid, the ***liquor folliculi***. With the formation of follicular cavity the size of the follicle increases.

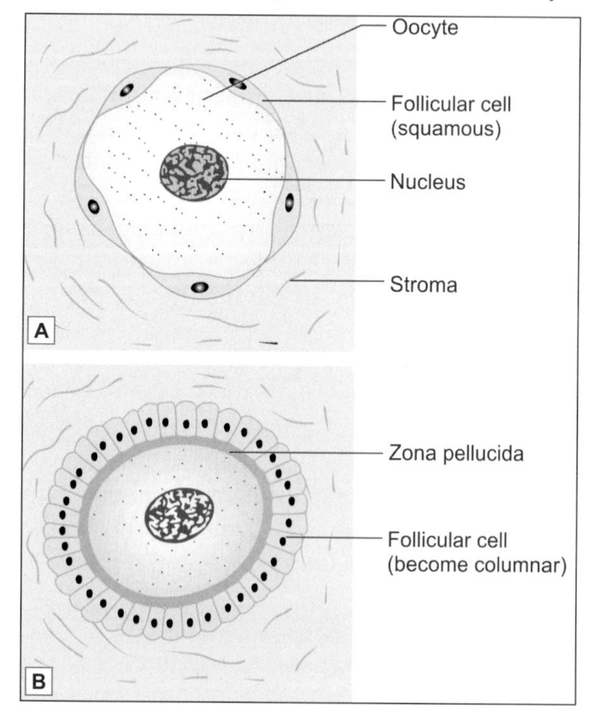

Fig. 20.4: A. Primordial follicle; **B.** Primary follicle (Schematic repesentation)

- ***Graafian follicle:*** With the further development, the follicular cavity rapidly increases in size. As a result, the wall of the follicle (formed by the granulosa cells) becomes relatively thin. The graffian follicle now measures about 10 mm or more in diameter and is seen bulging out of the cortex (Fig. 20.2). The oocyte now lies eccentrically in the follicle surrounded by some granulosa cells that are given the name of ***cumulus oophoricus*** (or ***cumulus oophorus***, or ***cumulus ovaricus***). The inner most layer of cumulus oophorus that lies directly adjacent to the zona pellucida is called ***corona radiata***. The granulosa cells that attach the oocyte to the wall of the follicle constitute the ***discus proligerus*** (Fig. 20.5).

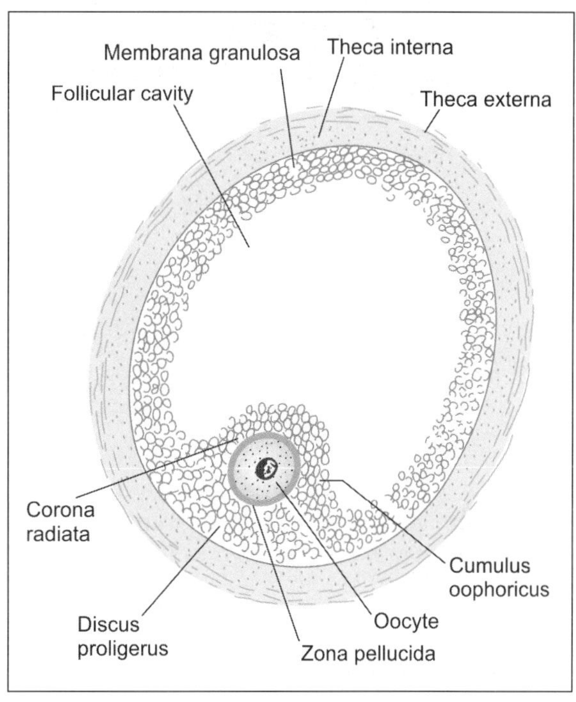

Fig. 20.5: Mature ovarian follicle (Schematic representation)

As the follicle expands the stromal cells surrounding the membrana granulosa become condensed to form a covering called the ***theca interna*** (theca = cover). The cells of the theca interna later secrete a hormone called ***oestrogen***, and they are then called the cells of the ***thecal gland***.

Outside the theca interna some fibrous tissues become condensed to form another covering for the follicle. This is the ***theca externa***. The theca interna and externa are collectively called the ***theca folliculi***.

Ovulation

The ovarian follicle is at first very small compared to the thickness of the ovarian cortex. As the follicle enlarges it becomes so big that it not only reaches the surface of the ovary, but forms a bulging in this situation. As a result the stroma and the theca on this side of the follicle are stretched and become very thin (Fig. 20.6).

An avascular area (***stigma***) appears over the most convex point of the follicle. At the same time the cells of the cumulus oophoricus become loosened by accumulation of fluid between

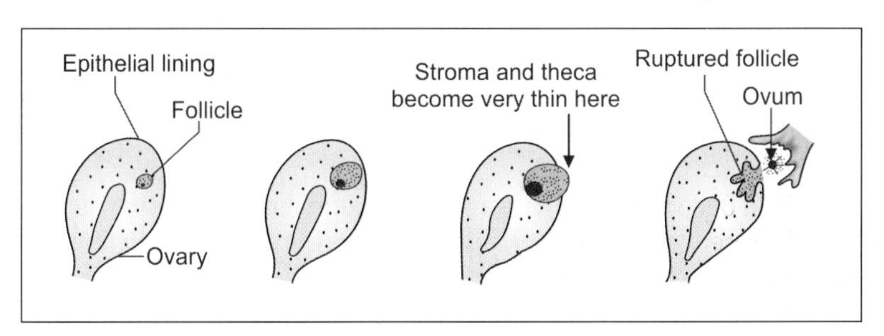

Fig. 20.6: Relationship of a growing ovarian follicle to the ovary (Schematic representation)

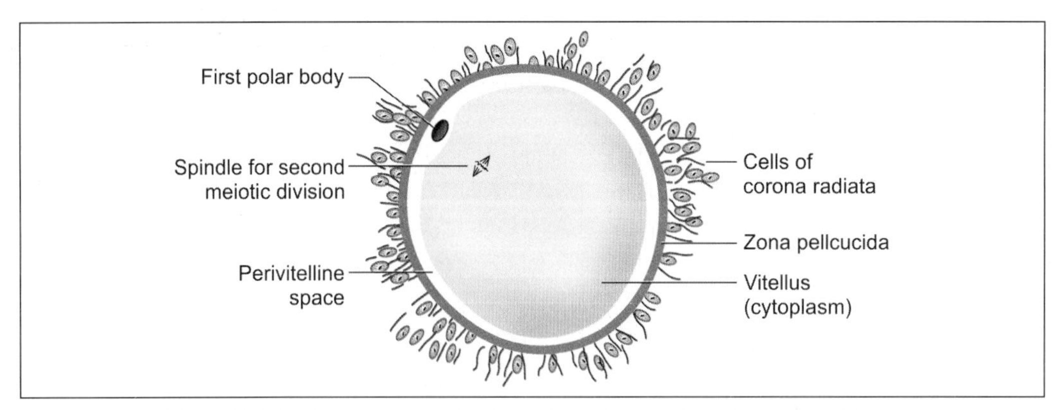

Fig. 20.7: Structure of ovum at the time of ovulation (Schematic representation)

them. The follicle ultimately ruptures and the ovum is shed from the ovary. The shedding of the ovum is called ***ovulation. The 'ovum' that is shed from the ovary is not fully mature. It is really a secondary oocyte surrounded by zona pellucida and corona radiata*** (Fig. 20.7).

Fate of the Ovum

The ovary is closely embraced by the fimbriated end of the uterine tube. Therefore, the ovum is easily carried into the tube partly by the follicular fluid discharged from the follicle and partly by the activity of ciliated cells lining the tube. The ovum slowly travels through the tube towards the uterus, taking three to four days to do so. If sexual intercourse takes place at about this time, the spermatozoa deposited in the vagina swim into the uterus and into the uterine tube. One of these spermatozoa may fertilize the ovum. If this happens, the fertilized ovum begins to develop into an embryo. It travels to the uterus and gets implanted in its wall. On the other hand, if the ovum (secondary oocyte) is not fertilized it dies in 12 to 24 hours. It passes through the uterus into the vagina and is discharged.

Corpus Luteum

The corpus luteum is an important structure. It secretes a hormone, ***progesterone***. The corpus luteum is derived from the ovarian follicle, after the latter has ruptured to shed the ovum, as follows (Figs 20.8 and 20.9).

- ***Corpus haemorrhagicum:*** When the follicle ruptures its wall collapses and becomes folded. Sudden reduction in pressure caused by rupture of the follicle results in bleeding into the follicle. The follicle filled with blood is called the ***corpus haemorrhagicum***. At this stage, the follicular cells are small and rounded.
- ***Corpus luteum:*** The cells now enlarge rapidly. As they increase in size their walls press against those of neighbouring cells so that the cells acquire a polyhedral shape (Figs 20.8 and 20.9).

  Their cytoplasm becomes filled with a yellow pigment called ***lutein***. They are now called ***luteal cells***. The presence of this yellow pigment gives the structure a yellow colour, and that is why it is called the corpus luteum (= yellow body).

  Some cells of the theca interna also enlarge and contribute to the corpus luteum. The cells of the corpus luteum contain abundant smooth ER and considerable amount of lipids.

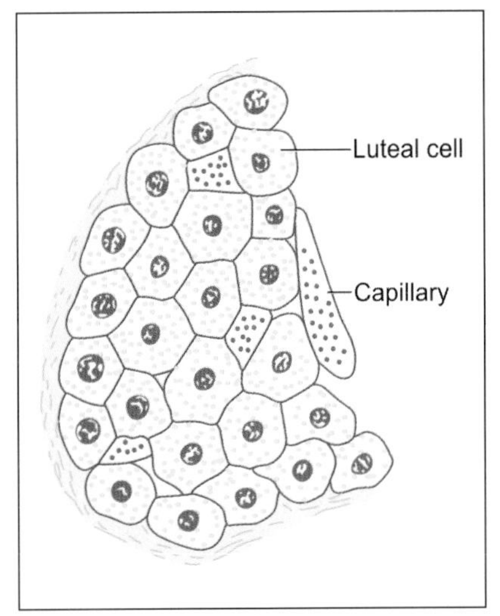

Fig. 20.8: Corpus luteum. Note the large hexagonal cells filled with yellow granules (Schematic representation)

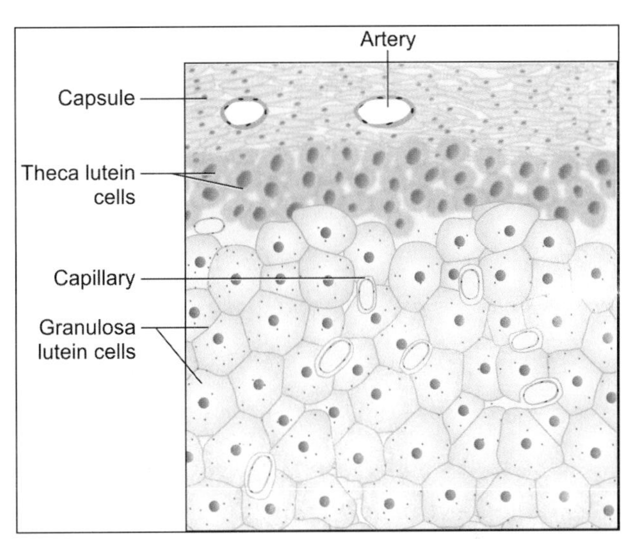

Fig. 20.9: Corpus luteum (high magnification) (Schematic representation)

The corpus luteum secretes progesterone. This secretion has to be poured into blood like secretions of endocrine glands. All endocrine glands are richly supplied with blood vessels for this purpose. ***The ovarian follicle itself has no blood vessels, but the surrounding theca interna is full of them. When the corpus luteum is forming, blood vessels from the theca interna invade it and provide it with a rich blood supply.***

The subsequent fate of the corpus luteum depends on whether the ovum is fertilised or not.

- If the ovum is not fertilised, the corpus luteum persists for about 14 days. During this period it secretes progesterone. It remains relatively small and is called the ***corpus luteum of menstruation***. At the end of its functional life, it degenerates and becomes converted into a mass of fibrous tissue called the ***corpus albicans*** (= white body) (Fig. 20.10).
- If the ovum is fertilised and pregnancy results, the corpus luteum persists for three to four months. It is larger than the corpus luteum of menstruation, and is called the ***corpus luteum of pregnancy***. The progesterone secreted by it is essential for the maintenance of pregnancy in the first few months. After the fourth month, the corpus luteum is no longer needed, as the placenta begins to secrete progesterone.

Fate of Ovarian Follicles

The series of changes that begin with the formation of an ovarian follicle, and end with the degeneration of the corpus luteum constitute what is called an ***ovarian cycle***.

In each ovarian cycle one follicle reaches maturity, sheds an ovum, and becomes a corpus luteum. At the same time, several other follicles also begin to develop, but do not reach maturity. It is interesting to note that, contrary to what one might expect, these follicles do not persist into the next ovarian cycle, but undergo degeneration. The ovum and granulosa cells of each follicle disappear. The cells of the theca interna, however, proliferate to form the ***interstitial glands***, also called the ***corpora atretica***. These glands are believed to secrete oestrogens.

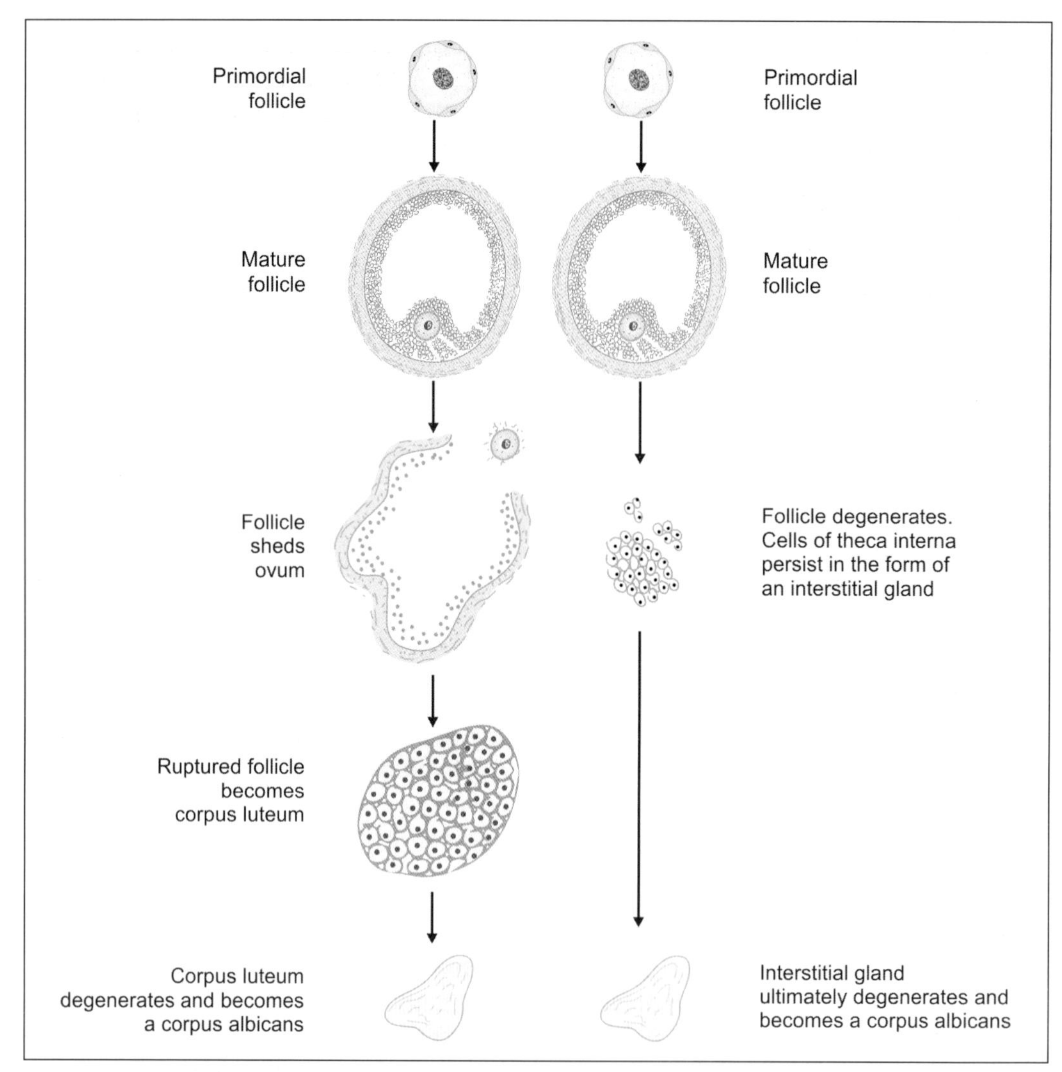

Fig. 20.10: Comparison of fate of ovarian follicles that shed an ovum and of those that do not (Schematic representation)

After a period of activity, each gland becomes a mass of scar tissue indistinguishable from the corpus albicans formed from the corpus luteum.

The cortex of an ovary (taken from a woman in the reproductive period) can show ovarian follicles (at various stages of maturation), corpora lutea, corpora albicantes, and corpora atretica.

The changes taking place during the ovarian cycle are greatly influenced by certain hormones produced by the hypophysis cerebri. The hormones produced by the theca interna and by the corpus luteum in turn influence other parts of the female reproductive system, notably the uterus, resulting in a cycle of changes referred to as the ***uterine cycle*** or ***menstrual cycle***.

UTERINE TUBES

Uterine tubes are paired muscular tubes and are also called ***fallopian tubes***. Each uterine tube has a medial or uterine end, attached to (and opening into) the uterus, and a lateral end that opens into the peritoneal cavity near the ovary. The tube has (from medial to lateral side),

- A ***uterine part*** that passes through the thick uterine wall
- A relatively narrow, thick walled part called the ***isthmus***
- A thin walled dilated part called the ***ampulla***
- Funnel shaped infundibulum. It is prolonged into a number of finger like processes or ***fimbriae***.

The wall of the uterine tube consists of following layers from within outwards.

Mucous Membrane

The mucous membrane shows numerous branching folds that almost fill the lumen of the tube (Plate 20.2). These folds are most conspicuous in the ampulla. Each fold has a highly cellular core of connective tissue. It is lined by columnar epithelium that rests on a basement membrane. Some of the lining cells are ciliated: ciliary action helps to move ova towards the uterus.

Other cells are secretory in nature and are also called as ***peg cells***. They contain secretory granules and are not ciliated. Their surface shows microvilli. A third variety of ***intercalary cells*** is also described.

Muscle Coat

The muscle coat has an inner circular layer and an outer longitudinal layer of smooth muscle. An additional inner longitudinal layer may also be present. The circular layer is thickest in the uterine part of the tube. The circular muscle is thickest in the isthmus. The pattern of mucosal folds is also different in this region. There is some evidence that the isthmus may have some control on the passage of a fertilised ovum through it.

Serosa

It consists of mesothelium supported by connective tissue.

Functions

The uterine tube conveys ova, shed by the ovary, to the uterus. Ova enter the tube at its fimbriated end. Spermatozoa enter the uterine tube through the vagina and uterus. Fertilisation normally takes place in the ampulla. When fertilisation occurs, the fertilised ovum travels towards the uterus through the tube. Secretions present in the tubes provide nutrition, oxygen and other requirements for ova and spermatozoa passing through the tube.

Clinical Correlation

Ectopic Tubal Pregnancy

The term ectopic tubal pregnancy is used for implantation of a fertilized ovum in the uterine tube. Though ectopic pregnancy may rarely occur in the uterine horn, cornu, ovary and abdominal cavity, tubal pregnancy is by far the most common form of ectopic gestation. The most frequent site of tubal pregnancy is the ampullary portion and the least common is interstitial pregnancy. Ectopic tubal pregnancy is a potentially hazardous problem because of rupture which is followed by intraperitoneal haemorrhage.

PLATE 20.2: Uterine Tube

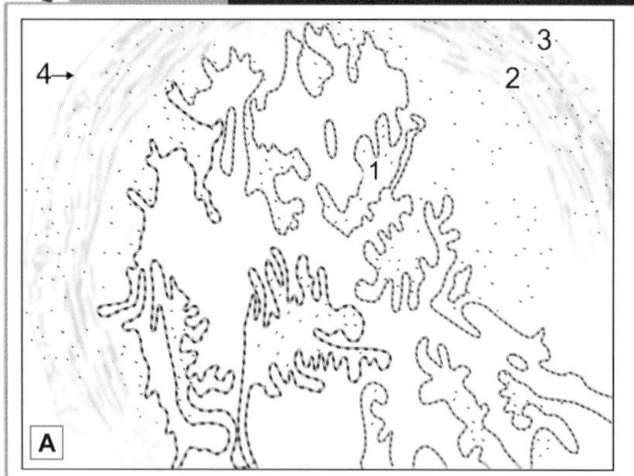

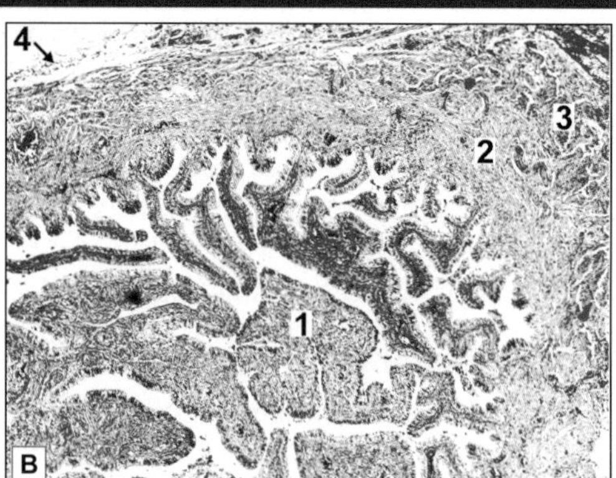

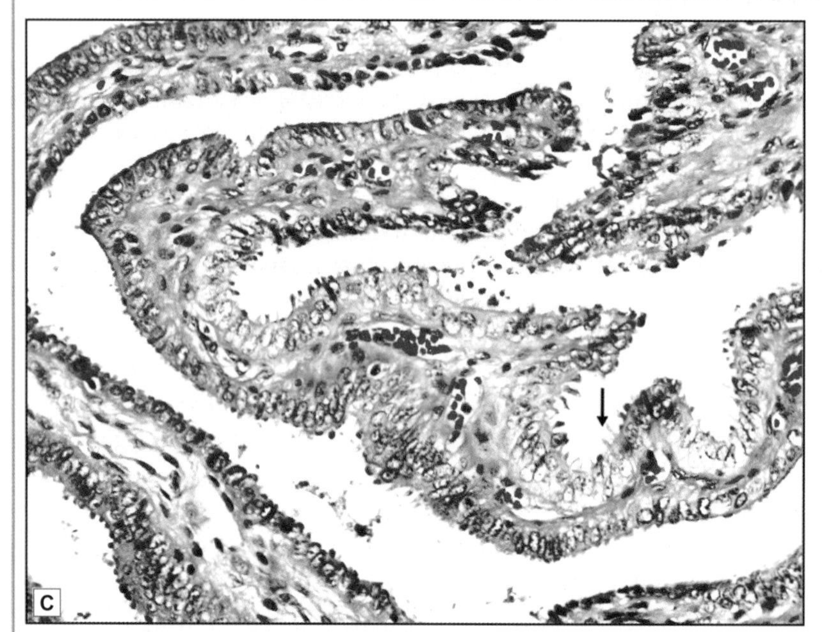

Uterine tube. A. As seen in drawing; B. Photomicrograph (low magnification); C. Photomicrograph (high magnification)

- The uterine tube is characterised by presence of numerous branching mucosal folds that almost fill the lumen of the tube
- The mucosa is lined by ciliated columnar epithelium
- The uterine tube has a muscular wall with an inner circular and outer longitudinal muscle layer.

Key

1. Mucous membrane with numerous branching folds
2. Inner circular muscle layer
3. Outer longitudinal muscle layer
4. Serosa

THE UTERUS

Uterus is a pear-shaped muscular organ located in the pelvic cavity.

The uterus consists of three parts: ***fundus, body*** and ***cervix***. The fundus is the upper dome-shaped part which is above the attachment of the fallopian tube. The body extends from the fundus to the isthmus; the isthmus is a narrow constricted part separating the body of the uterus from the cervix. Below the isthmus, the uterus becomes cylindrical in shape; this part is known as cervix.

The uterus has a very thick wall made up mainly of muscle (Plates 20.3 and 20.4). The lumen is small and is lined by mucous membrane. Part of the uterus is covered on the outside by peritoneum.

Myometrium

The muscle layer of the uterus is also called the ***myometrium***. It consists of bundles of smooth muscle amongst which there is connective tissue. Numerous blood vessels, nerves and lymphatics are also present in it.

The muscle fibres run in various directions and distinct layers are difficult to define. However, three layers, external, middle and internal are usually described. The fibres in the external layer are predominantly longitudinal. In the internal layer some bundles are longitudinal and others are circular. In the middle layer there is a mixture of bundles running in various directions.

The muscle cells of the uterus are capable of undergoing great elongation in association with the great enlargement of the organ in pregnancy (hypertrophy). New muscle fibres are also formed (hyperplasia). Contractions of the myometrium are responsible for expulsion of the fetus at the time of child birth.

Endometrium

The mucous membrane of the uterus is called the ***endometrium***. The endometrium consists of a lining epithelium that rests on a stroma. Numerous uterine glands are present in the stroma.

The lining epithelium is columnar. Before menarche (i.e., the age of onset of menstruation) the cells are ciliated, but thereafter most of the cells may not have cilia. The epithelium rests on a stroma that is highly cellular and contains numerous blood vessels. It also contains numerous simple tubular uterine glands. The glands are lined by columnar epithelium.

Menstrual Cycle

The endometrium undergoes marked cyclical changes that constitute the ***menstrual cycle***. The most prominent feature of this cycle is the monthly flow of blood from the uterus. This is called ***menstruation***. The menstrual cycle is divided (for descriptive convenience) into the following phases: ***postmenstrual, proliferative, secretory*** and ***menstrual***. The cyclical changes in the endometrium take place under the influence of hormones (oestrogen, progesterone) produced by the ovary. They are summarised below. (For details see the author's HUMAN EMBRYOLOGY).

- In the postmenstrual phase the endometrium is thin. It progressively increases in thickness being thickest at the end of the secretory phase.
- At the time of the next menstruation the greater part of its thickness (called the ***pars functionalis***) is shed off and flows out along with the menstrual blood. The part that remains is called the ***pars basalis***.
- The uterine glands are straight in the proliferative phase (Plate 20.3). As the endometrium increases in thickness the glands elongate, increase in diameter, and become twisted on themselves. Because of this twisting, they acquire a saw-toothed appearance in sections (Fig. 20.11 and Plate 20.4).
- At the time of menstruation the greater parts of the uterine glands are lost (along with the entire lining epithelium) leaving behind only their most basal parts.

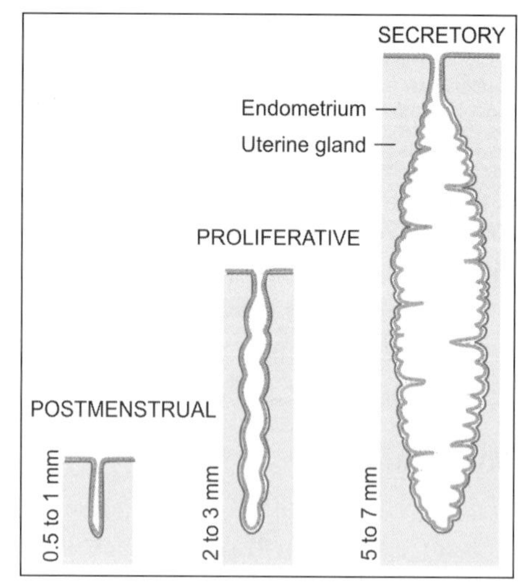

Fig. 20.11: Uterine glands at various stages of menstrual cycle. The thickness of endometrium is also indicated (Schematic representation)

- The lining epithelium is reformed (just after the cessation of menstruation) by proliferation of epithelial cells in the basal parts of the glands.
- The stroma and blood vessels of the endometrium also undergo cyclical changes.

Added Information

Hormones influencing Ovulation and Menstruation

We have seen that the changes taking place in the uterine endometrium during the menstrual cycle occur under the influence of:

- Oestrogens produced by the thecal gland (theca interna) and by the interstitial gland cells, and possibly by granulosa cells.
- Progesterone produced by the corpus luteum.

The development of the ovarian follicle, and of the corpus luteum, is in turn dependent on hormones produced by the anterior lobe of the hypophysis cerebri. These are:

- The ***follicle stimulating hormone*** (FSH) which stimulates the formation of follicles and the secretion of oestrogens by them; and
- The ***luteinising hormone*** (LH) which helps to convert the ovarian follicle into the corpus luteum, and stimulates the secretion of progesterone.

Secretion of FSH and LH is controlled by a ***gonadotropin releasing hormone*** (GnRH) produced by the hypothalamus. Production of LH is also stimulated by feed back of oestrogens secreted by follicular cells of the ovary. A sudden increase (surge) in the level of LH takes place near the middle of the menstrual cycle, and stimulates ovulation that takes place about 36 hours after the surge.

Apart from hormones, nervous and emotional influences may affect the ovarian and menstrual cycles. An emotional disturbance may delay or even prevent menstruation.

PLATE 20.3: Uterus (Proliferative Phase)

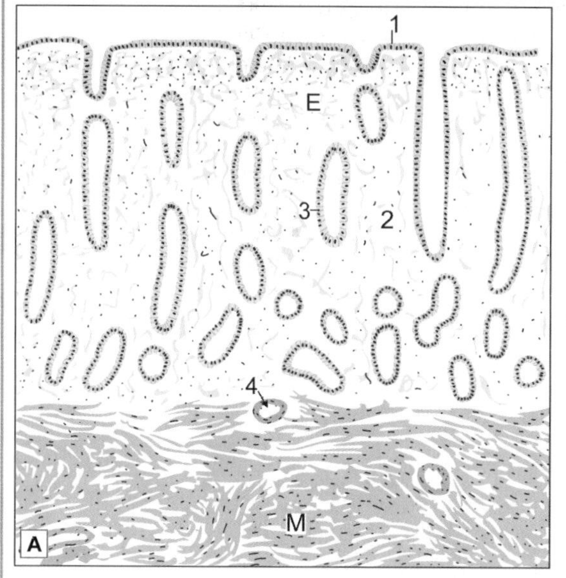

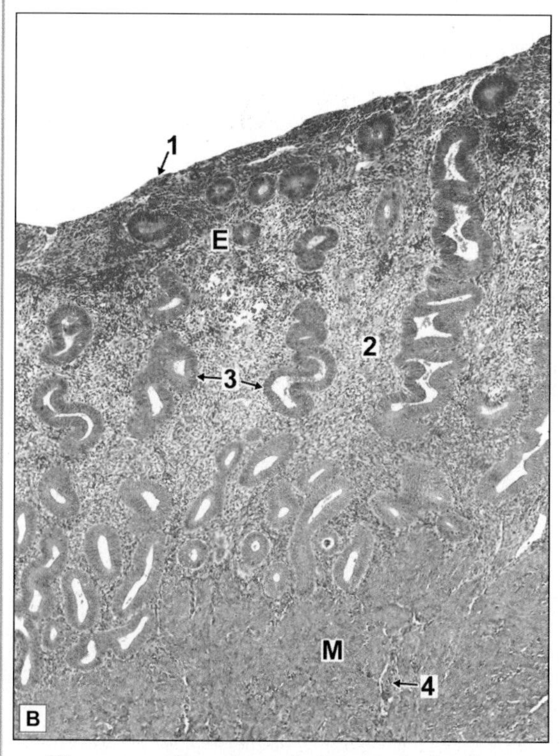

Uterus secretory phase. A. As seen in drawing; B. Photomicrograph

- The wall of the uterus consists of a mucous membrane (called the endometrium) and a very thick layer of muscle (the myometrium). The thickness of the muscle layer helps to identify the uterus easily
- The endometrium has a lining of columnar epithelium that rests on a stroma of connective tissue
- Numerous tubular uterine glands dip into the stroma
- The appearance of the endometrium varies considerably depending upon the phase of the menstrual cycle
  - The endometrium is thin and progressively increases in thickness
  - The uterine glands are straight and tubular in this phase.

Key

1. Columnar epithelium
2. Connective tissue
3. Uterine glands
4. Blood vessels

E. Endometrium
M. Myometrium

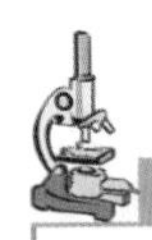

Plate 20.4: Uterus Secretory Phase

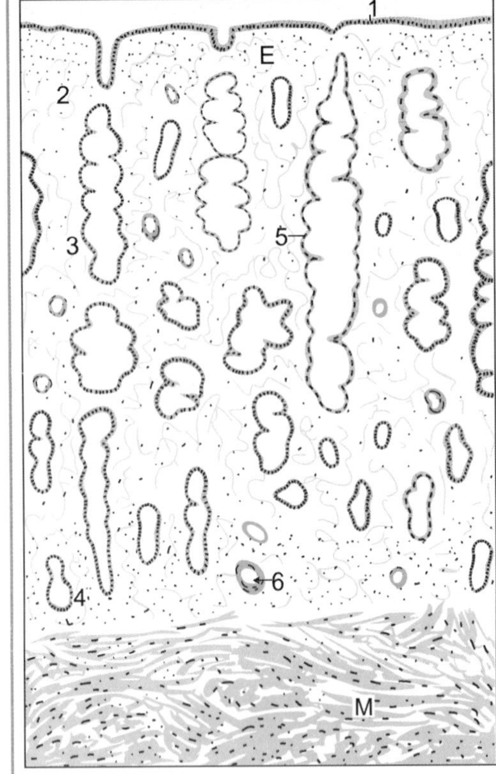

Uterus secretory phase (As seen in drawing)

In the secretory phase

- The thickness of the endometrium is much increased
- The uterine glands elongate, become dilated, and tortuous as a result of which they have saw-toothed margins in sections
- Blood vessels extend in the upper portion of endometrium.

Note: In this phase the appearance of the endometrium becomes so distinctive that the uterus cannot be confused with any other organ.

Key

1. Columnar epithelium
2. Stratum compactum
3. Stratum spongiosum
4. Stratum basale
5. Enlarged uterine gland
6. Blood vessel

E. Endometrium
M. Myometrium

Clinical Correlation

Dysfunctional Uterine Bleeding

Dysfunctional uterine bleeding (DUB) may be defined as excessive bleeding occurring during or between menstrual periods without a causative uterine lesion such as tumour, polyp, infection, hyperplasia, trauma, blood dyscrasia or pregnancy. DUB occurs most commonly in association with anovulatory cycles which are most frequent at the two extremes of menstrual life i.e. either when the ovarian function is just beginning (menarche) or when it is waning off (menopause).

Endometriosis

Endometriosis refers to the presence of endometrial glands and stroma in abnormal locations outside the uterus. Most common site being ovary followed by pouch of Douglas.

Adenomyosis

Adenomyosis is defined as abnormal distribution of histologically benign endometrial tissue within the myometrium along with myometrial hypertrophy. Adenomyosis is found in 15–20% of all hysterectomies. The possible underlying cause of the invasiveness and increased proliferation of the endometrium into the myometrium appears to be either a metaplasia or oestrogenic stimulation due to endocrine dysfunction of the ovary.

Cervix

The cervix is narrow lower part of uterus. The cavity of cervix (cervical canal) is narrow and communicates with the uterine cavity at its upper end and with vagina at its lower end. The upper and lower openings are referred to as internal and external os respectively. The lower portion of cervix which projects into vagina is called as ***portio vaginalis***.

The structure of the cervix of the uterus is somewhat different from that of the body. Here the mucous membrane (or ***endocervix***) has a number of obliquely placed ***palmate folds***. It contains deep branching glands that secrete mucous. The mucosa also shows small cysts that probably represent glandular elements that are distended with secretion. These cysts are called the ***ovula Nabothi***.

The mucous membrane of the upper two thirds of the cervical canal is lined by ciliated columnar epithelium, but over its lower one third the epithelium is non-ciliated columnar. Near the external os the canal is lined by stratified squamous epithelium. The part of the cervix that projects into vagina has an external surface that is covered by stratified squamous epithelium. The stroma underlying the epithelium of the cervix is less cellular than that of the body of the uterus and does not show muscle coat.

The lumen of the cervix is normally a narrow canal. It has tremendous capacity for dilation and, at the time of child birth, it becomes large enough for the fetal head to pass through.

VAGINA

The vagina is a fibromuscular elastic tube that extends from lower part of the cervix to the external genitalia. It is about 8 cm long. It is capable of considerable elongation and distension, this being helped by the rich network of elastic fibres in its wall. The wall of the vagina consists of following three layers (Plate 20.5).

Mucous Membrane

The mucous membrane shows numerous longitudinal folds, and is firmly fixed to the underlying muscle layer. It is lined by stratified squamous epithelium (nonkeratinised). The epithelial cells are rich in glycogen. (The glycogen content shows cyclical variation during the menstrual cycle).

The epithelium rests on dense connective tissue (lamina propria) that is highly vascular, many veins being present. The tissue is rich in elastic fibres. No glands are seen in the mucosa, the vaginal surface being kept moist by secretions of glands in the cervix of the uterus.

Muscle Coat

The muscle coat is made up of an outer layer of longitudinal fibres, and a much thinner inner layer of circular fibres. Many elastic fibres are present among the muscle fibres. The lower end of the vagina is surrounded by striated muscle fibres that form a sphincter for it. (The fibres belong to the bulbospongiosus muscle).

Adventitia

The muscle wall is surrounded by an adventitia made up of fibrous tissue containing many elastic fibres.

PLATE 20.5: Vagina

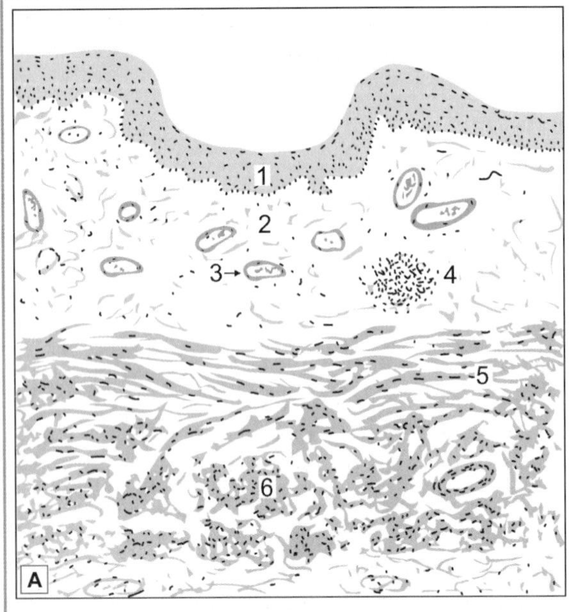

- The vagina is a fibromuscular structure consisting of an inner mucosa, a middle muscular layer and an outer adventitia
- The mucosa consists of stratified squamous non-keratinised epithelium and loose fibroelastic connective tissue lamina propria with many blood vessels and no glands
- The mucosa of vagina is rich in glycogen and hence the cells are pale stained which distinguishes it from oesophagus
- Muscular layer consists of smooth muscle fibres.

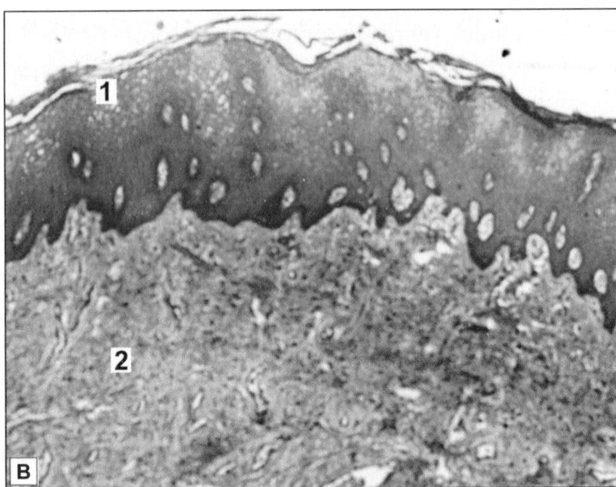

Key

1. Lining of stratified squamous epithelium
2. Lamina propria
3. Blood vessel
4. Lymphoid follicle
5. Muscle coat (longitudinal)
6. Muscle coat (circular)

Vagina. A. As seen in drawing B. Photomicrograph

Clinical Correlation

Vaginitis

The most common causes of vaginitis are *candida* (moniliasis) and *Trichomonas* (trichomoniasis). The hyphae of Candida can be seen in the vaginal smears. Similarly, the protozoa, Trichomonas, can be identified in smears. These infections are particularly common in pregnant and diabetic women and may involve both vulva and vagina. However, the adult vaginal mucosa is relatively resistant to gonococcal infection because of its structure.

FEMALE EXTERNAL GENITALIA

Labia Minora

These are folds of mucous membrane, made up of a core of connective tissue that is covered by stratified squamous epithelium. Modified sebaceous glands (not associated with hair follicles) are present.

Labia Majora

These are folds of skin containing hair, sebaceous glands, and sweat glands. The skin is supported on a core of connective tissue that contains abundant fat.

Clitoris

The clitoris may be regarded as a miniature penis, with the important difference that the urethra does not pass through it. Two ***corpora cavernosa*** and a ***glans*** are present. They contain erectile tissue. The surface of the clitoris is covered by mucous membrane (not skin) that is lined by stratified squamous epithelium. The mucosa is richly supplied with nerves.

MAMMARY GLAND

Although the mammary glands are present in both sexes they remain rudimentary in the male. In the female, they are well developed after puberty. Each breast is a soft rounded elevation present over the pectoral region. The skin over the centre of the elevation shows a darkly pigmented circular area called the ***areola***. Overlying the central part of the areola there is a projection called the ***nipple***.

Each mammary gland has an outer covering of skin deep to which there are several discrete masses of glandular tissue. These masses are separated (and covered) by considerable quantities of connective tissue and of adipose tissue (Plate 20.6). The fascia covering the gland is connected to overlying skin by fibrous processes called the ***suspensory ligaments*** (of Cooper). (In cancer of the breast these processes contract causing pitting of the overlying skin).

The glandular tissue (or mammary gland proper) is made up of 15 to 20 lobes. Each lobe consists of a number of lobules. The lobules are separated by moderately dense collagenous interlobular connective tissue. The lobule consists of many alveoli. The alveoli are embedded in loose cellular connective tissue (intralobular connective tissue).

The cells lining the alveoli vary in appearance in accordance with functional activity. In the 'resting' phase they are cuboidal. When actively producing secretion the cells become columnar. When the secretion begins to be poured into the lumen, distending them, the cells again become cuboidal, but are now much larger. The cells are filled with secretory vacuoles. Some distance from its termination each lactiferous duct shows a dilation called the ***lactiferous sinus***.

Ducts System

Each lobe drains into a ***lactiferous duct*** that opens at the summit of the nipple (Fig. 20.12). Beneath the nipple, the lactiferous duct dilates into ***lactiferous sinus***, which functions as a reservoir of milk. The smaller ducts are lined by columnar epithelium. In the larger ducts the epithelium has two or three layers of cells. Near their openings on the nipple the lining becomes stratified squamous. Between the epithelium and the basement membrane of the ducts, myoepithelial cells are present.

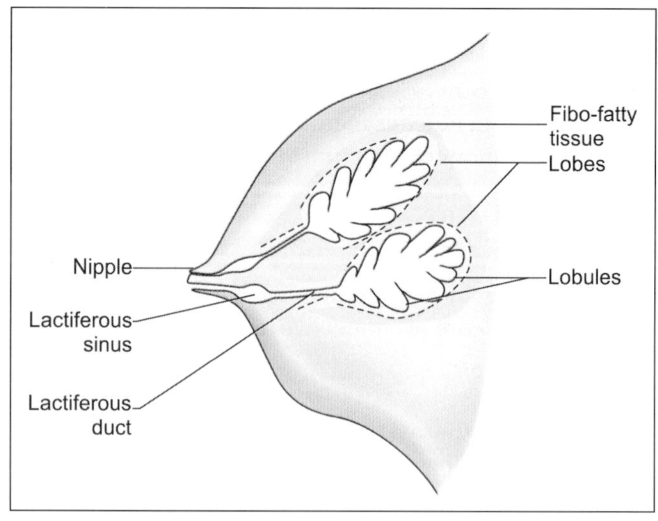

Fig. 20.12: Human female breast (Schematic representation)

Nipple and Areola

The nipple is covered by keratinised stratified squamous epithelium. It consists of dense connective tissue and smooth muscles arranged circularly and longitudinally. The pigmented skin around the nipple is called ***areola***. The skin of the areola lacks hair follicles. Circular smooth muscle is present in the dermis of the areola. Contraction of this muscle causes erection of the nipple. Many sebaceous glands and apocrine sweat glands are also present in the areola. At the periphery of the areola there are large sebaceous glands that are responsible for the formation of surface elevations called the ***tubercles of Montgomery***.

Added Information

- In the resting mammary gland, glandular epithelium is surrounded by an avascular zone containing fibroblasts. It has been claimed that this zone constitutes an ***epithelio-stromal junction*** that controls passage of materials to glandular cells.
- In the male, the mammary gland is rudimentary and consists of ducts that may be represented by solid cords of cells. The ducts do not extend beyond the areola.

Structure of Glandular Elements

The structure of the glandular elements of the mammary gland varies considerably at different periods of life as follows:

- Before the onset of puberty the glandular tissue consists entirely of ducts. Between puberty and the first pregnancy the duct system proliferates. At the end of each duct solid masses of polyhedral cells are formed, but proper alveoli are few or absent. The bulk of the breast consists of connective tissue and fat that widely separate the glandular elements.
- During pregnancy the ducts undergo marked proliferation and branching. Their terminal parts develop into proper alveoli. Each lobe is now a compound tubulo-alveolar gland. The ducts and alveoli are surrounded by very cellular periductal tissue. Towards the end of pregnancy the cells of the alveoli start secreting milk and the alveoli become distended.

  The development of breast tissue during pregnancy takes place under the influence of hormones produced by the hypophysis cerebri. Cells lining glandular tissue bear receptors for these hormones.

PLATE 20.6: Mammary Gland (Resting)

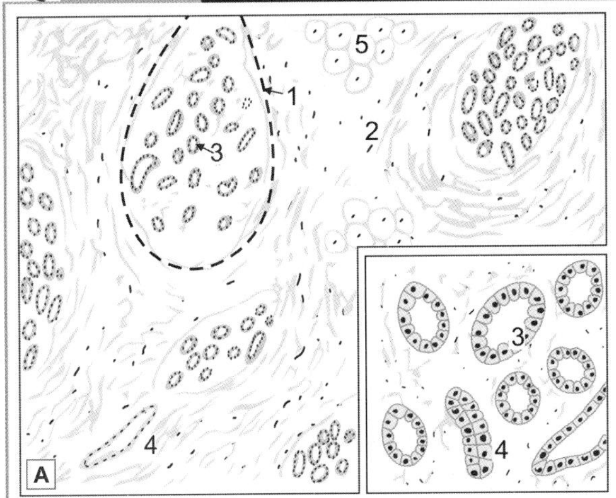

- Mammary gland consists of lobules of glandular tissue separated by considerable quantity of connective tissue and fat
- Non lactating mammary glands contain more connective tissue and less glandular tissue
- The glandular elements or alveoli are distinctly tubular. They are lined by cuboidal epithelium and have a large lumen so that they look like ducts. Some of them may be in form of solid cords of cells
- Extensive branching of duct system seen.

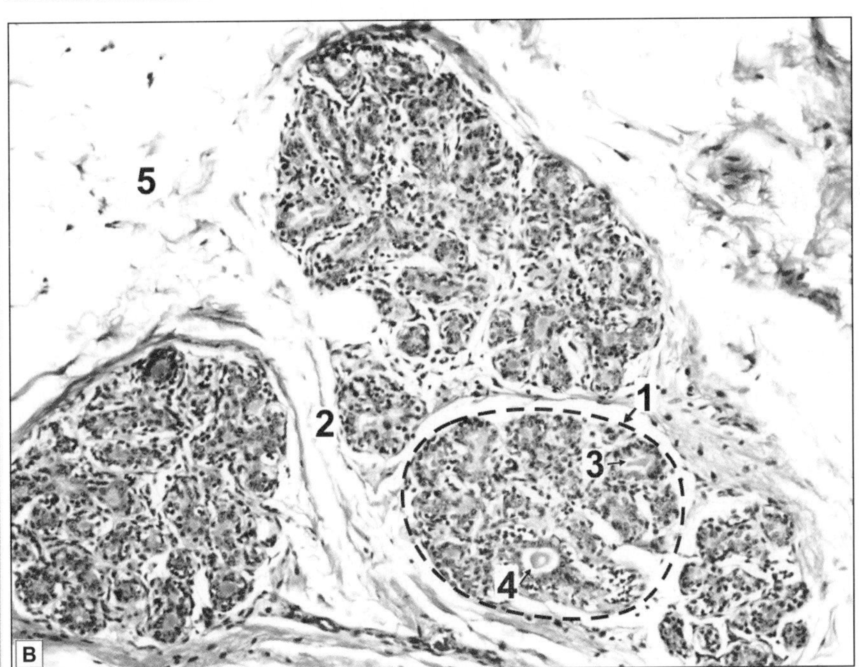

Mammary gland (resting) A. As seen in drawing; B. Photomicrograph

Key

1. Lobule
2. Connective tissue
3. Alveoli
4. Duct
5. Adipose tissue

PLATE 20.8: Mammary Gland (Lactating)

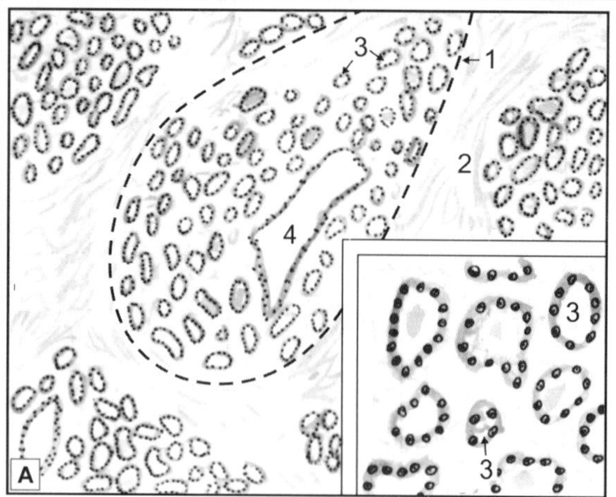

- In lactating mammary gland the glandular elements proliferate so that they become relatively more prominent than the connective tissue
- The interlobular connective tissue septum is very thin
- The lobules are formed by compactly arranged alveoli
- The alveoli are lined by simple cuboidal secretory epithelium and associated myoepithelial cells. Their lumen contains eosinophilic secretory material which appear vacuolated due to the presence of fat droplets.

Note: In fig. 20.8B observe the epithelial cells are arranged with abundant vacuolated cytoplasm containing lipo-proteinaceous material. Scant secretions may appear in the lumen.

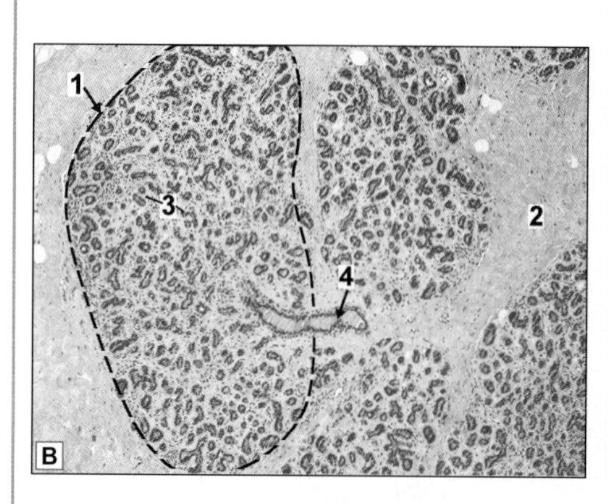

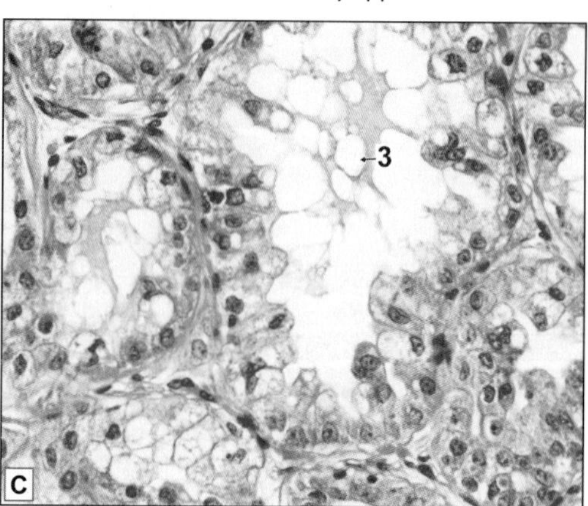

Mammary gland (Lactating). A. As seen in drawing; B. Photomicrograph (low magnification); C. Photomicrograph (high magnification)

Courtesy: Atlas of Histopathology. Ist Edition. Ivan Damjanov. Jaypee Brothers. 2012. p259

Key

1. Lobule
2. Connective tissue
3. Alveoli
4. Duct

- During lactation the glandular tissue is much more prominent than before, and there is a corresponding reduction in the volume of the connective tissue and fat (Plate 20.7).
- When lactation ceases the glandular tissue returns to the resting state. It undergoes atrophy after menopause (i.e., the age after which menstruation ceases).

Clinical Correlation

- **Fibroadenoma:** Fibroadenoma or adenofibroma is a benign tumour of fibrous and epithelial elements. It is the most common benign tumour of the female breast. Though it can occur at any age during reproductive life, most patients are between 15 and 30 years of age. Clinically, fibroadenoma generally appears as a solitary, discrete, freely mobile nodule within the breast.
- **Carcinoma of the breast:** Cancer of the breast is among the commonest of human cancers throughout the world. Its incidence varies in different countries but is particularly high in developed countries. The incidence of breast cancer is highest in the perimenopausal age group and is uncommon before the age of 25 years. Clinically, the breast cancer usually presents as a solitary, painless, palpable lump which is detected quite often by self-examination. Higher the age, more are the chances of breast lump turning out to be malignant. Thus, all breast lumps, irrespective of the age of the patient must be removed surgically.

Chapter 21

Endocrine System

Endocrine tissue is made up essentially of cells that produce secretions which are poured directly into blood. The secretions of endocrine cells are called ***hormones***. Along with the autonomic nervous system, the endocrine organs co-ordinate and control the metabolic activities and the internal environment of the body.

Endocrine tissues are highly vascular. The secretory pole of an endocrine cell is towards the wall of a capillary (or sinusoid). (In exocrine glands, the secretory pole is towards the surface over which secretions are discharged).

HORMONES

Hormones travel through blood to target cells whose functioning they may influence profoundly. A hormone acts on cells that bear specific receptors for it. Some hormones act only on one organ or on one type of cell, while other hormones may have widespread effects. On the basis of their chemical structure, hormones belong to four main types:

1. ***Amino acid derivatives***, for example adrenalin, noradrenalin and thyroxine.
2. ***Small peptides***, for example, encephalin, vasopressin and thyroid releasing hormone.
3. ***Proteins***, for example, insulin, parathormone and thyroid stimulating hormone.
4. ***Steroids***, for examples, progesterone, oestrogens, testosterone and cortisol.

DISTRIBUTION OF ENDOCRINE CELLS

Endocrine cells are distributed in three different ways:

- Some organs are entirely endocrine in function. They are referred to as ***endocrine glands*** (or ***ductless glands***). Those traditionally included under this heading are the hypophysis cerebri (or pituitary), the pineal gland, the thyroid gland, the parathyroid glands, and the suprarenal (or adrenal) glands.
- Groups of endocrine cells may be present in organs that have other functions. Several examples of such tissue have been described in previous Chapters. They include the islets of the pancreas, the interstitial cells of the testes, and the follicles and corpora lutea of the ovaries. Hormones are also produced by some cells in the kidneys, the thymus, and the placenta. Some authors describe the liver as being partly an endocrine gland.
- Isolated endocrine cells may be distributed in the epithelial lining of an organ. Such cells are seen most typically in the gut (endocrine cells of the gut). Similar cells are also present in the epithelium of the respiratory passages. Recent studies have shown that cells in many

other locations in the body produce amines that have endocrine functions. Many of these amines also act as neurotransmitters or as neuromodulators. These widely distributed cells are grouped together as the ***neuroendocrine system*** or the amine precursor uptake and decarboxylation ***(APUD)* cell system.**

HYPOPHYSIS CEREBRI

The hypophysis cerebri is also called the ***pituitary gland*** and is approximately the size of a pea. It is suspended from the floor of the third ventricle (of the brain) by a narrow funnel shaped stalk called the ***infundibulum***, and lies in a depression on the upper surface of the sphenoid bone, called ***sella turcica***.

The hypophysis cerebri is one of the most important endocrine glands. It produces several hormones some of which profoundly influence the activities of other endocrine tissues and is sometimes referred as "***master endocrine gland***". Its own activity is influenced by the hypothalamus, and by the pineal body.

SUBDIVISIONS OF THE HYPOPHYSIS CEREBRI

The hypophysis cerebri has, in the past, been divided into an anterior part, the ***pars anterior***; an intermediate part, the ***pars intermedia***; and a posterior part the ***pars posterior*** (or ***pars nervosa***) (Fig. 21.1 and Plate 21.1).

The pars anterior (which is also called the ***pars distalis***), and the pars intermedia, are both made up of cells having a direct secretory function. They are collectively referred to as the ***adenohypophysis***. An extension of the pars anterior surrounds the central nervous core of the infundibulum. Because of its tubular shape this extension is called the ***pars tuberalis***. The pars tuberalis is part of the adenohypophysis.

The pars posterior contains numerous nerve fibres. It is directly continuous with the central core of the infundibular stalk which is made up of nervous tissue. These two parts (pars posterior and infundibular stalk) are together referred to as the ***neurohypophysis***. The area in the floor of the third ventricle (tuber cinereum) immediately adjoining the attachment to it of the infundibulum is called the ***median eminence***. Some authorities include the median eminence in the neurohypophysis.

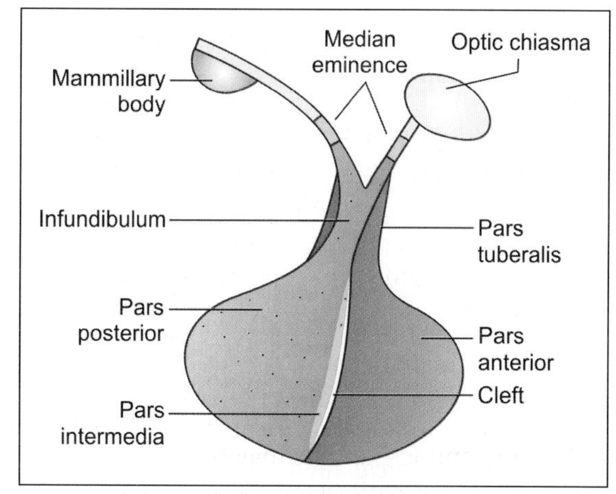

Fig. 21.1: Subdivisions of the hypophysis cerebri (Schematic diagram)

ADENOHYPOPHYSIS

Pars Anterior

The pars anterior consists of cords of cells separated by fenestrated sinusoids. Several types of cells, responsible for the production of different hormones, are present (Plate 21.2).

Using routine staining procedures the cells of the pars anterior can be divided into:

PLATE 21.1: Subdivisions of Hypophysis Cerebri

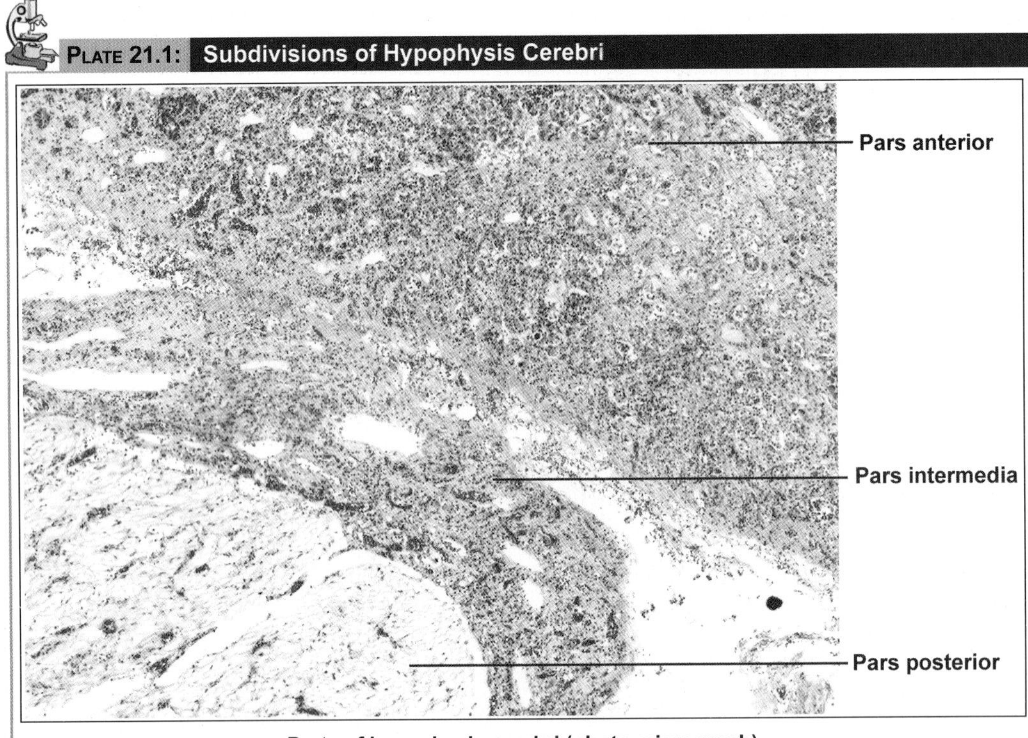

Parts of hypophysis cerebri (photo micrograph)

The hypophysis cerebri consists of three main parts:

- The pars anterior is cellular
- The pars intermedia is variable in structure. In this figure it appears to be cellular
- The pars posterior consists of fibres, and is lightly stained.

- ***Chromophil cells*** that have brightly staining granules in their cytoplasm
- ***Chromophobe cells*** in which granules are not prominent.

Chromophil Cells

Chromophil cells are further classified as ***acidophil*** when their granules stain with acid dyes (like eosin or orange G); or ***basophil*** when the granules stain with basic dyes (like haematoxylin). Basophil granules are also periodic acid Schiff (PAS) stain positive. The acidophil cells are often called ***alpha cells***, and the basophils are called ***beta cells*** (Plate 21.2).

Electron microscope (EM) examination shows that both acidophil and basophil cells contain abundant dense cored vesicles in the cytoplasm.

Both acidophils and basophils can be divided into sub-types on the basis of the size and shape of the granules in them. These findings have been correlated with those obtained by immunochemical methods—these methods allow cells responsible for production of individual hormones to be recognised. The following functional types of cells have been described.

PLATE 21.2: **Pars Anterior**

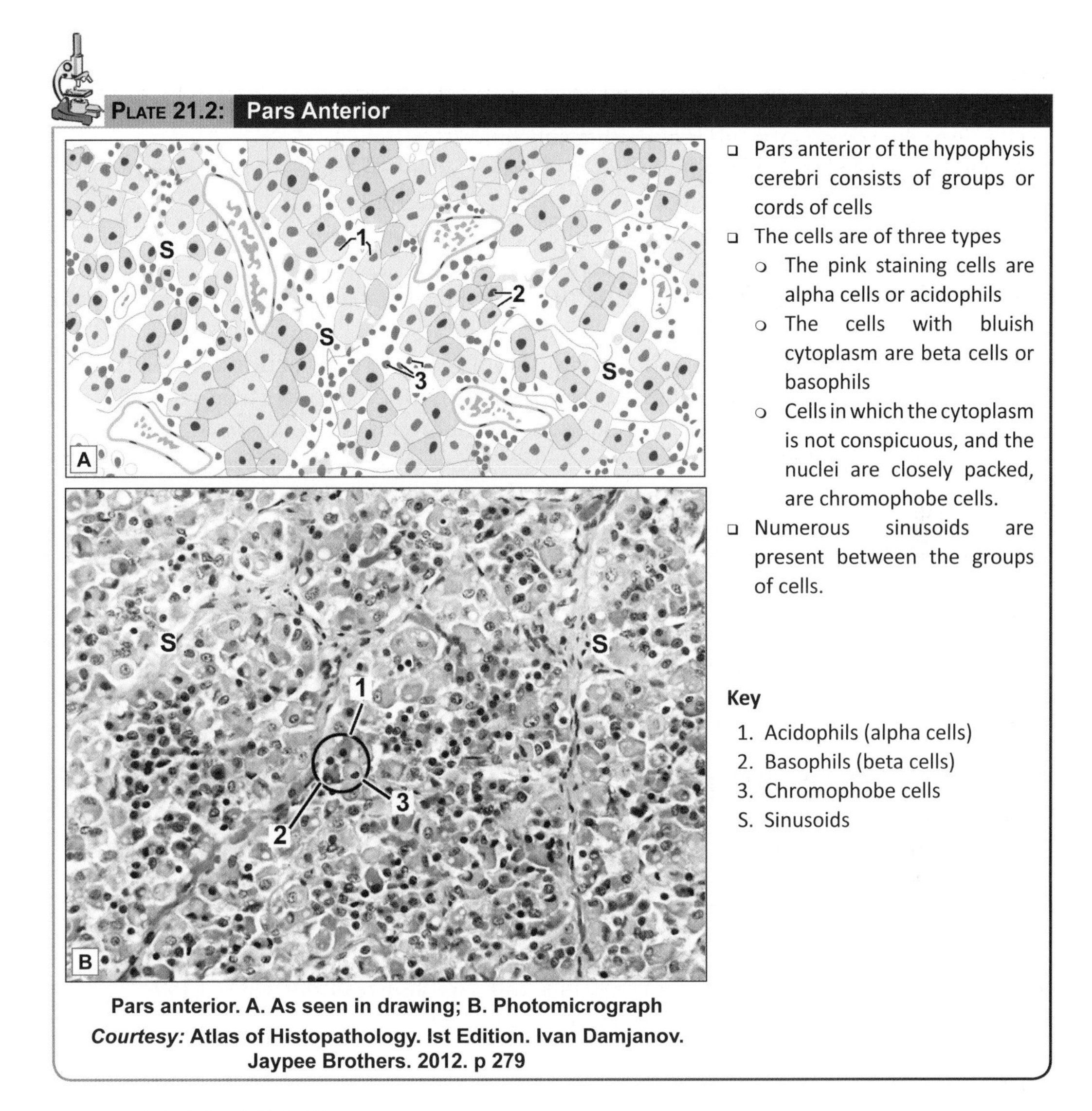

❑ Pars anterior of the hypophysis cerebri consists of groups or cords of cells
❑ The cells are of three types
  ❍ The pink staining cells are alpha cells or acidophils
  ❍ The cells with bluish cytoplasm are beta cells or basophils
  ❍ Cells in which the cytoplasm is not conspicuous, and the nuclei are closely packed, are chromophobe cells.
❑ Numerous sinusoids are present between the groups of cells.

Key

1. Acidophils (alpha cells)
2. Basophils (beta cells)
3. Chromophobe cells
S. Sinusoids

Pars anterior. A. As seen in drawing; B. Photomicrograph
Courtesy: **Atlas of Histopathology. Ist Edition. Ivan Damjanov. Jaypee Brothers. 2012. p 279**

Types of Acidophil Cells

❑ ***Somatotrophs*** produce the ***somatropic hormone*** [also called ***somatotropin (STH)***, or ***growth hormone (GH)***]. This hormone controls body growth, specially before puberty.
❑ ***Mammotrophs*** (or ***lactotrophs***) produce the ***mammotropic hormone*** [also called ***mammotropin***, ***prolactin (PRL)***, ***lactogenic hormone***, or LTH] which stimulates the growth and activity of the female mammary gland during pregnancy and lactation.

Types of Basophil Cells

❑ The ***corticotrophs*** (or ***corticotropes***) produce the ***corticotropic hormone*** (also called ***adreno-corticotropin*** or ACTH). This hormone stimulates the secretion of some hormones of the adrenal cortex. The staining characters of these cells are intermediate between those of acidophils and basophils. The cells are, therefore, frequently considered to be a variety of acidophils. In the human hypophysis their cytoplasm is weakly basophilic and PAS positive. The granules in the cells contain a complex molecule ***of pro-opio-melano-corticotropin***. This is broken down into ACTH and other substances.

Other corticotropic hormones that have been identified are β-lipotropin (β-LPH), α-melanocyte stimulating hormone (α-MSH)and β-endorphin.

- ***Thyrotrophs*** (or ***thyrotropes***) produce the ***thyrotropic hormone*** (***thyrotropin*** or TSH) which stimulates the activity of the thyroid gland.
- ***Gonadotrophs*** (***gonadotropes***, or ***delta basophils***) produce two types of hormones each type having a different action in the male and female.
  - In the female, the first of these hormones stimulates the growth of ovarian follicles. It is, therefore, called the ***follicle stimulating hormone*** (FSH). It also stimulates the secretion of oestrogens by the ovaries. In the male the same hormone stimulates spermatogenesis.
  - In the female, the second hormone stimulates the maturation of the corpus luteum, and the secretion by it of progesterone. It is called the ***luteinizing hormone*** (LH). In the male the same hormone stimulates the production of androgens by the interstitial cells of the testes, and is called the ***interstitial cell stimulating hormone*** (ICSH).

Chromophobe Cells

These cells do not stain darkly as they contain very few granules in their cytoplasm. Immunocytochemistry shows that they represent cells similar to the various types of chromophils mentioned above (including mammotrophs, somatotrophs, thyrotrophs, gonadotrophs or corticotrophs).

Added Information

- Somatotrophs constitute about 50%, mammotrophs about 25%, corticotrophs 15–20%, and gonadotrophs about 10% of the cell population of the pars anterior
- Somatotrophs are located mainly in the lateral parts of the anterior lobe. Thyrotrophs are concentrated in the anterior, median part; and corticotrophs in the posterior, median part. Gonadotrophs and mammotrophs are scattered throughout the anterior lobe
- The size of dense cored vesicles (seen by EM) is highly variable. It is about 400 nm in somatotrophs, about 300 nm in mammotrophs, 250–700 nm in corticotrophs and 150–400 nm in gonadotrophs.

Pars Tuberalis

The pars tuberalis consists mainly of undifferentiated cells. Some acidophil and basophil cells are also present.

Pars Intermedia

This is poorly developed in the human hypophysis. In ordinary preparations the most conspicuous feature is the presence of colloid filled vesicles (Fig. 21.2). These vesicles are remnants of the pouch of Rathke. Beta cells, other secretory cells, and chromophobe cells are present. Some cells of the pars intermedia produce the ***melanocyte stimulating hormone*** (MSH) which causes increased pigmentation of the skin. Other cells produce ACTH. ***Endorphins*** are present in the cytoplasm of secretory cells.

Note: The secretion of hormones from the adenohypophysis is under control of the hypothalamus as described later.

NEUROHYPOPHYSIS

Pars Posterior (Pars Nervosa)

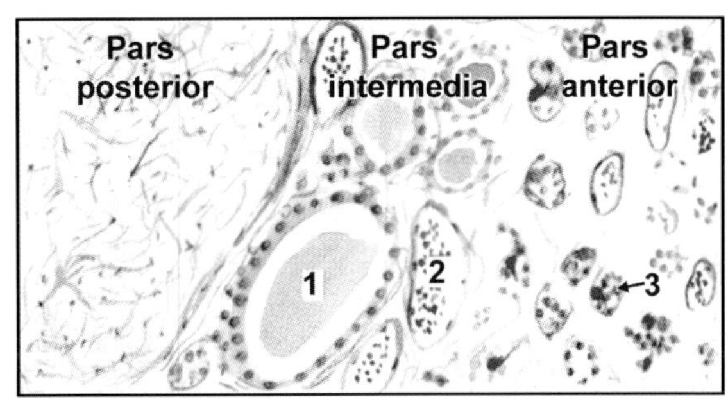

Fig. 21.2: Hypophysis cerebri. Pars posterior (left) and pars intermedia (right). 1–colloid; 2–capillary; 3–clumps of cells (including some acidophils and basophils) (Schematic representation)

The pars posterior consists of numerous unmyelinated nerve fibres which are the axons of neurons located in the hypothalamus (Fig. 21.2). Most of the nerve fibres arise in the supraoptic and paraventricular nuclei. Situated between these axons there are supporting cells of a special type called ***pituicytes***. These cells have long dendritic processes many of which lie parallel to the nerve fibres. The axons descending into the pars posterior from the hypothalamus end in terminals closely related to capillaries.

The pars posterior of the hypophysis is associated with the release into the blood of two hormones.

- ***Vasopressin:*** (Also called the ***antidiuretic hormone*** or ADH): This hormone controls reabsorption of water by kidney tubules.
- ***Oxytocin:*** It controls the contraction of smooth muscle of the uterus and also of the mammary gland.

These two hormones are not produced in the hypophysis cerebri at all. They are synthesised in neurons located mainly in the supraoptic and paraventricular nuclei of the hypothalamus. Vasopressin is produced mainly in the supraoptic nucleus, and oxytocin in the paraventricular nucleus. These secretions (which are bound with a glycoprotein called ***neurophysin***) pass down the axons of the neurons concerned. In axoplasm hormone is in the form of secretory vesicles (granules) and reaches the axon terminals in the pars posterior. The collection of these secretory granules at the terminal portion of axonal processing is called as ***Herring bodies*** (Fig. 21.3). Here they are released into the capillaries of the region and enter the general circulation.

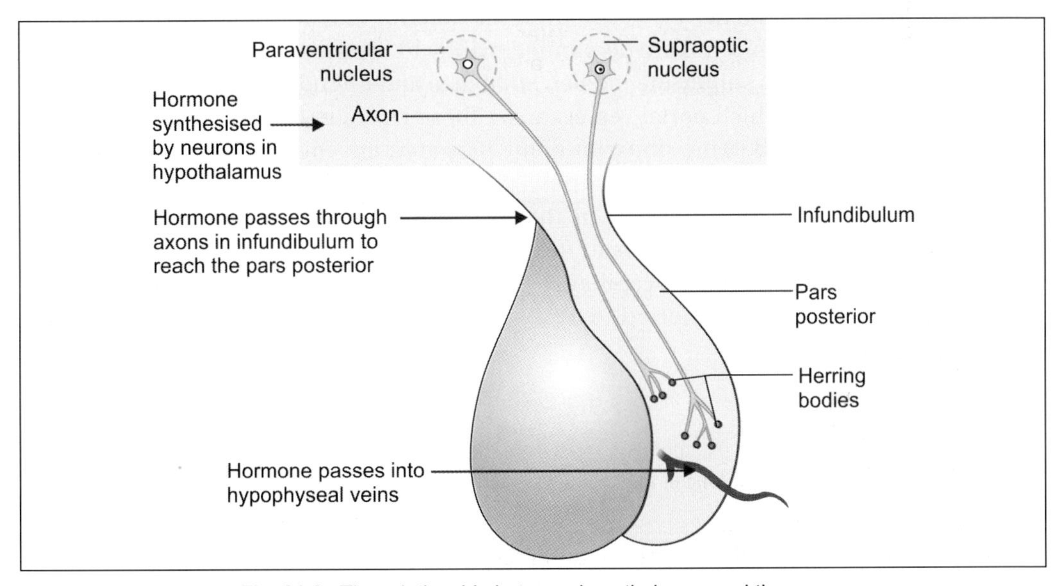

Fig. 21.3: The relationship between hypothalamus and the pars posterior of the hypophysis cerebri (Schematic representation)

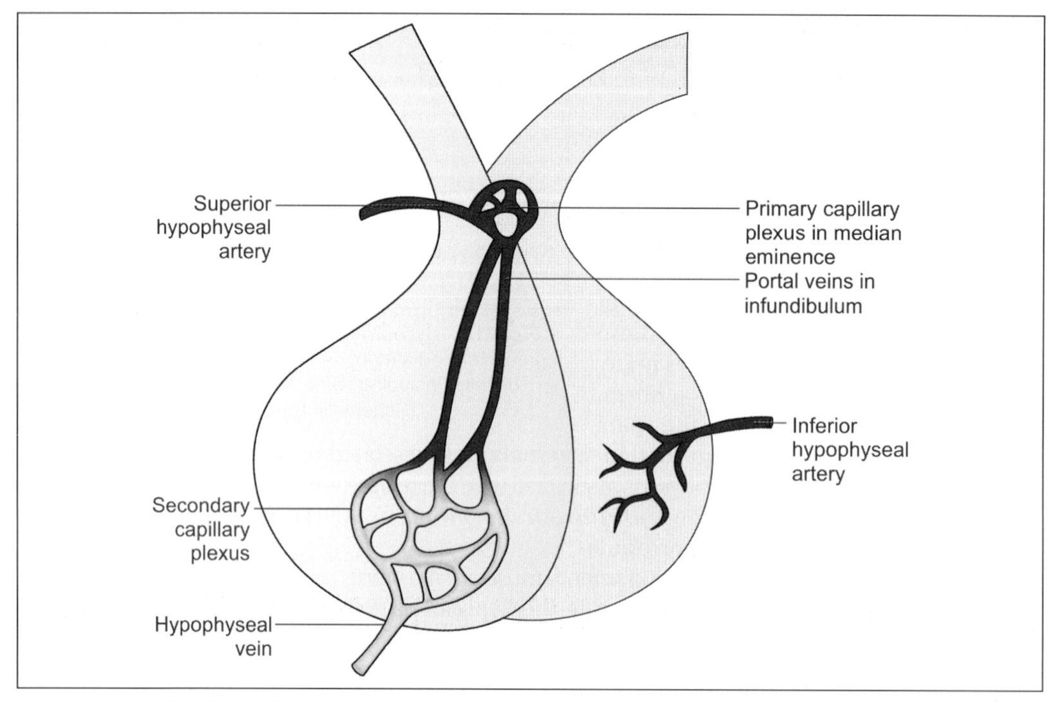

Fig. 21.4: Hypothalamo-hypophyseal portal system (Schematic representation)

BLOOD SUPPLY OF THE HYPOPHYSIS CEREBRI

The hypophysis cerebri is supplied by superior and inferior branches arising from the internal carotid arteries. Some branches also arise from the anterior and posterior cerebral arteries. The inferior hypophyseal arteries are distributed mainly to the pars posterior. Branches from the superior set of arteries supply the median eminence and infundibulum. Here they end in capillary plexuses from which portal vessels arise. These portal vessels descend through the infundibular stalk and end in the sinusoids of the pars anterior. The sinusoids are drained by veins that end in neighbouring venous sinuses.

The above arrangement is unusual in that two sets of capillaries intervene between the arteries and veins. One of these is in the median eminence and the upper part of the infundibulum. The second set of capillaries is represented by the sinusoids of the pars anterior. This arrangement is referred to as the ***hypothalamo-hypophyseal portal system*** (Fig. 21.4).

Clinical Correlation

The vessels descending through the infundibular stalk are easily damaged in severe head injuries. This leads to loss of function in the anterior lobe of the hypophysis cerebri.

CONTROL OF SECRETION OF HORMONES OF THE ADENOHYPOPHYSIS

The secretion of hormones by the adenohypophysis takes place under higher control of neurons in the hypothalamus, notably those in the median eminence and in the infundibular nucleus. The axons of these neurons end in relation to capillaries in the median eminence and in the upper part of the infundibulum.

Different neurons produce specific ***releasing factors*** (or releasing hormones) for each hormone of the adenohypophysis. (For details of hypothalamic nuclei and the releasing factors produced by them see the author's TEXTBOOK OF HUMAN NEUROANATOMY). These factors are released into the capillaries. Portal vessels arising from the capillaries carry these factors to the pars anterior of the hypophysis. Here they stimulate the release of appropriate hormones. Some factors inhibit the release of hormones. The synthesis and discharge of releasing factors by the neurons concerned is under nervous control. As these neurons serve as intermediaries between nervous impulses and hormone secretion they have been referred to as ***neuro-endocrine transducers***. Some cells called ***tanycytes***, present in ependyma, may transport releasing factors from neurons into the cerebrospinal fluid (CSF), or from CSF to blood capillaries. They may thus play a role in control of the adenohypophysis.

Added Information

Recent studies indicate that circulation in relation to the hypophysis cerebri may be more complex than presumed earlier. Some points of interest are as follows:

- The entire neurohypophysis (from the median eminence to the pars posterior) is permeated by a continuous network of capillaries in which blood may flow ***in either direction***. The capillaries provide a route through which hormones released in the pars posterior can travel back to the hypothalamus, and into CSF.
- Some veins draining the pars posterior pass into the adenohypophysis. Secretions by the adenohypophysis may thus be controlled not only from the median eminence, but by the entire neurohypophysis.
- Blood flow in veins connecting the pars anterior and pars posterior may be reversible providing a feed back from adenohypophysis to the neurohypophysis.

Clinical Correlation

- **Gigantism:** When growth hormone (GH) excess occurs prior to epiphyseal closure, gigantism is produced. Gigantism, therefore, occurs in prepubertal boys and girls and is much less frequent than acromegaly. The main clinical feature in gigantism is the excessive and proportionate growth of the child. There is enlargement as well as thickening of the bones resulting in considerable increase in height and enlarged thoracic cage.
- **Acromegaly:** Acromegaly results when there is overproduction of GH in adults following cessation of bone growth and is more common than gigantism. The term 'acromegaly' means increased growth of extremities (*acro*=extremity). There is enlargement of hands and feet, coarseness of facial features with increase in soft tissues, prominent supraorbital ridges and a more prominent lower jaw which when clenched results in protrusion of the lower teeth in front of upper teeth **(prognathism).**
- **Diabetes Insipidus:** Deficient secretion of ADH causes diabetes insipidus. The causes of ADH deficiency are—inflammatory and neoplastic lesions of the hypothalamo-hypophyseal axis, destruction of neurohypophysis due to surgery, radiation, head injury, and lastly, are those cases where no definite cause is known and are labelled as idiopathic. The main features of diabetes insipidus are excretion of a very large volume of dilute urine of low specific gravity (below 1.010), polyuria and polydipsia.

THYROID GLAND

Thyroid is a bi-lobed gland. Each lobe is situated on either side of trachea, below larynx, in lower neck. The two lobes are connected to each other by ***isthmus*** in front of trachea. Among endocrine glands, thyroid is unique as it stores large quantity of its hormonal secretion extracellularly as colloid in contrast to other endocrine glands which store very small quantities intracellularly as secretory granules only.

STRUCTURE OF THYROID GLAND

The thyroid gland is covered by a fibrous capsule. Septa extending into the gland from the capsule divide it into lobules. On microscopic examination each lobule is seen to be made up of an aggregation of ***follicles***. Each follicle is lined by ***follicular cells***, that rest on a basement membrane. The follicle has a cavity which is filled by a homogeneous material called ***colloid*** (which appears pink in haematoxylin and eosin stained sections) (Plate 21.3). The spaces between the follicles are filled by a stroma made up of delicate connective tissue in which there are numerous capillaries and lymphatics. The capillaries lie in close contact with the walls of follicles.

Apart from follicular cells the thyroid gland contains C-cells (or ***parafollicular cells***) which intervene (here and there) between the follicular cells and the basement membrane (Plate 21.3). They may also lie in the intervals between the follicles. Connective tissue stroma surrounding the follicles contain a dense capillary plexus, lymphatic capillaries and sympathetic nerves.

Follicular Cells

- The follicular cells vary in shape depending on the level of their activity (Fig. 21.5). Normally (at an average level of activity) the cells are cuboidal, and the colloid in the follicles is moderate in amount. When inactive (or resting) the cells are flat (squamous) and the follicles are distended with abundant colloid. Lastly, when the cells are highly active they become columnar and colloid is scanty. Different follicles may show different levels of activity (Fig. 21.5).
- With the EM a follicular cell shows the presence of apical microvilli, abundant granular endoplasmic reticulum, and a prominent supranuclear Golgi complex. Lysosomes, microtubules and microfilaments are also present. The apical part of the cell contains many secretory vacuoles (Fig. 21.6).
- The activity of follicular cells is influenced by the thyroid stimulating hormone (TSH or thyrotropin) produced by the hypophysis cerebri. There is some evidence to indicate that their activity may also be increased by sympathetic stimulation.

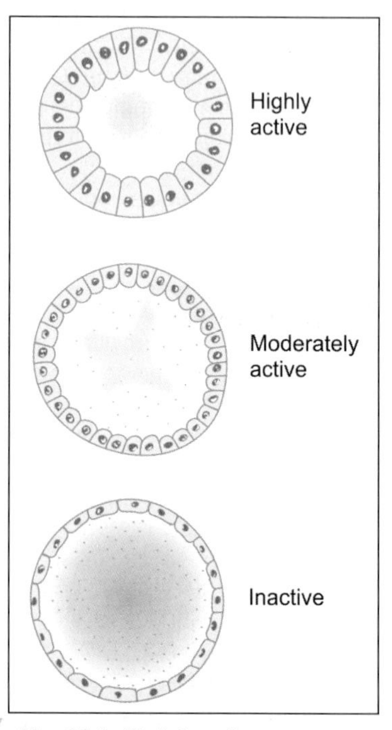

Fig. 21.5: Variations in appearance of thyroid follicles at different levels of activity (Schematic representation)

PLATE 21.3: Thyroid Gland

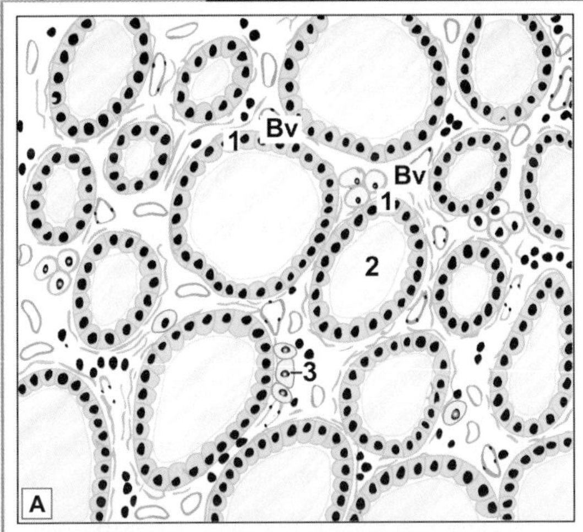

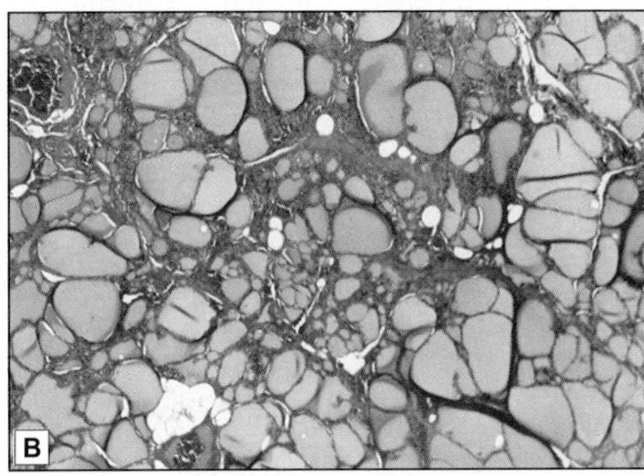

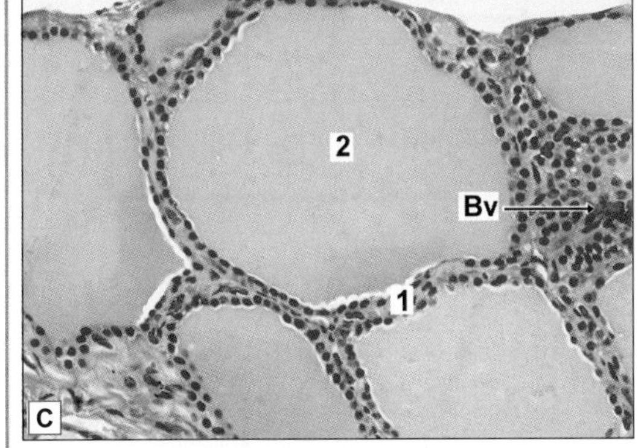

- The thyroid gland is made up of follicles lined by cuboidal epithelium
- In photomicrograph in low magnification it can be seen that follicles vary in shape and size
- Each follicle is filled with a homogenous pink colloid proteinaceous material composed primarily of thyroglobulin that has been produced by the follicular epithelial cells.
- Parafollicular cells are present in relation to the follicles and also as groups in the connective tissue
- In the intervals between the follicles there is some connective tissue
- Note the blood vessels between follicles.

Key

1. Follicles lined by cuboidal epithelium
2. Pink stained colloidal material
3. Parafollicular cells

Bv. Blood vessels

Thyroid gland. A. As seen in drawing; B. Photomicrograph (low magnification); C. Photomicrograph (high magnification).
Courtesy: **Atlas of Histopathology. Ist Edition. Ivan Damjanov. Jaypee Brothers. 2012. p281**

- The follicular cells secrete two hormones that influence the rate of metabolism. Iodine is an essential constituent of these hormones. One hormone containing three atoms of iodine in each molecule is called ***triodothyronine*** or T3. The second hormone containing four atoms of iodine in each molecule is called ***tetraiodothyronine***, T4, or thyroxine. T3 is much more active than T4.

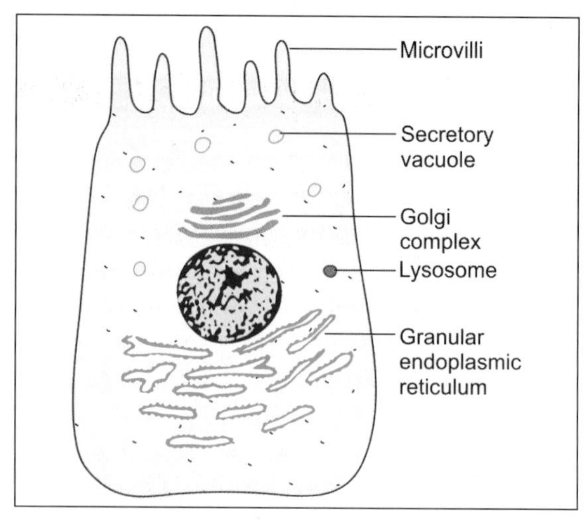

Fig. 21.6: Ultrastructure of a follicular cell of the thyroid gland (Schematic representation)

Synthesis of Thyroid Hormone

The synthesis and release of thyroid hormone takes place in two phases (Flow chart 21.1). In the first phase, thyroglobulin (a glycoprotein) is synthesised by granular endoplasmic reticulum and is packed into secretory vacuoles in the Golgi complex. The vacuoles travel to the luminal surface where they release thyroglobulin into the follicular cavity by exocytosis. Here the thyroglobulin combines with iodine to form colloid. Colloid is iodinated thyroglobulin.

In the second phase particles of colloid are taken back into the cell by endocytosis. Within the cell the iodinated thyroglobulin is acted upon by enzymes (present in lysosomes) releasing hormones T3 and T4 which pass basally through the cell and are released into blood.

Hormone produced in the thyroid gland is mainly T4 (output of T3 being less than 10%). In the liver, the kidneys (and some other tissues) T4 is converted to T3 by removal of one iodine molecule. T3 and T4 circulating in blood are bound to a protein [thyroxine binding globulin, TBG]. The bound form of hormone is not active.

Flow chart. 21.1: Some steps in the formation of hormones by the thyroid gland

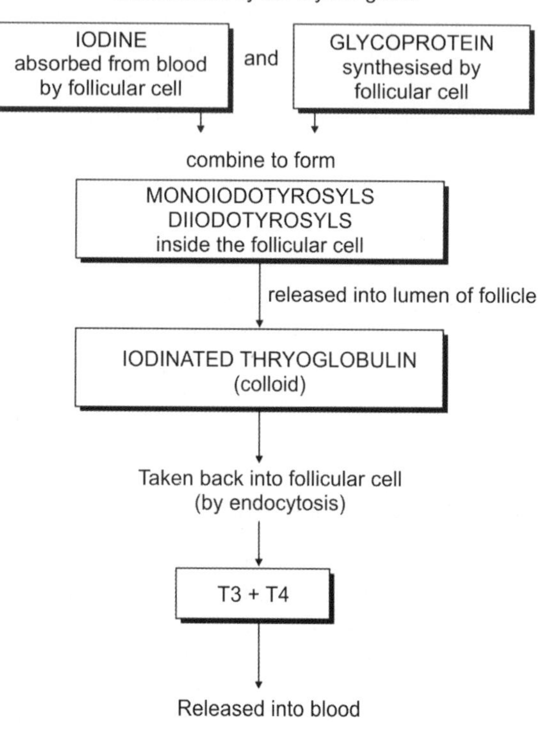

C-cells (Parafollicular Cells)

They are also called ***clear cells***, or ***light cells***. The cells are polyhedral, with oval eccentric nuclei. Typically, they lie between the follicular cells and their basement membrane. They may, however, lie between adjoining follicular cells; but they do not reach the lumen. In some species many parafollicular cells may lie in the connective tissue between the follicles and may be arranged in groups. With the EM the cells show well developed granular endoplasmic reticulum, Golgi complexes, numerous mitochondria, and membrane bound secretory granules.

Note: C-cells share features of the APUD cell system and are included in this system.

C-cells secrete the hormone ***thyro-calcitonin***. This hormone has an action opposite to that of the parathyroid hormone on calcium metabolism. This hormone comes into play when serum calcium level is high. It tends to lower the calcium level by suppressing release of calcium ions from bone. This is achieved by suppressing bone resorption by osteoclasts.

Clinical Correlation

- **Hyperthyroidism:** Hyperthyroidism, also called thyrotoxicosis, is a hypermetabolic clinical and biochemical state caused by excess production of thyroid hormones. The condition is more frequent in females and is associated with rise in both T3 and T4 levels in blood, though the increase in T3 is generally greater than that of T4.
- **Hypothyroidism:** Hypothyroidism is a hypometabolic clinical state resulting from inadequate production of thyroid hormones for prolonged periods, or rarely, from resistance of the peripheral tissues to the effects of thyroid hormones. The clinical manifestations of hypothyroidism, depending upon the age at onset of disorder, are divided into two forms:
  - **Cretinism** or congenital hypothyroidism is the development of severe hypothyroidism during infancy and childhood.
  - **Myxoedema** is the adulthood hypothyroidism.
- **Graves' disease:** Graves' disease, also known as Basedow's disease, primary hyperplasia, exophthalmic goitre, and diffuse toxic goitre, is characterised by a triad of features:
  - Hyperthyroidism (thyrotoxicosis)
  - Diffuse thyroid enlargement
  - Ophthalmopathy.

  The disease is more frequent between the age of 30 and 40 years and has five-fold increased prevalence among females.

PARATHYROID GLANDS

The parathyroid glands are so called because they lie in close relationship to the thyroid gland. Normally, there are two parathyroid glands, one superior and one inferior, on either side; there being four glands in all. Sometimes there may be as many as eight parathyroids.

STRUCTURE OF PARATHYROID GLANDS

Each gland has a connective tissue capsule from which some septa extend into the gland substance. Within the gland a network of reticular fibres supports the cells. Many fat cells (adipocytes) are present in the stroma (Plate 21.4).

The parenchyma of the gland is made up of cells that are arranged in cords. Numerous sinusoids lie in close relationship to the cells.

CELLS OF PARATHYROID GLANDS

The cells of the parathyroid glands are of two main types:
- ***Chief cells*** (or ***principal cells***)
- ***Oxyphil cells*** (or ***eosinophil cells***).

Chief Cells

The chief cells are much more numerous than the oxyphil cells.

With the light microscope the chief cells are seen to be small round cells with vesicular nuclei. Their cytoplasm is clear and either mildly eosinophil or basophil. Sometimes the cell accumulates glycogen and lipids and looks 'clear'. Three types of chief cells (light, dark, and clear) have been described.

With the EM active chief cells are seen to have abundant granular endoplasmic reticulum and well developed Golgi complexes. Small secretory granules are seen, specially in parts of the cytoplasm near adjacent blood sinusoids. These features become much less prominent in inactive cells. Both active and inactive cells contain glycogen, the amount of which is greater in inactive cells. In the normal parathyroid the number of inactive cells is greater than that of active cells.

The chief cells produce the parathyroid hormone (or parathormone). This hormone tends to increase the level of serum calcium by:
- Increasing bone resorption through stimulation of osteoclastic activity;
- Increasing calcium resorption from renal tubules (and inhibiting phosphate resorption);
- Enhancing calcium absorption from the gut.

Oxyphil Cells

The oxyphil cells are much larger than the chief cells and contain granules that stain strongly with acid dyes. Their nuclei are smaller and stain more intensely than those of chief cells. They are less numerous than the chief cells. The oxyphil cells are absent in the young and appear a little before the age of puberty.

With the EM it is seen that the granules of oxyphil cells are really mitochondria, large numbers of which are present in the cytoplasm. True secretory granules are not present. The functions of oxyphil cells are unknown.

Clinical Correlation

- **Hyperparathyroidism:** Hyperfunction of the parathyroid glands occurs due to excessive production of parathyroid hormone. It is classified into 3 types—primary, secondary and tertiary.
  - **Primary hyperparathyroidism** occurs from oversecretion of parathyroid hormone due to disease of the parathyroid glands.
  - **Secondary hyperparathyroidism** is caused by diseases in other parts of the body.
  - **Tertiary hyperparathyroidism** develops from secondary hyperplasia after removal of the cause of secondary hyperplasia.
- **Hypoparathyroidism:** Deficiency or absence of parathyroid hormone secretion causes hypoparathyroidism.

PLATE 21.4: Parathyroid Gland

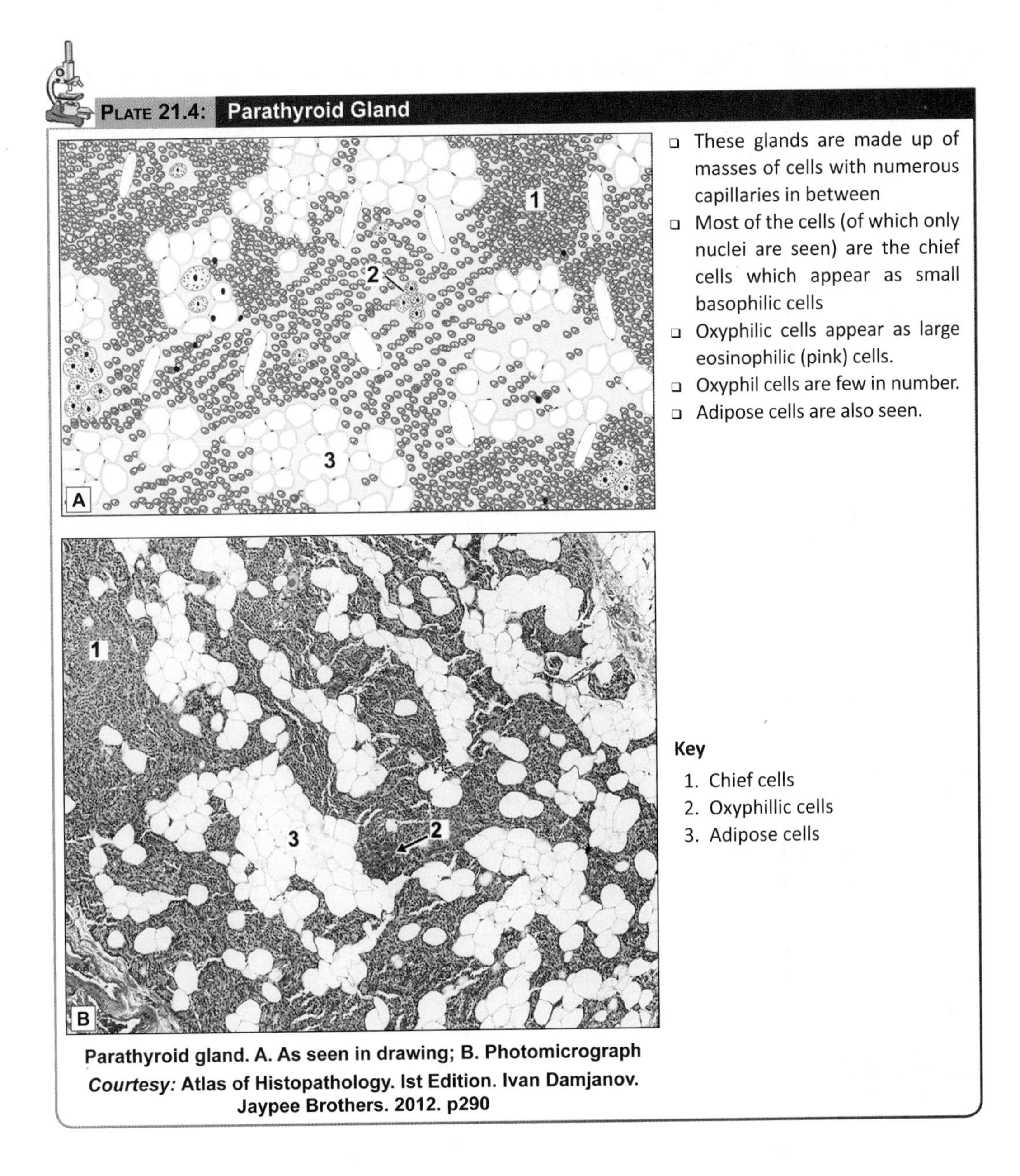

Parathyroid gland. A. As seen in drawing; B. Photomicrograph
Courtesy: Atlas of Histopathology. Ist Edition. Ivan Damjanov. Jaypee Brothers. 2012. p290

- These glands are made up of masses of cells with numerous capillaries in between
- Most of the cells (of which only nuclei are seen) are the chief cells which appear as small basophilic cells
- Oxyphilic cells appear as large eosinophilic (pink) cells.
- Oxyphil cells are few in number.
- Adipose cells are also seen.

Key

1. Chief cells
2. Oxyphillic cells
3. Adipose cells

SUPRARENAL GLANDS

As implied by their name the right and left suprarenal glands lie in the abdomen, close to the upper poles of the corresponding kidneys. In many animals they do not occupy a 'supra' renal position, but lie near the kidneys. They are, therefore, commonly called the ***adrenal glands***.

STRUCTURE OF SUPRARENAL GLANDS

Each suprarenal gland is covered by a connective tissue capsule from which septa extend into the gland substance. The gland is made up of two functionally distinct parts—a superficial part called the ***cortex***, and a deeper part called the ***medulla***. The volume of the cortex is about ten times that of the medulla.

Suprarenal Cortex

Layers of the Cortex

The suprarenal cortex is made up of cells arranged in cords. Sinusoids intervene between the cords. On the basis of the arrangement of its cells the cortex can be divided into three layers as follows (Plate 21.5):

- ***Zona glomerulosa:*** The outermost layer is called the ***zona glomerulosa***. Here the cells are arranged as inverted U-shaped formations, or acinus-like groups. The zona glomerulosa constitutes the outer one-fifth of the cortex. With the light microscope the cells of the zona glomerulosa are seen to be small, polyhedral or columnar, with basophilic cytoplasm and deeply staining nuclei.
- ***Zona fasciculata:*** The next zone is called the ***zona fasciculata***. Here the cells are arranged in straight columns, two cell thick. Sinusoids intervene between the columns. This layer forms the middle three-fifths of the cortex. With the light microscope the cells of the zona fasciculata are seen to be large, polyhedral, with basophilic cytoplasm and vesicular nuclei. The cells of the zona fasciculata are very rich in lipids which can be demonstrated by suitable stains. With routine methods the lipids are dissolved out during the processing of tissue, giving the cells an 'empty' or vacuolated appearance. These cells also contain considerable amounts of vitamin C.
- ***Zona reticularis:*** The innermost layer of the cortex (inner one-fifth) is called the ***zona reticularis***. It is so called because it is made up of cords that branch and anastomose with each other to form a kind of reticulum. The cells in this layer are smaller and more acidophilic than other two layer. With the light microscope the cells of the zona reticularis are seen to be similar to those of the zona fasciculata, but the lipid content is less. Their cytoplasm is often eosinophilic. The cells often contain brown pigment.

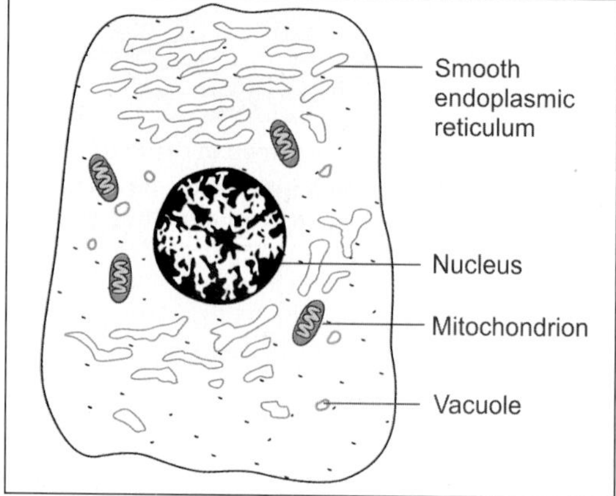

Fig. 21.7: Some features of ultrastructure of a cell from the adrenal cortex (Schematic representation)

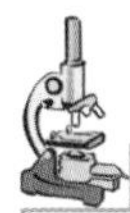

PLATE 21.5: Suprarenal Gland

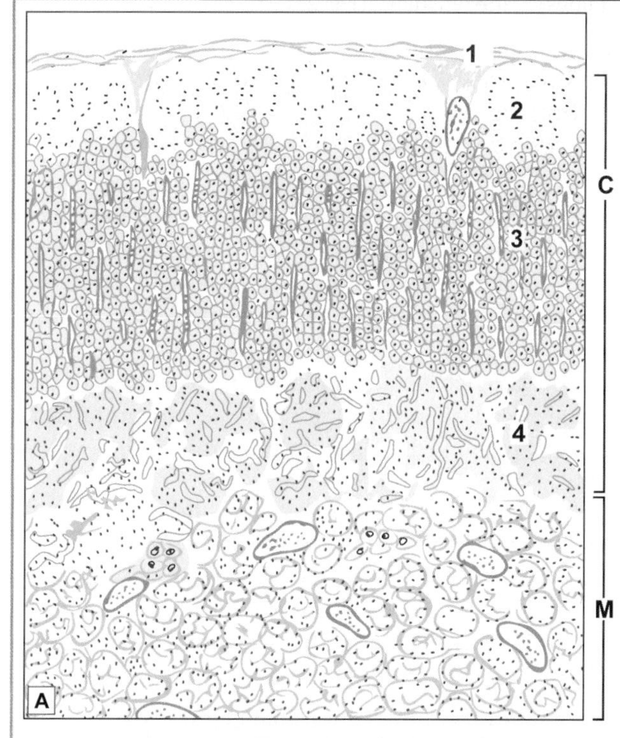

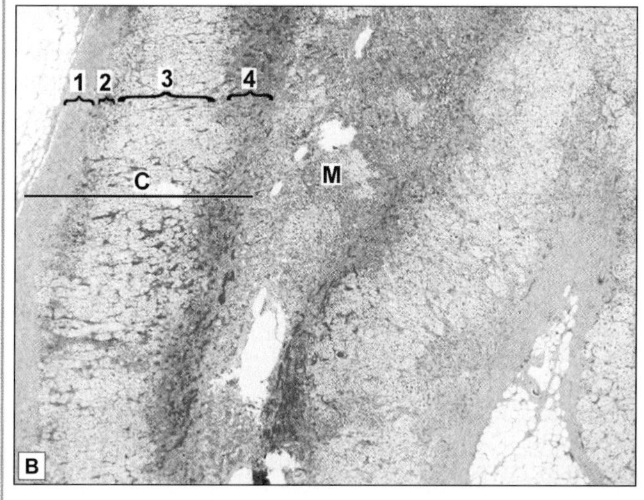

- The suprarenal gland is made up a large number of cells arranged in layers. It consists of an outer cortex and an inner medulla
- The cortex is divisible into three zones
  - The zona glomerulosa is most superficial. Here the cells are arranged in the form of inverted U-shaped structures or acinus-like groups
  - In the zona fasciculata the cells are arranged in straight columns (typically two cell thick). Sinusoids intervene between the columns
  - The zona reticularis is made up of cords of cells that branch and form a network
- The medulla is made up of groups of cells separated by wide sinusoids. Some sympathetic neurons are also present.

Key

1. Capsule
2. Zona glomerulosa
3. Zona fasciculata
4. Zona reticularis

C. Cortex
M. Medulla

Suprarenal gland A. As seen in drawing B. Photomicrograph
***Courtesy*: Atlas of Histopathology. Ist Edition. Ivan Damjanov. Jaypee Brothers. 2012. p292**

Note: With the EM the cells in all layers of the cortex are characterised by the presence of abundant agranular (or smooth) endoplasmic reticulum. The Golgi complex is best developed in cells of the zona fasciculata. Mitochondria are elongated in the glomerulosa, spherical in the fasciculata, and unusual with tubular cisternae (instead of the usual plates) in the reticularis (Fig. 21.7).

The suprarenal cortex is essential for life. Removal or destruction leads to death unless the hormones produced by it are supplied artificially. Increase in secretion of corticosteroids causes dramatic reduction in number of lymphocytes.

Hormones produced by the Suprarenal Cortex

- The cells of the zona glomerulosa produce the mineralocorticoid hormones ***aldosterone*** and ***deoxycorticosterone***. These hormones influence the electrolyte and water balance of the body. The secretion of aldosterone is influenced by renin secreted by juxta-glomerular cells of the kidney. The secretion of hormones by the zona glomerulosa appears to be largely independent of the hypophysis cerebri.
- The cells of the zona fasciculata produce the glucocorticoids ***cortisone*** and ***cortisol*** (***dihydro -cortisone***). These hormones have widespread effects, including those on carbohydrate metabolism and protein metabolism. They appear to decrease antibody responses and have an anti-inflammatory effect.

  The zona fascicularis also produces small amounts of ***dehydroepiandrosterone*** (DHA) which is an androgen.
- The cells of the zona reticularis also produce some glucocorticoids; and sex hormones, both oestrogens and androgens.

Suprarenal Medulla

Both functionally and embryologically the medulla of the suprarenal gland is distinct from the cortex. When a suprarenal gland is fixed in a solution containing a salt of chromium (e.g., potassium dichromate) the cells of the medulla show yellow granules in their cytoplasm. This is called the ***chromaffin reaction***, and the cells that give a positive reaction are called ***chromaffin cells (pheochromocytes)***. The cells of the suprarenal cortex do not give this reaction.

The medulla is made up of groups or columns of cells. The cell groups or columns are separated by wide sinusoids. The cells are columnar or polyhedral and have a basophilic cytoplasm.

Note: With the EM the cells of the adrenal medulla are seen to contain abundant granular endoplasmic reticulum (in contrast to the agranular endoplasmic reticulum of cortical cells), and a prominent Golgi complex (Fig. 21.8). The cells also contain membrane bound secretory vesicles. In some cells these vesicles are small and electron dense while in others they are large and not so dense. The former contain nor-adrenalin and the latter adrenalin.

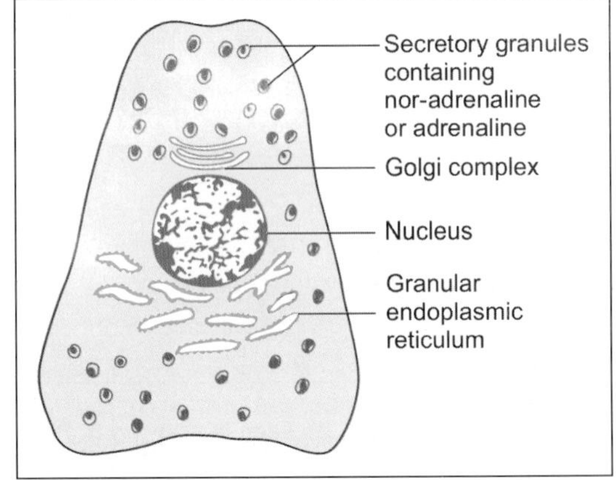

Fig. 21.8: Some features of ultrastructure of a cell from the adrenal medulla (Schematic representation)

The suprarenal medulla is now included in the APUD cell system. In contrast to the suprarenal cortex the medulla is not essential for life as its functions can be performed by other chromaffin tissues.

Hormone produced by Suprarenal Medulla

Functionally, the cells of the suprarenal medulla are considered to be modified postganglionic sympathetic neurons. Like typical postganglionic sympathetic neurons they secrete nor-adrenalin (nor-epinephrine) and adrenalin (epinephrine) into the blood. This secretion takes place mainly at times of stress (fear and anger) and results in widespread effects similar to those of stimulation of the sympathetic nervous system (e.g., increase in heart rate and blood pressure).

PINEAL GLAND

The pineal gland (or pineal body) is a small piriform structure present in relation to the posterior wall of the third ventricle of the brain. It is also called the ***epiphysis cerebri***. The pineal has for long been regarded as a vestigial organ of no functional importance. (Hence, the name pineal body). However, it is now known to be an endocrine gland of great importance.

MICROSCOPIC FEATURES

Sections of the pineal gland stained with haematoxylin and eosin reveal very little detail. The organ appears to be a mass of cells amongst which there are blood capillaries and nerve fibres. A distinctive feature of the pineal in such sections is the presence of irregular masses made up mainly of calcium salts. These masses constitute the ***corpora arenacea*** or ***brain sand*** (Fig. 21.9). The organ is covered by connective tissue (representing the piamater) from which septa pass into its interior.

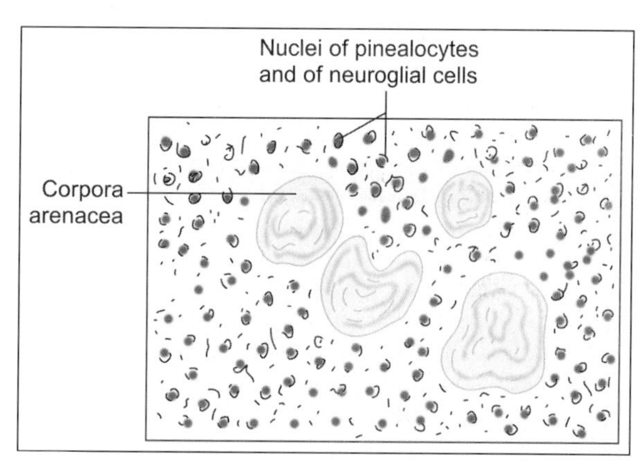

Fig. 21.9: Pineal body as seen with a light microscope (Schematic representation)

Pinealocytes

The organ is made up mainly of cells called ***pinealocytes***. Each cell has a polyhedral body containing a spherical oval or irregular nucleus. The cell body gives off long processes with expanded ***terminal buds*** that end in relation to the walls of capillaries, or in relation to the ependyma of the third ventricle.

The cell bodies of pinealocytes contain both granular and agranular endoplasmic reticulum, a well developed Golgi complex, and many mitochondria. An organelle of unusual structure made up of groups of microfibrils and perforated lamellae may be present (***canaliculate lamellar bodies***). The processes of pinealocytes contain numerous mitochondria. Apart from other organelles the terminal buds contain vesicles in which there are monamines and polypeptide hormones. The neurotransmitter gamma-amino-butyric acid is also present.

Hormone produced by Pinealocytes

The pinealocytes produce a number of hormones (chemically indolamines or polypeptides). These hormones have an important regulating influence (chiefly inhibitory) on many other endocrine organs. The organs influenced include the adenohypophysis, the neurohypophysis, the thyroid, the parathyroids, the adrenal cortex and medulla, the gonads, and the pancreatic islets. The hormones of the pineal body reach the hypophysis both through the blood and through the CSF. Pineal hormones may also influence the adenohypophysis by inhibiting production of releasing factors.

The best known hormone of the pineal gland is the amino acid ***melatonin*** (so called because it causes changes in skin colour in amphibia). Large concentrations of melatonin are present in the pineal gland. Considerable amounts of 5-hydroxytryptamine (serotonin), which is a precursor of melatonin, are also present. The presence of related enzymes has been demonstrated.

Interstitial Cells

The pinealocytes are separated from one another by neuroglial cells that resemble astrocytes in structure. They lie in proximity to blood vessel and pineolocytes.

Added Information

Cyclic Activity of Pineal Gland

The synthesis and discharge of melatonin is remarkably influenced by exposure of the animal to light, the pineal gland being most active in darkness. The neurological pathways concerned involve the hypothalamus and the sympathetic nerves. Because of this light mediated response, the pineal gland may act as a kind of biological clock which may produce circadian rhythms (variations following a 24-hour cycle) in various parameters.

The suprachiasmatic nucleus of the hypothalamus plays an important role in the cyclic activity of the pineal gland. This nucleus receives fibres from the retina. In turn it projects to the tegmental reticular nuclei (located in the brainstem). Reticulospinal fibres arising in these nuclei influence the sympathetic preganglionic neurons located in the first thoracic segment of the spinal cord. Axons of these neurons reach the superior cervical ganglion from where the ***nervus conarii*** arises and supplies the pineal gland.

It has often been stated in the past that the pineal gland degenerates with age. The corpora arenacea were considered to be signs of degeneration. Recent studies show that the organ does not degenerate with age. The corpora arenacea are now regarded as by-products of active secretory activity. It has been postulated that polypeptide hormones first exist in the form of complexes with a carrier protein called ***neuroepiphysin***. When hormones are released from the complex the carrier protein combines with calcium ions and is deposited as brain sand.

SOME OTHER ORGANS HAVING ENDOCRINE FUNCTIONS

PARAGANGLIA

Aggregations of cells similar to those of the adrenal medulla are to be found at various sites. They are collectively referred to as paraganglia because most of them are present in close relation to autonomic ganglia. The cells of paraganglia give a positive chromaffin reaction, receive a preganglionic sympathetic innervation, and have secretory granules containing catecholamines in their cytoplasm.

Like the cells of the adrenal medulla paraganglia are believed to develop from cells of the neural crest. Paraganglia are richly vascularised. They are regarded as endocrine glands that serve as alternative sites for the production of catecholamines in the foetus, and in early postnatal life, when the adrenal medulla is not yet fully differentiated. Because of their histochemical and ultrastructural features the cells of paraganglia are included in the APUD cell system. Most of the paraganglia retrogress with age, but some persist into adult life.

Cells similar to those of paraganglia are also present within some sympathetic ganglia. (They are called SIF or small intensely fluorescent cells). Here they are believed to act as interneurons.

Note: Some workers include the para-aortic bodies and carotid bodies amongst paraganglia.

PARA-AORTIC BODIES

These are two elongated bodies that lie, one on each side of the aorta, near the origin of the inferior mesenteric artery. The two masses may be united to each other by a band passing across the aorta.

These bodies have a structure similar to that of the adrenal medulla. The cells secrete noradrenalin. The aortic bodies retrogress with age.

CAROTID BODIES

These are small oval structures, present one on each side of the neck, at the bifurcation of the common carotid artery (i.e., near the carotid sinus).

The carotid bodies contain a network of capillaries in the intervals between which there are several types of cells.

Cells of Carotid Bodies

The most conspicuous cells of the carotid body are called ***glomus cells*** (or type I cells) (Fig. 21.10). These are large cells that have several similarities to neurons as follows:

- They give off dendritic processes.
- Their cytoplasm contains membrane bound granules which contain a number of neuropeptides. In the human carotid body the most prominent peptide present is encephalin. Others present include dopamine, serotonin, catecholamines, vasoactive intestinal peptide (VIP) and substance P.

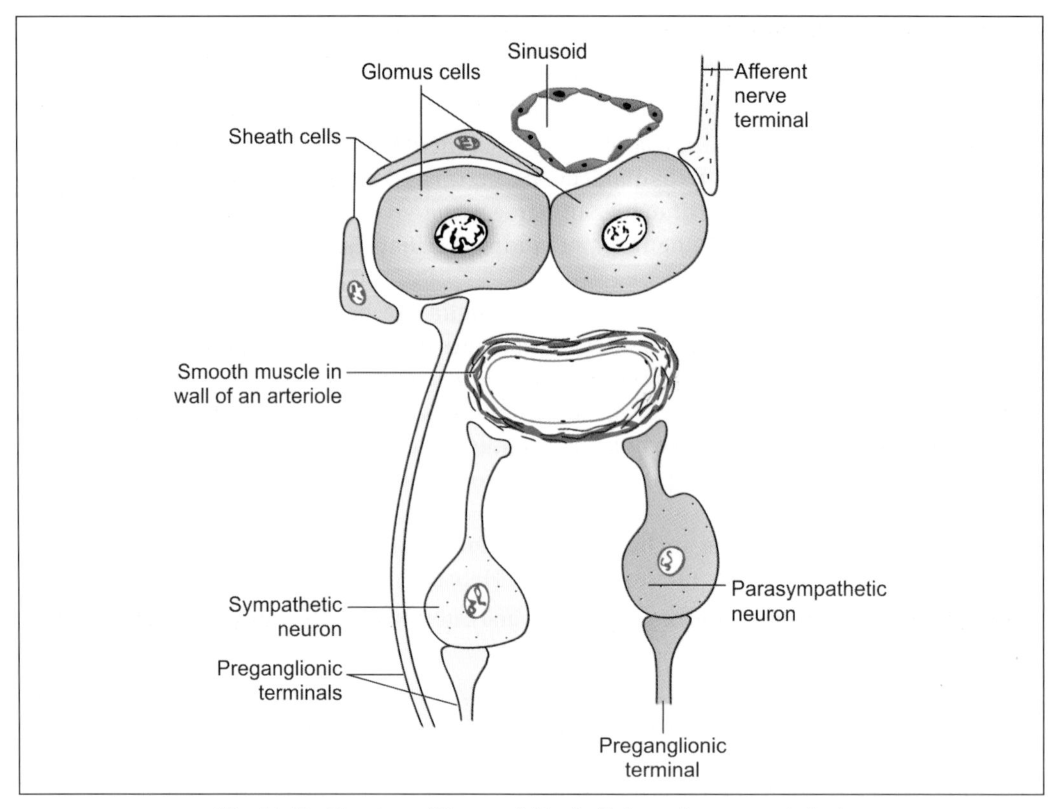

Fig. 21.10: Structure of the carotid body (Schematic representation)

- The cells are in synaptic contact with afferent nerve terminals of the glossopharyngeal nerve. Chemoreceptor impulses pass through these fibres to the brain. Some glomus cells also show synaptic connections with the endings of preganglionic sympathetic fibres, and with other glomus cells.
- The organisation of endoplasmic reticulum in them shows similarities to that of Nissl substance.
- They are surrounded by sheath cells that resemble neuroglial elements.

Because of these similarities to neurons, and because of the possibility that the cells release dopamine (and possibly other substances) they are sometimes described as neuroendocrine cells (and are included in the APUD cell category).

The exact significance of the glomus cells, and of their nervous connections, is not understood at present. They could possibly be sensory receptors sensitive to oxygen and carbon dioxide tension. Dopamine released by them may influence the sensitivity of chemoreceptor nerve endings. They may also serve as interneurons.

Apart from the glomus cells other cells present in the carotid bodies are as follows:
- Sheath cells (or type II cells) that surround the glomus cells.
- A few sympathetic and parasympathetic postganglionic neurons.
- Endothelial cells of blood vessels, and muscle cells in the walls of arterioles.
- Some connective tissue cells.

Nerve Supply of Carotid Bodies

The carotid body is richly innervated as follows:

- Afferent nerve terminals from the glossopharyngeal nerve form synapses with glomus cells.
- Preganglionic sympathetic and parasympathetic fibres end on the corresponding ganglion cells. Some preganglionic sympathetic fibres end by synapsing with glomus cells.
- Postganglionic fibres arising from the sympathetic and parasympathetic nerve cells within the carotid body innervate muscle in the walls of arterioles.

Functions of Carotid Bodies

The main function of the carotid bodies is that they act as chemoreceptors that monitor the oxygen and carbon dioxide levels in blood. They reflexly control the rate and depth of respiration through respiratory centres located in the brainstem. In addition to this function the carotid bodies are also believed to have an endocrine function.

Note: The precise mechanism by which the carotid bodies respond to changes in oxygen and carbon dioxide tension is not understood. It is not certain as to which cells, or nerve terminals are responsible for this function.

Added Information

Diffuse Neuroendocrine or APUD Cell System

Apart from the discrete endocrine organs considered in this chapter there are groups of endocrine cells scattered in various parts of the body. These cells share some common characteristics with each other, and also with the cells of some discrete endocrine organs. All these cells take up precursor substances from the circulation and process them (by decarboxylation) to form amines or peptides. They are, therefore, included in what is called the ***APUD cell system***. These peptides or amines serve as hormones. Many of them also function as neurotransmitters. Hence, the APUD cell system is also called the ***diffuse neuroendocrine system***. The cells of this system contain spherical or oval membrane bound granules with a dense core. There is an electronlucent halo around the dense core.

Some of the cells included in the APUD or diffuse neuroendocrine systems also give a positive chromaffin reaction. They were earlier referred to as cells of the ***chromaffin system***. However, they appear to be closely related functionally to other cells that are not chromaffin, and the tendency is to consider all these cells under the common category of diffuse neuroendocrine cells.

The diffuse neuroendocrine system is regarded as representing a link between the autonomic nervous system on the one hand, and the organs classically recognized as endocrine on the other, as it shares some features of both. The effects of the amines or peptides produced by the cells of the system are sometimes 'local' (like those of neurotransmitters) and sometimes widespread (like those of better known hormones).

The list of cell types included in the APUD cell system, as well as of their secretions, is long. As stated above even some discrete endocrine glands are now included under this heading. Some cells are enumerated:

- Various cells of the adenohypophysis.
- Neurons in the hypothalamus that synthesize the hormones of the neurohypophysis (oxytocin, vasopressin); and the cells that synthesize releasing factors controlling the secretion of hormones by the adenohypophysis.

Contd...

Contd...

- The chief cells of the parathyroid glands producing parathyroid hormone.
- The C- cells (parafollicular cells) of the thyroid, producing calcitonin.
- Cells of the adrenal medulla (along with some outlying chromaffin tissues) that secrete adrenalin and noradrenalin. These include the SIF cells of sympathetic ganglia.
- Cells of the gastro-entero-pancreatic endocrine system which includes cells of pancreatic islets producing insulin, glucagon and some other amines. It also includes endocrine cells scattered in the epithelium of the stomach and intestines producing one or more of the following: 5-hydroxytryptamine, glucagon, dopamine, somatostatin, substance P, motilin, gastrin, cholecystokinin, secretin, vasoactive intestinal polypeptide (VIP), and some other peptides.
- Glomus cells of the carotid bodies producing dopamine and noradrenalin.
- Melanocytes of the skin producing promelanin.
- Some cells in the pineal gland, the placenta, and modified myocytes of the heart called ***myoendocrine cells.***
- Renin producing cells of the kidneys.

Chapter 22

Special Senses: Eye

INTRODUCTION

The eyes are peripheral organs for vision and are located in the bony orbit. Each eyeball is like a camera. It has a ***lens*** that produces images of objects that we look at. The images fall on a light sensitive membrane called the ***retina***. Cells in the retina convert light images into nervous impulses that pass through the optic nerve, and other parts of the visual pathway, to reach visual areas in the cerebral cortex. It is in the cortex that vision is actually perceived.

STRUCTURE OF EYEBALL

The wall of an eyeball consists of three layers:

- Outer fibrous coat that includes sclera and cornea
- Middle vascular coat that includes choroid, ciliary body and iris
- Inner retina.

The space between the iris and the cornea is called the ***anterior chamber***, while the space between the iris and the front of the lens is called the ***posterior chamber***. These chambers are filled with a fluid called the ***aqueous humour***. The part of the eyeball behind the lens is filled by a jelly-like substance called the ***vitreous body***. The main parts of eyeball (as seen in section) are shown in Fig. 22.1.

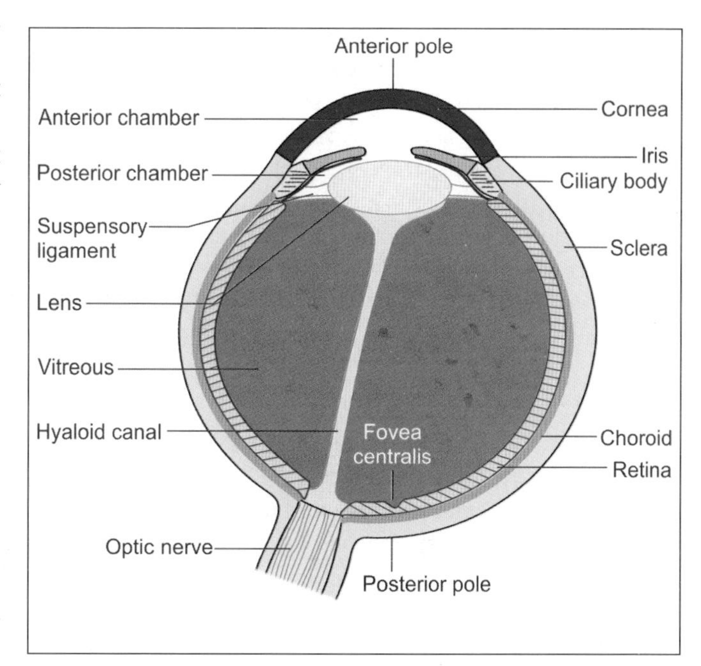

Fig. 22.1: Section across the eyeball to show its main parts (Schematic representation)

OUTER FIBROUS COAT

Sclera

The outer wall of the eyeball is formed (in its posterior five-sixths) by a thick white opaque membrane called the ***sclera***. The sclera consists of white fibrous tissue (collagen). Some elastic fibres, and connective tissue cells (mainly fibroblasts) are also present. Some of the cells are pigmented.

Externally, the sclera is covered in its anterior part by the ocular conjunctiva, and posteriorly by a fascial sheath (or ***episclera***). The deep surface of the sclera is separated from the choroid by the ***perichoroidal space***. Delicate connective tissue present in this space constitutes the ***suprachoroid lamina*** (or ***lamina fusca***).

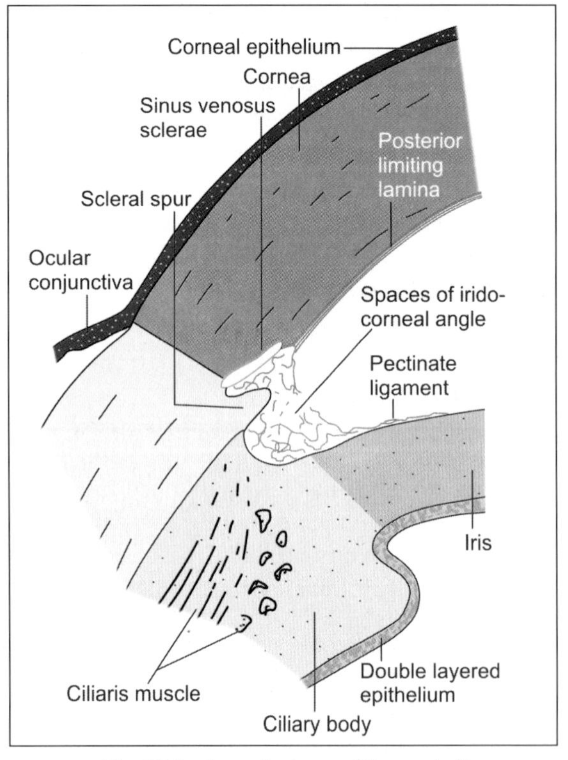

Fig. 22.2: Some features of the eyeball to be seen at the junction of the cornea with the sclera (Schematic representation)

Anteriorly, the sclera becomes continuous with the cornea at the ***corneoscleral junction*** (also called ***sclerocorneal junction*** or ***limbus***). A circular channel called the ***sinus venosus sclerae*** (or ***canal of Schlemm***) is located in the sclera just behind the corneoscleral junction (Fig. 22.2). A triangular mass of scleral tissue projects towards the cornea just medial to this sinus. This projection is called the ***scleral spur***.

The optic nerve is attached to the back of the eyeball a short distance medial to the posterior pole. Here the sclera is perforated like a sieve, and the area is, therefore, called the ***lamina cribrosa***. Bundles of optic nerve fibres pass through the perforations of the lamina cribrosa.

Functions

- The sclera (along with the cornea) collectively forms the ***fibrous tunic*** of the eyeball and provides protection to delicate structures within the eye.
- It resists intraocular pressure and maintains the shape of the eyeball.
- Its smooth external surface allows eye movements to take place with ease.
- The sclera also provides attachment to muscles that move the eyeball.

Cornea

In the anterior one-sixth of the eyeball the sclera is replaced by a transparent disc called the ***cornea***. The cornea is convex forwards. It is colourless and avascular but has a very rich nerve supply. The cornea is made up of five layers (Plate 22.1).

- ***Corneal epithelium:*** The outermost layer is of non-keratinised stratified squamous epithelium (corneal epithelium). The cells in the deepest layer of the epithelium are columnar; in the middle layers they are polygonal; and in the superficial layers they are flattened. The cells are arranged with great regularity.

  With the EM the cells on the superficial surface of the epithelium show projections either in the form of microvilli or folds of plasma membrane. These folds are believed to play an important role in retaining a film of fluid over the surface of the cornea. The corneal epithelium regenerates rapidly after damage.
- ***Bowman's membrane:*** The corneal epithelium rests on the ***anterior limiting lamina*** (also called ***Bowman's membrane***). With the light microscope this lamina appears to be structureless, but with the EM it is seen to be made up of fine collagen fibrils embedded in matrix. It gives great stability and strength to the cornea.
- ***Corneal stroma:*** Most of the thickness of the cornea is formed by the ***substantia propria*** (or ***corneal stroma***). The substantia propria is made up of type 1 collagen fibres embedded in a ground substance containing sulphated glycosaminoglycans.

  They are arranged with great regularity and form lamellae. The fibres within one lamellus are parallel to one another, but the fibres in adjoining lamellae run in different directions forming obtuse angles with each other. The transparency of the cornea is due to the regular arrangement of fibres, and because of the fact that the fibres and the ground substance have the same refractive index.

  Fibroblasts are present in the substantia propria. They appear to be flattened in vertical sections through the cornea, but are seen to be star-shaped on surface view. They are also called ***keratocytes*** or ***corneal corpuscles***.
- ***Descemet's membrane:*** Deep to the substantia propria there is a thin homogeneous layer called the ***posterior limiting lamina*** (or ***Descemet's membrane***). It is a true basement membrane.

  At the margin of the cornea the posterior limiting membrane becomes continuous with fibres that form a network in the angle between the cornea and the iris (***irido-corneal angle***). The spaces between the fibres of the network are called the ***spaces of the irido-corneal angle***. Some of the fibres of the network pass onto the iris as the ***pectinate ligament*** (Fig. 22.2).
- ***Endothelium:*** The posterior surface of the cornea is lined by a single layer of flattened cells that constitute the ***endothelium of the anterior chamber***. This layer is in contact with the aqueous humour of the anterior chamber.

  The endothelial cells are adapted for transport of ions. They possess numerous mitochondria. They are united to neighbouring cells by desmosomes and by occluding junctions. The cells pump out excessive fluid from cornea, and thus ensure its transparency.

VASCULAR COAT OR UVEA

Deep to the sclera there is a vascular coat (uvea) that consists of the choroid, the ciliary body and the iris.

PLATE 22.1: Cornea

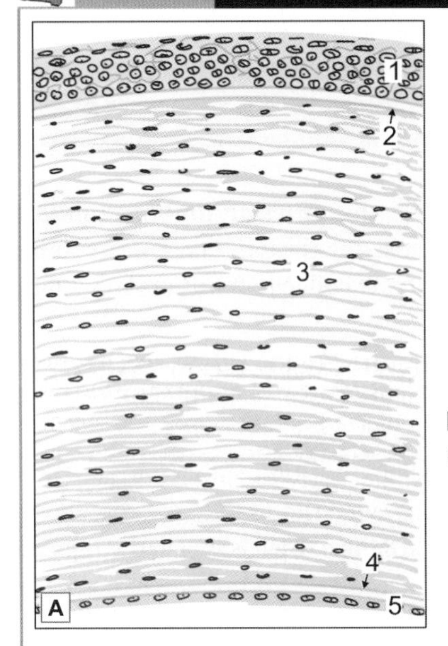

The cornea is made up of five layers

- The outermost layer is of non-keratinised stratified squamous epithelium (corneal epithelium)
- The corneal epithelium rests on the structureless anterior limiting lamina (also called Bowman's membrane)
- Most of the thickness of the cornea is formed by the substantia propria (or corneal stroma) made up of collagen fibres embedded in a ground substance
- Deep to the substantia propria there is a thin homogeneous layer called the posterior limiting lamina
- The posterior surface of the cornea is lined by a single layer of flattened or cuboidal cells.

Note: The structure of the cornea is fairly distinctive and its recognition should not be a problem.

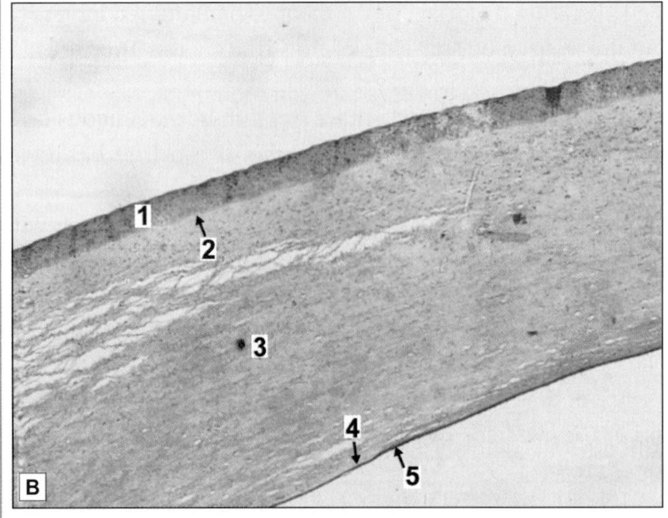

Key

1. Stratified sqaumous corneal epithelium
2. Anterior limiting membrane (Bowman's)
3. Substantia propria
4. Posterior limiting membrane (Descemet's)
5. Posterior cuboidal epithelium

Cornea. A. As seen in drawing; B. Photomicrograph

Choroid

The choroid consists of

- ***Choroid proper***
- ***Suprachoroid lamina*** that separates the choroid proper from the sclera
- ***Basal lamina*** (***membrane of Bruch***) which intervenes between the choroid proper and the retina (Fig. 22.3).

Choroid Proper

The choroid proper consists of a network of blood vessels supported by connective tissue in which many pigmented cells are present, giving the choroid a dark colour. This colour darkens the interior of the eyeball. The pigment also prevents reflection of light within the eyeball. Both these factors help in formation of sharp images on the retina.

Added Information

The choroid proper is made up of an outer ***vascular lamina*** containing small arteries and veins, and lymphatics; and an inner ***capillary lamina*** (or ***choroidocapillaris***). The connective tissue supporting the vessels of the vascular lamina is the ***choroidal stroma***. Apart from collagen fibres it contains melanocytes, lymphocytes and mast cells. The capillary lamina is not pigmented. Nutrients diffusing out of the capillaries pass through the basal lamina to provide nutrition to the outer layers of the retina.

Suprachoroid Lamina

The suprachoroid lamina is also called the ***lamina fusca***. It is non-vascular. It is made up of delicate connective tissue containing collagen, elastic fibres, and branching cells containing pigment. A plexus of nerve fibres is present. Some neurons may be seen in the plexus.

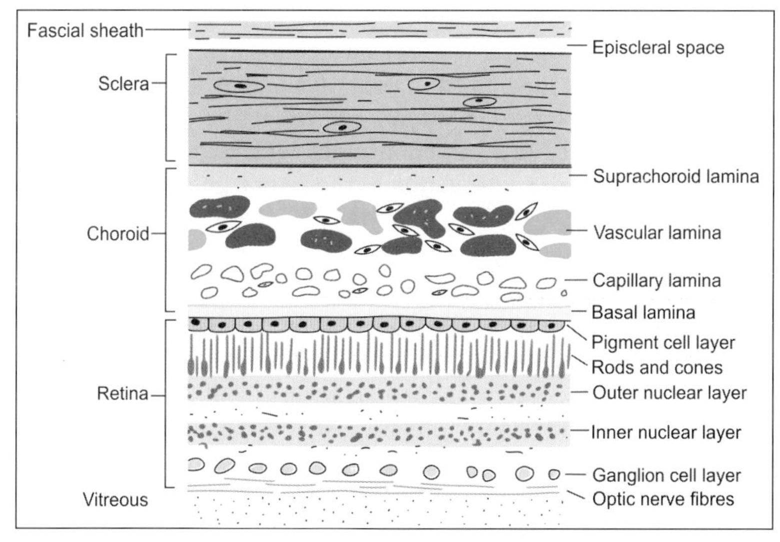

Fig. 22.3: Various layers of the eyeball (Schematic representation)

Basal Lamina

With the light microscope the basal lamina (or ***membrane of Bruch***) appears to be a homogeneous layer. However, with the EM the membrane is seen to have a middle layer of elastic fibres, on either side of which there is a layer of delicate collagen fibres. The basal lamina is said to provide a smooth surface on which pigment cells and receptors of the retina can be arranged in precise orientation.

Ciliary Body

The ciliary body represents an anterior continuation of the choroid. It is a ring-like structure continuous with the periphery of the iris. It is connected to the lens by the suspensory ligament.

The ciliary body is made up of vascular tissue, connective tissue and muscle. The muscle component constitutes the ***ciliaris muscle***. The ciliaris muscle is responsible for producing alterations in the convexity of the lens (through the suspensory ligament) enabling the eye to see objects at varying distances from it. In other words the ciliaris is responsible for accommodation.

The inner surface of the ciliary body is lined by a double layered epithelium. The outer cell layer is pigmented, whereas the inner cell layer (facing the posterior chamber) is non-pigmented. The cells of the inner layer secrete aqueous humour. The anterior part of the inner surface of the ciliary body has short processes towards the lens, known as ***ciliary processes***.

Iris

The iris is the most anterior part of the vascular coat of the eyeball. It forms a diaphragm placed immediately in front of the lens. At its periphery it is continuous with the ciliary body. In its centre, there is an aperture the ***pupil***. The pupil regulates the amount of light passing into the eye.

The iris is composed of a stroma of connective tissue containing numerous pigment cells, and in which are embedded blood vessels and smooth muscle. Some smooth muscle fibres are arranged circularly around the pupil and constrict it. They form the ***sphincter pupillae***. Other fibres run radially and form the ***dilator pupillae***. The posterior surface of the iris is lined by a double layer of epithelium continuous with that over the ciliary body. This epithelium represents a forward continuation of the retina. The cells of this epithelium are deeply pigmented.

RETINA

This is the inner coat of eyeball and lines its posterior ¾ surface. The retina contains photo-receptors (rods and cones) which are essential for vision. Retina has a specialized area where vision is most acute, called as fovea centralis or macula (Fig. 22.5). This area contains only cones which are essentially bare (the over-lying layers are pushed to the side). The retina also has a 'blind spot', the optic disc, where the optic nerve leaves the eye and there are no photoreceptor cells.

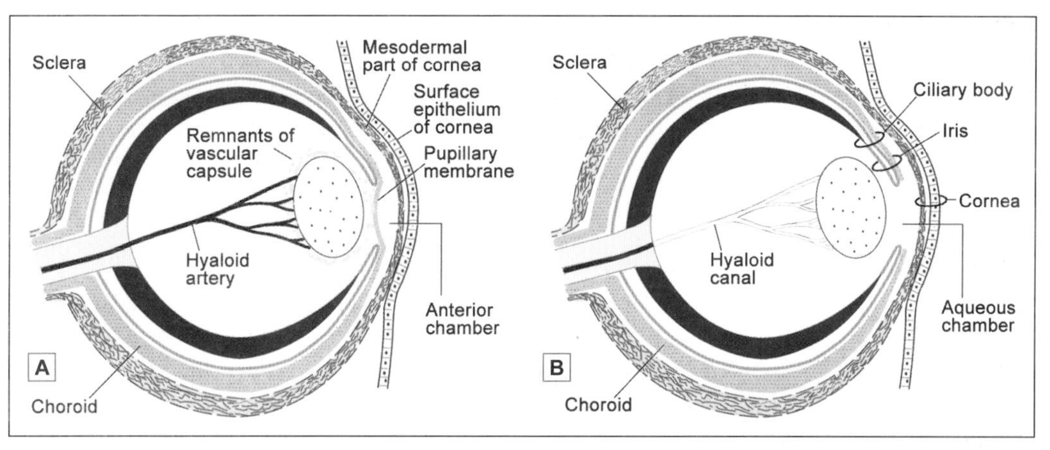

Fig. 22.4: Some features of the developing eye. **A.** Early stage; **B.** Later stage (Schematic representation)

Embryological Considerations

To understand the structure of the retina brief reference to its development is necessary (Figs 22.4A and B). The retina develops as an outgrowth from the brain (diencephalon). The proximal part of the diverticulum remains narrow and is called the ***optic stalk***. It later becomes the optic nerve. The distal part of the diverticulum forms a rounded hollow structure called the ***optic vesicle***. This vesicle is invaginated by the developing lens (and other surrounding tissues) so that it gets converted into a two layered ***optic cup***. At first, each layer of the cup is made up of a single layer of cells. The outer layer persists as a single layered epithelium that becomes pigmented. It forms the ***pigment cell layer*** of the retina. Over the greater part of the optic cup the cells of the inner layer multiply to form several layers of cells that become the ***nervous layer of the retina.*** In the anterior part, both layers of the optic cup remain single layered. These two layer sline (a) the inner surface of the ciliary body forming the ***ciliary part of the retina***; and (b) the posterior surface of the iris forming the ***iridial part of the retina***.

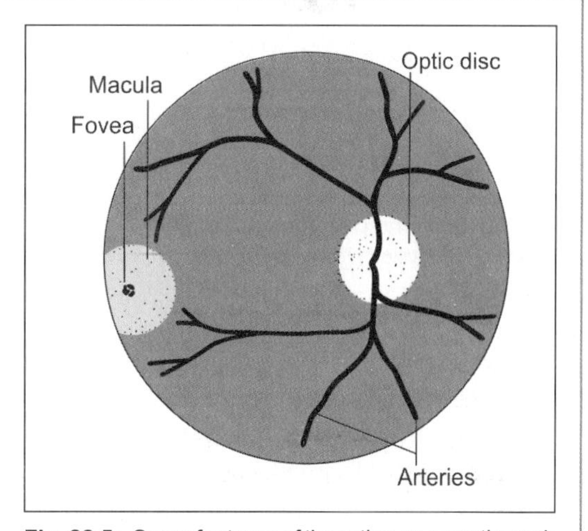

Fig. 22.5: Some features of the retina as seen through an ophthalmoscope (Schematic representation) Note the arteries emerging through the optic disc. Veins are omitted for the sake of clarity

Opposite the posterior pole of the eyeball the retina shows a central region about 6 mm in diameter. This region is responsible for sharp vision. In the centre of this region an area about 2 mm in diameter has a yellow colour and is called the macula lutea (Fig. 22.5). In the centre of the macula lutea there is a small depression that is called the fovea centralis. The floor of the fovea centralis is often called the foveola. This is the area of clearest vision.

The optic nerve is attached to the eyeball a short distance medial to the posterior pole. The nerve fibres arising from the retina converge to this region, where they pass through the lamina cribrosa. When viewed from the inside of the eyeball this area of the retina is seen as a circular area called the optic disc.

PLATE 22.2: Eye ball

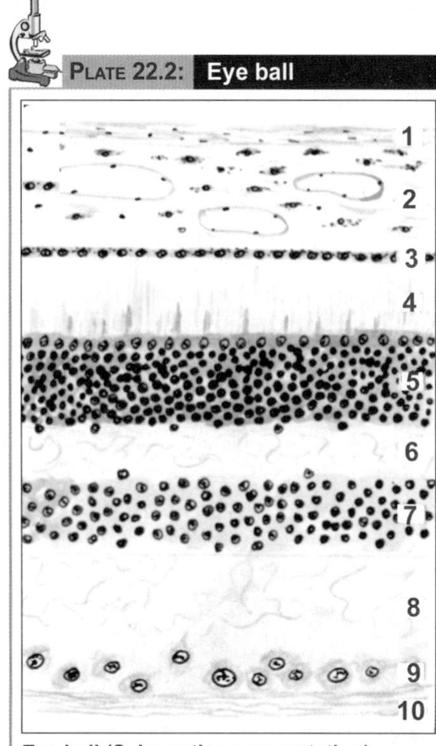

Eye ball (Schematic representation)

The wall of the eye ball is made up of several layers as follows:

1. Sclera, made up of collagen fibres
2. Choroid, containing blood vessels and pigment cells. The remaining layers are subdivisions of the retina
3. Pigment cell layer
4. Layer of rods and cones
5. Outer nuclear layer
6. Outer plexiform layer
7. Inner nuclear layer
8. Inner plexiform layer
9. Layer of ganglion cells
10. Layer of optic nerve fibres.

The appearance is not likely to be confused with any other tissue.

Basic Structure of the Retina

When we examine sections through the retina (stained by haematoxylin and eosin, Fig. 22.6) a number of layers can be distinguished. The significance of the layers becomes apparent, however, only if we study the retina using special methods. A highly schematic presentation of the layers of the retina, and of the cells present in them is shown in Fig. 22.6. The retina can be said to have an external surface that is in contact with the choroid, and an internal surface that is in contact with the vitreous. Beginning from the external surface the following layers can be made out (Plate 22.2).

- ***Pigment Cell Layer:*** It is the outermost layer of retina which is separated from choroid by Bruch's membrane. This consists of a single layer of low cuboidal cells containing melanin pigment. Processes from pigment cells extend into the next layer. This layer performs the following functions:
  - It absorbs and prevents reflection of light that has passed through the neural layers of the retina.
  - The pigment cells phagocytose the shed membranous discs of the outer segment of rods and cones.
  - These cells also produce melanin.
- ***Layer of Rods and Cones:*** The rods are processes of rod cells, and cones are processes of cone cells. The peripheral process is rod shaped in the case of rod cells, and cone shaped in the case of cone cells.

- ***External Nuclear Layer:*** The external nuclear layer contains the cell bodies and nuclei of rod cells and of cone cells.

  These cells are photoreceptors that convert the stimulus of light into nervous impulses. Each rod cell or cone cell can be regarded as a modified neuron. It consists of a cell body, a peripheral (or external) process, and a central (or internal) process. The peripheral processes lie in the layer of rods and cones described above. The nuclei of these cells are arranged in several layers in the form of external nuclear layer. This layer is darkly stained.

  Between second and third layer there is presence of a pink linear marking called as outer limiting membrane or lamina. This results because of zonula adherens of the glial cells (Muller cells) with the cell bodies of photoreceptor cells. The Muller cells are supporting cells of retina. They have long slender body that is radially oriented in retina. The central process of each rod cell or cone cell is an axon. It extends into the external plexiform layer where it synapses with dendrites of bipolar neurons.
- ***External Plexiform Layer:*** The external plexiform layer (or ***outer synaptic zone***) consists only of nerve fibres that form a plexus. The axons of rods and cones synapse here with dendrites of bipolar neurons and horizontal cells. This layer stains lightly.

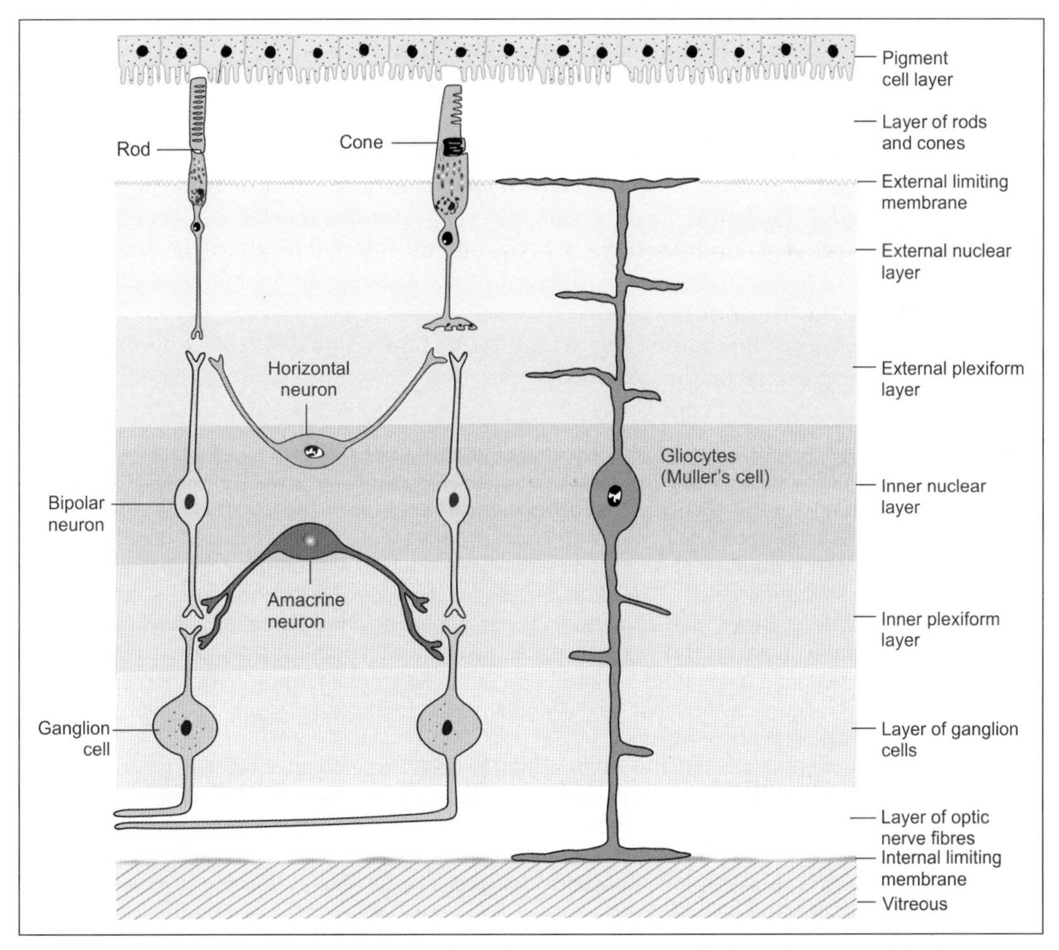

Fig. 22.6: Layers of the retina and the main structures therein (Schematic representation)

- ***Internal Nuclear Layer:*** The internal nuclear layer contains the cell bodies and nuclei of three types of neurons.
  - The ***bipolar cells*** give off dendrites that enter the external plexiform layer to synapse with the axons of rod and cone cells; and axons that enter the internal plexiform layer where they synapse with dendrites of ganglion cells. The bipolar cells are oriented perpendicular to the layers of retina.
  - The ***horizontal cells*** give off processes that run parallel to the retinal surface. These processes enter the outer plexiform layer and synapse with rods, cones, and dendrites of bipolar cells. The horizontal cells are oriented parallel to the layers of retina.
  - The ***amacrine cells*** also lie horizontally in the retina. Their processes enter the inner plexiform layer where they synapse with axons of bipolar cells, and with dendrite of ganglion cells.

Note: Apart from bipolar, horizontal and amacrine neurons, the internal nuclear layer also contains the nuclei of retinal gliocytes or ***cells of Muller*** (Fig. 22.6). These cells give off numerous protoplasmic processes that extend through almost the whole thickness of the retina. Externally, they extend to the junction of the layer of rods and cones with the external nuclear layer. Here the processes of adjoining gliocytes meet to form a thin ***external limiting membrane***. Internally, the gliocytes extend to the internal surface of the retina where they form an ***internal limiting membrane***. This membrane separates the retina from the vitreous. The retinal gliocytes are neuroglial in nature. they support the neurons of the retina and may ensheath them. They probably have a nutritive function as well.

- ***Internal Plexiform Layer:*** The ***internal plexiform layer*** (or ***inner synaptic zone***) consists of synapsing nerve fibres. The axons of bipolar cells synapse with dendrites of ganglion cells; and both these processes synapse with processes of amacrine cells. The internal plexiform layer also contains some horizontally placed ***internal plexiform cells***; and also a few ganglion cells.
- ***Layer of Ganglion Cells:*** The layer of ganglion cells contains the cell bodies of ganglion cells. We have seen that dendrites of these cells enter the internal plexiform layer to synapse with processes of bipolar cells and of amacrine cells. Each ganglion cell gives off an axon that forms a fibre of the optic nerve.
- ***Layer of Optic Nerve Fibres:*** The layer of optic nerve fibres is made up of axons of ganglion cells. The fibres converge on the optic disc where they pass through foramina of the lamina cribrosa to enter the optic nerve.

Added Information

- Horizontal neurons are of two types, ***rod horizontals*** and ***cone horizontals***, depending on whether they synapse predominantly with rods or cones. Each horizontal cell gives off one long process, and a number of short processes (7 in case of rod horizontal cells, and 10 in case of cone horizontal cells). The short processes are specific for the type of cell: those of rod horizontals synapse with a number of rod spherules, and those of cone horizontals synapse with cone pedicles. The long processes synapse with both rods and cones (which are situated some distance away from the cell body of the horizontal neuron). The long and short processes of horizontal cells cannot be distinguished as dendrites or axons, and each process probably conducts in both directions.
- The term amacrine is applied to neurons that have no true axon. Like the processes of horizontal cells those of amacrine neurons also conduct impulses in both directions. Each cell gives off one or two thick processes that divide further into a number of branches. The amacrine cells are believed to play a very important role in the interaction between adjacent areas of the retina resulting in production of sharp images. They are also involved in the analysis of motion in the field of vision.

Appearance of the Retina in Sections

Having considered the structures comprising the various layers of the retina it is now possible to understand the appearance of the retina as seen in sections stained by haematoxylin and eosin (Plate 22.2). The inner and outer nuclear layers can be made out even at low magnification. The outer nuclear layer is thicker, and the nuclei in it more densely packed than in the inner nuclear layer. We have seen that this (outer nuclear) layer contains the nuclei of rods and cones. The cone nuclei are oval and lie in a single row adjoining the layer of rods and cones. The remaining nuclei are those of rods.

The nuclei in the inner nuclear layer belong (as explained above) to bipolar cells, horizontal cells, amacrine cells, and gliocytes.

The layer of ganglion cells is (at most places) made up of a single row of cells of varying size. The cell outlines are indistinct, but the nuclei can be made out. They are of various sizes. On the whole they are larger and stain more lightly than nuclei in the inner and outer nuclear layers.

The layer of pigment cells resembles a low cuboidal epithelium. All the nuclei in this layer are of similar size, and lie in a row.

The remaining layers (layers of rods and cones, inner and outer plexiform layers, and the layer of optic nerve fibres) are seen as light staining areas in which no detail can be made out. The layer of rods and cones may show vertical striations.

We will now consider the individual cells of the retina in greater detail.

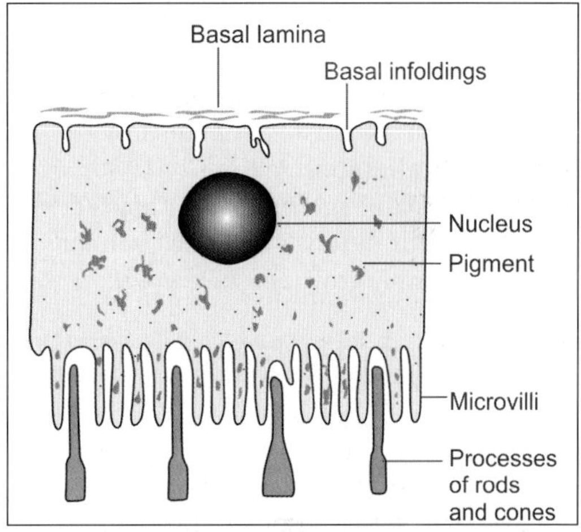

Fig. 22.7: Some features of a pigment cell of the retina (Schematic representation)

Blood-retina Barrier

The blood vessels that ramify in the retina do not supply the rods and cones. These are supplied by diffusion from choroidal vessels. The endothelial cells of capillaries in the retina are united by tight junctions to prevent diffusion of substances into the rods and cones. This is referred to as the blood-retina barrier.

Pigment Cells

Pigment cells appear to be rectangular in vertical section, their width being greater than their height (Fig. 22.7). In surface view they are hexagonal. The nucleus is basal in position. The pigment in the cytoplasm is melanin. With the

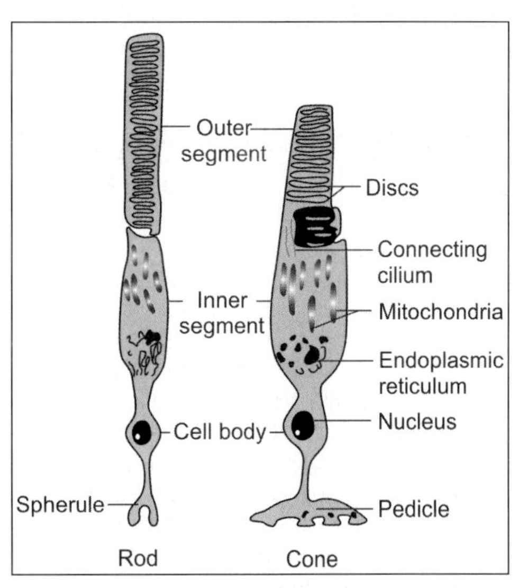

Fig. 22.8: The main parts of rods and cones (Schematic representation)

EM it can be seen that the surface of the cell shows large microvilli that contain pigment. These microvilli project into the intervals between the processes of rods and cones. Each pigment cell is related to about a dozen rods and cones. The plasma membrane at the base of the cell shows numerous infoldings.

The functions attributed to pigment cells include the following.

- The absorption of excessive light and avoidance of back reflection.
- They may play a role in regular spacing of rods and cones and may provide mechanical support to them.
- They have a phagocytic role. They 'eat up' the ends of rods and cones (which are constantly growing: see below).

Rods and Cones (Fig. 22.8)

There are about seven million cones in each retina. The rods are far more numerous. They number more than 100 million. The cones respond best to bright light (photopic vision). They are responsible for sharp vision and for the discrimination of colour. Rods can respond to poor light (scotopic vision) and specially to movement across the field of vision.

Each rod is about 50 μm in length and about 2 μm thick. Cones are about 40 μm in length and 3-5 μm thick.

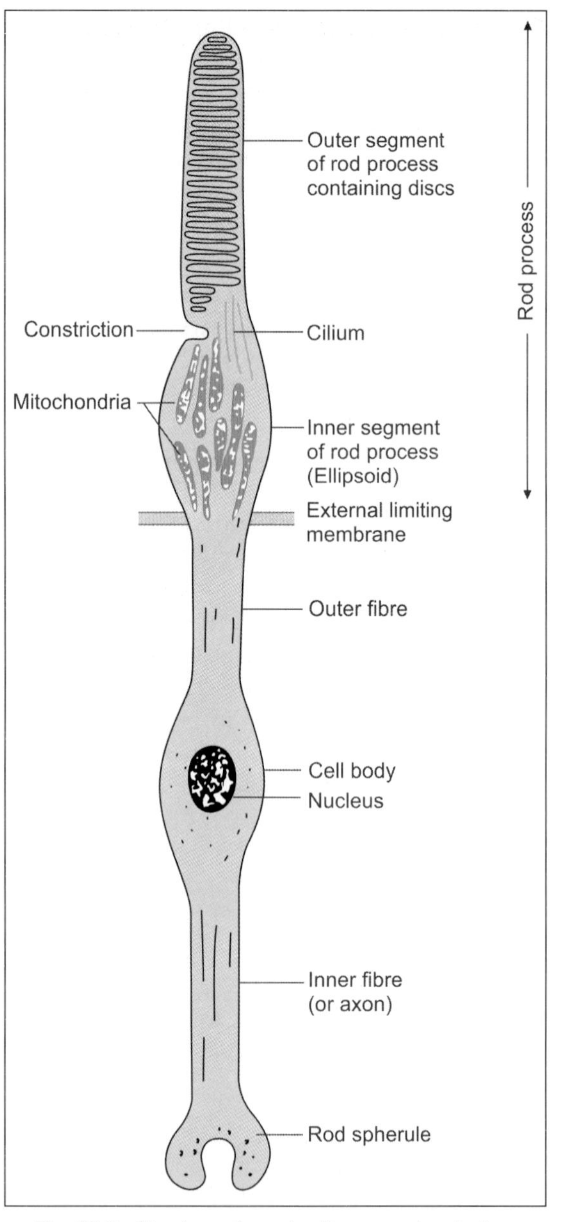

Fig. 22.9: Structure of a rod cell as seen by electron microscope (Schematic representation)

Ultrastructure of Rod and Cone Cells

The ultrastructure of rod cells and of cone cells is similar and is, therefore, considered together. Each rod or cone cell consists of a cell body containing the nucleus, and of external and internal processes, an inner fibre and spherule (Fig. 22.9). The parts of cone cells are almost same except the terminal part which is called pedicle instead of spherule.

The cell body (lying in the external nuclear layer) gives off two 'fibres,' inner and outer. The ***outer fibre*** passes outwards up to the external limiting membrane and becomes continuous with the rod process, or the cone process. The process itself can be divided into an ***inner segment***, and an ***outer segment***. The outer segment is the real photo-receptor element. It contains a large number of membranous discs stacked on one another. It is believed that the discs are produced by the cilium (see below) and gradually move towards the tip of the outer segment. Here old discs are phagocytosed by pigment cells.

The outer segments of rods and cones contain photo-sensitive pigments that are concerned with the conversion of light into nerve impulses. The pigments are believed to be bound to the membranes of the sacs of the outer segments. The pigment in the rods is ***rhodopsin***, and that in the cones is ***iodopsin***. Cones are believed to be of three types, red sensitive, green sensitive, and blue sensitive. Iodopsin has, therefore, to exist in three forms, one for each of these colours. However, the three types of cones cannot be distinguished from one another on the basis of their ultrastructure.

The inner segment of the rod or cone process is wider than the outer segment. It contains a large number of mitochondria that are concentrated in a region that is called the ***ellipsoid***.

At the junction of the inner and outer segments of the rod or cone process there is an indentation of the plasma membrane on one side, so that the connection becomes very narrow. This narrow part contains a fibrillar ***cilium*** in which the microfibrils are orientated as in cilia elsewhere. This cilium is believed to give rise to the flattened discs of the outer segment.

We have seen that the part of the rod cell between the cell body and the external limiting membrane is the outer fibre. The length of the outer fibre varies from rod to rod, being greatest in those rods that have cell bodies placed 'lower down' in the external nuclear layer. The outer fibre is absent in cones, the inner segment of the cone process being separated from the cone cell body only by a slight constriction.

The ***cell bodies*** of rod cells and of cone cells show no particular peculiarities of ultrastructure.

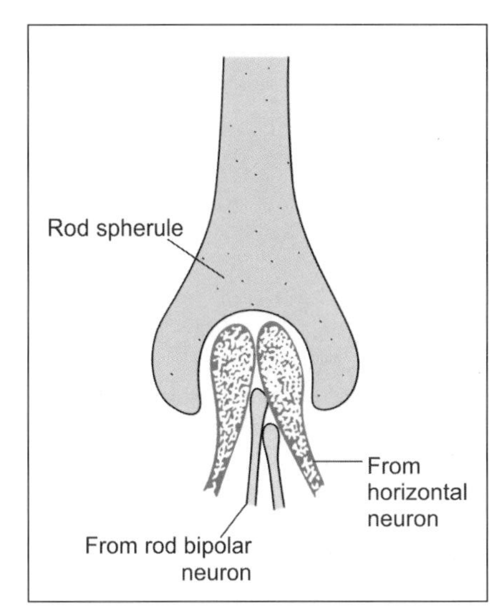

Fig. 22.10: Rod spherule synapsing with terminals of rod bipolar cells and horizontal cells (Schematic representation)

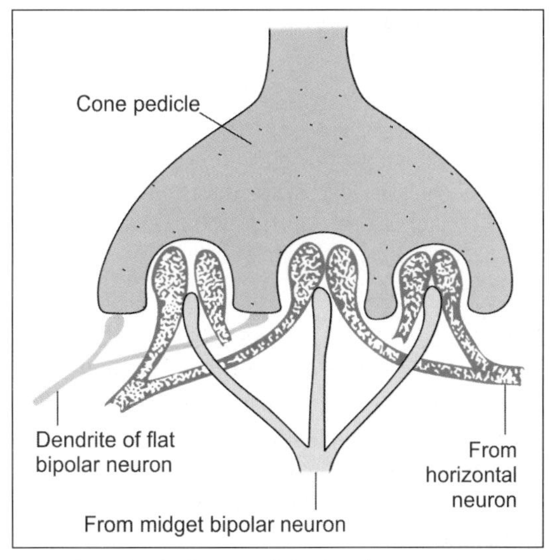

Fig. 22.11: Cone pedicle showing a number of synaptic areas, each area receiving three terminals (Schematic representation)

The ***inner fibres*** of rod and cone cells resemble axons. At its termination each rod axon expands into a spherical structure called the ***rod spherule***, while cone axons end in expanded terminals called ***cone pedicles*** (Figs 22.10 and 22.11). The rod spherules and cone pedicles form complex synaptic junctions with the dendrites of bipolar neurons, and with processes of horizontal cells. Each rod spherule synapses with processes of two bipolar neurons, and with processes of horizontal neurons.

Each cone pedicle has numerous synapses with processes from one or more bipolar cells, and with processes of horizontal cells. In many situations the cone pedicle bears several invaginations that are areas of synaptic contacts. Each such area receives one process from a bipolar dendrite; and two processes, one each from two horizontal neurons. Such groups are referred to as ***triads***. Each cone pedicle has 24 such triads. Apart from triads the cone pedicle bears numerous other synaptic contacts in areas intervening between the triads. These areas synapse with dendrites of diffuse bipolar cells. Some pedicles also establish synaptic contacts with other cone pedicles.

Clinical Correlation

Retinal Detachment

Retinal detachment is the separation of the neurosensory retina from the retinal pigment epithelium. It may occur spontaneously in older individuals past 50 years of age or may be secondary to trauma in the region of head and neck. There are 3 pathogenetic mechanisms of retinal detachment:

- Pathologic processes in the vitreous or anterior segment
- Collection of serous fluid in the sub-retinal space
- Accumulation of vitreous under the retina through a hole or a tear in the retina.

Retinitis Pigmentosa

Retinitis pigmentosa is a group of systemic and ocular diseases of unknown etiology, characterised by degeneration of the retinal pigment epithelium. The earliest clinical finding is night blindness due to loss of rods and may progress to total blindness.

Retinoblastoma

This is the most common malignant ocular tumour in children. It may be present at birth or recognised in early childhood before the age of 4 years. About 60% cases of retinoblastoma are sporadic and the remaining 40% are familial. Familial tumours are often multiple and multifocal and transmitted as an autosomal dominant trait by retinoblastoma susceptibility gene (RB) located on chromosome 13. Such individuals have a higher incidence of bilateral tumours and have increased risk of developing second primary tumour, particularly osteogenic sarcoma. Clinically, the child presents with leukokoria, i.e. white pupillary reflex.

LENS

The lens of the eye is a transparent biconvex avascular structure. It is suspended between the iris and the vitreous by the zonules, which connect the lens with the ciliary body. It is surrounded by an elastic capsule which is a semipermeable membrane. The posterior surface of the lens is more curved than the anterior.

The lens consists of three parts:

- Lens capsule
- Lens epithelium
- Lens substance (Fig. 22.12).

Lens Capsule

The ***lens capsule*** is a transparent homogeneous and highly elastic collagenous basement membrane. It is made up mainly of type IV collagen and glycoproteins. The capsule is thicker in front than behind. It is secreted by the lens epithelium.

Lens Epithelium

Deep to the capsule the lens is covered on its anterior surface by a lens epithelium. The cells of the epithelium are cuboidal. However, towards the periphery of the lens the cells become progressively longer. Ultimately they are converted into long fibres that form the substance of the lens.

The cells of epithelium are metabolically active, contain Na^+-K^+-ATPase and generate adenosine triphosphate (ATP) to meet the energy demand of the lens. The cells show high mitotic activity and form new cells which migrate toward the equator. The lens epithelial cells continue to divide and develop into the lens fibres.

Lens Fibres

The ***lens fibres*** develop from the lens epithelial cells that continue to divide and get elongated and transformed into lens fibres. They are mainly composed of soluble proteins called *crystallins.*

The fibres formed earlier lie in the deeper plane (nucleus of the lens), the newer ones occupy a more superficial plane.

When the lens is examined from the front, or from behind, three faint lines are seen radiating from the centre to the periphery. In the fetus these lines form a 'Y' that is upright on the front of the lens, and inverted at the back (Fig. 22.13). The lines become more complex in the adult. These lines are called ***sutural lines***. They are made up of amorphous material. The

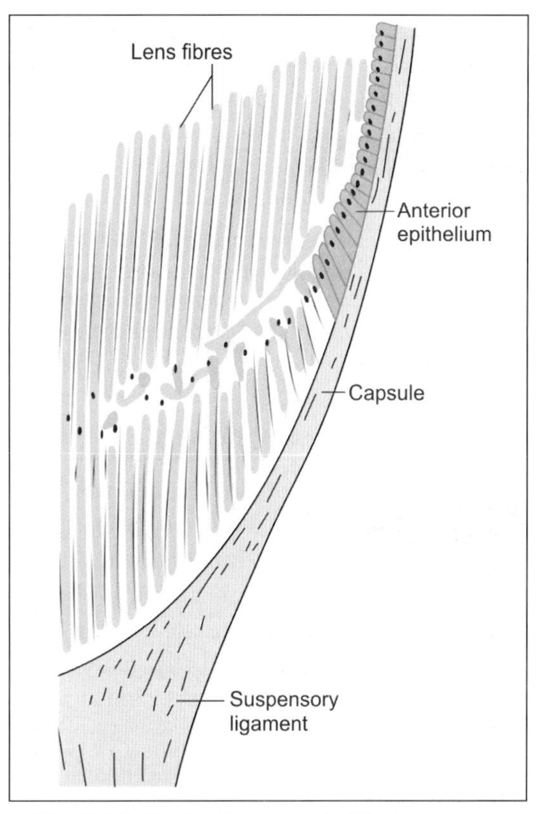

Fig. 22.12: Section through part of the lens near its margin (Schematic representation)

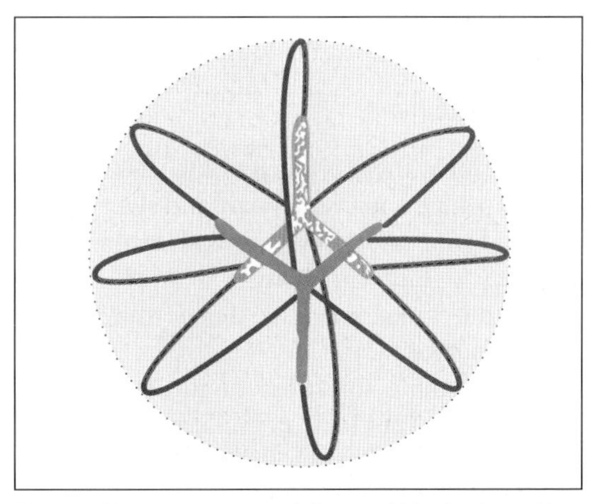

Fig. 22.13: Arrangement of fibres within the lens. Note the Y-shaped lines on the front and back of the lens (Schematic representation)

ends of lens fibres are attached at these lines. Each lens fibre starts on one surface at such a line, and follows a curved course to reach the opposite surface where it ends by joining another such line.

Clinical Correlation

Cataract

The cataract is the opacification of the normally crystalline lens which leads to gradual painless blurring of vision. The various causes of cataract are: senility, congenital (e.g. Down syndrome, rubella, galactosaemia), traumatic (e.g. penetrating injury, electrical injury), metabolic (e.g. diabetes, hypoparathyroidism), drug associated (e.g. long-term corticosteroid therapy), smoking and heavy alcohol consumption.

ACCESSORY VISUAL ORGANS

The accessory visual organs include the extraocular muscles and related fascia, the eyebrows, the eyelids, the conjunctiva, and the lacrimal gland. The structure of extraocular muscles corresponds to that of skeletal muscle elsewhere in the body; and the structure of eyebrows is similar to that of hair in other parts of the body. The remaining structures are considered below.

Eyelids

Eyelids are two movable skin folds that protect the eye from injury and keep the cornea clean and moist. The basic structure of an eyelid is shown in Fig. 22.14.

- Anteriorly, there is a layer of true skin with which a few small hair and sweat glands are associated. The skin is thin.
- Deep to the skin there is a layer of delicate connective tissue that normally does not contain fat.
- Considerable thickness of the lid is formed by fasciculi of the palpebral part of the orbicularis oculi muscle (skeletal muscle).
- The 'skeleton' of each eyelid is formed by a mass of fibrous tissue called the ***tarsus***, or ***tarsal plate***.

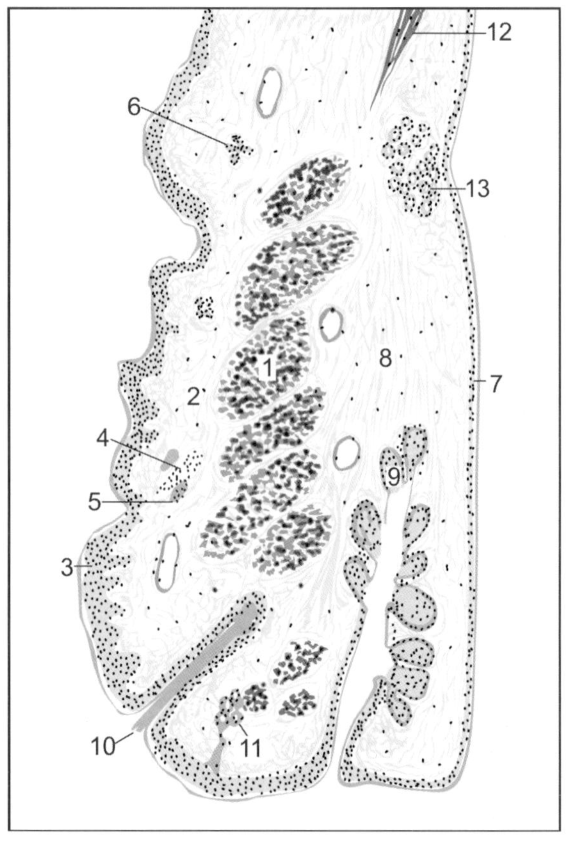

Fig. 22.14: Eyelid (Schematic representation). 1–core of skeletal muscle; 2–dense connective tissue; 3–skin; 4–hair follicle; 5–Sebaceous gland; 6–Sweat gland; 7–Palpebral conjunctiva; 8–Tarsal plate; 9–Tarsal glands; 10–Eyelash; 11–Ciliary gland; 12–Levator palpebrae superioris; 13–Accessory lacrimal glands

- On the deep surface of the tarsal plate there are a series of vertical grooves in which ***tarsal glands*** (or ***Meibomian glands***) are lodged. Occasionally, these glands may be embedded within the tarsal plate. Each gland has a duct that opens at the free margin of the lid. The tarsal glands are modified sebaceous glands. They produce an oily secretion a thin film of which spreads over the lacrimal fluid (in the conjunctival sac) and delays its evaporation.
- Modified sweat glands, called ***ciliary glands*** (or ***glands of Moll***), are present in the lid near its free edge. Sebaceous glands present in relation to eyelashes constitute the ***glands of Zeis***. They open into hair follicles. Accessory lacrimal glands are often present just above the tarsal plate (***glands of Wolfring***).
- The inner surface of the eyelid is lined by the palpebral conjunctiva.

Clinical Correlation

Stye (Hordeolum)

Stye or 'external hordeolum' is an acute suppurative inflammation of the sebaceous glands of Zeis, the apocrine glands of Moll and the eyelash follicles.

Chalazion

Chalazion is a very common lesion and is the chronic inflammatory process involving the meibomian glands. It occurs as a result of obstruction to the drainage of secretions. The inflammatory process begins with destruction of meibomian glands and duct and subsequently involves tarsal plate.

Conjunctiva

The conjunctiva is a thin transparent membrane that covers the inner surface of each eyelid (***palpebral conjunctiva***) and the anterior part of the sclera (***ocular conjunctiva***). At the free margin of the eyelid the palpebral conjunctiva becomes continuous with skin; and at the margin of the cornea the ocular conjunctiva becomes continuous with the anterior epithelium of the cornea. When the eyelids are closed the conjunctiva forms a closed ***conjunctival sac***. The line along which palpebral conjunctiva is reflected onto the eyeball is called the ***conjunctival fornix***: superior, or inferior.

Conjunctiva consists of an epithelial lining that rests on connective tissue. Over the eyelids this connective tissue is highly vascular and contains much lymphoid tissue. It is much less vascular over the sclera.

The epithelium lining the palpebral conjunctiva is typically two layered. There is a superficial layer of columnar cells, and a deeper layer of flattened cells. At the fornix, and over the sclera, the epithelium is three layered there being an additional layer of polygonal cells between the two layers mentioned above. The three layered epithelium changes to stratified squamous at the sclerocorneal junction.

Lacrimal Gland

The lacrimal gland is a compound tubuloalveolar gland and consists of a number of lobes that drain through about twenty ducts. It is a tear-secreting gland. The structure of the lacrimal gland is similar to that of a serous salivary gland (Fig. 22.15).

Sections of the lacrimal gland can be distinguished from those of serous salivary glands because of the following features.

- The acini are larger, and have wider lumina.
- All cells appear to be of the same type. They are low columnar in shape and stain pink with haematoxylin and eosin.
- The profiles of the acini are often irregular or elongated.
- The walls of adjacent acini within a lobule may be pressed together, there being very little connective tissue between them. However, the acini of different lobules are widely separated by connective tissue. Myoepithelial cells are present as in salivary glands.

Small ducts of the lacrimal gland are lined by cuboidal or columnar epithelium. Larger ducts have a two layered columnar epithelium or a pseudostratified columnar epithelium.

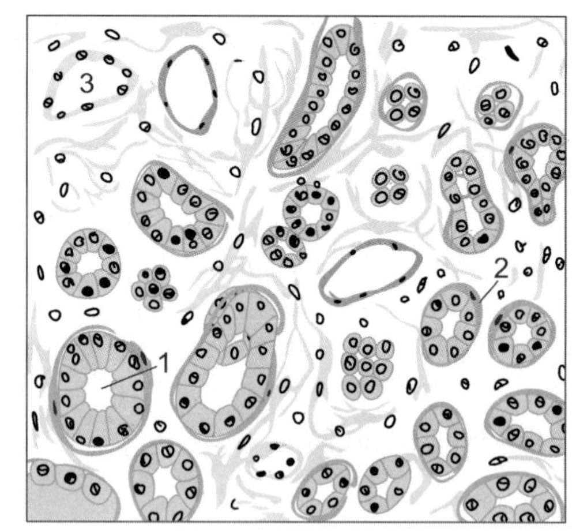

Fig. 22.15: Lacrimal gland (Schematic representation) 1–lumen of acinus; 2–myoepithelial cell; 3–duct

EM studies on the human lacrimal gland reveal that the secretory cells may be of several types, including both mucous and serous cells.

The ducts of the lacrimal gland open into the lateral part of the superior conjunctival fornix. Lacrimal fluid keeps the conjunctiva moist. Accessory lacrimal glands are present near the superior conjunctival fornix (***glands of Krause***).

Added Information

Density of Rods and Cones in Retina

The density of rods and cones in different parts of the retina is shown schematically in Fig. 22.16. Note the following points:

- The density of cones is greatest in the fovea (about 1.5 million/mm^2). Their density decreases sharply in proceeding to the margin of the central area, but thereafter the density is uniform up to the ora serrata (about 5000/mm^2).
- The density of rods is greatest at the margin of the central area (about 1.5 million/mm^2). It decreases sharply on proceeding towards the margin of the central area. There are no rods in the foveola. The density of rods also decreases in passing towards the ora serrata (where it is about 30,000/mm^2).

Bipolar Neurons

Bipolar cells of the retina are of various types. The terminology used for them is confusing as it is based on multiple criteria. The main points to note are as follows:

- The primary division is into bipolars that synapse with rods (rod-bipolars), and those that synapse with cones (cone-bipolars).
- As there are three types of cones, responding to the colours red, green and blue we can distinguish three corresponding types of cone bipolars (red cone bipolar, green cone bipolar, blue cone bipolar).
- When a photoreceptor (rod or cone) is exposed to light it releases neurotransmitter at its synapse with the bipolar cell. Some bipolars respond to neurotransmitter by depolarisation (and secretion of neurotransmitter at their synapses with ganglion cells). These are called ON-bipolars as they are 'switched on' by light. Other bipolars respond to release of neurotransmitter by hyperpolarisation. In other words they are 'switched off' by light and are called OFF-bipolars.

Contd...

- On the basis of structural characteristics, and the synapses established by them, cone bipolars are divided into three types: ***midget***, ***blue cone*** and ***diffuse***.
  - A midget bipolar establishes synapses with a single cone (which may be red or green sensitive). Some midget bipolars synapse with indented areas on cone pedicles forming triads (Fig. 22.11). These are ON-bipolars. Other midget bipolars establish 'flat' synapses with the cone pedicle (and are also referred to as flat-bipolars). These are OFF-bipolars.
  - A blue cone bipolar connects to one blue cone, and establishes triads. It may be of the ON or OFF variety.
  - Diffuse cone bipolars establish synapses with several cone pedicles. They are not colour specific.

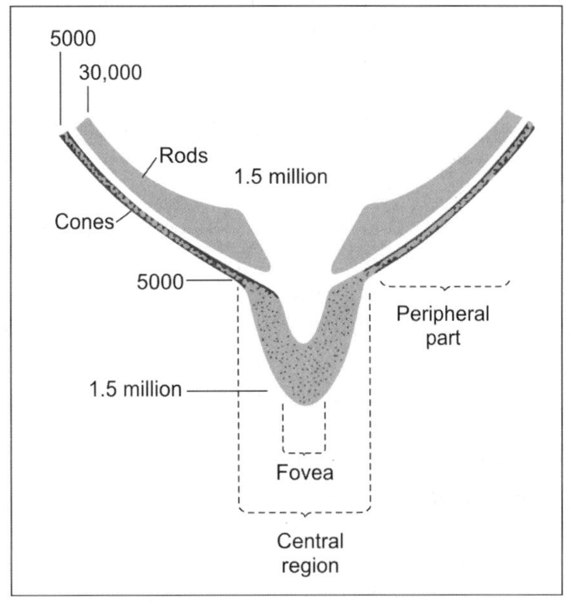

Fig. 22.16: Scheme to show the relative number of rods and cones in different parts of the retina. The figures represent number of receptors per mm^2.The diagram is not drawn to scale

Axons of rod bipolar neurons synapse with up to four ganglion cells, but those of one midget bipolar neuron synapse with only one (midget) ganglion cell, and with amacrine neurons.

Ganglion Cells

We have seen that the dendrites of ganglion cells synapse with axons of bipolar cells, and also with processes of amacrine cells. The axons arising from ganglion cells constitute the fibres of the optic nerve.

Ganglion cells are of two main types. Those that synapse with only one bipolar neuron are ***mono-synaptic***, while those that synapse with many bipolar neurons are ***polysynaptic***. Monosynaptic ganglion cells are also called ***midget ganglion cells***. Each of them synapses with one midget bipolar neuron. We have seen that midget bipolars in turn receive impulses from a single cone. This arrangement is usual in the central region of the retina, and allows high resolution of vision to be attained.

Polysynaptic ganglion cells are of various types. Some of them synapse only with rod bipolars (***rod ganglion cells***). Others have very wide dendritic ramifications that may synapse with several hundred bipolar neurons (***diffuse ganglion cells***). This arrangement allows for summation of stimuli received through very large numbers of photoreceptors facilitating vision in poor light. On physiological grounds ganglion cells are also classified as 'ON' or 'OFF' cells.

Horizontal Neurons

Horizontal neurons establish numerous connections between photoreceptors (Fig. 22.17). Some of them are excitatory, while others are inhibitory. In this way these neurons play a role in integrating the activity of photorecepors located in adjacent parts of the retina. As they participate in synapses between photoreceptors and bipolar neurons horizontal neurons may regulate synaptic transmission between these cells.

Contd...

Amacrine Neurons

Different types of amacrine neurons are recognised depending upon the pattern of branching. We have seen that the processes of amacrine neurons enter the internal plexiform layer where they may synapse with axons of several bipolar cells, and with the dendrites of several ganglion cells (Fig. 22.18). They also synapse with other amacrine cells. At many places an amacrine process synapsing with a ganglion cell is accompanied by a bipolar cell axon. The two are referred to as a ***dyad.***

Internal plexiform cells (present in the internal plexiform layer) represent a third variety of horizontally oriented neurons in the retina.

Apart from integration of impulses from rods and cones horizontal, amacrine and internal plexiform cells act as 'gates' that can modulate passage of inputs from rods and cones to ganglion cells. In this connection it is to be noted that processes of amacrine neurons are interposed between processes of bipolar cells and ganglion cells, while processes of horizontal cells are interposed between photoreceptors and bipolar cells.

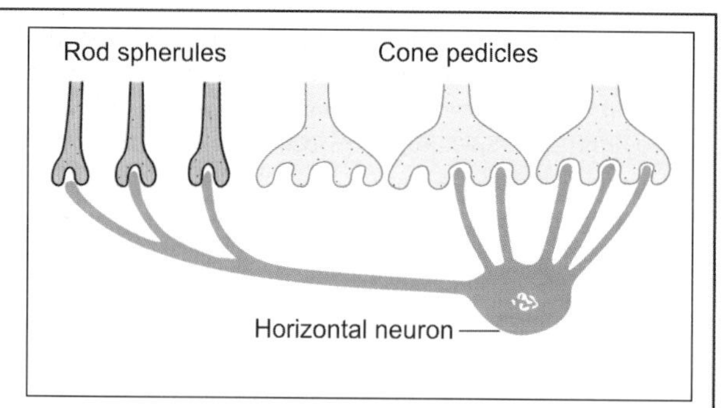

Fig. 22.17: Connections of a cone horizontal neuron (Schematic representation)

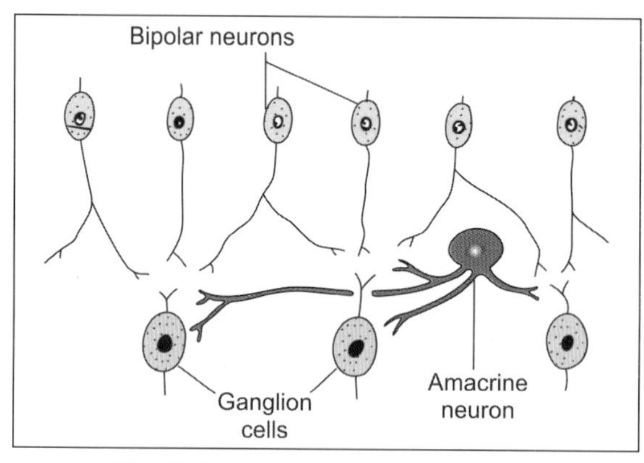

Fig. 22.18: Connections of an amacrine neuron (Schematic representation)

Some Further Remarks About Connections of Retinal Neurons

- While there are well over a hundred million photoreceptors in each retina there are only about one million ganglion cells, each giving origin to one fibre of the optic nerve. (The bipolar cells are intermediate in number between photoreceptors and ganglion cells). In passing from the photoreceptors to the ganglion cells there has, therefore, to be considerable convergence of impulses (Fig. 22.19). Each ganglion cell would be influenced by impulses originating in several photoreceptors. On functional considerations it would be expected that such convergence would be most marked near the periphery of the retina; and that it would involve the rods much

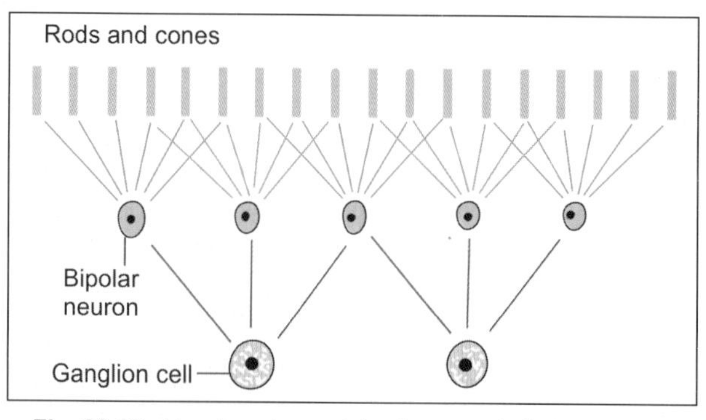

Fig. 22.19: How impulses arising in several photoreceptors concentrate on one ganglion cell (Schematic representation)

Contd...

more than the cones. It has been estimated that in the peripheral parts of the retina one ganglion cell may be connected to as many as 300 rods or to 10 cones. Convergence leads to summation of impulses arising in many photoreceptors and allows vision even in very dim light. It would also be expected that convergence would be minimal in the macula, and absent in the foveola, to allow maximal resolution.

- The second highly important fact about intra-retinal connections is the presence of numerous arrangements for interaction of adjacent regions of the retina as follows:
  - Firstly, cone pedicles establish numerous contacts with other cone pedicles and with adjacent rod spherules.
  - Except in the fovea, most photoreceptors are connected to more than one bipolar cell. In turn each bipolar cell is usually connected to more than one ganglion cell.
  - The vertically arranged elements of the retina (photoreceptors, bipolar cells, ganglion cells) are intimately interconnected to adjacent elements through horizontal neurons and amacrine neurons.

Mechanism of Firing of Bipolar Neurons

- When no light falls on the retina photoreceptors are depolarised. Exposure to light causes hyperpolarisation.
- When a photoreceptor is depolarised it releases inhibitor at its junction with a bipolar neuron. This prevents the bipolar neuron from firing. Release of inhibitor is controlled by voltage gated calcium channels.
- Hyperpolarisation of photoreceptor, caused by exposure to light, leads to closure of Ca++ gates and release of inhibitor is stopped. This causes the bipolar neuron to fire. As explained earlier, this description applies to ON-bipolars.
- Rhodopsin, present in photoreceptors, is a complex of a protein ***opsin*** and ***cis-retinal*** that is sensitive to light. When exposed to light cis-retinal is transformed to trans-retinal. This leads to decrease in concentration of cyclic GMP that in turn leads to closure of sodium channels. Closure of sodium channels results in hyperpolarisation of photo-receptor (Flow chart 22.1).

Flow chart. 22.1: Mechanism of firing of bipolar neurons

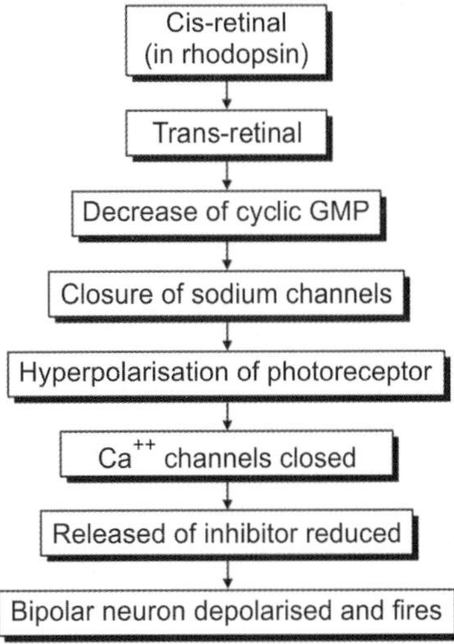

Chapter 23

Special Senses: Ear

INTRODUCTION

Ear is the peripheral sense organ concerned with hearing and equilibrium. Anatomically speaking, the ear is made up of three main parts called the ***external ear***, the ***middle ear***, and the ***internal ear***. The external and middle ears are concerned exclusively with hearing. The internal ear has a ***cochlear part*** concerned with hearing; and a ***vestibular part*** which provides information to the brain regarding the position and movements of the head.

The main parts of the ear are shown in Figure 23.1.

External Ear

The external ear consists of ***auricle*** or ***pinna*** and ***external acoustic meatus (external auditory canal).***

The part of the ear that is seen on the surface of the body (i.e., the part that the lay person calls the ear) is anatomically speaking, the auricle or pinna. Leading inwards from the auricle there is a tube called the external acoustic meatus.

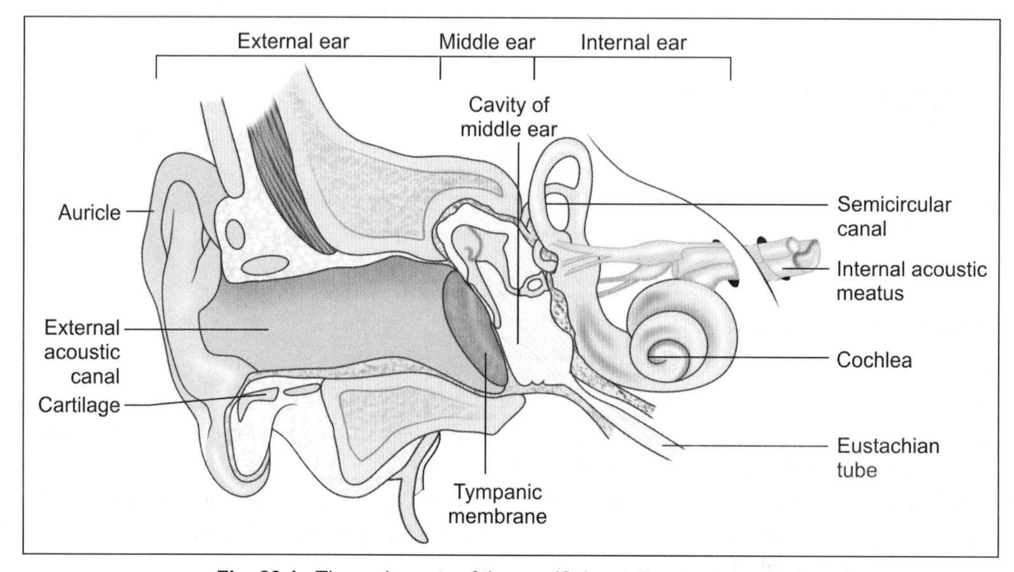

Fig. 23.1: The main parts of the ear (Schematic representation)

The inner end of the external acoustic meatus is closed by a thin membranous diaphragm called the ***tympanic membrane***. This membrane separates the external acoustic meatus from the middle ear.

Middle Ear

The ***middle ear*** is a small space placed deep within the petrous part of the temporal bone. It is also called the ***tympanum***.

Medially, the middle ear is closely related to parts of the internal ear. It is lined with mucous membrane.

The cavity of the middle ear is continuous with that of the nasopharynx through a passage called the ***auditory tube***. Within the cavity of the middle ear there are three small bones or ***ossicles***: the ***malleus***, the ***incus***, and the ***stapes***. They form a chain that is attached on one side to the tympanic membrane, and at the other end to a part of the internal ear.

Internal Ear

The ***internal ear*** is in the form of a complex system of cavities lying within the petrous temporal bone. It has sense organs for both hearing and balance. It has a central part called the ***vestibule***. Continuous with the front of the vestibule there is a spiral shaped cavity called the ***cochlea***. Posteriorly, the vestibule is continuous with three ***semicircular canals***.

THE EXTERNAL EAR

The Auricle (Pinna)

The auricle consists of a thin plate of elastic cartilage covered on both sides by true skin (Plate 23.1). The skin is closely adherent to the cartilage on its lateral surface while it is comparatively loose on medial surface. Epithelium is squamous keratinising. Hair follicles, sebaceous glands, and sweat glands are present in the skin, adipose tissue is present only in lobule.

Clinical Correlation

- ***Grafts in rhinoplasty***: The conchal cartilage is frequently used to correct depressed nasal bridge. The composite grafts of the skin and cartilage can be used for repair of defects of ala of nose.
- ***Grafts in tympanoplasty***: Tragal and conchal cartilage and perichondrium and fat from lobule are often used during tympanoplasty operations.

The External Acoustic Meatus

Dimensions: External auditory canal (EAC) measures about 24 mm and extends from the concha to the tympanic membrane.

EAC is usually divided into 2 parts: (1) cartilaginous and (2) bony. Its outer one-third (8 mm) is cartilaginous and its inner two-third (16 mm) is bony.

- ***Cartilaginous EAC:*** It is a continuation of the cartilage that forms the framework of the pinna. The skin of the cartilaginous canal is thick and contains hair follicles, ceruminous and pilosebaceous glands that secrete wax.

PLATE 23.1: Pinna

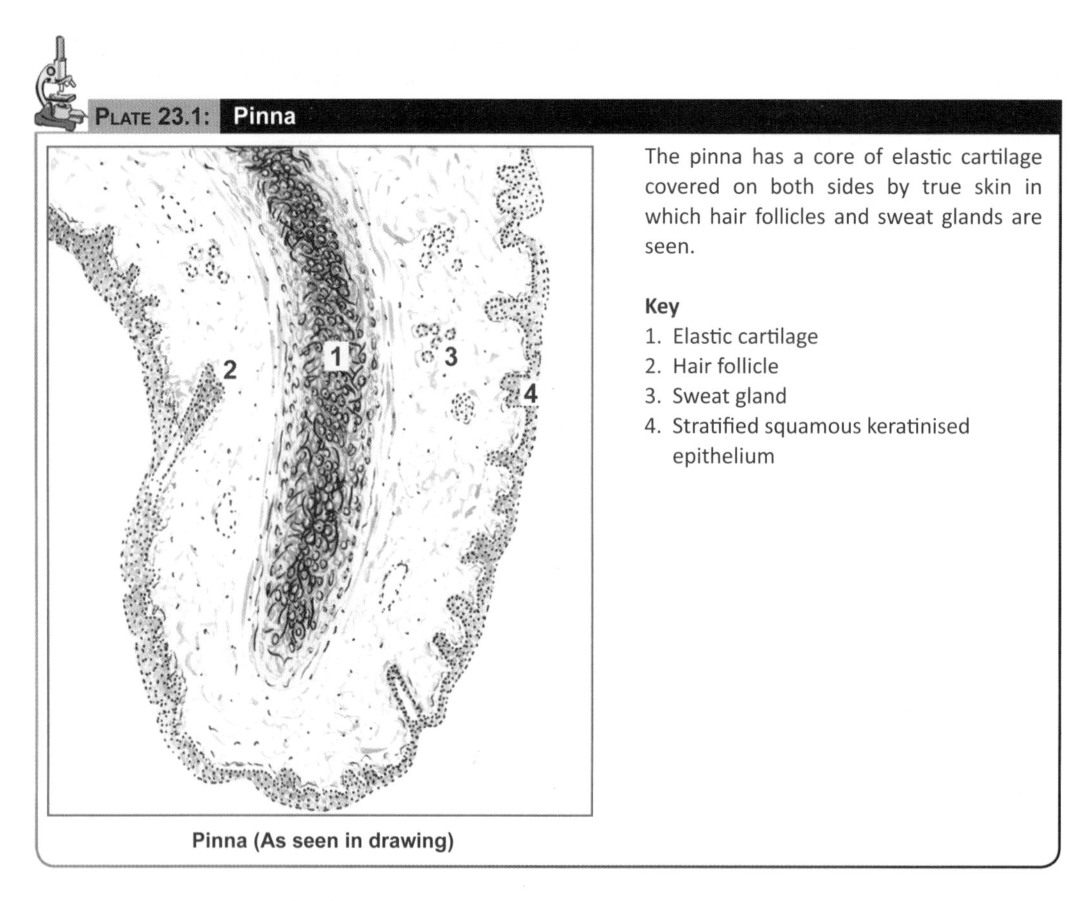

Pinna (As seen in drawing)

The pinna has a core of elastic cartilage covered on both sides by true skin in which hair follicles and sweat glands are seen.

Key

1. Elastic cartilage
2. Hair follicle
3. Sweat gland
4. Stratified squamous keratinised epithelium

Note: The ceruminous glands secrete the wax of the ear. They are modified sweat glands lined by a columnar, cuboidal or squamous epithelium.

Clinical Correlation

- ***Fissures of Santorini*:** Transverse slits in the floor of cartilaginous EAC called "fissures of Santorini" provide passages for infections and neoplasms to and from the surrounding soft tissue (especially parotid gland). The parotid and mastoid infections can manifest in the EAC.
- ***Hair follicles*** are present only in the outer cartilaginous canal and therefore furuncles (staphylococcal infection of hair follicles) are seen only in the cartilaginous EAC.

- ***Bony EAC:*** It is mainly formed by the tympanic portion of temporal bone but roof is formed by the squamous part of the temporal bone.
  - Skin of the bony EAC is thin and continuous over the tympanic membrane skin is devoid of subcutaneous layer, hair follicles and ceruminous glands.
  - ***Isthmus:*** Approximately 6 mm lateral to tympanic membrane, bony EAC has a narrowing called the isthmus.

Clinical Correlation

- Foreign body impacted medial to bony isthmus of EAC are difficult to remove.
- ***Foramen of Huschke*:** In children and occasionally in adults, anteroinferior bony EAC may have a deficiency that is called foramen of Huschke. Foramen of Huschke permits spread of infections to and from EAC and parotid.

Note: The skin of EAC has a unique self-cleansing mechanism. This migratory process continues from the medial to lateral side. The sloughed epithelium is extruded out as a component of cerumen.

The Tympanic Membrane

- ***Dimensions:*** Its dimensions are: 9–10 mm height and 8–9 mm width. It is 0.1 mm thick.
- ***Position:*** Tympanic membrane (TM) is a partition wall between the EAC and the middle ear. It is positioned obliquely. It forms angle of 55° with deep EAC.
- ***Structure:*** Tympanic membrane consists of the following three layers:
  - ***Outer epithelial layer:*** It is continuous with the EAC skin.
  - ***Middle fibrous layer:*** The middle layer is made up of fibrous tissue, which is lined on the outside by skin (continuous with that of the external acoustic meatus), and on the inside by mucous membrane of the tympanic cavity.

    The fibrous layer contains collagen fibres and some elastic fibres. The fibres are arranged in two layers. In the outer layer they are placed radially, while in the inner layer they run circularly.
  - ***Inner mucosal layer:*** The mucous membrane is lined by an epithelium which may be cuboidal or squamous. It is said that the mucosa over the upper part of the tympanic membrane may have patches of ciliated columnar epithelium, but this is not borne out by EM studies.
- ***Otoscopy:*** Normal tympanic membrane is shiny and pearly-grey in colour. Its transparency varies from person to person.

MIDDLE EAR

The Tympanic Cavity

The walls of the tympanic cavity are formed by bone which is lined by mucous membrane. The mucous membrane also covers the ossicles. The lining epithelium varies from region to region. Typically it is cuboidal or squamous. At places it may be ciliated columnar. The ossicles of the middle ear consist of compact bone, but do not have marrow cavities.

The Auditory Tube (Eustachian Tube)

It is a channel connecting the tympanic cavity with the nasopharynx. The length of eustachian tube (ET) is 36 mm. Its lateral third is bony and medial 2/3 (i.e. 24 mm) is fibrocartilaginous.

The bone or cartilage is covered by mucous membrane which is lined by ciliated columnar epithelium. Near the pharyngeal end of the tube the epithelium becomes pseudostratified columnar. Goblet cells and tubuloalveolar mucous glands are also present. A substantial collection of lymphoid tissue, present at the pharyngeal end, forms the ***tubal tonsil***.

THE INTERNAL EAR

Introduction

The internal ear is in the form of a complex system of cavities within the petrous temporal bone. Because of the complex shape of these intercommunicating cavities the internal ear is also called the ***labyrinth***.

It consists of a bony labyrinth contained within the petrous part of temporal bone.

Note: The basic structure of the labyrinth is best understood by looking at a transverse section through a relatively simple part of it e.g., a semicircular canal (Fig. 23.2). The space bounded by bone is ***bony labyrinth.*** Its wall is made up of bone that is more dense than the surrounding bone. Its inner surface is lined by periosteum.

Lying within the bony labyrinth there is a system of ducts which constitute the ***membranous labyrinth***. The space within the membranous labyrinth is filled by a fluid called the ***endolymph***. The space between the membranous labyrinth and the bony labyrinth is filled by another fluid called the ***perilymph***.

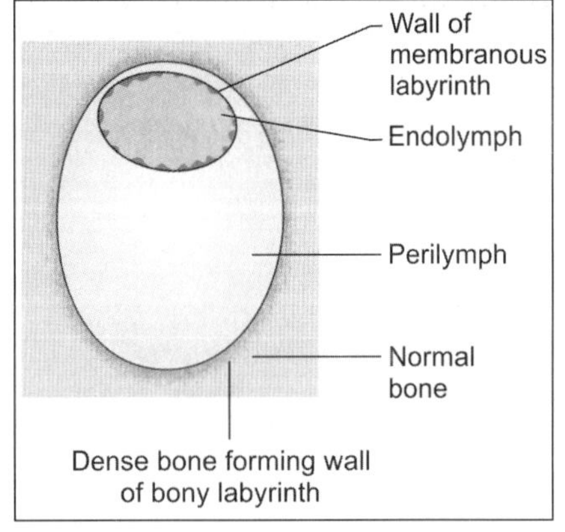

Fig. 23.2: Basic structure of internal ear as seen in a section through a semicircular canal (Schematic representation)

Bony Labyrinth

The bony labyrinth consists of three parts:

- Vestibule
- Semicircular canals
- The bony cochlea

Vestibule

Vestibule is the central part (Fig. 23.3). It is continuous anteriorly with the ***cochlea***; and posteriorly with three ***semicircular canals***.

Semicircular Canal

There are three semicircular canals (SCCs): lateral (horizontal), posterior and superior (anterior). Each canal occupies 2/3rd of a circle and has a diameter of 0.8 mm. They lie in planes at right angles to one another. Each canal has two ends: ampullated and non-ampullated. All the three ampullated ends and non-ampullated end of lateral SCC open independently and directly into the vestibule.

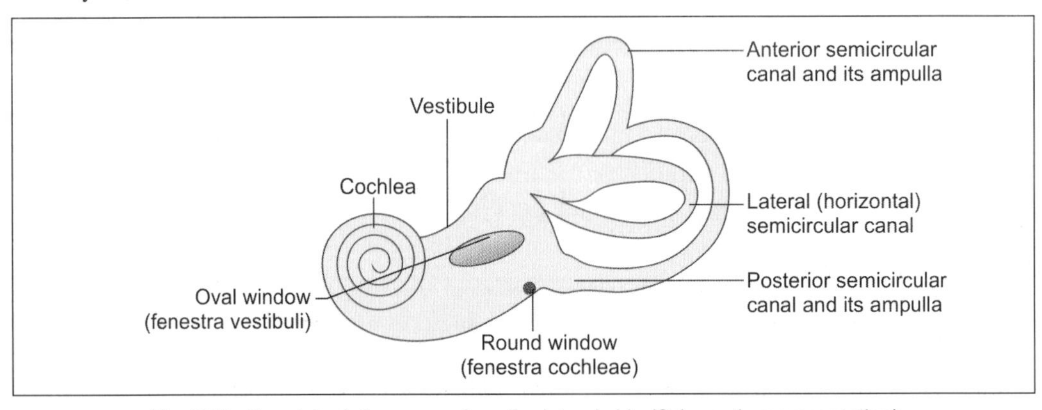

Fig. 23.3: Bony labyrinth as seen from the lateral side (Schematic representation)

Note: The non-ampullated ends of posterior and superior canals join and form a crus commune (4 mm length), which then opens into the medial part of vestibule. So, the three SCCs open into the vestibule by five openings.

The Bony Cochlea

The cochlea has a striking resemblance to a snail shell. It is basically a tube that is coiled on itself for two and three-fourth turns. The 'turns' rest on a solid core of bone called the ***modiolus***.

Note: Because of the spiral nature of the cochlea the mutual relationships of the structures within it differ in different parts of the cochlea. A structure that is 'inferior' in the upper part of the canal becomes 'superior' in the lower part. For descriptive convenience the structures lying next to the modiolus are described as 'inner' and those away from it as 'outer'. These terms as used here are not equivalents of 'medial' and 'lateral' as normally used. The words 'superior' and 'inferior' indicate relationships as they exist in the lowest (or basal) turn of the cochlea.

In sections, through the middle of the cochlea the cochlear canal is cut up six times as shown in Plate 23.2. The cochlear canal is partially divided into two parts by a bony lamina that projects outwards from the modiolus. This bony projection is called the ***spiral lamina***. Passing from the tip of the spiral lamina to the opposite wall of the canal there is the ***basilar membrane***.

The spiral lamina and the basilar membrane together divide the cochlear canal into three parts—scala vestibuli, scala tympani and scala media (membranous cochlea). The lower most channel is the ***scala tympani***. When traced proximally, the scala tympani opens in the medial wall of the middle ear through an aperture the ***fenestra cochleae (round window)***, which is closed by the ***secondary tympanic membrane***.

The part of the cochlear canal above the basilar membrane is further divided into two parts by an obliquely placed ***vestibular membrane*** (***of Reissner***). The part above the vestibular membrane is the ***scala vestibuli***. When traced proximally it becomes continuous with the vestibule (Fig 23.4). At the apex of the cochlea the scala vestibuli becomes continuous with the scala tympani called helicotrema. Both scala vestibuli and scala tympani are filled with perilymph. The triangular space between the basilar and vestibular membranes is called the ***duct of the cochlea***. This duct represents the membranous labyrinth of the cochlea and contains endolymph.

The vestibular membrane consists of a basal lamina lined on either side by squamous cells. Some of the cells show an ultrastructure indicative of a fluid transport function. The cells of the membrane form a barrier to the flow of ions between endolymph and perilymph so that these two fluids have different concentrations of electrolytes.

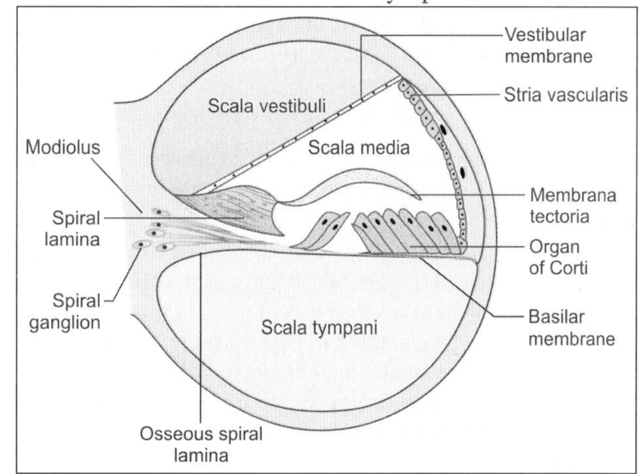

Fig. 23.4: Structure of cochlear canal to show scala vestibuli and tympani (Schematic representation)

PLATE 23.2: Cochlea

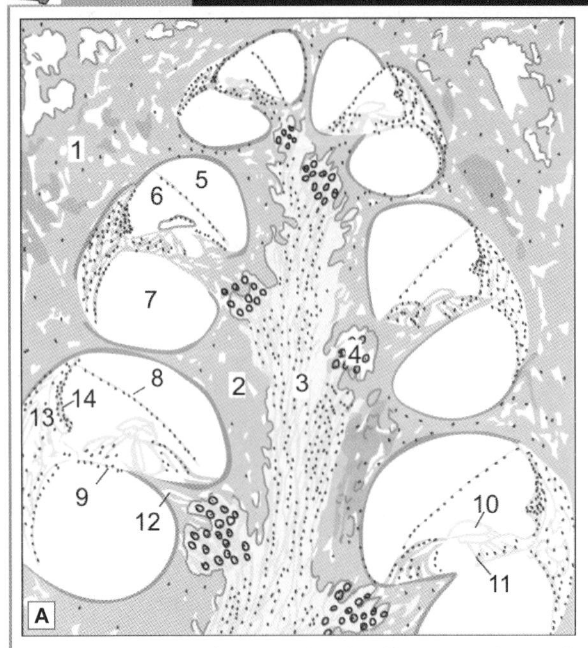

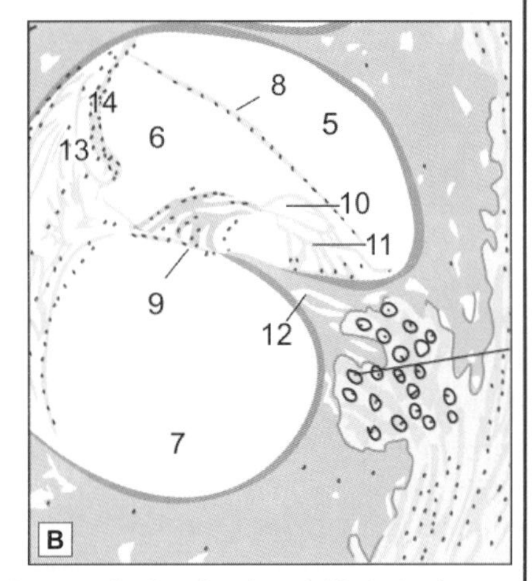

Cochlea. A. As seen in drawing (low magnification); B. As seen in drawing (magnified view)

Key

1. Petrous temporal bone
2. Modiolus
3. Canal for passage of cochlear nerve fibres
4. Spiral ganglion
5. Scala vestibuli
6. Scala media
7. Scala tympani
8. Vestibular membrane
9. Basilar membrane
10. Membrana tectoria
11. Organ of Corti
12. Spiral lamina
13. Spiral ligament
14. Stria vascularis

A low power view to show the general structure of the cochlea:

- The cochlea is embedded in the petrous temporal bone
- It is in the form of a spiral canal and is, therefore, cut up six times
- The cone-shaped mass of bone surrounded by these turns of the cochlea is called the modiolus which contains a canal through which fibres of the cochlear nerve pass
- A mass of neurons belonging to the spiral ganglion lies to the inner side of each turn of the cochlea
- The parts to be identified in each turn of the cochlea are the scala vestibuli, scala media, the scala tympani, the vestibular membrane, the basilar membrane, the membrana tectoria, and the organ of Corti, and the spiral lamina
- Outer wall of the cochlear turn is the spiral ligament and it is lined by avascularised epithelium (stria vascularis).

The basilar membrane is divisible into two parts. The part supporting the organ of Corti is the ***zona arcuata***. The part lateral to the zona arcuata is the ***zona pectinata***. The zona arcuata is made up of a single layer of delicate filaments of collagen. The zona pectinata is made up of three layers of fibres.

Note: ***Aqueduct of cochlea:*** The scala tympani is connected with the subarachnoid space through the aqueduct of cochlea. It is thought to regulate perilymph and pressure in bony labyrinth.

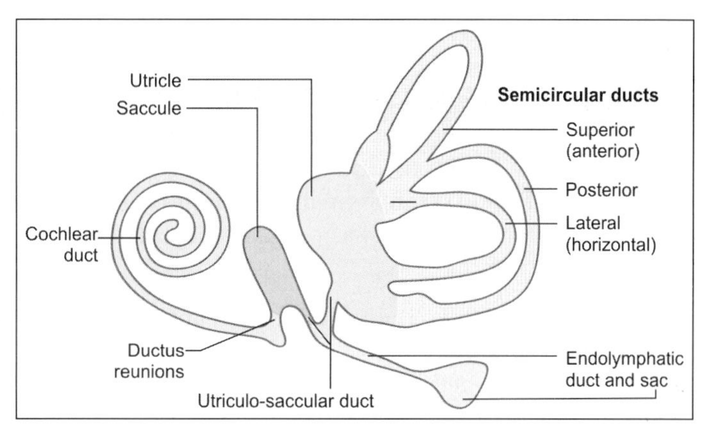

Fig. 23.5: Parts of the membranous labyrinth (Schematic representation)

Membranous Labyrinth

Membranous labyrinth (Fig. 23.5) consists of cochlear duct, utricle, saccule, three semicircular ducts, and endolymphatic duct and sac.

The parts of the membranous labyrinth are shown in Figure 23.6. Within each semicircular canal the membranous labyrinth is represented by a ***semicircular duct***. The part of the membranous labyrinth present in the cochlea is called the ***duct of the cochlea (membranous cochlea or scala media)***. The part of the membranous labyrinth that lies within the vestibule is in the form of two distinct membranous sacs called the ***saccule*** and the ***utricle***.

- ***Utricle:*** The utricle, which is oblong and irregular, has anteriorly upward slope at an approximate angle of 30°. It lies in the posterior part of bony vestibule and receives the five openings of the three semicircular ducts. The utricle (4.33 mm^2) is bigger than saccule (2.4 mm^2) and lies superior to saccule. The utricle is connected with the saccule through utriculosaccular duct. Its sensory epithelium, which is called macula, is concerned with linear acceleration and deceleration.
- ***Saccule:*** The saccule lies anterior to the utricle opposite the stapes footplate in the bony vestibule. Its sensory epithelium, macula responds to linear acceleration and deceleration. The saccule is connected to the cochlea through the thin reunion duct.

The wall of the membranous labyrinth is trilaminar. The outer layer is fibrous and is covered with ***perilymphatic cells.*** The middle layer is vascular. The inner layer is epithelial, the lining cells being squamous or cuboidal. Some of the cells (called ***dark cells***) have an ultrastructure indicative of active ionic transport. They probably control the ionic composition of endolymph.

Inner Ear Fluids

Perilymph fills the space between bony and membranous labyrinth while endolymph fills the entire membranous labyrinth.

Perilymph

It resembles extracellular fluid and is rich in sodium ions. The aqueduct of cochlea provides communication between scala tympani and subarachnoid space. Perilymph percolates through the arachnoid type connective tissue present in the aqueduct of cochlea.

- ***Source*:** There are two theories:
  - Filtrate of blood serum from the capillaries of spiral ligament.
  - CSF reaching labyrinth via aqueduct of cochlea.

Endolymph

It resembles intracellular fluid and is rich in potassium ions. Protein and glucose contents are less than in perilymph.

- ***Source*:** They are believed to be following:
  - Stria vascularis
  - Dark cells of utricle and ampullated ends of semicircular ducts.
- ***Absorption*:** There are following two opinions regarding the absorption of endolymph:
  - ***Endolymphatic sac:*** The longitudinal flow theory believes that from cochlear duct endolymph reaches saccule, utricle and endolymphatic duct and is then absorbed by endolymphatic sac.
  - ***Stria vascularis:*** The radial flow theory believes that endolymph is secreted as well as absorbed by the stria vascularis.

The Spiral Organ of Corti

This sensory organ of the hearing is situated on the basilar membrane. It is spread like a ribbon along the entire length of basilar membrane (Figs 23.4 and 23.6).

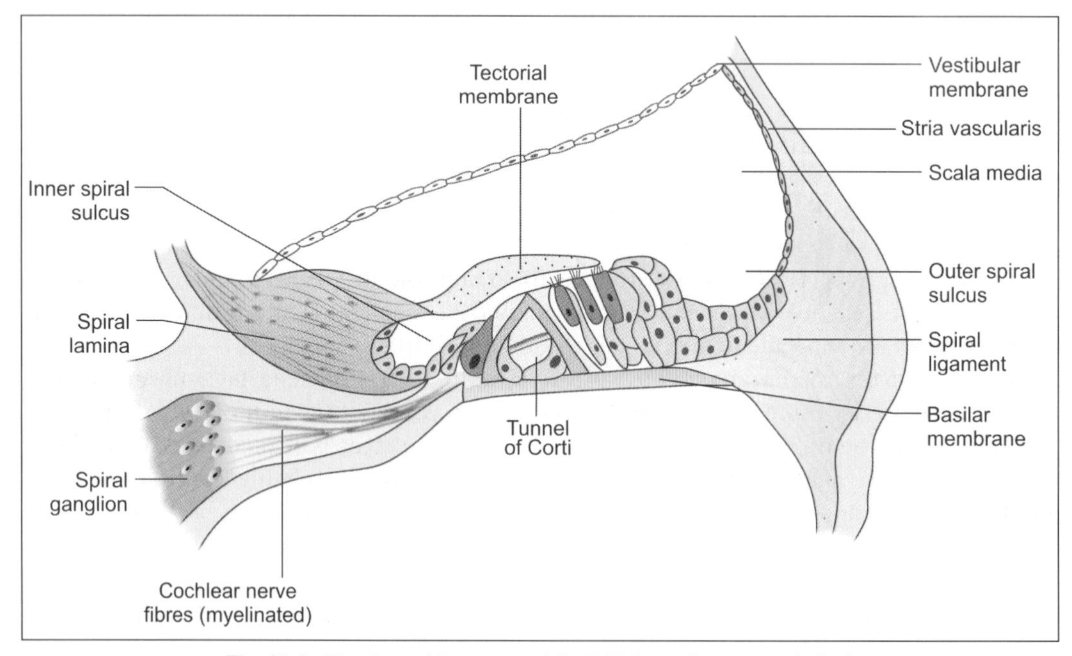

Fig. 23.6: Structure of the organ of Corti (Schematic representation)

The spiral organ of Corti is so called because (like other structures in the cochlea) it extends in a spiral manner through the turns of the cochlea. It is made up of epithelial cells that are arranged in a complicated manner. The cells are divisible into the true receptor cells or ***hair cells***, and supporting elements which are given different names depending on their location. The cells of the spiral organ are covered from above by a gelatinous mass called the ***membrana tectoria***. It consists of delicate fibres embedded in a gelatinous matrix. This material is probably secreted by cells lining the vestibular lip of the limbus lamina spiralis.

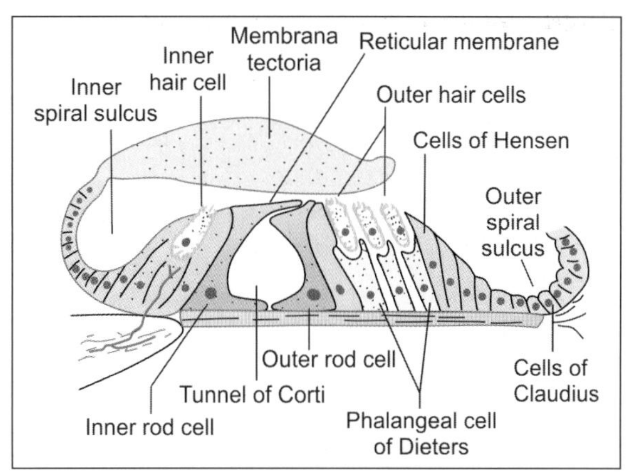

Fig. 23.7: The cells in the organ of Corti (Schematic representation)

In Figure 23.7 a tunnel can be seen called the tunnel of Corti, which is situated between the inner and outer rod cells and contains a fluid called cortilymph. The base of the tunnel lies over the basilar membrane.

To the internal side of the inner rod cells there is a single row of ***inner hair cells***. The inner hair cell is supported by tall cells lining the tympanic lip of the internal spiral sulcus.

On the outer side of each external rod cell there are three or four ***outer hair cells***. The outer hair cells do not lie directly on the basilar membrane, but are supported by the ***phalangeal cells*** (***of Dieters***) which rest on the basilar membrane. To the outer side of the outer hair cells and the phalangeal cells, there are tall supporting cells (***cells of Hensen***). Still more externally the outer spiral sulcus is lined by cubical cells (***cells of Claudius***).

A narrow space the ***cuniculum externum*** intervenes between the outermost hair cells and the cells of Hensen. A third space, the ***cuniculum medium*** (or ***space of Nuel***) lies between the outer rod cell and the outer hair cells. The spaces are filled with perilymph (or ***cortilymph***).

We will now examine some of the structures mentioned above in greater detail.

Rod Cells (Fig. 23.7)

Each rod cell (or ***pillar cell***) has a broad ***base*** (or ***footplate***, or ***crus***) that rests on the basilar membrane; an elongated middle part (***rod*** or ***scapus***); and an expanded upper end called the ***head*** or ***caput***.

The bases of the rod cells are greatly expanded and contain their nuclei. The bases of the inner and outer rod cells meet each other forming the base of the tunnel of Corti. The heads of these cells also meet at the apex of the tunnel. Here a convex prominence on the head of the outer rod cell fits into a concavity on the head of the inner rod cell. The uppermost parts of the heads are expanded into horizontal plates called the ***phalangeal processes***. These processes join similar processes of neighbouring cells to form a continuous membrane called the ***reticular lamina***.

Table 23.1: Difference between inner hair cells (IHCs) and outer hair cells (OHCs)		
	Inner hair cells	**Outer hair cells**
Cells numbers	3500	12000
Rows	One	Three or four
Shape	Flask	Cylindrical
Nerve supply	Mainly afferent fibres	Mainly efferent fibres
Development	Early	Late
Function	Transmit auditory stimuli	Modulate function of inner hair cells
Ototoxicity	More resistant	More sensitive and easily damaged
High intensity noise	More resistant	More sensitive and easily damaged
Generation of otoacoustic emissions	No	Yes

The Hair Cells

These important receptor cells of hearing transduce sound energy into electrical energy. The hair cells are so called because their free 'upper' or apical ends bear a number of 'hair'. The hair are really stereocilia. Each cell is columnar or piriform. The hair cells are distinctly shorter than the rod cells. Their apices are at the level of the reticular lamina. Their lower ends (or bases) do not reach the basilar membrane. They rest on phalangeal cells. The plasma membrane at the base of each hair cell forms numerous synaptic contacts with the terminations of the peripheral processes of neurons in the spiral ganglion. Some efferent terminals are also present. The apical surface of each hair cell is thickened to form a ***cuticular plate*** the edges of which are attached to neighbouring cells.

There are two types of hair cells—***inner and outer***. Differences between inner and outer hair cells are given in Table 23.1.

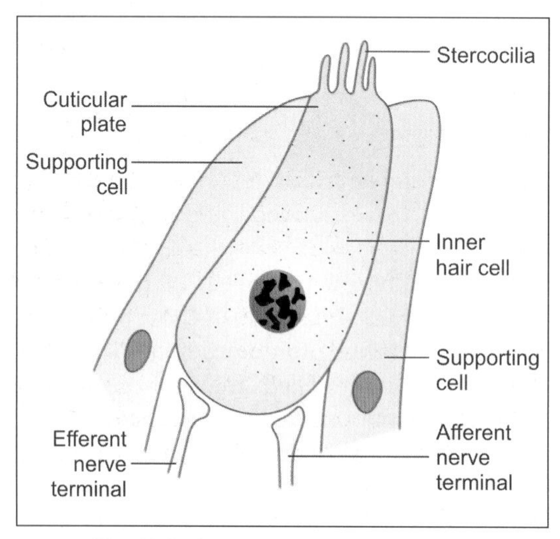

Fig. 23.8: Structure of the inner hair cell (Schematic representation)

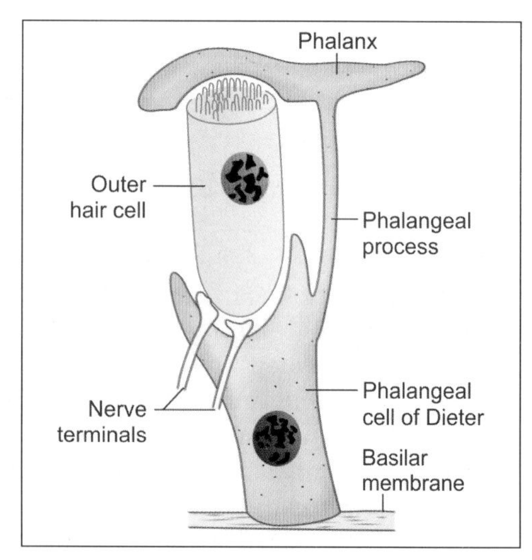

Fig. 23.9: Structure of the outer hair cell (Schematic representation)

- ***Inner hair cells:*** Inner hair cells (IHCs) form a single row and are richly supplied by afferent cochlear fibres. These are flask-shaped cells and relatively short (Fig. 23.8). They are very important in the transmission of auditory impulses. Their nerve fibres are mainly afferent.
- ***Outer hair cells:*** Outer hair cells (OHCs) are arranged in three or four rows and mainly receive efferent innervation from the olivary complex. These are long cylindrical cells which modulate the function of inner hair cells (Fig. 23.9). Their nerve fibres are mainly efferent. The lower end of each outer hair cell fits into a depression on the upper end of a phalangeal cell, but the inner hair cells do not have such a relationship. The 'hair' of the outer hair cells are somewhat longer and more slender than those on inner hair cells. They are arranged as a shallow 'U' rather than a 'V'. Occasionally, the outer hair cells may have more than three rows of hair, and the rows may assume the shape of a 'W' (instead of a 'V').

Added Information

With the EM the 'hair' of hair cells are seen to be similar to microvilli. Each hair has a covering of plasma membrane within which there is a core of microfilaments. Each hair is cylindrical over most of its length, but it is much narrowed at its base. The hair can, therefore, bend easily at this site.

The hair on each hair cell are arranged in a definite manner. When viewed from 'above' they are seen to be arranged in the form of the letter 'V' or 'U'. Each limb of the 'V' has three rows of hairs. The hairs in the three rows are of unequal height being tallest in the 'outer' row, intermediate in the middle row, and shortest in the 'inner' row. The 'V' formed by the hairs of various hair cells are all in alignment, the apex of the 'V' pointing towards the 'outer' wall of the cochlear canal. At the point corresponding to the apex of the 'V' there is a centriole lying just under the apical cell membrane, but a true kinocilium is not present (unlike hair cells of ampullary crests).

The above description applies to both inner and outer hair cells.

The direction of the 'V' is of functional importance. Like hair cells of the maculae and cristae, those of the cochlea are polarised. Bending of stereocilia towards the apex of the 'V' causes depolarisation, while the reverse causes hyperpolarisation. Ionic gradients associated with depolarisation and hyperpolarisation are maintained because apices of hair cells and surrounding cells are tightly sealed by occluding junctions.

The Outer Phalangeal Cells and Reticular Lamina

These are the cells that support the outer hair cells. They lie lateral to the outer rod cells. Their bases rest on the basilar membrane. Their apical parts have a complicated configuration. The greater part of the apex forms a cup-like depression into which the base of an outer hair cell fits. Arising from one side (of the apical part) of the cell there is a thin rod-like ***phalangeal process***. This process passes 'upwards', in the interval between hair cells, to reach the level of the apices of hair cells. Here the phalangeal process expands to form a transverse plate called the ***phalanx***. The edges of the phalanges of adjoining phalangeal cells unite with each other to form a membrane called the ***reticular lamina*** (The reticular lamina also receives contributions from the heads of hair cells). The apices of hair cells protrude through apertures in this lamina.

The cell edges forming the reticular lamina contain bundles of microtubules embedded in dense cytoplasm. Adjacent cell margins are united by desmosomes, occluding junctions and gap junctions. The reticular lamina forms a barrier impermeable to ions except through the cell membranes. It also forms a rigid support between the apical parts of hair cells thus ensuring that the hair cells rub against the membrana tectoria when the basilar membrane vibrates.

The Cochlear Duct

We have seen that the cochlear duct is a triangular canal lying between the basilar membrane and the vestibular membrane. We may now note some further details (Figs 23.4 and 23.6).

- The endosteum on the outer wall of the cochlear canal is thickened. This thickened endosteum forms the outer wall of the duct of the cochlea. The basilar and vestibular membranes are attached to this endosteum. The thickened endosteum shows a projection in the region of attachment of the basilar membrane: this projection is called the ***spiral ligament***. A little above the spiral ligament the thickened endosteum shows a much larger rounded projection into the cochlear duct: this is the ***spiral prominence***. The spiral prominence forms the upper border of a concavity called the ***outer spiral sulcus***.

  Between the spiral prominence and the attachment of the vestibular membrane the thickened endosteum is covered by a specialised epithelium that is called the ***stria vascularis***. The region is so called because there are capillaries ***within*** the thickness of the epithelium (This is the only such epithelium in the whole body). The epithelium of the stria vascularis is made up of three layers of cells: marginal, intermediate and basal. The cells of the marginal layer are called ***dark cells***. They are in contact with the endolymph filling the duct of the cochlea. These cells have a structure and function similar to that of the dark cells already described in the planum semilunatum. These dark cells may be responsible for the formation of endolymph. The basal parts of the dark cells give off processes that come into intimate contact with the intraepithelial capillaries. The capillaries are also in contact with processes arising from cells in the intermediate and basal layers of the stria vascularis.
- We have seen that the spiral lamina is a bony projection into the cochlear canal. Near the attachment of the spiral lamina to the modiolus there is a spiral cavity in which the ***spiral ganglion*** is lodged. This ganglion is made up of bipolar cells. Central processes arising from these cells form the fibres of the cochlear nerve. Peripheral processes of the ganglion cells pass through canals in the spiral lamina to reach the spiral organ of Corti (described below).
- The periosteum on the upper surface of the spiral lamina is greatly thickened to form a mass called the ***limbus lamina spiralis*** (or ***spiral limbus***).

The limbus is roughly triangular in shape. It has a flat 'lower' surface attached to the spiral lamina; a convex 'upper' surface to which the vestibular membrane is attached; and a deeply concave 'outer' surface. The concavity is called the ***internal spiral sulcus***. This sulcus is bounded above by a sharp ***vestibular lip*** and below by a ***tympanic lip*** which is fused to the spiral lamina.

SPECIALISED END ORGANS IN THE MEMBRANOUS LABYRINTH

The internal ear is a highly specialised end organ that performs the dual functions of hearing and of providing information about the position and movements of the head. The impulses in question are converted into nerve impulses by a number of structures that act as transducers. These are ***spiral organ*** (***of Corti***) for hearing and maculae (singular = ***macula***) present in the utricle and saccule for changes in position of the head (Fig. 23.10).

Information about angular movements of the head is provided by end organs called the ***ampullary crests*** (or ***cristae ampullae***). One such crest is present in each semicircular duct. One end of each semicircular duct is dilated to form an ***ampulla***, and the end organ lies within this dilatation. These end organs are described below.

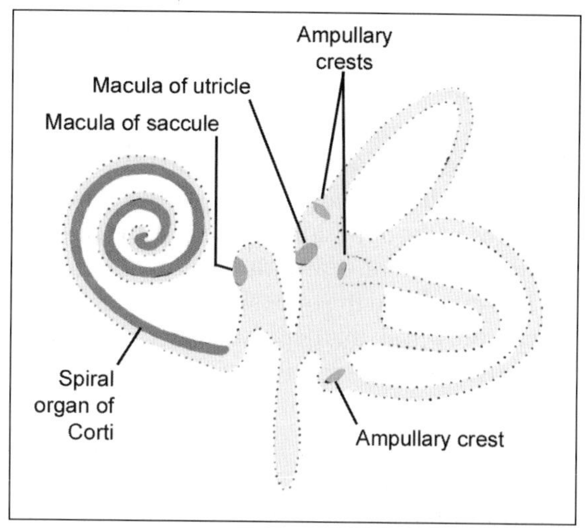

Fig. 23.10: End organs in the membranous labyrinth (Schematic representation)

Ampullary Crests (Fig. 23.11)

One ampullary crest is present in the ampullated end of each of the three semicircular ducts. Each crest is an elongated ridge projecting into the ampulla, and reaching almost up to the opposite wall of the ampulla. The long axis of the crest lies at right angles to that of the semicircular duct. The crest is lined by a columnar epithelium in which two kinds of cells are present. These are ***hair cells*** which are specialised mechano-receptors, and ***supporting*** (or ***sustentacular***) ***cells***.

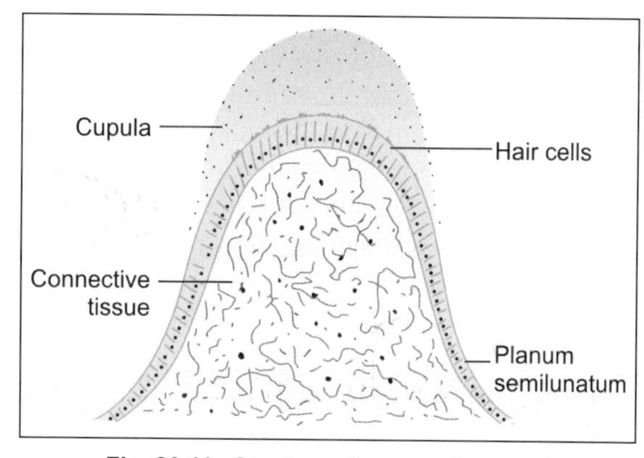

Fig. 23.11: Structure of an ampullary crest (Schematic representation)

The Hair Cells

The hair cells occupy only the upper half of the epithelium. The luminal surface of each hair cell bears 'hairs'. When examined by EM the 'hair' are seen to be of two types as follows:

- There is one large kinocilium which is probably non-motile.
- There are a number of stereocilia (large microvilli).

These 'hair' extend into a gelatinous (protein polysaccharide) material which covers the crest and is called the ***cupula***. The hair processes of the hair cells are arranged in a definite pattern the orientation being specific for each semicircular duct. This orientation is of functional importance.

Each hair cell is innervated by terminals of afferent fibres of the vestibular nerve. Efferent fibres that can alter the threshold of the receptors are also present.

Hair cells can be divided into two types depending on their shape and on the pattern of nerve endings around them. Type I hair cells (inner hair cell) are flask shaped. They have a rounded base and a short neck. The nucleus lies in the expanded basal part (outer hair cell).

The basal part is surrounded by a goblet shaped nerve terminal (or ***calix***). Type II hair cells are columnar. Both types of hair cells receive nerve terminals which are afferent (non-granular) as well as efferent (granular).

Both in the ampullae of semicircular ducts, and in the maculae of the utricle and saccule, each hair cell is polarised with regard to the position of the kinocilium relative to the stereocilia. Each hair cell (in an ampulla) can be said to have a side that is towards the utricle, and a side that faces in the opposite direction. In the lateral semicircular duct, the kinocilia lie on the side of the cells which are towards the utricle; while in the anterior and posterior semicircular ducts the kinocilia lie on the opposite side. When stereocilia are bent towards the kinocilium the cell is hyperpolarised. It is depolarised when bending is away from the kinocilium. Depolarisation depends on the opening up of Ca^{++} channels.

The Supporting Cells

The supporting (or sustentacular) cells are elongated and may be shaped like hour glasses (narrow in the middle and wide at each end). They support the hair cells and provide them with nutrition. They may also modify the composition of endolymph.

Functioning of Ampullary Crests

The ampullary crests are stimulated by movements of the head (specially by acceleration). When the head moves, a current is produced in the endolymph of the semicircular ducts (by inertia). This movement causes deflection of the cupula to one side distorting the hair cells. It appears likely that distortion of the crest in one direction causes stimulation of nerve impulses, while distortion in the opposite direction produces inhibition. In any given movement the cristae of some semicircular ducts are stimulated while those of others are inhibited. Perception of the exact direction of movement of the head depends on the precise pattern formed by responses from the various cristae.

Added Information

Planum Semilunatum

On each side of each ampullary crest the epithelium of the semicircular duct shows an area of thickened epithelium that is called the planum semilunatum. The importance of this area is that (amongst other cells) it contains certain dark cells that have an ultrastructure similar to cells (elsewhere in the body) that are specialised for ionic transport. The cells bear microvilli and have deep infoldings of their basal plasma membrane. The areas between the folds are occupied by elongated mitochondria (Compare with structure of cells of the distal convoluted tubules of the kidney). The dark cells are believed to control the ionic content of the endolymph. Similar cells are also present elsewhere in the membranous labyrinth. The planum semilunatum may secrete endolymph.

The Maculae

They lie in otolith organs (utricle and saccule). Macula of the utricle is situated in its floor in a horizontal plane in the dilated superior portion of the utricle. Macula of saccule is situated in its medial wall in a vertical plane. The macula utriculi (approximately 33,000 hair cells) are larger than saccular macula (approximately 18,000 hair cells). The striola, which is a narrow

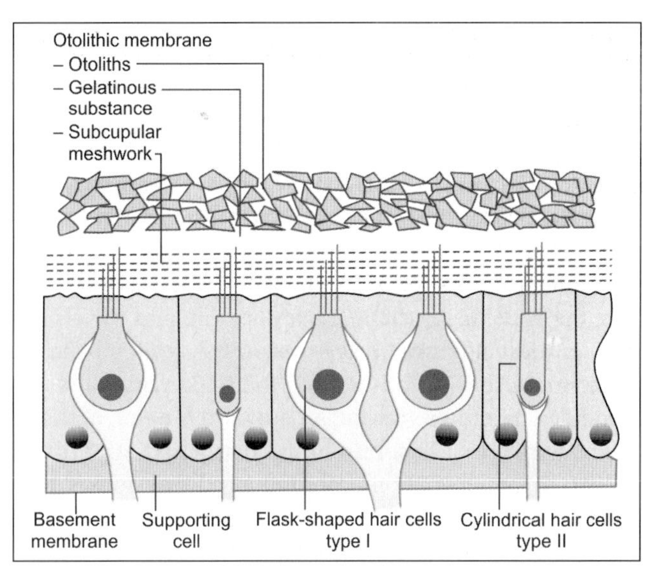

Fig. 23.12: Macula of otolith organs utricle and saccule (Schematic representation)

curved line in centre, divides the macula into two areas. They appreciate position of head in response to gravity and linear acceleration. A macula consists mainly of two parts: a sensory neuroepithelium and an otolith membrane.

- ***Sensory neuroepithelium:*** It is made up of type 1 and type 2 cells, which are similar to the hair cells of the ampullary cristae. Type I cells are in higher concentration in the area of striola and change orientation (mirror-shaped) along the line of striola with opposite polarity. The kinocilia face striola in the utricular macula, whereas in saccule, they face away from the striola. The polarity and curvilinear shape of striola offer CNS wide range of neural information of angles in all the three dimensions for optimal perception and compensatory correction. During tilt, translational head movements and positioning, visual stimuli combined with receptors of neck muscles, joint and ligaments play an important part.
- ***Otolithic membrane:*** The otoconial membrane consists of a gelatinous mass, a subgelatinous space and the crystals of calcium carbonate called otoliths (otoconia or statoconia) (Fig. 23.12). The otoconia, which are multitude of small cylindrical and hexagonally shaped bodies with pointed ends, consists of an organic protein matrix together with crystallized calcium carbonate. The otoconia (3–19 μm long) lie on the top of the gelatinous mass. The cilia of hair cells project into the gelatinous layer. The linear, gravitational and head tilt movements result into the displacement of otolithic membrane, which stimulate the hair cells lying in different planes.

The maculae give information about the position of the head and are organs of ***static balance***. In contrast the ampullary crests are organs of ***kinetic balance***. The macula of the saccule may be concerned with the reception of low frequencies of sound. Impulses arising from the ampullary crests and the maculae influence the position of the eyes. They also have an influence on body posture (through the vestibular nuclei).

SOME ELEMENTARY FACTS ABOUT THE MECHANISM OF HEARING

Sound waves travelling through air pass into the external acoustic meatus and produce vibrations in the tympanic membrane. These vibrations are transmitted through the chain of ossicles to perilymph in the vestibule. In this process the force of vibration undergoes considerable amplification because (a) the chain of ossicles acts as a lever; and (b) the area of the tympanic membrane is much greater than that of the footplate of the stapes (increasing the force per unit area).

Movement of the stapes (towards the vestibule) sets up a pressure wave in the perilymph. This wave passes from the vestibule into the scala vestibuli, and travels through it to the apex of the cochlea. At this point (called the ***helicotrema***) the scala vestibuli is continuous with the scala tympani. The pressure wave passes into the scala tympani and again traverses the whole length of the cochlea to end by causing an outward bulging of the secondary tympanic membrane. In this way vibrations are set up in the perilymph, and through it in the basilar membrane. Movements of the basilar membrane produces forces that result in friction between the 'hairs' of hair cells against the membrana tectoria. This friction leads to bending of the 'hairs'. This bending generates nerve impulses that travel through the cochlear nerve to the brain.

The presence of efferent terminals on the hair cells probably controls the afferent impulses reaching the brain. It can also lead to sharpening of impulses emanating from particular segments of the spiral organ by suppressing impulses from adjoining areas.

It has to be remembered that the transverse length of the basilar membrane is not equal in different parts of the cochlear canal. The membrane is shortest in the basal turn of the cochlea, and longest in the apical turn (quite contrary to what one might expect). Different segments of the membrane vibrate most strongly in response to different frequencies of sound thus providing a mechanism for differentiation of sound frequencies. Low frequency sounds are detected by hair cells in the organ of Corti lying near the apex of the cochlea, while high frequency sounds are detected by hair cells placed near the base of the cochlea.

The intensity of sound depends on the amplitude of vibration. For further details of the mechanism of hearing consult a textbook on physiology.

INDEX

Page numbers preceded by '**A**' refer to Atlas and '**P**' refer to plate number followed by page number.
Page numbers followed by ***f*** refer to figure and ***t*** refer to table

B

C

D

E

F

G

H

I

J

K

L

R

S

T

U

V

W

X

Z